W9-ATR-866

HERPETOLOGY

HERPETOLOGY

AN INTRODUCTORY BIOLOGY OF AMPHIBIANS AND REPTILES

George R. Zug

Department of Vertebrate Zoology
National Museum of Natural History
Smithsonian Institution
Washington, D.C.

ACADEMIC PRESS
A Division of Harcourt Brace & Company
San Diego New York Boston
London Sydney Tokyo Toronto

Find Us on the Web! http://www.apnet.com

This book is printed on acid-free paper. ♾

ACADEMIC PRESS
525 B Street, Suite 1900
San Diego, California 92101-4495

United Kingdom Edition published by
Academic Press Limited
24–28 Oval Road, London NW1 7DX

Library of Congress Cataloging-in-Publication Data

Zug, George R., date
Herpetology : An Introductory Biology of Amphibians and Reptiles / George R. Zug.
p. cm.
Includes bibliographical references and index.
ISBN 0-12-782620-3
1. Herpetology. I. Title.
QL641.Z84 1993
597.6–dc20 92-30758
CIP

Printed in the United States of America
98 99 MM 8 7 6 5 4

Dedicated to my mentors:

A. F. and G. B. Zug,

A. Schwartz,

W. J. Riemer,

and C. F. Walker

CONTENTS

PREFACE

It is an admirable feature of herpetologists that they are able to cross the boundaries between different aspects of their subject, which remains, perhaps more than other branches of zoology, a single coherent discipline.

A. d'A. Bellairs and C. B. Cox, 1976

The classification followed in this book is a mixture of the old and the new. Vertebrate zoologists will recognize that the division of the subphylum Vertebrata into seven classes (Cyclostomata, Chondrichthyes, Osteichthyes, Amphibia, Reptilia, Aves, and Mammalia) does not recognize single evolutionary lineages or groups of equivalent taxonomic rank. In a strictly phylogenetic sense, the Tetrapoda (Amphibia through Mammalia) are a subgroup of bony fishes specialized for terrestrial life. However, that classification level is not of immediate concern to us as herpetologists, our concern, rather, is with the classification arrangement within the Tetrapoda and its accurate depiction of evolutionary lineages. Breaking with traditional classifications, the following cladistic classification more accurately reflects our knowledge of tetrapod evolution but drastically alters the ranks of many long-used taxa and, in some cases, changes the content of taxa.

- Tetrapoda
  - Amniota
    - Mammalia
    - Reptilia
      - Testudines
      - Sauria
        - Archosauria
          - Crocodylia
          - Aves
        - Lepidosauria
          - Sphenodontida
          - Squamata
  - Amphibia
    - Lissamphibia
      - Gymnophiona
      - Salientia
      - Caudata

Note particularly the difference in taxonomic rank of Amphibia and Reptilia and the inclusion of birds in the Reptilia. These rankings and relationships do not surprise the professional zoologist, but they are unsettling after having used another classification for years. Due to this and to the lack of agreement on category names for the new rankings, the formal names of the taxa are not preceded with category titles for any category higher than family. Because herpetology is defined by its history of study, I use "reptiles" in a vernacular sense to include all the Reptilia, while generally excluding the Ornithosuchia (the lineage that includes the dinosaurs and birds).

My major goals are to portray the diversity of living amphibians and reptiles in most aspects of their biology and to introduce students (upper-level undergraduates) to the breadth of herpetology today. Obviously, not all aspects of herpetology can be covered in depth in an introductory text, so a bibliography is included for each chapter. Each bibliography contains the references I consulted in writing the chapter as well as summary and review articles that provide information beyond that presented herein.

George R. Zug

ACKNOWLEDGMENTS

The preparation of any book requires the assistance of many individuals. I thank all of you—those who provided direct assistance and those whose ideas and discoveries were extracted from research articles and seminars. A few individuals deserve a special thanks: Pat Zug and Ron Heyer for constant support and picking up the slack when I failed to do so, Ron Crombie for his amazingly encyclopaedic recall of the herpetological literature, and Josh Seiff and Rob Wilson for literature retrieval and map production. Seven reviewers—C. Gans, L. M. Hardy, W. R. Heyer, G. Middendorf, J. Mitchell, W. Witt, and K. Witt—graciously volunteered to read the text in its earliest and roughest stages; their critical reviews and tolerances are much appreciated. I also thank the other reviewers—M. Benabib, R. Crombie, C. Crumly, K. de Queiroz, C. Ernst, O. Flores-V., J. Iverson, G. Mayer, R. W. McDiarmid, J. Slowinski, K. Troyer, V. Wallach, P. Weldon, and A. Wynn—for their assistance on one or more chapters. Their suggestions have made this a better text. Undoubtedly errors remain, since I do not always follow good advice. For these, I am alone responsible.

I thank the following photographers for permitting me to reproduce their materials in original or modified form: R. W. Barbour, A. Bauer, A. Cardoso, R. E. Clark, Jr., C. K. Dodd, Jr., C. Gans, J. W. Lang, R. W. McDiarmid, K. Miyata, E. O. Moll, K. Nemuras, R. A. Nussbaum, C. A. Ross, T. Schwaner, R. G. Tuck, Jr., H. I. Uible, and R. W. Van Devender.

PART I

DIVERSITY AND HISTORY

Life, life an endless march, an endless army . . .

W. Whitman, 1891–1892

Living amphibians and reptiles display an amazing variety of shapes and sizes. Some forms, such as the legless and wormlike caecilians and amphisbaenians, are easily mistaken for animals other than amphibians and reptiles, whereas frogs and turtles are always recognized for what they are. The variety of body forms is matched by equally diverse behaviors, ecologies, and physiologies. All of this variety reflects a complex and long evolutionary history for both amphibians and reptiles and their adaptations to a multitude of environments and life-styles.

The great variety of adaptations is matched by high species richness or diversity. There are approximately 4300 species of amphibians and over 6000 species of reptiles. Neither matches the diversity of the bony fishes, but both are comparable to the diversity of mammals (ca. 5000 species). This great diversity demonstrates that amphibians and reptiles are not degenerate groups hanging on at the edge of survival in this "Age of Mammals." Their life-style is different but not inferior to that of mammals. In fact, they have been able to adapt and radiate widely in several environments that have proved more hostile for mammals. "Cold-bloodedness," a supposed maladapted characteristic of amphibians and reptiles, is less energy demanding and allows a lower metabolism and discontinuous growth thereby permitting survival in many hostile environments. However, "cold-blooded" is a poor descriptor for the ectothermic amphibians and reptiles. Ectothermy denotes the physiological condition in which an animal's body temperature depends on heat derived from an external (**ecto-**) source. Ectothermy is not an adaptation; rather it is the original (ancestral) state of life. Its importance lies at two extremes: (1) what it permits an animal to do for immediate survival; and (2) evolutionarily, what adaptive modes have been possible within the boundaries imposed by this physical state. The latter aspect is emphasized in the subsequent four chapters through a general introduction to the anatomy of amphibians and reptiles and their present and past diversity.

CHAPTER 1

Amphibians

Many living amphibians have a two-phase life-style. Indeed, they are the only vertebrates with a free-living aquatic developmental stage and a terrestrial juvenile/adult stage. The larval stage highlights the ancestral origin from fish, and the adult stage demonstrates the mastery of the land. Their name derives from the Greek *amphibios* for "double life," recognizing their duplex life-style.

Living amphibians share a number of other unique traits. These traits distinguish them from fish and the other limbed vertebrates (tetrapods) and indicate that the three dissimilar groups of modern amphibians possessed a common ancestor. All share a reliance on cutaneous respiration, a pair of sensory papillae in the inner ear, doubled transmission channels in the middle ear, specialized visual cells in the retina, pedicellate teeth, two types of skin glands, as well as several other unique traits.

Three respiratory surfaces are used by amphibians, usually two of them operating simultaneously. Aquatic amphibians, particularly the larvae, use gills; terrestrial ones use lungs. However, in both air and water, the skin serves as a major surface for the transfer of oxygen and carbon dioxide. The skin has proved so effective that some terrestrial amphibians have lost or reduced their lungs, and some aquatic ones have eliminated their gills, depending entirely on the skin for the exchange of gases. All lunged species use a force-pump mechanism for moving air in and out of the lungs.

The double-channeled auditory system of amphibians has one channel that is common to all tetrapods, the columella-basilar papilla channel. The other channel, opercular-amphibian papilla, allows the reception of low-frequency sounds ($<$1000 Hz). The possession of double receptors may not seem peculiar for the frogs since they are vocal animals. However, for the largely mute salamanders, a dual hearing system seems peculiar and redundant.

Salamanders and frogs have green rods in the retina, presumably absent in the degenerate-eyed caecilians. This type of visual cell is found in no other animals, and its particular function is as yet unknown.

The teeth of modern amphibians are two-part structures: an elongate base (pedicel) anchored in the jaw bone and a crown protruding above the gum. Each tooth is usually constricted where the crown attaches to the pedicel. As the crowns wear down, they break free at the constriction and are replaced by a new crown emerging from within the pedicel. Only a few living amphibians lack pedicellate teeth.

Two types of skin glands are present in all living amphibians: mucous and granular (poison) glands. The mucous glands secrete mucus, which keeps the skin surface moist for cutaneous respiration. Although the structure of the poison glands is identical in all species, the toxicity of the secretions is highly variable, ranging from barely irritating to lethal.

Living amphibians possess other unique traits. All have fat bodies that develop from the germinal ridge of the embryo and retain an association with the gonads in adults. Frogs and salamanders are the only vertebrates able to raise and lower their eyes. The bony orbit of all amphibians opens into the roof of the mouth with a special muscle stretched across this opening; this muscle elevates the eye. The ribs of amphibians do not encircle the body.

MODERN GROUPS OF AMPHIBIANS

Three major lineages of amphibians exist today: frogs, salamanders, and caecilians (Fig. 1.1). Although these three groups share a number of unique traits for tetrapods, the question remains whether they evolved from the same

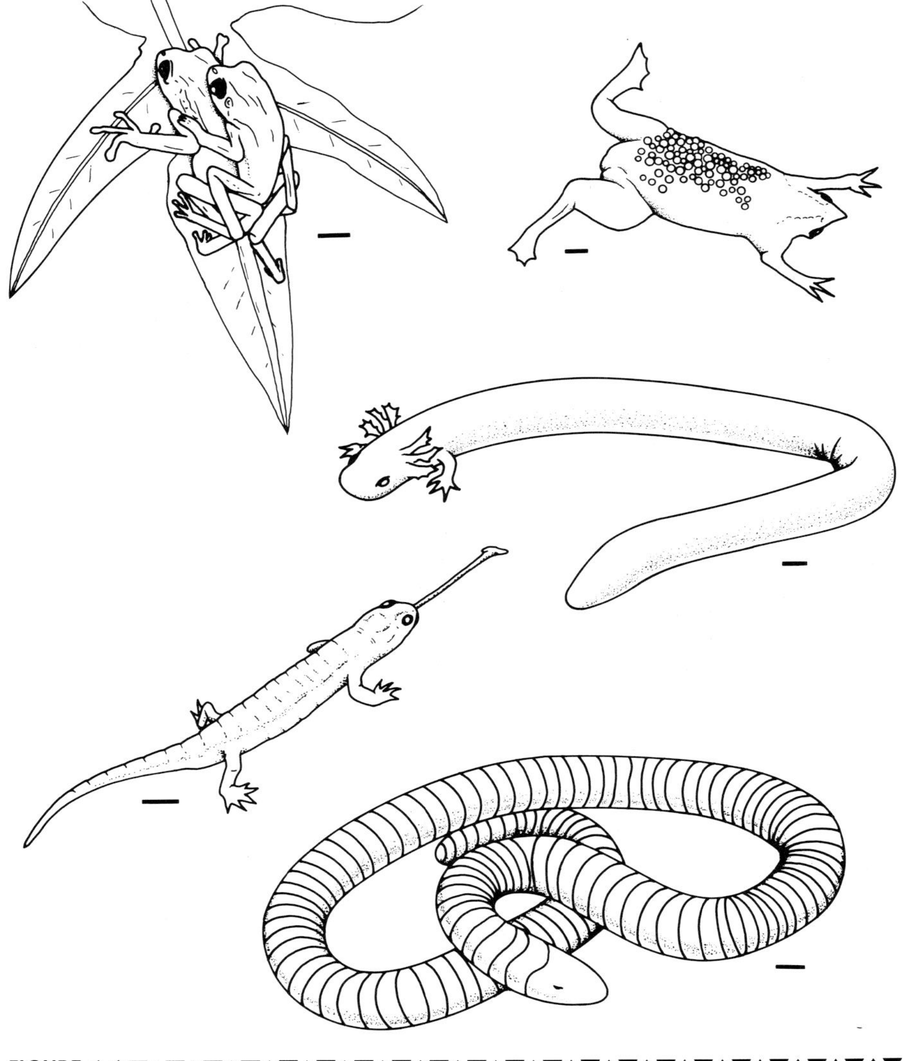

FIGURE 1.1 ▼▲▼

A sampling of adult body forms (habitus) in living amphibians. *Phyllomedusa* (Hylidae), *Pipa* (Pipidae), *Siren* (Sirenidae), *Gymnophis* (Caeciliaidae), and *Hydromantoides* (Plethodontidae), clockwise from upper left. Scale: bar = 1 cm.

ancestral amphibians or from different ones. If the unique traits are primitive, that is, shared by many groups of ancient amphibians, the traits provide no evidence of relationship among modern amphibians other than that all three are amphibians. If, on the other hand, the traits are uniquely derived (= independently evolved) and were confined to a single group of ancient amphibians, then frogs, salamanders, and caecilians are closely related phylogenetically, although long separated in their divergence from a common ancestor. Today, proponents argue either for a common ancestor for the three or for the salamanders' origin from a different ancestor. The issue remains unresolved, because the evidence can be interpreted differently. The consensus currently favors a shared ancestor for the three extant lineages and places them in a single taxonomic group, the Lissamphibia. The pros and cons for this interpretation are examined later (see Chapter 2).

Uncertainty about the relationships of the three extant amphibian groups is not surprising. Morphologically they are very different. The caecilians are limbless; even the internal girdle (pectoral and pelvic) elements are absent. The body is an elongate cylinder, regularly encircled by grooves forming segments; the head is blunt and cone-shaped with an underhung lower jaw and nearly invisible degenerate eyes; and the tail is very short and blunt (if present). They are usually shades of pinkish browns and grays. All these features create the appearance of an earthworm. Like earthworms, most caecilians are fossorial, living and burrowing in the soil; a few are aquatic. The salamanders are lizardlike in body form; typically they have long, low-slung bodies with moderate-length limbs and long tails. The head is broad with distinct eyes and separated from the body by a neck. The species are nearly evenly divided between terrestrial and aquatic forms. A few of the latter have reduced limbs and appear eellike. The body form of frogs is dominated by large hindlimbs, an adaptation for their jumping (saltatory) locomotion. The entire body form is adapted for jumping, although not all species jump or hop. Nonetheless, the shortened body with broad head, no neck, no tail, and well-developed limbs matches the physical needs of a creature propelling itself with synchronous extensions of the hindlimbs. A few frog species are entirely aquatic; most are semiterrestrial to terrestrial. Clearly, a frog cannot be mistaken for either a salamander or a caecilian, and neither of the latter two is likely to be confused. The following sections introduce the main groups of living amphibians; more detailed descriptions of each amphibian family are presented in Part VI, Systematics and Classification Sections, Chapters 14 and 15.

Caecilians

The caecilians (Gymnophiona) are not very diverse, with approximately 160 species divided among six families (Table 1.1). So little is known of their biology and relationships that the families are not introduced here (see Chapter 14). The caecilians are tropical amphibians, occurring in tropical America, Africa, and Asia, and typically in areas of abundant rainfall. They vary in total length from 7–8 cm to 1.5 m; most are moderate-sized and less than 50 cm. All species have internal fertilization, assisted by a protrusible portion (phallodeum) of the male's cloaca. The fertilized eggs may be deposited externally and develop through a

TABLE 1.1 Classification of Extant Caecilians and Salamanders

Amphibia
  Lissamphibia
    Gymnophiona (Apoda)—caecilians
      Family Caeciliaidae[a]
      Family Ichthyophiidae
      Family Rhinatrematidae
      Family Scolecomorphidae
      Family Typhlonectidae
      Family Uraeotyphlidae
    Caudata (Urodela)—salamanders
      Cryptobranchoidea
        Family Cryptobranchidae
        Family Hynobiidae
      Salamandroidea
        Family Ambystomatidae
        Family Amphiumidae
        Family Dicamptodontidae
        Family Plethodontidae
        Family Proteidae
        Family Salamandridae
      Meantes (Trachystomata, Sirenoidea)
        Family Sirenidae

Note: The names in parentheses are alternate, usually more recent, names for the groups.

[a] This peculiar spelling is dictated by a 1988 opinion of the International Commission on Zoological Nomenclature. The spelling was proposed to remove the homonymy of the insect Caeciliidae and the amphibian Caeciliidae.

free-living larval stage. In some species, the external eggs have direct development; the free-living stage is skipped and the young hatch as miniatures of the adults. In others, the fertilized eggs are retained in the female's oviducts; the developing fetuses feed on the oviducal wall, by scraping its epithelial lining with specialized embryonic teeth.

Salamanders

The living salamanders (Caudata) are moderately speciose with nearly 400 species. These species cluster into three evolutionary lineages (Table 1.1): sirens (Meantes); primitive salamanders (Cryptobranchoidea); and advanced salamanders (Salamandroidea). The sirens are eellike salamanders with external gills.

They live in slow-moving and still waters of southeastern North America and range in total length from 25 cm for the dwarf sirens to nearly 1 m for the sirens. Sirens lack hindlimbs and pelvic girdles, and their forelimbs are reduced to small, slender appendages. Although abundant in plant-choked waterways, many aspects of their biology remain unknown, such as site of fertilization, which is assumed to be external owing to the absence of specialized reproductive structures in the cloacae of males and females. The cryptobranchoid salamanders comprise two dissimilar families. The giant salamanders (Cryptobranchidae) are the largest living salamanders with the Oriental *Andrias* reaching total lengths of 1.5 m and the North American *Cryptobranchus* to 75 cm. These large salamanders live in cool mountain streams. Their smaller (10–20 cm total length/TL) relatives, the Asiatic brook salamanders (Hynobiidae), are also predominantly aquatic animals, living in the cool streams and ponds of Asia from the Ural Mountains eastward to the Pacific. Both cryptobranchoid families reproduce with external fertilization and have aquatic eggs and free-living larvae.

The salamandroids are the most diverse (330+ species) and widespread group of salamanders. All have internal fertilization in which a male deposits a spermatophore (a packet of sperm on a mucoid pedicel) externally and the female picks up the sperm packet and stores it in a special pocket (spermatheca) in her cloaca. The six families possess a variety of body forms and life-styles. The amphiumas (Amphiumidae) are a group of three large (1 m TL), eellike salamanders of the rivers and swamps of southeastern North America. They are voracious carnivores that might be mistaken for sirens except they possess tiny fore- and hindlimbs and lack external gills. The mudpuppies and waterdogs (Proteidae) are also aquatic, but smaller (11–43 cm TL) with well-developed limbs and large bushy external gills. The mole salamanders (Ambystomatidae and Dicamptodontidae) of North America are heavy-bodied salamanders with robust limbs. The ambystomatids are principally subterranean species that return to water in the winter or spring for reproduction. A few species are aquatic exceptions; these species are the axolotls, which may be absolute or facultative paedomorphs, that is, larvae have the ability to reproduce. Many salamander species possess larval characteristics as adults, and this paedomorphosis (see Chapter 7) seems to have served as a major mechanism in the evolution of salamanders. The dicamptodontids are structurally similar to the ambystomatids and strongly aquatic.

The newts (Salamandridae) and lungless salamanders (Plethodontidae) are the two most speciose families of the advanced salamanders. The newts are predominantly Eurasian, although a few species occur in eastern (*Notophthalmus*) and western (*Taricha*) North America. Many newts have a tripartite life history. The larvae are aquatic and metamorphose into terrestrial juveniles, the efts. Efts wander about on land for two to eight years before maturing and returning to streams or lakes as aquatic adults, the newts. Most salamandrids are less than 12 cm (TL), even though a few reach lengths to 20 cm; however, the most striking feature of all salamandrids is their toxic skin secretion, often associated with bright colors or bold color patterns in the most toxic species or life stage.

In contrast, the plethodontids tend toward more somber colors or cryptic color patterns. With a single European exception, the plethodontids are confined to the Americas. As their common name implies, all plethodontid salamanders lack lungs; their skin is the major respiratory surface. Lunglessness would seem to be a major handicap, yet the plethodontids are the most speciose salamanders with over 250 species—60% of the living species of salamanders. Furthermore, they are the only salamanders to have successfully invaded the tropics, and although most tropical species occur in cool mountain forests, a few species do live in hot lowland forests. Many of the plethodontids are terrestrial, and many have terrestrial eggs and direct development, skipping a free-living, aquatic larval stage. The plethodontids show a variety of body shapes and sizes, from heavy-bodied species to elongate, tiny-limbed forms, from dwarf to moderate-sized species (15–125 mm snout–vent length/SVL).

Frogs

Frogs (Anura) are the most successful living amphibians. Many species have dense populations and occupy broad geographical ranges. In total, there are over 3800 species of living frogs. Their diversity and the ancientness of many lineages make frogs difficult to classify (Table 1.2). The frogs are divided into 20 or more families, and these are grouped into two or three suborders. The more primitive families contain only a few living species: Ascaphidae, one species (*Ascaphus truei*); Leiopelmatidae, three species; Discoglossidae, 14 species. The tailed frog (*A. truei*) is a small (±45 mm SVL), mountain-stream frog of northwestern North America. The leiopelmatids are small frogs (<50 mm SVL) living in forested areas of New Zealand. The discoglossids are Eurasian frogs of small to moderate sizes.

A second group of frogs contains two families of highly specialized frogs (Pipidae, Rhinophrynidae) and two families (Pelobatidae, Pelodytidae) that are structurally intermediate between the primitive and advanced frogs. The Rhinophrynidae contains a single toadlike species from southernmost Texas, Mexico, and northern Central America. *Rhinophrynus dorsalis* spends much of its life underground eating ants and termites, emerging only to breed. The Pipidae (25+ species) consists of the clawed frogs and their relatives of sub-Saharan Africa and the Surinam toad and its relatives of northern South America. All pipids are highly aquatic frogs, dorsoventrally flattened with webbed hindfeet and muscular hindlimbs for swimming. The pelobatids (80+ species) and pelodytids (2 species) have a typical frog appearance and most are small- to moderate-sized (<80 mm SVL) species. The pelobatids occur in Europe and southeast Asia and also in North America (spadefoot toads, *Scaphiopus*); the pelodytids occur in western Europe and the Causasia.

The advanced frogs include a dozen or more families, some with only one species and others with hundreds. Although there is little agreement among the frog specialists on the classification of the frogs at higher levels, there is near

TABLE 1.2
Classification of Extant Frogs

- Amphibia
  - Lissamphibia
    - Salientia—frogs
      - Proanura
      - Anura
        - Archaeobatrachia
          - Family Ascaphidae
          - Family Discoglossidae
          - Family Leiopelmatidae
        - Mesobatrachia
          - Pipoidea
            - Family Pipidae
            - Family Rhinophrynidae
          - Pelobatoidea
            - Family Pelobatidae
            - Family Pelodytidae
        - Neobatrachia
          - Bufonoidea
            - Family Brachycephalidae
            - Family Bufonidae
            - Family Centrolenidae
            - Family Dendrobatidae
            - Family Heleophrynidae
            - Family Hylidae
            - Family Leptodactylidae
            - Family Myobatrachidae
            - Family Pelodryadidae
            - Family Pseudidae
            - Family Rhinodermatidae
            - Family Sooglossidae
          - Microhyloidea
            - Family Microhylidae
          - Ranoidea
            - Family Hyperoliidae
            - Family Ranidae
            - Family Rhacophoridae

unanimity that the advanced frogs divide into three major evolutionary lineages: microhyloids, ranoids, and bufonoids. (Names ending in -idae are anglicized to -id ending for families, and those ending in -oidea to -oid for suprafamilies.) The microhyloids are a single family (Microhylidae) of small- to moderate-sized burrowing and terrestrial frogs. Many are robust, tear-shaped animals with microcephalic heads, and others have a typical frog or treefrog shape. They occur worldwide in the tropics and in the subtropics of eastern Asia and North America.

The ranoids are a more diverse group of four to six families. The Ranidae is often labeled the true frogs, and the namesake genus *Rana* is one of the most widespread genera of frogs, occurring on all continents and consisting of over 200 species ranging from strongly aquatic to highly terrestrial frogs. The ranids include several dozen other species in a variety of body forms; these are mainly terrestrial or semiaquatic species. The African grass or reed frogs (Hyperoliidae) are small- to moderate-sized frogs; all have distinct enlarged toepads and most are brightly colored. The Rhacophoridae is also principally treefrogs, ranging in size from less than 20 mm to over 100 mm (SVL). Their distribution centers on tropical Asia, although a few species occur in Africa. The sooglossids include three species of small frogs confined to the granitic Seychelle islands. Their relationships are uncertain, and they may be ranoids or bufonoids.

The bufonoids include nearly a dozen families with a large variety of body forms and life-styles. The namesake Bufonidae include the true toads with squat, warty bodies; their natural geographic distribution encompasses all continents except Australia. Their warts and neck glands are concentrations of poison glands and protect them from predators. The Dendrobatidae or poison-dart frogs have a smooth skin with many small glands. Their poisonous secretions are, nonetheless, highly toxic, particularly so in the brightly colored species. The dendrobatids are small terrestrial frogs of Middle and South America. Three bufonoid families are treefrogs: Centrolenidae, Hylidae, and Pelodryadidae. The centrolenids are called the glass frogs, because many have transparent belly skin making the viscera visible. These tropical American frogs are predominantly small and lay their eggs on rocks or leaves overhanging streams; the hatching tadpoles drop into the water below. The widespread treefrogs of the Americas and temperate Eurasia are the Hylidae. Even though many are arboreal and have expanded digit tips, others are terrestrial and lack expanded tips. Hylids range in size from small to large and display numerous reproductive modes. The Australian treefrogs (Pelodryadidae) are closely related to the hylids and share many of the latter's characteristics and diverse adaptations. The Americas and Australia share another pair of bufonoid families, Leptodactylidae and Myobatrachidae, respectively. These frogs are mainly, although not exclusively, terrestrial and semiterrestrial. Both families range from $<$20 mm to $>$110 mm (SVL), and both have species with the typical tadpole lifestage as well as those with terrestrial eggs and direct development. There are several more families of bufonoid frogs (Table 1.2); each has only a few species and, with the exception of the African Heleophrynidae, is confined to South America.

GENERAL ANATOMY

Skin

The skin is the cellular envelope forming the boundary between the animal and the external environment. As the interface between the internal and external environments, the skin serves multiple roles. Foremost are its roles in support and protection. The skin holds the other tissues and organs in place, yet is sufficiently elastic and flexible to permit expansion, movement, and growth. As a protective barrier, it prevents the invasion of microbes and other small animals, resists mechanical invasion and abrasion, and buffers the internal environment from the extremes of the external environment. The skin also serves in physiological regulation (e.g., heat and osmotic regulation), sensory detection (chemo- and mechanoreception), respiration, and coloration.

The skin (Fig. 1.2) consists of an external layer, the epidermis, separated from the internal layer, the dermis, by a thin basement membrane. The epidermis is of variable thickness, two to three cell layers thick in larval amphibians and five to seven layers thick in juveniles and adults. In both larvae and adults, the epidermis is an ever-growing tissue. The innermost layer of cells (stratum germinativum) divides continuously to replace the worn outer layer of epidermal cells. The outer cell layer is alive in larvae, but in most adults, cells slowly flatten, keratinize, and die as they are pushed outward. This layer of dead, keratinous cells (stratum corneum) shields the inner layers of living cells from injury. The dermis is a thicker layer, containing many cell types and structures (pigment cells, mucous and granular glands, blood vessels, nerves) embedded in a connective tissue matrix. The innermost layer of dermis is a densely knit connective tissue (stratum compactum), and the outer layer (stratum spongiosum) is a looser matrix of connective tissue, blood vessels, nerve endings, glands, and other cellular structures. In caecilians and salamanders, the stratum compactum is tightly linked with the connective tissue sheaths of the muscles and bones. In contrast, much of the body skin is loosely attached in frogs.

Glands and Specialized Structures

Amphibians possess several types of epidermal glands. Mucous and granular glands occur in all postmetamorphic (i.e., juvenile and adult) amphibians and are numerous and widespread on the head, body, and limbs. Both types are multicellular, flask-shaped glands with the bulbous, secretory portion lying within the stratum spongiosum of the dermis; their narrow necks extend upward through and open on the surface of the epidermis. Although occurring over the entire body, the glands are not evenly distributed; their role determines their density and location. Mucous glands are the most abundant glands, about 10 mucous glands for every granular one. The mucous glands are especially dense dorsally, and they continuously secrete a clear, slimy mucus that maintains a thin, moist film over the skin. The granular glands tend to be concentrated on the head and shoulders, where their poisonous or noxious secretions will be immediately encountered by predators. The granular glands may be aggregated into macroglands, such as the parotoid glands of some frogs and salamanders. Usually these macroglands contain more complex individual glands.

Larvae have a greater variety of epidermal glands. Most are single-celled (unicellular) glands, although many may be concentrated in a single region. For

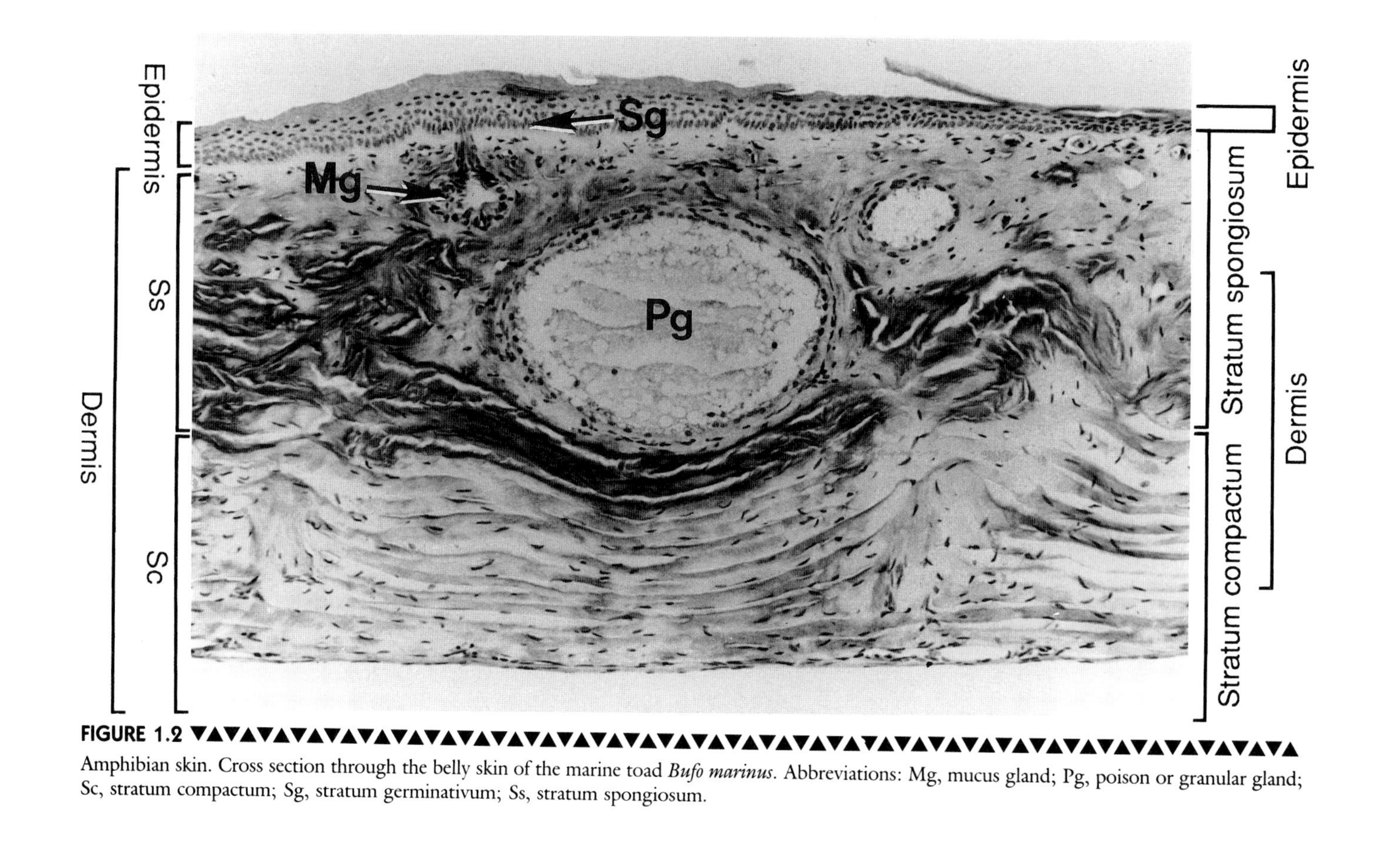

FIGURE 1.2 Amphibian skin. Cross section through the belly skin of the marine toad *Bufo marinus*. Abbreviations: Mg, mucus gland; Pg, poison or granular gland; Sc, stratum compactum; Sg, stratum germinativum; Ss, stratum spongiosum.

example, the hatching glands are clustered on the dorsal forepart of the head; their secretion dissolves the gelatinous egg coat, allowing the larva to escape from the egg. Unicellular mucous glands are widespread and secrete a protective mucous coat over the surface of the living epidermis. This mucous coat also serves as a lubricant to enhance the flow of water over the larva when swimming. Merkel and flask cells are scattered throughout the larval epidermis, but they are nowhere abundant. The function of these cells remains uncertain. Flask cells may be involved in salt and water balance.

The skin of amphibians ranges from smooth to rough. Some of the integumentary projections are strictly epidermal, but most involve both the epidermis and dermis. Integumentary annuli of caecilians and costal grooves of salamanders match the segmentation of the axial musculature and vertebral column. Each primary annulus (caecilians) and each costal groove (salamanders) lies directly over the myosepta (connective tissue sheet) between the muscle masses, thus the number of annuli equals the number of trunk vertebrae. In caecilians, this annular pattern may be complicated by the development of secondary and tertiary grooves; the secondary grooves appear directly above the myosepta. The warts, papillae, flaps, tubercles, and ridges in frogs and salamanders may be aggregations of glands or simply thickenings in the underlying dermis and epidermis.

Although amphibians lack epidermal scales, they do possess keratinous structures. The clawlike toe tips of pipid frogs, the spade of pelobatid frogs, and the rough, spiny skin of some frogs and salamanders are keratinous. These structures persist year-round; other keratinous structures are seasonal and usually associated with reproduction. Many male salamanders and frogs have keratinous nuptial pads on their thumbs at the beginning of the mating season; some even develop keratinous spines or tubercles on their arms or chests. At the end of the mating season, these specialized mating structures are usually shed and then reappear in subsequent breeding seasons.

Dermal scales exist only in caecilians, although not in all species. These scales are small, flat, bony plates that are buried deeply in pockets within the annular grooves. Whether these scales are homologues of fish scales remains uncertain. A few species of unrelated frogs, for example, some species of *Ceratophrys* and *Megophrys*, have osteoderms (bony plates) embedded in or immediately above the dermis. Also in some species of frogs, the dorsal skin of the head is compacted and the connective tissue of the dermis is co-ossified with the skull bones.

Ecdysis

Adult amphibians shed their skin in a cyclic pattern of several days to a few weeks. This shedding (ecdysis, sloughing, or molting) involves only the stratum corneum and is commonly divided into several phases. At its simplest, the shedding cycle contains epidermal germination and maturation phases, preecdysis, and actual ecdysis. These phases are controlled hormonally, although timing and mechanisms differ between species and amphibian groups. The stratum germinativum produces new cells and they move outward and upward in a conveyer-belt-like fashion as new cells are produced beneath them. Once these new cells lose contact with the basement membrane, they cease dividing and begin to mature with the appearance and loss of subcellular organelles. Preecdysis is signaled by development of mucous lakes between the maturing cells and the stratum corneum. The lakes expand and coalesce; the cellular connections break between the dead cells of the stratum corneum and underlying and maturing cells. Externally, the skin commonly splits middorsally over the head first, and the split continues down the middle of the back. Using its limbs, the frog or salamander emerges from the old skin and then often eats it. During the preecdytic and/or the ecdytic phase, the epidermal cells beneath the mucous lakes complete their keratinization and die.

The shedding process of larval amphibians is not well known. In the mudpuppy *Necturus maculosus* and likely in most other larvae, the skin is shed as single cells or small pieces. The shed skin is not keratinized and may still be living when shed. The epidermal scales do mature as they are pushed to the surface, however keratinization is not part of maturation.

Coloration

The color of amphibians derives from the presence of pigment cells (chromatophores) in the skin. Three classes of chromatophores exist: melanophores with black, brown, or red pigment; white or reflective iridophores; and xanthophores with yellow, orange, or red pigment. Each of the three cell types is structurally different as well as containing a different pigment. The three classes of chromatophores as a unit produce an animal's external coloration, for example, blue of iridophores and yellow of xanthophores produce a green skin.

Melanophores have a central cell body with long, attenuate processes radiating outward. Melanophores occur individually in the epidermis or as part of a chromatophore unit in the dermis. The epidermal melanophores (melanocytes) are common in larvae and are often lost or their number greatly reduced at metamorphosis. The chromatophore unit in the dermis contains a basal melanophore, an iridophore, and a terminal xanthophore; the processes of the melanophore extend upward and over the iridophore. The color produced by the unit depends largely on the color of the pigment in the xanthophore and the reflectivity of the iridophore (see Chapter 6). Melanophores are largely responsible for lightening or darkening of the color produced by the interaction of light in the other two chromatophores. Color changes can occur quickly (less than a minute) by dispersal or contraction of the melanin within the melanophores' processes or slowly (weeks to months) by the increase or decrease of pigment concentration within the chromatophores or by pigment deposited in adjacent cells. The short-term color changes are controlled by hormonal or nervous stimulation.

Muscles and Skeleton

The phyletic transition from fish to amphibian was accompanied by major reorganizations within the musculoskeletal system. As the ancestral amphibians shifted their activities from an aquatic to a terrestrial environment, the buoyant support of water was lost and the pull of gravity required a strengthening of the vertebral column to support the viscera. Simultaneously, the ancestral lineage was shifting from undulatory locomotion to limbed locomotion. The new functions and demands on the musculoskeletal system required a more tightly linked vertebral column, elaboration of the limbs and girdles, and modification of the cranium for capture and ingestion of terrestrial food.

While the postmetamorphic stages of all three amphibian groups are predominantly terrestrial, each group has had a long and independent evolutionary history. Many structural differences appeared during this long divergence, and these differences are nowhere more apparent than in the composition and organization of the musculoskeletal system. The diversity within and between each group permits only a general survey.

Head and Hyoid

The cranial skeleton of vertebrates contains elements from three units: chondrocranium, splanchnocranium, and dermocranium. The chondrocranium or neurocranium comprises the skeleton surrounding the brain (braincase) and the sense organs (olfactory, optic, and otic capsules). The splanchnocranium is the branchial (visceral) arch skeleton, including the upper and lower jaws, hyobranchium, gill arches, and their derivatives. Most elements from these two cranial skeletons appear first as cartilage. The cartilaginous precursors define the position of the later developing bony element. The bone formed by replacement of cartilage is labeled replacement or endochondral bone. The dermocranium contains the roofing elements lying external to the chondro- and splanchnocranial elements. These roofing elements have no cartilaginous precursors, instead ossifications develop in the dermis, hence this type of bone is called dermal (or membrane) bone.

All three crania are represented by numerous skeletal elements in fish and the fish ancestors of amphibians. The earliest amphibians showed a loss of elements from each of the crania and a firmer articulation of the elements remaining. The reduction has continued in the modern orders, which have lost additional elements and often different ones in each group. This loss has been least in the caecilians, where the skull is a major digging tool and must remain sturdy and firmly knit, often with the fusion of adjacent elements.

In extant amphibians, much of the chondrocranium remains cartilaginous throughout life (Fig. 1.3). Only the sphenoethmoid (orbitosphenoid in salamanders), forming the inner wall of the orbit, and the fused prootic and exoccipital, forming the rear of the skull, ossify. Within the skull proper, the bony elements

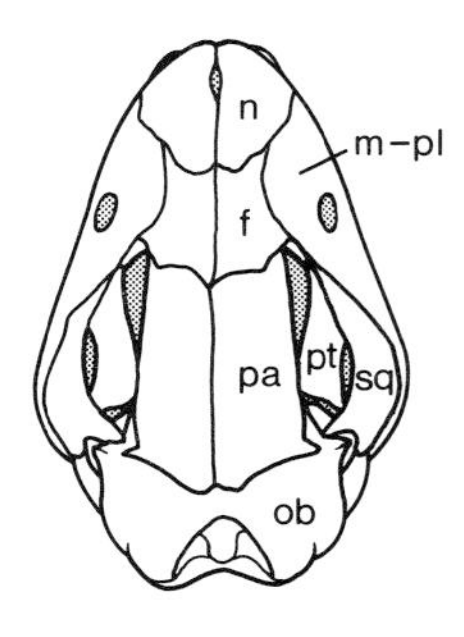

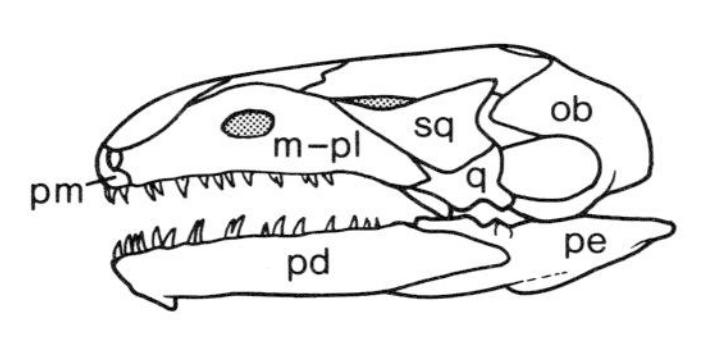

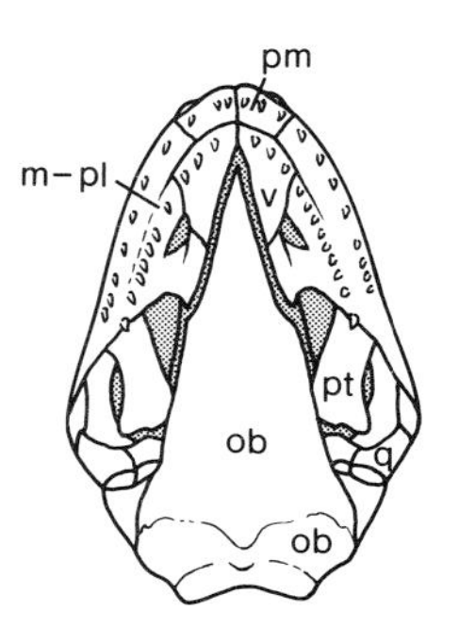

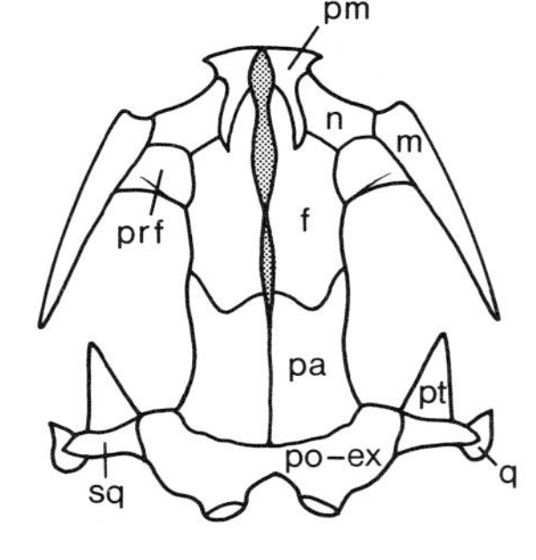

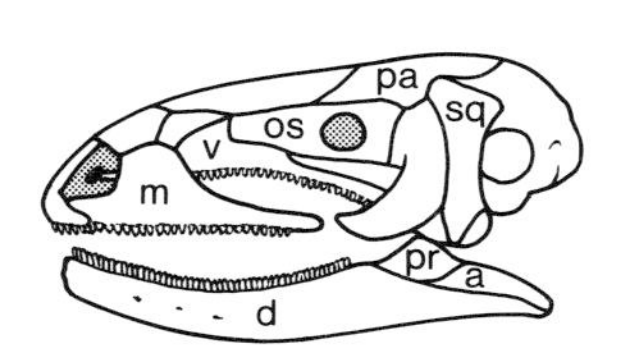

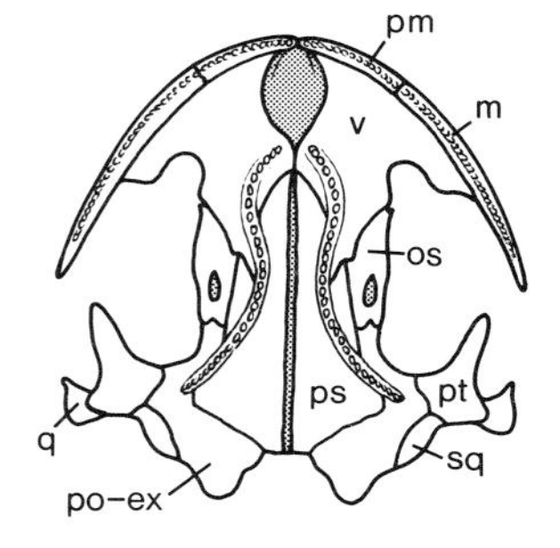

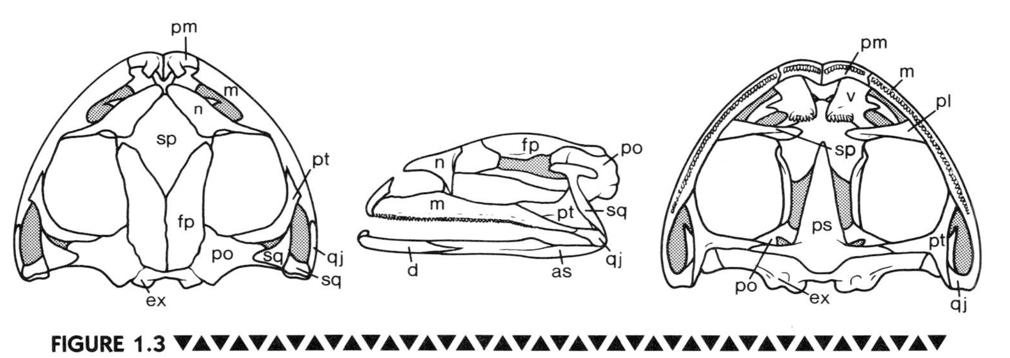

FIGURE 1.3 ▼▲▼

Cranial skeletons of representatives of the three lineages of amphibians. Dorsal, lateral, and ventral views (left to right) of the caecilian *Epicrionops petersi*, the salamander *Salamandra salamandra*, and the frog *Gastrotheca walkeri*, top to bottom. Abbreviations: a, articular; as, angulosplenial; d, dentary; ex, exoccipital; f, frontal; fp, frontoparietal; m, maxillary; mm, mentomeckelian bone; n, nasal; o, occipital; ob, os basale; os, orbitosphenoid; pa, parietal; pl, palatine; pd, pseudodentary; pe, pseudoangular; pm, premaxillary; po, prootic; pr, prearticular; prf, prefrontal; ps, parasphenoid; pt, pterygoid; pv, prevomer; q, quadrate; qj, quadratojugal; sp, sphenethmoid; sq, squamosal; v, vomer. [Adapted from Duellman and Trueb (1986).]

of the splanchnocranium are the columella (ear) and the quadrate (upper jaw); Meckel's cartilage forms the core of the mandible (lower jaw) and ossification in its anterior end forms the mentomeckelian bone and in its posterior end the articular. The dermal bones form the major portion of the adult skull, linking the various cranial elements and forming a protective sheath over the cartilaginous elements, the brain, and the sense organs. The skull is roofed (anterior to posterior) by the premaxillae, nasals, frontals, and parietals; each side of the skull contains the maxilla, septomaxilla, prefrontal, and squamosal. Dermal bones also sheath the skull ventrally, creating the primary palate (roof of mouth), composed of the vomers, palatines, pterygoids, quadratojugals, and a parasphenoid (the only unpaired dermal bone in the amphibian skull). The dermal bones of the mandible are the dentary, angular, and prearticular; all encase Meckel's cartilage. Teeth occur commonly on the premaxillae, maxillae, vomers, palatines, and dentaries.

The jaws of vertebrates arose evolutionarily from the first visceral (branchial) arch. The second visceral (hyomandibular) arch supported the jaws and bore gills, and the third and subsequent visceral arches comprised the major gill arches. Remnants of many of these arches remain in modern amphibians. The jaws consist mostly of dermal bones; only the mentomeckelian, articular, and quadrate are bony remnants of the first arch. The quadrate becomes part of the skull proper and the dorsalmost element of the hyomandibular arch becomes the columella for the transmission of sound waves from the external eardrum (tympanum) to the inner ear. The ventral portion of the second arch persists as part of the hyoid apparatus. The subsequent two to four visceral arches may persist, at least in part, as gill arches in larvae and some gilled adults (e.g., Proteidae) and also as elements of the hyoid in juveniles and adults. Some elements from the more posterior visceral arch become structural supports in the glottis, larynx, and trachea.

The composition and architecture of the hyoid is highly variable within and between each group of living amphibians. However, in all, the hyoid lies in the floor of the mouth and forms the structural support for the tongue. In some species, the components of the hyoid can be traced accurately to their visceral arch origin; in others, their origin from a specific arch element is uncertain. The hyoid elements in primitive salamanders (Fig. 1.4) retain an architecture similar to that of the fish visceral arches but with the loss of arch elements. In the more advanced salamanders, the number of hyoid elements is further reduced. The hyoid remains cartilaginous in caecilians (Fig. 1.4) without segmentation of hyoid arms into individual elements. The anuran hyoid is a single cartilaginous plate with two to four processes and has little resemblance to its visceral arch precursor.

The cranial musculature contains two main functional groups: jaw movement and respiratory-swallowing muscles. The jaw muscles fill the temporal area of the skull, extending from the parietal-prootic-squamosal area to the mandible.

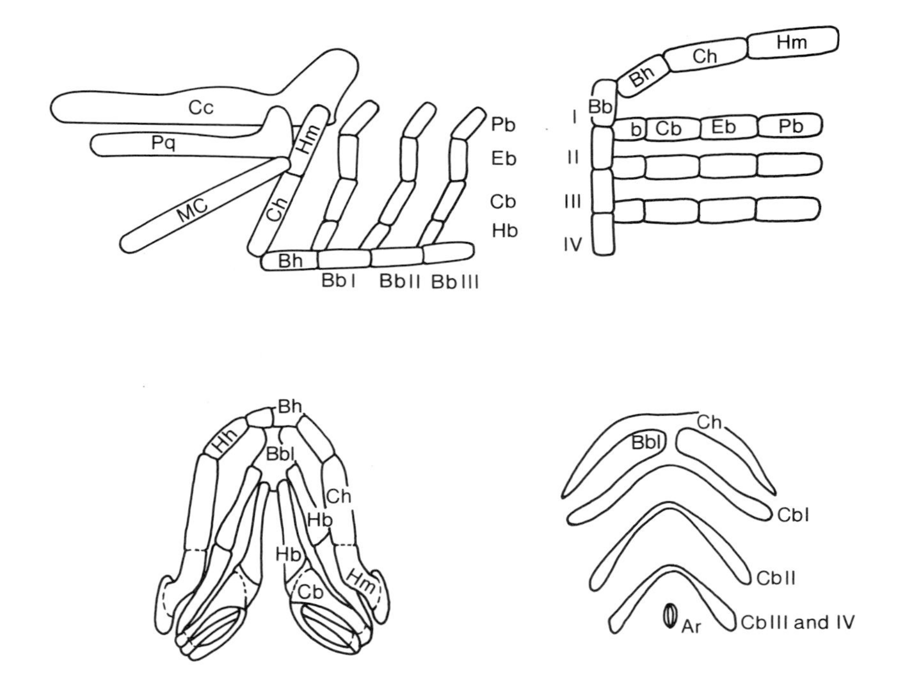

FIGURE 1.4 ▼▲▼

The hypobranchial skeleton of amphibians. Basic organization of the hypobranchial skeleton for vertebrates, lateral view (upper left) and dorsal view (upper right) with left side removed. Hypobranchial skeletons of the salamander *Cryptobranchus*, dorsal view (lower left), and the caecilian *Ichthyophis*, ventral view (lower right). Abbreviations: Ar, arytenoid cartilage; Bb, basibranchial; Bh, basihyal; Cb, ceratobranchial; Cc, chondrocranium; Ch, ceratohyal; Eb, epibranchial; Hb, hyobranchial; Hh, hyohyal; Hm, hyomandibular; MC, Meckel's cartilage; Pb, pharyngobranchial; Pq, palatoquadrate. [Adapted from Duellman and Trueb (1986).]

The muscles attaching to the dorsal surface of mandible close the mouth; those attaching to the lateral and ventral surface of the mandible open the mouth. The respiratory-swallowing muscles form the floor of the mouth, the throat, and neck. These muscles move and support the gills and/or the hyoid and tongue.

Vertebral Column

The amphibian vertebral column must combine rigidity and strength to support the head, limb girdles, and viscera with flexibility to permit lateral and dorsoventral flexure of the column. These seemingly conflicting roles are met by the presence of sliding and rotating articular facets on the ends of each vertebra and overlapping sets of muscular slips linking adjacent vertebrae.

Each vertebra consists of a ventral cylinder (centrum) and a dorsal arch (neural arch) that may have a dorsal projection (neural spine). The anterior end of the centrum articulates with the posterior end of the preceding centrum. These central articular surfaces are variously shaped: convex anterior and concave posterior central surfaces in opisthocoelous vertebrae; concave anterior and convex posterior surfaces in procoelous ones; both anterior and posterior surfaces concave in amphicoelous ones. Intervertebral discs, usually of fibrocartilage, lie between central surfaces of adjacent vertebrae. A pair of flat processes extend from the anterior (prezygapophyses) and posterior (postzygapophyses) edges of the neural arch, forming another set of articulations between adjacent vertebrae. Articular surfaces for the ribs lie on the sides of each vertebra, dorsally a diapophysis at the base of the neural arch and a parapophysis on the side of the centrum. Ribs are much shorter in amphibians than in other tetrapods (i.e., Reptilia and Mammalia) and do not extend more than halfway down the sides.

The first postcranial vertebra (atlas) is modified to create a mobile attachment between the skull and the vertebral column. The atlantal condyles on the anterior surface articulate with the paired occipital condyles of the skull. The succeeding vertebrae of the trunk match the general pattern just described. The number and shape of the vertebrae differ in the three amphibian groups. Salamanders have 10–60 presacral vertebrae: a single atlas or cervical vertebra and variable number of trunk vertebrae. The trunk vertebrae are all very similar, possess well-developed zygapophyses, neural spines, and usually bicapitate (two-headed) ribs. Rather than exiting intervertebrally between neural arches of adjacent vertebrae as in other vertebrates, the spinal nerves of salamanders often exit through foramina in the neural arches. Postsacral vertebrae are of variable number but always present and differentiated into two to four precaudal (cloacal) and numerous caudal vertebrae. Caecilians have 60–285 vertebrae: a single atlas, numerous trunk vertebrae, no sacral vertebrae, and a few irregular bony nodules representing the precaudal vertebrae. The trunk vertebrae are robust with large centra and neural spines; most bear bicapitate ribs. Frogs have five to eight presacral vertebrae. The atlas (presacral I) lacks transverse processes, usually present on all the other presacral vertebrae. Ribs are absent in most frogs, present only in leiopelmatid, discoglossid, and pipid frogs and only then on the presacrals II–IV. The sacral vertebra has large transverse processes (called sacral diapophyses, although whether they are true diapophyses is uncertain) and articulates posteriorly with an elongate urostyle, which represents a rod of fused postsacral vertebrae.

The musculature of the vertebral column consists of epaxial (dorsal trunk) and hypaxial (flank or ventral trunk) muscles. The epaxial muscles consist largely of longitudinal slips linking various sets of adjacent vertebrae. These muscles lie principally above rib attachments (apophyses) and attach to the neural arches and spines. They are the muscular components providing the rigidity and strength to the vertebral column. The hypaxial muscles support the viscera and contain the oblique muscle series of the flanks and the rectus muscle series midventrally along the abdomen.

Girdles and Limbs

The limbs of amphibians and other tetrapods are specialized fins for terrestrial locomotion. The girdle and limb components (appendicular muscles and skeleton) of tetrapod vertebrates derive from the girdle and fin components of their fish ancestors. Several opposite trends are evident in the evolution of limbs from fins. The anterior (pectoral) girdle loses its articulation with the skull and a number of its elements. In contrast, the posterior (pelvic) girdle elaborates and enlarges its elements and articulates with the vertebral column. Within the limbs, the number of skeletal elements are reduced, and a series of highly flexible joints appear between the proximal to distal limb segments: propodial segment (humerus or femur); epipodial (radius and ulna or fibula and tibia); mesopodial (carpal or tarsal elements); metapodial (metacarpals or metatarsals); and phalanges (Fig. 1.5). These trends largely reflect the change in function of the appendages from steering and stability in fish locomotion to support and propulsion in amphibian locomotion.

The girdles provide the internal support for the limbs and translate limb movement into locomotion. Primitively, the amphibian pectoral girdle contained dermal and endochondral elements. The endochondral coracoid and scapula form the two arms of a V-shaped strut with a concave facet (glenoid fossa) at their juncture; the glenoid fossa is the articular surface for the head of the humerus. The dermal elements (cleithral elements and a clavicle) strengthen the endochondral girdle. A dermal interclavicle—the only unpaired pectoral element—provides midventral strengthening to the articulation of clavicles and coracoids of the two sides. This midventral articulation includes the sternum posteriorly. The pelvic girdle and limbs (fore and hind) contain only endochondral elements. Three paired elements form the pelvic girdle. A ventral plate contains the pubes anteriorly and the ischia posteriorly; an ilium projects upward on each side from the edge of the puboishial plate and articulates with the diapophyses of the sacral vertebra. A concave facet (acetabulum) lies at the juncture of the three pelvic elements and is the articular surface for the head of the femur.

The girdles are anchored to the trunk by axial muscles. Since the pectoral girdle lacks an attachment to the axial skeleton, a series of muscles form a sling extending from the back of the skull across the anterior trunk vertebrae to insert on the scapula and humerus. The pelvic girdle has a bony attachment to the vertebral column, hence its muscular sling is less extensive. The muscles of the limbs divide into a dorsal extensor and a ventral flexor unit. Within each unit, most of the muscles are single joint muscles and cross only a single joint such as girdle to humerus or humerus to ulna.

Caecilians have lost all components of the appendicular skeleton and musculature. Limbs and girdles are present in most salamanders, although they may be reduced in size and have lost distal elements, as in the dwarf siren. All frogs

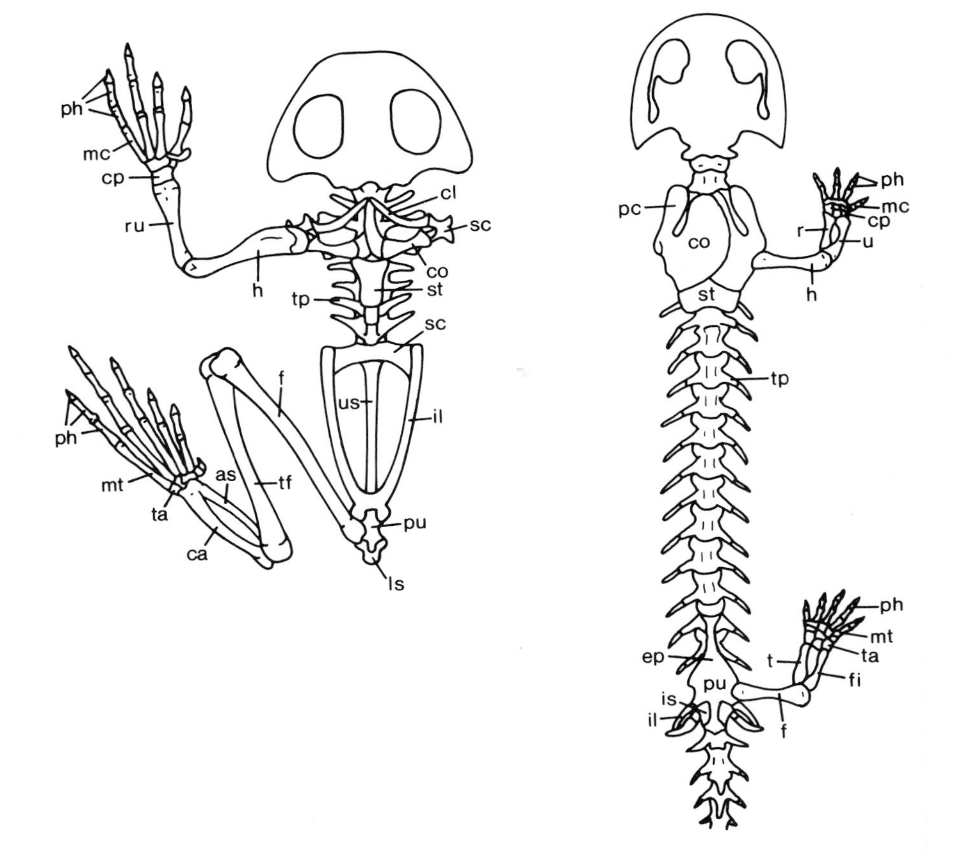

FIGURE 1.5 Postcranial skeletons (ventral view) of a gray treefrog (left, *Hyla versicolor*) and a hellbender (right, *Cryptobranchus alleganiensis*). Abbreviations: as, astragulus; ca, calcaneum; cl, clavicle; co, coracoid; cp, carpals; ep, epipubis; f, femur; h, humerus; il, ilium; is, ischium; mc, metacarpals; mt, metatarsals; ph, phalanges; pc, procoracoid; pu, pubis; r, radius, ru, radioulna; sc, scapula; sd, sacral diapophysis; st, sternum; sv, sacral vertebrae; t, tibia; ta, tarsals; tf, tibiofibula; tp, transverse process; u, ulna; us, urostyle. [Adapted from Cope (1898).]

possess well-developed limbs and girdles. Salamanders and frogs have only four (or fewer in some salamanders) digits on the forefeet; the missing digit is the fifth (postaxial or outer) digit. The hindfeet of anurans and salamanders usually retain all digits; if one is lost, it is also the fifth digit.

Reduction and loss are common features of the salamander skeleton. The pectoral girdle remains largely cartilaginous and contains only three elements: scapula, procoracoid, and coracoid. These three are usually indistinguishably fused and ossified only in the area of the glenoid fossa. The left and right halves of the girdle overlap and do not articulate with one another. A small, diamond-shaped, cartilaginous sternum lies on the ventral midline posterior to the girdle halves and is grooved anteriorly for a sliding articulation with the edges of the coracoids. The humerus, radius, and ulna have ossified shafts, but their ends remain cartilaginous. The carpals are often entirely cartilaginous as well or have a small ossification node in the center of the larger carpal elements. Reduction by loss and fusion of adjacent carpals is common in salamanders. The phalanges ossify, but their number in each digit is reduced. The common phalangeal formula for modern amphibians is 1-2-3-2 or 2-2-3-3, compared to the 2-3-4-5-4 formula for the early amphibians.

The salamander pelvic girdle has a more robust appearance than the pectoral girdle. The ilia and ischia are ossified, although the pubes remain largely cartilaginous. The two halves of the girdle are firmly articulated, and a Y-shaped cartilaginous rod (ypsiloid cartilage) extends forward and likely supports the viscera. The hindlimb elements show the same pattern of ossification as those of the forelimbs; the hindfoot is typically 1-2-3-3-2 and the loss of the fifth toe is not uncommon, for example, in *Hemidactylium*.

The appendicular skeleton of frogs is robust and well ossified. The saltatory locomotion of anurans, in both jumping and landing, requires a strong skeleton. The pectoral girdle contains a scapula capped by a bony cleithrum and a cartilaginous suprascapula and, ventrally, a clavicle and a coracoid; an omosternum (or episternum) and a sternum extend anteriorly and posteriorly, respectively, from the midline of the girdle. Two types of girdles occur in anurans: arciferal and firmisternal girdles. In both types, the clavicles articulate firmly on the midline. In the firmisternal girdle, the coracoids are joined firmly through the fusion of their epicoracoidal caps. In contrast, the epicoracoidal caps overlap in arciferal girdles and can slide pass one another. The two girdle types are quite distinct in many species, although in others, the girdle structure is intermediate. The humerus is entirely ossified and has an elevated, spherical head. The epipodial elements fuse into a single bony element, the radioulna. The carpal elements are bony and reduced in number by fusion. The phalangeal formula is rarely reduced from 2-2-3-3.

The anuran pelvic girdle is unlike that of any other tetrapod. The puboishiadic plate is compressed into a bony, vertical semicircular block on the midline; the ischia lie posterodorsally and the pubes form the ventral edge. The ilia complete the anterior portion of the pelvic block, and each ilium also projects forward as an elongate blade that attaches to the sacral diapophysis. The hindlimb elements are elongate and proportionately much longer than the forelimb. The epipodial elements are also fused into a single bone, the tibiofibula, which is typically as long or longer than the femur. Two mesopodial elements, fibulare and tibiale, are greatly elongated, giving frogs a long ankle. Most of the other mesopodial elements are lost or greatly reduced in size. With the exception of two species, frogs have five toes and seldom deviate from a 2-2-3-4-3 phalangeal formula.

Nerves and Sense Organs

The nervous system of vertebrates has four morphologically distinct, but integrated units: central nervous system, peripheral nervous system, autonomic nervous system, and sense organs. The first three of these units are composed principally of neurons or nerve cells, each of which consists of a cell body and one or more processes (axons and dendrites) of various lengths. The appearance of nervous system structures depends on the organization of and the parts of the neurons within the structure. For example, nerves are bundles of axons, and the gray matter of the brain results from concentrations of cell bodies. The sense organs show a greater diversity of structure and organization, ranging from single-cell units for mechanoreception to the more multicellular eyes and ears. Neurons or parts of neurons are important components of sense organs, but most sense organs require and contain a variety of other cell and tissue types to make them functional organs.

Nervous Systems

The central nervous system includes the brain and the spinal cord. Both derive embryologically and evolutionarily from a middorsal neural tube. The anterior end of this tube enlarges to form the brain, which serves as the major center for the coordination of neuromuscular activity and for the integration of and response to all sensory input. The brain is divided during development by a flexure into the forebrain and hindbrain (Fig. 1.6). The fore- and hindbrain are each further partitioned, structurally and functionally, into distinct units. From anterior to posterior, the forebrain consists of telencephalon, diencephalon, and mesencephalon; the metencephalon and myelencephalon (medulla) form the hindbrain. Twelve pairs (ten in extant amphibians) of cranial nerves arise from the brain: olfactory (I) from the telencephalon; optic (II) from the diencephalon; oculomotor (III), trochlear (IV), and abducens (VI) from the mesencephalon; and trigeminal (V), facial (VII), auditory (VIII), glossopharyngeal (IX), and vagus (X) from the medulla. The accessory (XI) and hypoglossal (XII) cranial nerves also originate from the myelencephalon in other vertebrates, but apparently a shortening of the cranium places them outside the skull in amphibians, hence they become spinal nerves.

The embryonic flexure disappears in amphibians as subsequent embryonic growth straightens the brain. The morphology of the brain is similar in the three living groups, although the brain is shortened in frogs and more elongate in salamanders and caecilians. The telencephalon contains elongate and swollen cerebral hemispheres dorsally encompassing the ventral olfactory lobes. The cerebral hemispheres comprise half of the total amphibian brain. The small, unpaired diencephalon lies behind the hemispheres and merges smoothly into the mesencephalon's bulbous optic lobes. Internally, the diencephalon is divided into the epithalamus, thalamus, and hypothalamus. A small pineal organ (epiphy-

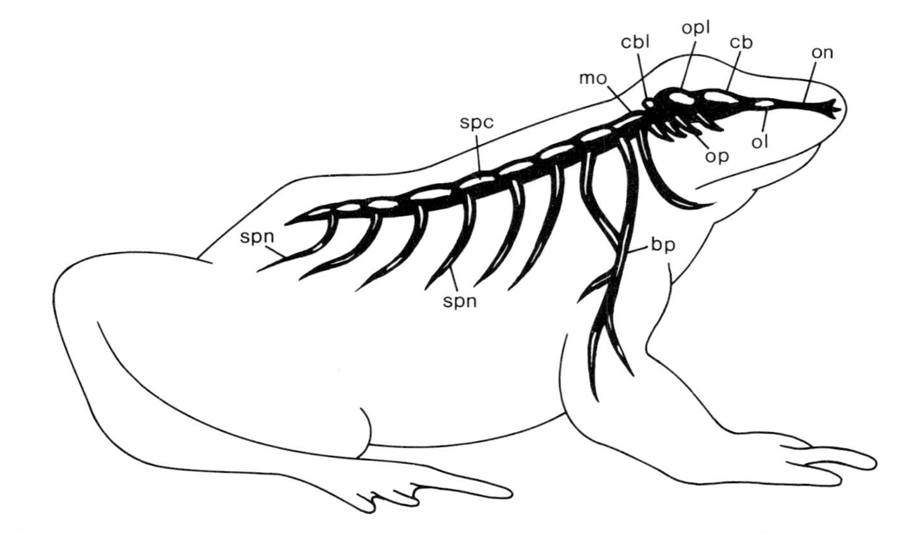

FIGURE 1.6

Brain and spinal cord of a frog; a diagrammatic lateral view. Abbreviations: bp, brachial plexus; cb, cerebrum; cbl, cerebellum; mo, medulla oblongata; ol, olfactory lobe; on, olfactory nerve; op, optic nerve; opl, optic lobe; spc, spinal cord; spn, spinal nerve.

sis) projects dorsally from the epithalamus; a parietal process, lying anterior to the epiphysis, is absent in extant amphibians. The ventral hypothalamus holds anteriorly the optic chiasma, where the optic nerves cross as they enter the brain, and posteriorly the infundibular area, from which projects the hypophysis or pituitary gland. Behind the optic lobes, the hindbrain is a flattened triangular area tapering gradually into the spinal cord. Neither the cerebellum (base of the triangle abutting the optic lobes) nor the medulla is enlarged.

The spinal cord is a flattened cylinder of nerve cells extending caudad through the vertebrae. Spinal nerves exit the cord in a regular, segmental fashion as bilateral pairs. Each spinal nerve has a dorsal (sensory) and a ventral (motor) root, which fuse near their origins and soon divide into dorsal, ventral, and communicating nerve branches. The neurons of the first two branches innervate the body wall (somatic = skin, muscle, and skeleton); the neurons of the communication branch join the central nervous system to the autonomic system innervating the viscera (visceral = digestive, urogenital, circulatory, endocrine, and respiratory organs).

The spinal cord extends to the end of the vertebral column in salamanders and caecilians, but in anurans, the cord ends at the 6–7th vertebrae and a bundle of spinal nerves (cauda equina) continues caudad through the neural canal. The diameter of the cord is nearly uniform from brain to base of tail, except for slight expansion in the region of the limbs. The organization of the spinal nerves is similar in all living amphibians. The dorsal root contains somatic and visceral sensory neurons and some visceral motor neurons; the somatic motor and some visceral motor neurons comprise the ventral root.

The nerves and their ganglia (aggregations of neuron cell bodies) outside the skull and vertebral column comprise either the peripheral or autonomic nerve systems. The peripheral system contains the somatic sensory neurons and axons of motor neurons; the autonomic system contains the visceral sensory and some motor neurons. The latter are generally associated with the involuntary activity of the smooth muscles and glands of the viscera. Both the peripheral and autonomic systems appear similar in the three amphibian groups, but neither system has been studied extensively and the autonomic system least so. Importantly, the peripheral nerves transmit the animal's perception of the outside world to the central nervous system and then transmit messages to the appropriate organs for the animal's response.

Sense Organs

The sense organs provide the animal with information about itself and its surroundings. Both sets of sense organs, the internal ones monitoring the internal environment and the external ones monitoring the external environment, are integrated directly to the central nervous system or to it through the autonomic and peripheral networks. The eyes, ears, and nose are obvious external receptors. Heat and pressure receptors of the skin are less obvious, as are also internal receptors, for example, proprioceptors of joints and muscles.

Cutaneous Sense Organs

The skin contains a variety of receptors that register the environment's impingement on the animal's exterior. Pain and temperature receptors consist of free and encapsulated nerve endings, lying in the dermis and a few extending into the epidermis. Mechanoreceptors, sensitive to pressure and touch, are similarly positioned in the skin. The pressure receptors may also sense temperature.

The lateral line system is the most evident of the cutaneous sense organs. Superficially it appears as the series of dots or dashed lines on the head and body of the aquatic larvae and some strongly aquatic adults (cryptobranchid, amphiumid, proteid, and sirenid salamanders, typhlonectid caecilians, and pipid frogs). The mechanoreceptor organs or neuromasts are arranged singly or in compact linear arrays (called stitches) to form the various lines (canals) traversing the head and trunk. Each neuromast contains a small set of cilia projecting from its outer surface; the cilia bend in only one axis, hence sensing water pressure changes only along that axis. They are sensitive to very light currents and are used to locate food. Their importance as sense organs is attested by the numerous lines of neuromasts on the head and their universal presence in aquatic amphibians, being reduced only in species living in rapidly flowing water.

Recently, ampullary organs were discovered on the heads of some larval salamanders and caecilians. These electroreceptors are less numerous, lying in rows parallel to the neuromasts. Like the neuromast, the ampullary organs provide the larva with a sense of its surroundings, identifying both stationary and moving objects in the electrical field around the larva.

Ears

The ears of tetrapods, whether frogs or birds, are structurally similar and serve two functions: hearing, the reception of sound waves; and balance, monitoring the position and movement of the animal's head. The receptors for both functions are neuromasts located in the inner ear. These neuromasts differ somewhat from those of the lateral line system, but they similarly record fluid movements along a single axis by the deflection of terminal cilia.

Ears are paired structures, one on each side of the head just above and behind the articulation of the lower jaw. Each ear consists of an inner, middle, and outer unit. The inner ear is a fluid-filled membranous sac, containing the sensory receptors and suspended in a fluid-filled cavity of the bony or cartilaginous otic capsule. The middle ear contains the bone and muscular links that transfer vibrations from the ear drum (tympanum) to the inner ear. An outer ear is usually no more than a slight depression of the tympanum or may be absent. Salamanders, caecilians, and a few frogs lack tympana. In these amphibians, low-frequency sounds may be transmitted via the appendicular and cranial skeleton to the inner ear.

Unlike amniotes, the amphibian middle ear has two auditory pathways: the tympanum–columella path for airborne sounds and the forelimb–opercular path for seismic sounds. Both pathways reach the inner ear through the fenestra ovalis of the otic capsule. The tympanum–columella path is shared with the other tetrapods. In amphibians, it is a single bony rod (columella) extending between the external eardrum and the fenestra ovalis of the inner ear. In most frogs, the columella lies within an air-filled cavity, and in salamanders and caecilians the columella is embedded in muscles. The limb–opercular path is unique to frogs and salamanders. Sound waves are transmitted from the ground through the forelimb skeleton onto the tensed opercular muscle joining the shoulder girdle to the operculum lying in the fenestra ovalis.

The membranous inner ear consists basically of two sacs joined by a broad passage. The dorsal sac or utriculus has three semicircular canals projecting outward from it. One of these canals lies horizontally, the other two are vertical, and all three are perpendicular to one another, thus oriented and recording movement in three different planes. The neuromasts are not scattered throughout the canals and sacs, instead they lie in patches, one patch in each semicircular canal and one or more patches in the utriculus and the ventral sac, the sacculus. The ventral sacculus also contains several outpocketings, the amphibian papilla, basilar papilla, lagena, and endolymphatic duct. The two papillae contain patches of neuromasts specialized for acoustic reception.

Eyes

Eyes vary from large and prominent to small and inconspicuous in extant amphibians. All have a pair of laterally placed eyes. Most terrestrial-arboreal salamanders and frogs have moderate to large eyes; fossorial and aquatic species usually have small eyes; and eyes are degenerate and beneath the skin in caecilians and cave-dwelling salamanders, even lying beneath bone in a few caecilians.

The structure of the eye is similar in all vertebrates. It is a hollow sphere lined internally with a heavily pigmented sensory layer, the retina. The retina is supported by the sclera, a dense connective tissue sheath forming the outside wall of the eyeball. The cornea is the transparent part of the outer sheath lying over a gap in the retina and allows light to enter the eye. In postmetamorphic amphibians, eyelids and a nictitating membrane slide across the exposed cornea to protect and moisten it. A spherical lens lies behind the cornea and is anchored by a corona of fibers extending peripherally to the cornea–scleral juncture. The amount of light passing through the lens and onto the retina is regulated by a delicate, pigmented iris lying behind the cornea. Its central opening, the pupil, is opened (dilated) or closed (contracted) by peripherally placed muscles. The eye retains its spherical shape by presence of fluid, vitreous humor in the cavity behind the lens and aqueous humor in front of the lens.

Light enters the vitreous chamber through the iris and is focused on the retina by the lens. The organization of the retina's several layers differs from what might be expected. The sensory or light-registering surfaces are not the innermost surface of the eye. Instead, the innermost layer consists of transmission axons carrying impulses to the optic nerve; the next layer contains connector neurons transferring impulses from the adjacent receptor cell layer; and the deepest layer contains the pigment cells adjacent to the sclera. The actual receptor surfaces of the sensory cells face inward (not toward the incoming light) against and in the pigment layer. Amphibians have four kinds of light receptors: red and green rods, single and double cones. The rods are the color receptors. They possess specialized pigments that are sensitive (i.e., change chemical state) to a narrow range of wavelengths. Amphibians are the only vertebrates with two types of rods, and the green rods are unique to amphibians, although these rods are absent in taxa with degenerate eyes. The visual pigment of the cone is sensitive to all wavelengths of light, hence cones register only the presence or absence of light.

Nasal Organs

Olfaction or smelling is performed by two bilaterally paired structures: the nose and the vomeronasal (Jacobson's) organ. The nose opens to the exterior through the external naris and internally into the buccal cavity via the choana (internal naris). Between these openings is the large olfactory (principal) cavity and several accessory chambers extending laterally and ventrally; the vomeronasal organ is in one of the accessory chambers. A nasolacrimal duct extends from the anterior corner of each eye to the principal cavity. The surfaces of the chambers are lined with ciliated epithelium, containing support and mucous cells. The ciliated neuroepithelium occurs in three patches: the largest patch occupying the roof, medial wall, and anterior end of the principal cavity; a small, protruding patch on the middle of the floor; and another small patch in the vomeronasal organ chamber. The neuroepithelium of the principal cavity is innervated by neurons from the olfactory bulb of the brain, and the vomeronasal organ by a separate olfactory branch to the brain. Olfaction is a chemosensory reaction. The actual receptor site on the cell is unknown but may be either on the base of each cilium or near the cilium's junction with the cell body.

The nose of salamanders is relatively simple with a large main cavity partially divided by a ventrolateral fold. Aquatic salamanders have the simplest and smallest nasal cavities, but a large vomeronasal organ. Frogs, in general, have a complex nasal cavity of three chambers and a large vomeronasal organ. Caecilians have a simple nasal cavity similar to salamanders, but with a major modification, a sensory tentacle. The size, position, and structure of the tentacle vary among the different species; however, in all, the tentacle arises from a combination of nasal and orbital tissues as a tubular evagination from the corner of the eye. The tentacle's exterior sheath is flexible, but nonretractable. The tentacle proper can be extruded and retracted into its sheath, where odor particles are transported via the nasolacrimal duct to the vomeronasal organ.

Internal Sense Organs

The major internal sense organs are the proprioceptor organs embedded in the muscles, tendons, ligaments, and joints. These organs record the tension and stress on the musculoskeletal system and allow the brain to coordinate the movement of limbs and body during locomotor and stationary behaviors. The proprioceptors show a structural diversity from simple nerve endings and netlike endings to specialized corpuscles.

Taste buds or gustatory organs are present in all amphibians, although they have been little studied and nearly exclusively in frogs. There are two types: papillary discs and nonpapillary organs. The discs occur only on the tongue and always on the surface of the tongue's fungiform papillae. The nonpapillary organs occur throughout the buccal cavity, except on the tongue. Each type of taste bud is a composite of receptor and support cells. The buds are highly sensitive to salts, acids, quinine (bitter), and pure water.

Heart and Vascular Network

The circulatory system is a transport system carrying nutrients and oxygen to all body tissues and removing waste products and carbon dioxide. This system contains four components: blood, the transport medium; vascular and lymphatic vessels, the distribution network; and the heart, the pump or propulsive mechanism.

Blood

Amphibian blood is a colorless fluid (the plasma) with three major types of blood cells (erythrocytes, leucocytes, and thrombocytes). The blood cells are typically nucleated, although in salamanders a small portion of each of the three types lack nuclei. The erythrocytes carry oxygen to and carbon dioxide from the tissues; both gases attach to the respiratory pigment hemoglobin. The leucocytes consist of a variety of cell types, most of which are involved in maintenance duties such as removing cell debris and bacteria or producing antibodies. The thrombocytes serve as clotting agents. Only the erythrocytes are confined to vascular vessels; the other blood cells and the plasma leak through the walls of the vascular vessels and bathe the cells of all tissues. They may reenter the vascular vessels directly or collect in the lymphatic vessels that empty into the vascular system.

Arterial and Venous Circulation

The vascular vessels form a closed network of ducts, beginning and ending at the heart (Fig. 1.7). Blood leaves the heart through the arteries. These divide into smaller and smaller vessels (arterioles); eventually the vessels (capillaries) are only slightly larger than the blood cells flowing through them. Here within the capillary beds, the plasma and some blood cells leak through to the lymphatic system. Beyond the capillaries, the vessels join, becoming increasingly larger (venules to veins), and the veins return the blood to the heart.

Blood exits the heart through the ventral aorta, which bifurcates and divides again after a short distance. The first pair of vessels remain ventral aortae; each aorta then branches successively into a pulmocutaneous artery, a systemic arch, and a carotid arch. The branch to the skin and lung serves the respiratory surfaces for gaseous exchange. The systemic arch curves dorsally and fuses on the midline with its bilateral mate to form the dorsal aorta. Vessels branch from the dorsal aorta as it extends posteriorly and these branches provide blood to all the viscera and limbs. The branches of the carotid arch carry blood to the tissues and organs of the head and neck. The venous system shares this general distributional pattern but in reverse. A pair of common jugular veins drain the numerous veins of the head and neck; the subclavian veins gather blood from the smaller veins of the forelimb and skin; and the pulmonary veins drain the lungs. A single postcaval vein is the major efferent vessel for the viscera and hindlimbs. All of these veins, except the pulmonary, empty into the sinus venosus, which opens directly into the right atrium. The sizes, shapes, and branching patterns within the vascular network are nearly as variable within groups as between unrelated taxa. There are a few distinctive conditions, such as the retention of aortic arches IV and V in salamanders giving them double systemic arches rather than a single one (IV) as in frogs.

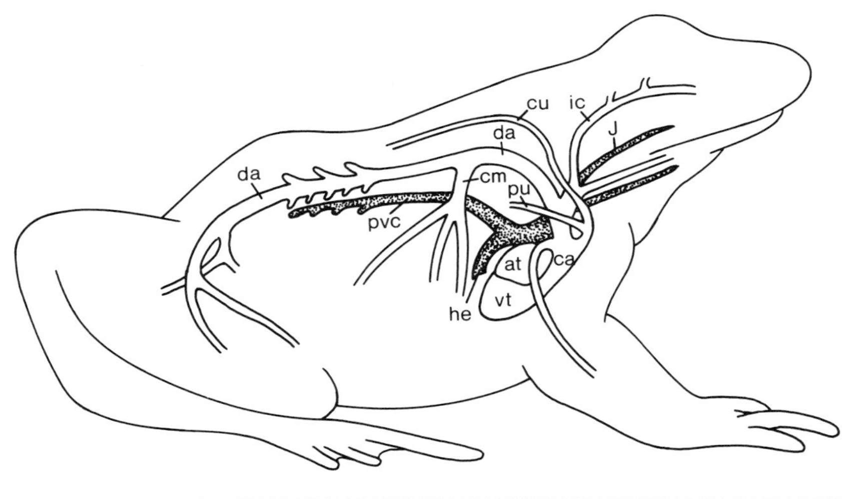

FIGURE 1.7 ▼▲▼

Circulatory system of a frog; a diagrammatic lateral view; arteries as open vessels, veins in gray. Abbreviations: Heart—at, atrium; vt, ventricle. Arteries—ca, conus arteriosus; cm, coeliacomesenteric; cu, cutaneous; da, dorsal aorta; ic, internal carotid; pu, pulmonary. Veins—he, hepatic; j, jugular; pvc, posterior vena cava.

Lymphatic Network

The lymphatic network is an open system, containing both vessels and open cavities (sinuses within the muscles and visceral mesenteries, and beneath the skin). It is a one-way network, collecting the plasma that has leaked out of the capillaries and returning it to the vascular system. The lymph sinuses are the major collection sites, and the subcutaneous ones are especially large in frogs. The sinuses are drained by lymphatic vessels that empty into veins. In amphibians and fishes, lymph hearts lie at venous junctions; they are contractile structures with backflow valves, thereby speeding the flow of lymph into the veins. Frogs and salamanders have 10–20 lymph hearts; the elongate caecilians have more than a hundred.

Heart

Heart structure is highly variable in amphibians. All share a three-chambered heart (two atria and one ventricle), but the actual morphology of the chambers and the pattern of blood flow through the chambers differ. The differences are associated with the relative importance of cutaneous and pulmonary respiration. Even differences in an amphibian's physiological state modifies flow pattern, for example, a hibernating frog might mix pulmonary and systemic blood in the ventricle whereas an active frog will not. The atria are thin-walled sacs separated by an interatrial septum. The sinus venosus empties into the right atrium, and pulmonary veins into the left atrium. Both atria empty into the thick, muscular-walled ventricle, which pumps the blood into the conus arteriosus (base of the ventral aorta), whence blood flows into the various arteries and arches. Although the ventricle is not divided by a septum, oxygenated and unoxygenated blood can be directed into different arterial pathways. Such segregation is possible owing to the volume and position of the blood in the ventricle, the nature of ventricular contraction, the spiral fold of the conus arteriosus, the branching pattern of the arteries from the conus, and the relative resistance of the pulmonary and systemic pathways.

Digestive and Respiratory Structures

The digestive and pulmonary systems are linked by a common embryological origin, similar functions, and shared passageways. The lungs and respiratory tubes form as an outpocketing of the principal regions. Both systems are intake ports and processors for the fuels of living: oxygen (respiratory), water, and food (digestive).

Digestive Structures

The digestive system has two major components: the digestive tube with specialized regions and the digestive glands. The digestive tube or tract extends from the mouth to the anus, which empties into the cloaca. From beginning to end, the regions are the buccal (oral) cavity, pharynx, esophagus, stomach, and small and large intestines. The general morphology of these regions is similar within the amphibians, differing mainly in their lengths: short in anurans, long in caecilians.

The mouth opens directly into the buccal cavity and is bordered by flexible, immobile lips. The buccal cavity is continuous posteriorly at the angle of the jaw with the pharynx. Its roof is the primary palate, its floor the tongue. The tongue is variously developed in amphibians, from little more than a bump on the floor in pipid frogs to elaborate projectile tongues in some salamanders and many advanced frogs. In its least-developed form, the tongue is a small muscular pad lying on a simple hyoid skeleton. The more mobile tongues (not necessarily protrusible, as in caecilians) have a more elaborate hyoid skeleton and associated musculature with a glandular pad attached to the muscular base.

Amphibian teeth are typically simple structures; each tooth has an exposed bicuspid crown anchored to a base (pedicel) in the jaw. Caecilians and a few frogs have unicuspid curved teeth. Salamanders and caecilians have teeth on all the jawbones; most frogs lack teeth on the lower jaw and a few lack teeth on the upper jaw.

The pharynx is the antechamber for directing the food into the esophagus and air into the lungs. A muscular sphincter controls the movement of food in the thin-walled esophagus and peristalsis moves food downward into the stomach. The stomach is an enlarged and expandable region of the digestive tube.

Its thick muscular walls and secretory lining initiate the first major digestive breakdown of food. The food bolus passes from the stomach through the pyloric valve into the narrower and thin-walled small intestine. The forepart of the small intestine is the duodenum, which receives the digestive juices from the liver and pancreas. The small intestine of amphibians has only a small amount of internal folding and villi to increase surface area for nutrient absorption. It is continuous with a slightly broader large intestine in caecilians, salamanders, and some frogs; in the more advanced frogs, a valve separates the large and small intestines. The large intestine empties into the cloaca, which is a saclike cavity receiving the products and by-products of the digestive, urinary, and reproductive systems. The cloaca exits to the outside through the vent.

Digestive Glands

A variety of glands occur within the digestive tract. The lining of the buccal cavity contains unicellular and multicellular glands. The multicellular ones secrete mucus that lubricates the surface, and although numerous and widespread in terrestrial amphibians, they are less abundant in aquatic taxa such as pipid frogs and neotenic salamanders. The intermaxillary gland opens in the middle of the palate and secretes a sticky compound that helps prey adhere to the tip of the tongue. Numerous unicellular and multicellular glands are present in the lining of the remainder of the digestive tract; most secrete mucus, and a few secrete digestive enzymes and acid in the stomach.

The liver and pancreas are major secretory structures derived from the embryonic gut and lying astride the stomach and duodendum. The liver is the largest of the digestive glands, serving as a nutrient storage organ and producer of bile. The bile drains from the liver into the gallbladder, then via the bile duct into the duodenum, where it assists in the breakdown of food. The pancreas is a smaller, diffuse gland. It secretes digestive fluids into the duodenum and also produces the hormone insulin.

Respiratory Structures

Lungs

The respiratory passage includes the external nares, olfactory chambers, internal nares, buccopharyngeal cavity, glottis, larynx, trachea, bronchial tubes, and lungs. The glottis is a slitlike opening on the floor of the pharynx and is a valve controlling airflow in and out of the exclusively respiratory passages. The glottis opens directly into a boxlike larynx. This voice box occurs in all amphibians although is anatomically most complex in frogs. The larynx exits into the trachea; the latter bifurcates into the bronchi then into the lungs. Bronchi are absent in all frogs except the pipids. The lungs are highly vascularized, thin-walled sacs. Internally, they are weakly partitioned by thin connective tissue septa. This weak partitioning and the small size, even absence, of the lungs emphasize the use of multiple respiratory surfaces in the amphibians. Lung ventilation is triphasic via a buccopharyngeal force-pump mechanism. Inhalation begins with nares open, glottis closed, and depression of the buccopharyngeal floor drawing air into this cavity. The glottis then opens, and elastic recoil of the lungs forces the pulmonary air out and over the new air in the buccopharyngeal pocket. The nares close, the buccopharyngeal floor contracts, thus pumping air into the lungs, and the glottis closes, keeping the air in the lungs under supra-atmospheric pressure. Similar, but faster and shallower, throat movements occur regularly in frogs and salamanders; these smelling movements rapidly flush air in and out of the olfactory chambers.

Other Respiratory Surfaces

In amphibians, lungs are only one of several respiratory structures. A few caecilians have a small third lung budding off the trachea. The buccopharyngeal cavity is heavily vascularized in many amphibians and is a minor gaseous exchange surface.

Gills are major respiratory structures in larvae and a few adult salamanders. Three pairs of external gills, developing and projecting from the outside of the pharyngeal arches, occur in salamanders and caecilians. External and internal gills occur sequentially in anuran larvae; the former arise early, remain largely rudimentary, and are replaced quickly by the latter.

In most adults and larvae, the skin is the major respiratory surface and is highly vascularized. Gaseous exchange in all vertebrates requires a moist surface; drying alters the cell surfaces and prevents diffusion across cell membranes.

Urinary and Reproductive Organs

The urinary and reproductive systems are intimately related in their location along the midline of the dorsal body wall and their frequent sharing of ducts. The structures of each system are paired.

Kidneys and Urinary Ducts

The kidneys remove nitrogenous waste from the bloodstream and maintain water balance by regulating the removal or retention of water and salts. The functional unit of the kidney is the nephron or kidney tubule. Each nephron consists of a renal corpuscle and a convoluted tubule of three segments, each of variable length in different species. The corpuscle encloses a ball of capillaries, and the major filtration occurs here. Filtration (selective secretion) may also occur in the tubule, but resorption (return to bloodstream) of salts and water is the major activity as the filtrate passes through the tubule. The tubules of adjacent nephrons empty into collecting ducts, which in turn empty into larger ducts and eventually into the urinary duct draining each kidney.

Primitively and embryologically, the kidney developed from a ridge of mesomeric tissue along the entire length of the body cavity. In modern amphibians, a holonephric kidney exists embryologically but never becomes functional. Instead the functional kidney (pronephros) of embryos and larvae arises from the anterior part of the "holonephric" ridge. The pronephros begins to degenerate as the larva approaches metamorphosis and a new kidney, the opisthonephros, develops from the posterior part of the ridge. The tubules of the anterior end of the male's opisthonephric kidney take on the additional role of sperm transport. In the primitive salamanders, this new role causes the anterior end of the kidney to narrow and the tubules to lose their filtration role; in caecilians, the kidney remains unchanged; and in anurans and advanced salamanders, the kidney shortens into a compact, ellipsoidal organ with the loss of the anterior end. A single urinary duct (archinephric duct) drains each kidney; it receives urine from the collecting ducts and empties into the cloaca. The urinary drainage shows two principal patterns in amphibians. Only the archinephric duct drains the kidney in the caecilians and primitive salamanders; whereas in frogs and advanced salamanders, the archinephric duct drains the anterior portion of the kidney and an accessory duct the posterior half. The bladder has a single and separate duct (urethra) emptying into the cloaca; fluids enter and exit the bladder through this duct.

Gonads and Genital Ducts

The female and male gonads, ovaries and testes, develop from the same embryological organs (Fig. 1.8). The undifferentiated organs arise on the dorsal body wall between the kidneys. The germ cells migrate into each organ and initiate the reorganization and consolidation of the pregonadal tissue into an external cortex and internal medulla. Later when sexual differentiation occurs, the cortex is elaborated into an ovary in females, and the medulla into a testis in males.

Structurally, these two gonads are quite different. The ovary is a thin-walled sac with the germ cells sandwiched between the inner and outer ovarian walls. The germ cells divide, duplicating themselves and producing ova. A single layer of follicle cells from the wall epithelium encases each ovum, providing support and nourishment. This unit (the follicle) of ovum and follicle cells grows and protrudes into the ovarian lumen. The numerous developing follicles are the visible portion of the ovaries in gravid females. The testis is a mass of convoluted seminiferous tubules encased in a thin-walled sac. Small amounts of interstitial tissue fill the spaces between the tubules. The developmental cycle (gametogenesis) of ova and spermatozoa is presented in Chapter 7.

Spermatozoa collect in the lumens of the seminiferous tubules and then move sequentially through progressively larger collecting ducts into kidney collecting ducts before emptying into the archinephric duct. Because of its dual role in urine and sperm transport, the archinephric duct is called the urogenital (Wolffian) duct. The oviducts (Müllerian ducts) are paired tubes, one on each side of the dorsal body wall, lateral to each ovary. Each arises *de novo* as a fold

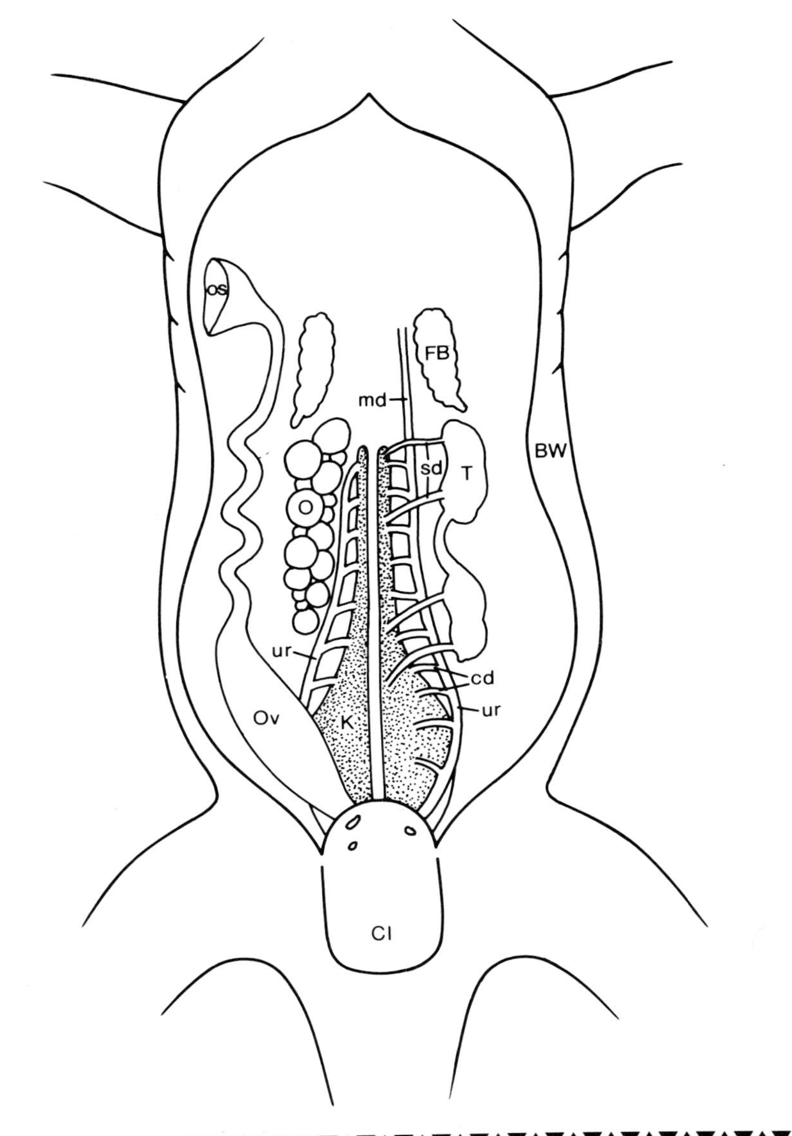

FIGURE 1.8

Female (left) and male (right) reproductive tracts of a salamander; a diagrammatic ventral view. Abbreviations: BW, body wall; Cl, cloaca; FB, fat bodies; K, kidney; ur, urogenital duct. Female — O, ovary; os, ostium; Ov, oviduct (Müllerian duct). Male—cd, collecting ducts; md, Müllerian duct; sd, sperm ducts; T, testis.

of the peritoneum (or by a splitting of the archinephric duct in salamanders). The anterior end remains open as an ostium; ova are shed into the body cavity and move to and through the ostium into the oviduct for their passage to the outside. The oviduct extends posteriorly and empties into the cloaca. Oviducts form in both males and females, degenerating although not disappearing in many male amphibians; this nonfunctional duct is Bidder's duct. Similarly some males retain a part of the gonadal cortex (Bidder's organ) attached to the anterior end of the testis.

Endocrine Glands

The endocrine system comprises numerous glands scattered throughout the head and trunk. The glands are an integrative system, initiating and coordinating the body's reactions to internal and external stimuli. But unlike the nervous system, the glands do not communicate directly with one another and their target organs. Instead, they rely on vascular and neural pathways to transmit their chemical messengers. Unlike other organ systems, the endocrine system is a composite of unrelated anatomical structures from the other systems, for example, the pituitary of the nervous and digestive systems, the gonads of the reproductive system, or the pancreas of the digestive system. Only a few of the many glands and their functions are mentioned, and these are described only superficially. The commonality of all the endocrine organs is their secretion of one or more chemical messengers (hormones) that stimulate or arrest the action of one or more target organs (including other endocrine glands/tissues). Hormones work in both the short term and continually to maintain a stable internal environment and in the long term and cyclically to control periodic behaviors, such as reproduction.

Pituitary Gland

The pituitary gland or hypophysis is the master gland of the body. Structurally, it contains two parts: the dorsal part arising from the ventral portion of the diencephalon (neurohypophysis) and a ventral part derived from the roof of the buccal cavity (adenohypophysis). The neurohypophysis and adenohypophysis interdigitate and are joined by neural and vascular connections. The chain of action begins with the release of neurohormones by the brain cells. These hormones reach the neurohypophysis through blood vessels or secretory axons of neurons ending in the neurohypophysis. In turn, the neurohypophysis produces hormones that stimulate the adenohypophysis (e.g., GnRH, gonadotropin-releasing hormone) or act directly on the target organs (ADH, antidiuretic hormone; MSH, melanophore-stimulating hormone). The adenohypophysis secretes six major hormones: adrenocorticotropin, two gonadotropins (FSH, LH), prolatin, somatotropin, and thyrotropin. These hormones control growth, metamorphosis, reproduction, water balance, and a variety of other life processes.

Pineal Complex

The pineal complex consists of the pineal (epiphysis) and a frontal (parapineal) organ, each arising embryologically from the roof of the diencephalon. These two organs are light receptors as well as endocrine glands. As light receptors, they record the presence or absence of light and, as glands, they produce and release melatonin. These two functions are associated with cyclic activities, both daily cycles or circadian rhythms and seasonal cycles. Frogs possess both a pineal organ lying inside the skull and a frontal organ piercing the skull and lying beneath the skin on top of the head. Caecilians and salamanders have only the pineal organ, which may extend upward to, but does not pierce, the skull roof.

Thyroid and Parathyroid Glands

These two glands occur near one another behind the larynx in the throat. Both arise as outpocketings of the pharyngeal pouches. The thyroid is a bilobular gland and the parathyroid is a pair of glands. Thyroid hormones are the principal initiators and controllers of metamorphosis. The parathyroid hormones are largely responsible for calcium regulation, determining when bone is deposited and resorbed and the regulation of calcium absorption–excretion by the kidneys and intestine.

Pancreas

The pancreas contains two types of secretory tissues. One type secretes digestive enzymes and the other (isles of Langerhans) secretes the hormone insulin. Insulin is critical for regulating carbohydrate metabolism; it stimulates the liver and adipose tissue to remove glucose from the bloodstream through glycogen production and fat synthesis, respectively. Insulin is important also for striated muscle activity by increasing the movement of glycogen into the muscle cells.

Gonads

The gonads are major target organs for the gonadotropins of the pituitary. In addition to initiating gametogenesis, the gonadotropins stimulate the production of estrogens and androgens (the female and male sex hormones) by gonadal tissues. Estrogens and androgens are steroids, and several closely related estrogens or androgens are produced in each animal. Stimulation and inhibition of the reproductive structures are obvious actions of the sex hormones, but they interact also with a variety of other tissues, such as the skin to produce secondary sexual characteristics and the brain to stimulate the production of gonadotropins. Estrogens are produced largely by the follicle cells in the ovarian follicles and corpora lutea. Androgens derive principally from the interstitial cells (cells of Leydig) lying in the spaces between the seminiferous tubules.

Adrenals

The adrenals are paired, elongate glands lying on the ventral surface of the kidneys. Each adrenal is an admixture of two tissues: the interrenal (cortical) cells form the main matrix of the gland and the adrenal (medullary) cells form strands and islets within the interrenal matrix. These two tissues have different embryological origins and distinctly different functions. The chromaffin cells produce adrenalin and noradrenalin, both of which affect blood flow to brain, kidney, liver, and striated muscles, mainly during stress reactions. The interrenal tissue produces a variety of steroid hormones. One group of interrenal hormones affects sodium and potassium metabolism, another group carbohydrate metabolism, and a third group, the androgens, reproductive processes.

ASPECTS OF LARVAL ANATOMY

The diversity of larval morphologies equals the diversity of the adult stages. In amphibians, larvae are free-living embryos. Most larvae also feed during this free-living developmental period; however, some do not eat and depend on the yolk stores of the original egg for nourishment. Caecilian and salamander larvae resemble their adult stages in general appearance and anatomical organization (Fig. 1.9). Their transition (metamorphosis) from an embryonic larva to nonembryonic juvenile is gradual with only minor reorganization. In contrast, the anuran larva (tadpole) undergoes a major reorganization in its metamorphosis from embryo to juvenile because the tadpole is anatomically so different from the juvenile/adult.

The larvae of the three amphibian groups are aquatic with very few exceptions and share anatomical characteristics associated with an aquatic existence. All have a thin, fragile skin of two to three epidermal layers. The skin is also heavily vascularized owing to its role as a major respiratory surface, a role shared with the gills. All amphibian larvae develop pharyngeal slits and external gills, usually three pairs projecting from the outside of the pharyngeal arches; they persist and function throughout the larval period of salamanders, primitive anurans, and caecilians. In advanced frogs, external gills are resorbed and replaced by internal gills, lamellar structures on the walls of the pharyngeal slits. All the larvae have lidless eyes and large, nonvalvular nares. They have muscular trunks and tails for undulatory swimming, and the tails have dorsal and ventral fins. The skeleton is entirely or mainly cartilaginous. All have well-developed lateral line systems.

Caecilian and salamander larvae are miniature adults, differing mainly by their smaller size, pharyngeal slits and gills, tail fins, a rudimentary tongue, and specialized larval dentition. In contrast, the body plan of the anuran tadpole bears little similarity to the adult's. The general tadpole can be described as a sac

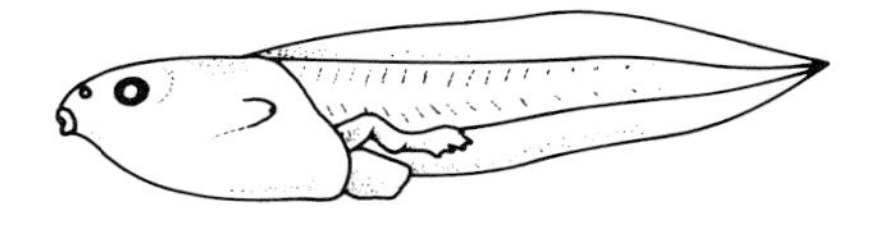
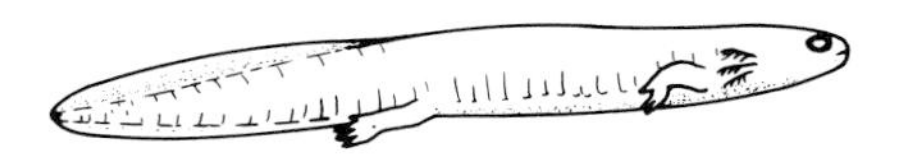

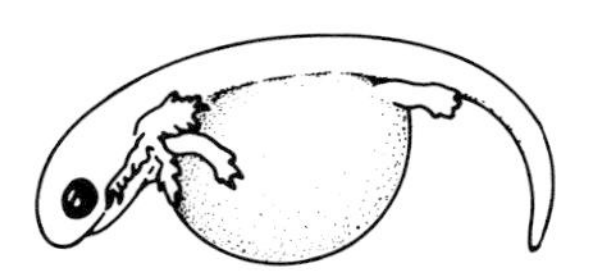

FIGURE 1.9 ▼▲▼
The body forms of some amphibian larvae. Salamanders (left) and frogs (right) arranged by habitat type: aquatic/pond types (top), *Ambystoma tigrinum*, *Sphaenorhynchus orophilus*; aquatic/stream types (middle), *Eurycea bislineata*, *Rana holsti*; terrestrial/direct developers (bottom), *Plethodon glutinosus*, *Eleutherodactylus nubicola*.

of guts with mouth and eyes at one end and a muscular tail at the other end; limbs do not appear until late in the developmental process and then only the hindlimbs are visible externally.

The general tadpole body form has been modified into hundreds of different shapes and sizes, each specially adapted to a specific aquatic habitat and feeding behavior. This diversity has been variously partitioned. Orton's tadpole divisions recognize four basic body plans or morphotypes, each defining an evolutionary grade and to a limited extent phylogenetic relationships. Another approach is to examine the relationship between the tadpole's morphology and ecological niche. Such an analysis recently defined 18 ecomorphological guilds, which with their subcategories recognize 33 body types; these morphotypes define a tadpole's adaptive zone not its phylogenetic relationships. Both classifications emphasize external and oral-pharyngeal morphology.

Most tadpoles have a large, fleshy disc encircling their mouth. Depending on the manner of feeding and the type of food, this oral disc ranges in position from ventral horizontal (suctorial, to anchor in swift water and scrape food off rocks) to dorsal horizontal (grazing on surface film in calm water) and in shape from round to dumbbell. The margin of the disc is variously covered with papillae, and these have a variety of shapes. The actual function of the papillae remains uncertain, although chemosensory, tactile, and water-current detection are possibilities. Tadpoles lack teeth on their jaws; instead many tadpoles have keratinous beaks and parallel rows of keratinous denticles on the oral disc above and below the mouth. The beaks serve to cut large food items into smaller pieces; the denticle rows act as scrapers or raspers to remove food from rock or plant surfaces. The oral-pharyngeal cavity is large; its structures trap and guide food into the esophagus, and it is also a pump to move water through the cavity and across the gills. The gills are initially visible externally, but at hatching or shortly thereafter, an operculum grows posteriorly from the back of the head to fuse to the trunk, enclosing the gills and the developing forelimbs. To permit water flow, a single or pair of spiracles remain open on the posterior margin of the operculum. Since the operculum covers the gill region, the head and body form a single globular mass. The muscular tail is usually 1 1/2 to 2 times the length of the body and is the sole locomotor mechanism. Hindlimbs appear at about a third of the way through the larval period but are largely nonfunctional until shortly before metamorphosis. At that time, they become functional, concurrently with the forelimbs breaking through the operculum.

CHAPTER 2

Origin and Evolution of Amphibians

ORIGIN OF TERRESTRIAL VERTEBRATES

Sometime in the Middle to Late Devonian (360–370 million years ago), a fish ancestor gave rise to the first tetrapods. Soon thereafter in a geological sense, the tetrapods split into two lineages, amphibians and anthracosaurs. Reptiles would soon evolve from one descendant lineage of the early anthracosaurs. These evolutionary events occurred in landscapes that appeared alien compared to the earth's present ones. Plants, like animals, were only beginning to break away from a completely aquatic existence. The uplands were desert barrens of bare rock and soil. Plants grew only in valleys and along the coasts where water occurred in abundance.

The transition from fish to tetrapod occurred in water. The earliest tetrapods were likely highly aquatic creatures. They were no more adapted to survive on land than their fish ancestors. The proposition is no longer widely accepted that the tetrapods evolved from fish using modified fins to escape from shrinking pools in drying river beds. Although the issue will never be resolved with absolute certainty, limbs likely arose in an aquatic environment, perhaps for stalking prey in heavy vegetation, perhaps as props to permit aerial respiration in shallow, stagnant pools. Another unresolved controversy is whether tetrapods evolved in a marine/brackish or a freshwater environment. The consensus favors a freshwater origin owing to the kidney structure and physiology of the tetrapod kidney and to the preponderance of early tetrapod fossils from nonmarine sediments. Some early amphibians are from marine sediments, and proponents of a marine origin interpret the kidney data to support a brackish environment (estuarine) for the transition from fish to tetrapod.

Just as disagreement exists over the aqueous environment and the selective factors for the origin of tetrapods, several groups of fishes have been proposed for the ancestor of tetrapods (see History of Extant Amphibians, this chapter). The consensus favors the lobe-finned fishes, specifically the rhipidistian crossopterygians, and evidence continues to mount supporting this fish group as the tetrapod ancestor. One of the earliest tetrapods, *Ichthyostega*, appears much like the rhipidistian *Eusthenopteron*, even though the former had legs and the latter fins. This transitional tetrapod may not have been the actual ancestor of amphibians and anthracosaurs (Fig. 2.1), but their ancestor was structurally similar. The extant coelacanth *Latimeria* is not a rhipidistian, hence not a living ancestor of tetrapods.

Before proceeding further, the contents of some amphibian groups require clarification. A plethora of group names are used for the various early tetrapods. Even "tetrapod" has two different meanings. In the general sense and as used above, a tetrapod is any vertebrate with four limbs. In the specific sense, tetrapod equals the category Tetrapoda and that includes all vertebrates with four limbs, except the Ichthyostegalia (Fig. 2.1). Amphibia is another potential source of confusion, because until recently, Amphibia included many anthracosaurian groups (Anthracosauroideae, Batrachosauria, Seymouriamorpha), the temnospondyls, and the lepospondyls. Anthracosaurians represent a separate lineage and the one that gave rise to reptiles and mammals, therefore to include anthracosaurs with the other amphibians makes Amphibia paraphyletic (see Fig. 13.2). Both temnospondyls and lepospondyls were used previously as formal group names. The members of the lepospondyls shared features associated with small body size and aquatic behavior, but not shared because of genetic relatedness. Thus, the lepospondyls (= Aistopoda + Microsauria + Nectridea) are a composite group of unrelated tetrapods. The content of temnospondyls equals the content of the present Amphibia without the lepospondyls. Labyrinthodont amphibian also has had long and frequent use. Its use encompassed a broad spectrum of early tetrapods, including the temnospondyls, anthracosaurs, and

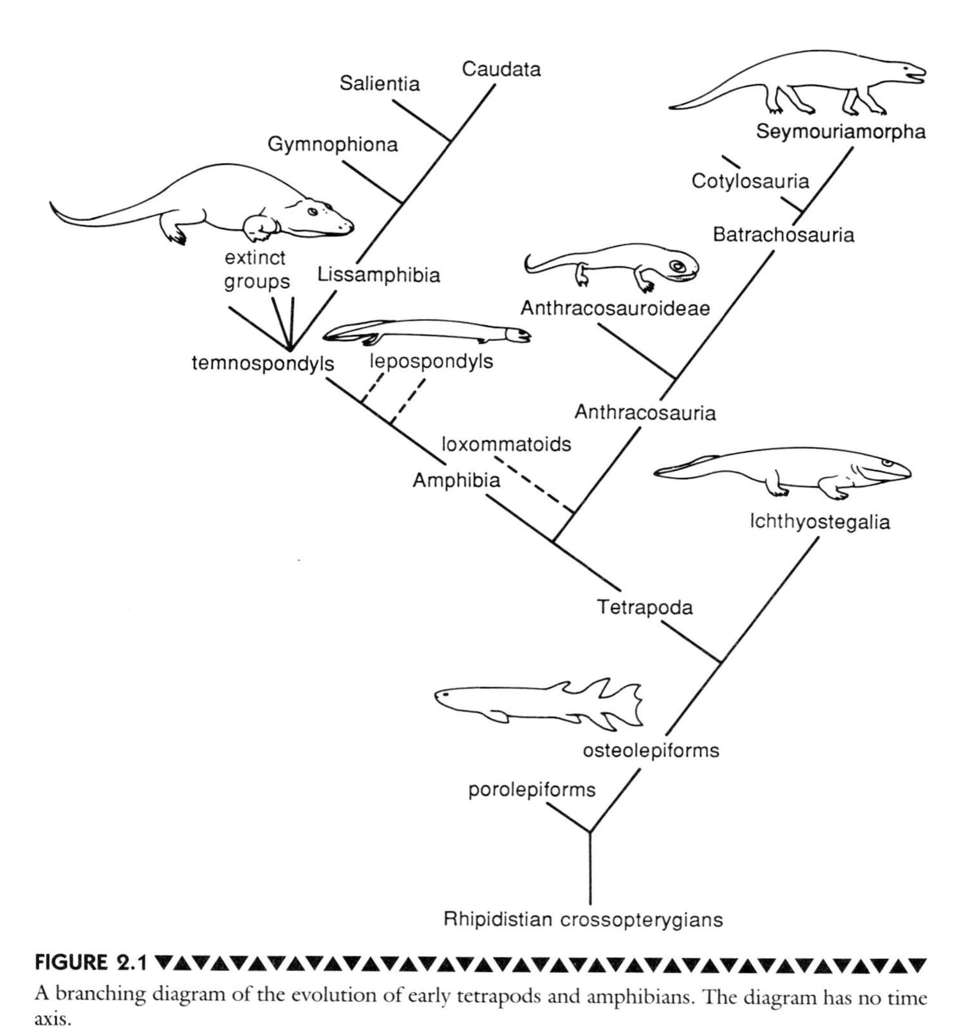

FIGURE 2.1 ▼▲▼
A branching diagram of the evolution of early tetrapods and amphibians. The diagram has no time axis.

ichthyostegalians. Labyrinthodonts are an unnatural group based on shared primitive (ancestral) characters, and they are not a monophyletic group. The formal groups (Table 2.1) are presumed to be monophyletic, and each includes an ancestor and all its descendants.

FISH TO TETRAPOD

Whatever the driving force, one group of fishes began the transmutation from an entirely aquatic life-style to a terrestrial one. The early stage began in

TABLE 2.1
Hierarchical Classification Derived from the Sister Relationships of the Tetrapod Vertebrates

- Tetrapoda
  - Amphibia
    - Lissamphibia
      - Gymnophiona
      - Caudata
      - Salientia
  - Anthracosauria
    - Anthracosauroideae
    - Batrachosauria
      - Seymouriamorpha
      - Cotylosauria
        - Diadectomorpha
        - Amniota

Note: Category titles are not assigned to the hierarchical ranks above family, and not all category levels or steps are included.

water and the prototetrapod was no more able to survive on land than a fish. The changes affected most aspects of anatomy and physiology. These changes did not occur all at once. Some structural changes were linked, others were not.

Terrestriality requires aerial respiration, but lungs appeared early in the evolution of bony fishes, long before any group of fishes displays any other terrestrial adaptation. Indeed, lungs preceded and are the structural predecessors of swim bladders in the advanced fishes. Lungs likely developed as accessory respiratory structures for gaseous exchange in anoxic or low-oxygen waters. The lung structure of the rhipidistian and the earliest tetrapods is unknown. Presumably it was a ventral outpocketing off the pharynx, probably with a short trachea leading to either an elongated or a bilobed sac. The internal surface was likely only lightly vascularized, for cutaneous respiration was also possible. Respiration (i.e., ventilation) depended on water pressure. The fish would rise to the surface, gulp air, and dive. With the head lower than the body, water pressure compresses the buccal cavity and forces the air rearward into the lungs, since water pressure is less on the body higher in the water column. Reverse air flow occurs as the fish surfaces head first. This mechanism is still used by air-breathing fish for ventilation. In the early tetrapods, this passive-pump mechanism was replaced by a buccal-force pump. Air enters through the mouth with the floor depressed, the mouth closes, the floor contracts (elevates) driving air into the lungs, and the glottis closes holding the pulmonary air at supra-atmospheric pressure. Exhalation results from the elastic recoil of the body wall driving air outward. Gills were present in the rhipidistian fish, but presumably absent in adult ichthyostegalians. The loss of the gills may have occurred in a situation where aquatic respiration was negligible and closure of the gill openings improved the inhalation cycle.

The next step toward terrestriality involved the transformation of fins to limbs. The cause remains debatable, but lobed fins seem a prerequisite. A lobed fin projects well outward from the body wall and contains internal skeletal and muscular elements that permit it to serve as a strut or prop. Since the limbs evolved for locomotion in water, probably for a slow progression along the bottom, the limbs were not required to support the body because buoyancy reduced body weight. At first, the fin-limbs probably acted like oars, rowing the body forward with the fin tips pushing against the bottom. Minimal modifications of the fin-limb are necessary for bottom-walking in this manner. The next stage would involve bending of the fin-limb to allow the tip to make broader contact with the substrate. These bends or joints would be the sites of the future elbow/knee and wrist/ankle. As the flexibility of the joints increased, the limb segments developed increased mobility and their skeletal and muscular components lost the simple architecture of the fin elements. Probably at this stage, the fin rays were lost and replaced by short, robust digits, and the pectoral girdle lost its connection with the skull, allowing the head to be lifted and to retain a forward orientation as the limbs extended and retracted. The ichthyostegalians represent this stage. Their limb movements, although in water, must have matched the basic terrestrial walking pattern of extant salamanders, that is, the extension-retraction and rotation of the proximal segment, the rotation of the middle segment (forearm or crus), and flexure of the distal segment (foot). As tetrapods became increasingly terrestrial and buoyancy no longer counteracted the pull of gravity, the vertebral column became a sturdier arch with stronger intervertebral links, muscular as well as skeletal ones. The limb girdles also became supportive, the pelvic girdle by a direct connection to the vertebral column and the pectoral girdle through a strong muscular sling connected to skin and trunk.

While in the water, the fluidity and resistance of water assists in grasping and swallowing food. Once the tetrapod begins to feed in shallow or no water, inertial feeding becomes important. The food item is stationary in inertial feeding and the mouth/head of the tetrapod must move forward over the food. Several modifications of the skull may be associated with this feeding behavior. With the pectoral girdle and skull independent, the skull could move left and right, and up and down on occipital condyles-atlas/axis articulation. The snout and jaws elongated. The intracranial joint locked and the primary palate became a broader and solid bony plate.

As tetrapod behavior became more terrestrial, the sense organs shifted from aquatic to aerial perception. Lateral line and electric organs function only in water and occur only in the aquatic phase of the life cycle or in aquatic species.

Hearing and middle ears appeared. The eyes changed to sharpen their focus for aerial vision. The nasal passages became a dual channel, air passages for respiration and the surfaces for olfaction.

The skin of larval amphibians and fish is very similar. The epidermis is only two to three layers thick and protected by a mucous coat, secreted by numerous unicellular mucous cells. Adult amphibians show the modifications that occurred in the transition from an aquatic to a terrestrial existence. The epidermis increases its thickness to five or seven layers; the basal two layers are living and equivalent to the fish or larval epidermis. The external layers undergo keratinization and the mucoid cuticle persists between the basal and keratinized layers. Increased keratinization likely appeared as a protection against abrasion, since terrestrial habitats and the low body posture of the early tetrapods exposed the body to constant contact with the substrate and the probability of greater and frequent surface damage.

The preceding changes represent the major anatomical alterations that occurred in the transition from fish to tetrapod. Many physiological modifications also occurred; some of these are described in Chapter 10. However, some aspects, like reproduction, remained fishlike: external fertilization, eggs encased in gelatinous capsules, free-living embryos or larvae, and larvae with gills. Metamorphosis from the aquatic larval to semiaquatic adult stage was a new developmental feature.

HISTORY OF EXTINCT AMPHIBIANS

Radiation of Early Tetrapods

Tetrapods in the Late Devonian were largely aquatic, but beginning to conquer the land. Vegetation completely covered the lowland coastal areas and floodplains, for the plants were no longer confined to water or the margins of streams, lakes, and seas. Herbs and shrubs were the dominant plants, but trees had appeared and in some places formed forests. Plants were even beginning to invade the upland areas.

The vertebrates began their invasion of land with the ichthyostegalians. These earliest tetrapods were aquatic creatures, although likely shallow-water inhabitants. They had stout limbs and could move about on land; however, any terrestrial forays probably occurred only in humid, heavily vegetated habitats. It is not unlikely that they were feeding along the shore's edge, because arthropods had successfully invaded land earlier and in the Late Devonian were undergoing a terrestrial radiation. Nonetheless, the ichthyostegalians look similar to their rhipidistian ancestors and like them were probably fish predators. The ichthyostegalians were robust animals, about 1 m long (TL, total length). The body and head retained the fusiform shape of fish; their tails had dorsal fins and in some species both dorsal and ventral fins. The head and mouth were large. The jaws were heavily toothed and large tusks projected from the margin of the palate.

The ichthyostegalians disappeared from the fossil record at the end of the Devonian (Fig. 2.2). The next tetrapods appeared in the Upper Mississippian and appeared en masse. Several groups of anthracosaurs, three groups of amphibians (temnospondyls), loxommatoids, and aistopodans were present. Most of these tetrapods were highly aquatic, although the lowlands were now clothed by an even greater diversity of plants and plant communities. The aistopodans were elongate, limbless amphibians. None exceeded 70 cm (TL). Presumably they were aquatic and semiaquatic; their fragile skulls do not match the needs of burrowing animals. The loxommatoids appear to have been moderate-sized, reduced-limbed animals, which are known principally from skulls. They are recognized by an anterior elongation of each orbit that probably housed a large gland. Of the three temnospondyl groups, the trimerorhachoids included limbed and reduced-limbed amphibians and likely contained both semiaquatic and aquatic members. The trimerorhachoids were especially long-lived, appearing in the middle of the Mississippian and surviving to the end of the Permian. The colosteid temnospondyls had a much shorter appearance in the mid-Carboniferous. The colosteids were small, aquatic amphibians. The third group, edopoids, contained larger, more robust amphibians. They persisted through the Carboniferous into the early Permian. The earlier edopoids were mainly aquatic forms, but the later ones became increasingly terrestrial. Two groups of anthracosaurs, the embolomeres and eoherpetontid, were part of the early tetrapod community. Both groups were mainly aquatic animals. The embolomeres were the largest tetrapods, ranging from 1 to 4 m (TL). All had heavy crocodilelike skulls and short, robust limbs.

In the Pennsylvanian, many of the preceding groups were joined by new groups of tetrapods: geophyrostegid and limnoscelid anthracosaurs, eryopoid amphibians, nectrideans, and three groups of microsaurs. This fauna remained a moisture-loving one and predominantly a lowland fauna. The climate was generally hot and wet and supported diverse and dense plant communities. The uplands also bore a thick plant cover. These tetrapods were primarily aquatic forms, although a few had become terrestrial. The limnoscelids were one of the terrestrial groups. They were moderate-sized (1–2 m TL), robust-bodied and -limbed, and long-tailed tetrapods. Structurally, they possessed some features that would appear in reptiles. The eryopoids included aquatic to terrestrial, small to large amphibians. The heavy-bodied *Eryops* is characteristic of this group, although it was larger (nearly 2 m TL) than most eryopoids. The nectrideans were aquatic forms with short bodies and long, laterally compressed tails. The heads of some were arrow-shaped with large, laterally projecting horns. The microsaurs were small (most <50 cm TL), salamanderlike tetrapods. They commonly had long bodies and tails, and short limbs and were probably predominantly aquatic-semiaquatic.

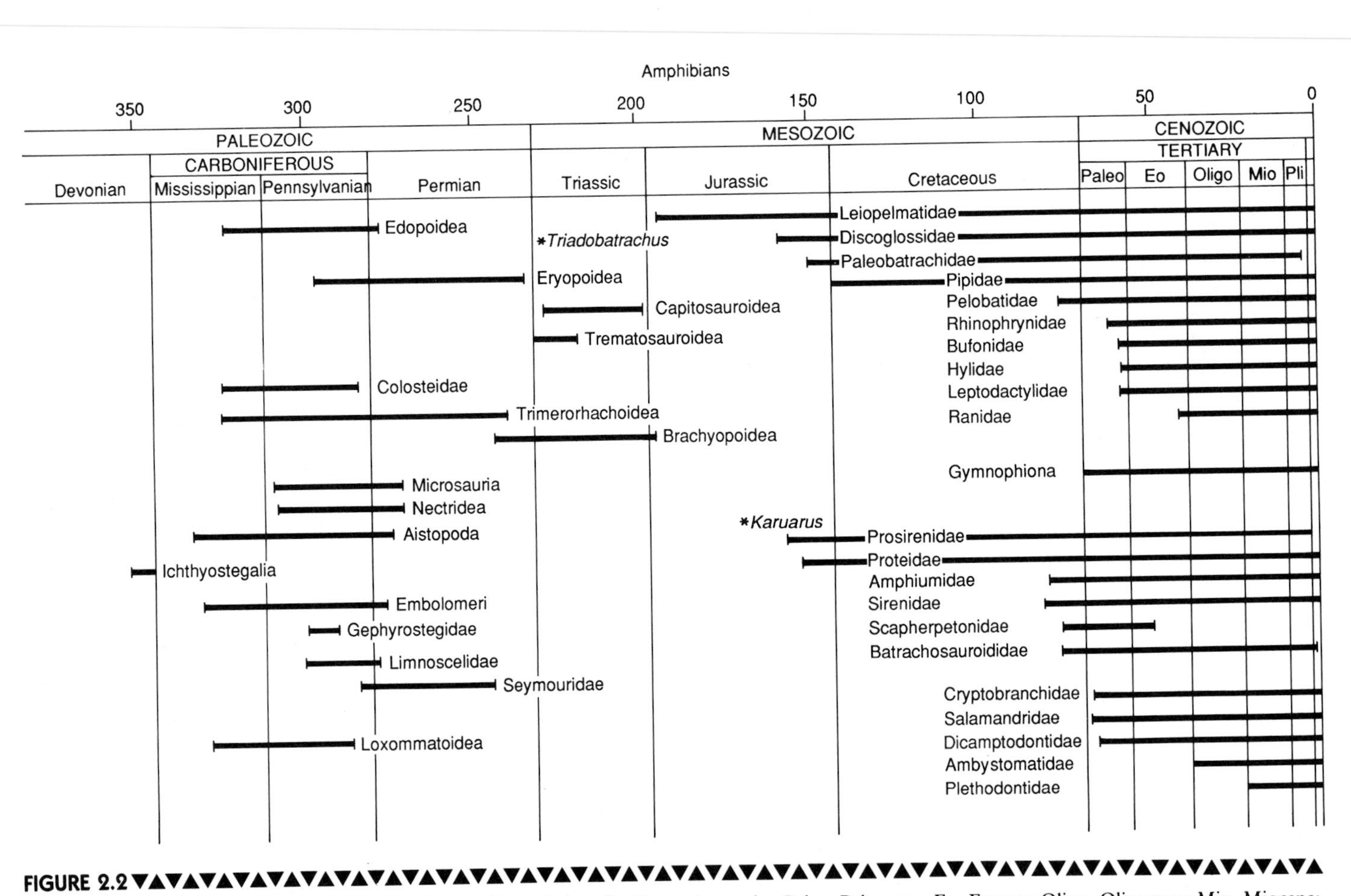

FIGURE 2.2 ▼▲

Geological occurrence of extinct and living amphibians. Abbreviations for Cenozoic epochs: Paleo, Paleocene; Eo, Eocene; Oligo, Oligocene; Mio, Miocene; Pli, Pliocene; Pleistocene is the narrow, unlabeled epoch at the top of the chart.

Amphibians of Late Paleozoic

Many Late Carboniferous groups persisted into the Early Permian, but not beyond. Only the trimerorhachoids and eryopoids survived into the Late Permian, and they were joined by the seymourids for most of their Permian tenure. A few other small anthracosaurs appeared briefly. The loss of amphibian and anthracosaurian diversity occurred concurrently with the diversification of reptiles and synapsids (reptilelike mammals) and a shift to an arid climate. The aquatic habitats shrunk and many disappeared. Plant cover was reduced, and the drier upland vegetation spread into the lowlands.

The trimerorhachoids had greatly reduced limbs and were probably highly aquatic. The eryopoids as a group remained diverse in size, body form, and habits (aquatic to terrestrial). The surviving embolomeres had become shallow-water denizens in contrast to their deep-water ancestors of the Carboniferous. They did not survive beyond the Early Permian. The seymourids were a much more successful group of semiterrestrial and terrestrial anthracosaurs. Generally, they had large heads with well-developed jaws, robust bodies, and strong limbs. Most were within the 1-m size range (Fig. 2.3).

FIGURE 2.3
Seymouria, an Early Permian anthracosaur from Texas. Scale: bar = 5 cm. (R. S. Clarke)

Amphibians of Early Mesozoic

In the Triassic, the reptiles and synapsids had become the dominant terrestrial vertebrates. A few anthracosaur groups survived into the earliest Triassic but soon disappeared. In contrast, the amphibians experienced a mini-diversity explosion with the appearance of at least seven different groups of temnospondyls and the first lissamphibian (*Triadobatrachus*). This radiation differed from earlier ones by emphasizing small size (<30 cm TL). The trematosaurs of the Early Triassic were long-snouted creatures and some of them were marine, an anomaly for amphibians. Only two of the temnospondyl groups occurred throughout the Triassic, the capitosauroids and brachyopoids (Fig. 2.4). Although never common, they persisted throughout this period. The first frog, *Triadobatrachus,* appeared in the Early Triassic, but is known only from one fossil deposit. It is clearly a frog, although it possessed 14 body vertebrae and a short tail of six vertebrae. Its pelvic girdle and skull are very froglike.

FIGURE 2.4 ▼▲▼

A death assemblage of *Buettneria perfecta*, a Late Triassic temnospondyl from New Mexico. Scale: bar = 10 cm. (R. S. Clarke)

HISTORY OF EXTANT AMPHIBIANS

The perennial disagreement concerning the relationships of the three extant groups of amphibians may now be nearing resolution, although it is unlikely that unanimous agreement will ever be attained. Nonetheless, the preponderance of evidence points to caecilians, salamanders, and frogs as each others' closest relatives and their derivation from the same ancestral lineage. Importantly, much of this evidence is uniquely derived characters; such characters are the strongest evidence for genetic relatedness. Thus, the living amphibian groups are appropriately contained in the same taxon: Lissamphibia. A major reason for some uncertainty is that the earliest fossils of each are recognizable as a frog, salamander, or caecilian. No "intermediates" have been discovered bridging the morphological gap between any of the Paleozoic amphibian groups and the lissamphibians.

Three main hypotheses exist. The Holmgren hypothesis proposes that salamanders arose from the lungfish lineage, and frogs and amniotes from crossopterygian fish. The Jarvik hypothesis has salamanders, frogs, and amniotes arising from rhipidistian crossopterygians, but the salamanders from one group of rhipidistians (porolepiforms), and frogs and amniotes from another (osteolepiforms); lungfish are unrelated to the tetrapod lineage. The monophyletic hypothesis proposes the origin of all tetrapods from a single group of crossopterygians (osteolepiforms), and among the living tetrapods, the three living amphibian groups are a separate lineage, well removed from the amniote lineage.

The Holmgren hypothesis is based on supposed differences in the early development of the limb skeleton. Recent investigations have not confirmed differences in the mesenchyme aggregations, which are the precursors to the limb skeleton. In fact, the embryonic sequence for the differentiation of bony limb elements is identical for frogs and salamanders, furthermore it more closely matches the organization of the fin skeleton in crossopterygian fishes than in lungfishes. The evidence for lungfish being the ancestors of tetrapods or any group of tetrapods is equivocal. The structural divergences of the tetrapods from fish are easily traceable to structures of the rhipidistian fishes. Conversely, derivation of tetrapod structures from lungfishes requires a major reorganization of the anatomy of Paleozoic lungfishes, but no such reorganization is evident in modern lungfishes.

The Jarvik biphyly of modern amphibians derived mainly from the anatomy of rhipidistian snouts and the supposed distinctiveness of salamander and frog snouts. Recent studies show snout anatomy of frogs and salamanders to be more similar than Jarvik's earlier studies indicated. Further, the abundance of shared derived traits of salamanders and frogs advocate their origin from a common ancestor (monophyly).

The origin of the lissamphibians likely occurred in the mid-Permian or earlier. The fossil evidence now suggests that the ancestor was a dissorophid.

Presumably, this ancestral stock was a small, semiaquatic salamanderlike amphibian. They would have had external fertilization, a larval developmental stage, and many other physiological and anatomical features shared by today's lissamphibians. There are no fossils available to show the times and manner of divergence of the three modern groups. Often caecilians are depicted as diverging first owing to their extreme structural divergence from frogs and salamanders. But if occurrence of fossils is used as an indicator, frogs diverged first. This issue remains unresolved.

Caecilians

Caecilians are represented by a single Paleocene fossil vertebra from Brazil. This vertebra is most similar to the vertebrae of the African genus *Geotrypetes* (Caeciliaidae) and has been named *Apodops*. The African–South American affinity of *Geotrypetes* and *Apodops* favors the Gondwanan origin and evolution of the caecilians. Also *Apodops*' ready recognition as a member of the extant Caeciliaidae suggests the origin of the modern gymnophionan families in the Late Cretaceous or earlier.

Salamanders

Five groups of salamanders occur as fossils: the extinct karauroids and prosirenoids, and the extant sirenoids (Meantes), cryptobranchoids, and salamandroids. The history of these five groups is linked to the Northern Hemisphere (Holarctic) and to the ancient continent of Laurasia.

The karauroids are represented by a single fossil from the Late Jurassic of Kazakhstan. The fossil of *Karaurus* (Fig. 2.5) is fortunately nearly complete, and is definitely a salamander. It is extremely primitive and is postulated to be an early offshoot of the lineage leading to the three living groups. *Karaurus* was small (about 120 mm SVL) and terrestrial judging from its body form and the dermal sculpturing (skin fused to bone) on the skull bones.

The next two groups to appear in the fossil record (see Fig. 2.2) were the prosirenoids and the salamandroids. The prosirenoids are another early offshoot of the lineage leading to modern salamanders (i.e., cryptobranchoids and salamandroids). The prosirenoids are sufficiently different that some researchers have suggested that they were not salamanders. However, bicipital ribs, spinal nerve foramina in vertebrae, and other traits link them to the other salamander groups. In contrast, the salamandroids are the advanced salamanders and the ones comprising the majority of living salamanders.

The first prosirenoid, *Albanerpeton*, appeared in the Late Jurassic and persisted into the Miocene; during that time, several species arose and disappeared. Prosirenoid fossils derive principally from Europe, although one species occurred in the Late Cretaceous of North America. All were small salamanders (<12 cm

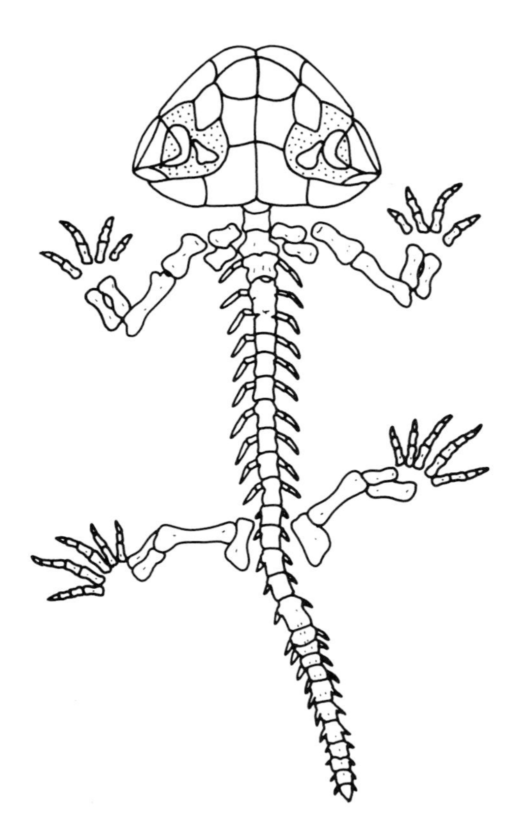

FIGURE 2.5 *Karuarus*, the earliest known salamander (Late Jurassic; ca. 15 cm TL). [Drawn as a partial reconstruction from a photograph in Carroll (1988).]

TL) and presumably terrestrial or semiterrestrial. Their limbs were usually well developed, and the robust lower jaw possessed a unique interlocking symphysis. *Prosiren* was a short-bodied, weak-limbed salamander from the Early Cretaceous of North America. It is unrelated to the living sirenids and is not their ancestor.

Salamandroids also appeared in the Late Jurassic. These were the Proteidae, which is potentially a composite group of salamanders sharing a few characteristics that may be associated with paedomorphic origins rather than a shared evolutionary history. The Jurassic *Comonecturoides* is the earliest known salamander of North America. In general, this extinct salamander appears similar to the extant mudpuppies (*Necturus*); however, the fossil material is too fragmentary to be certain of this relationship. Definite proteid fossils occur first in the Late Paleocene of North America and mid-Miocene of Europe. These fossils represent the extant *Necturus* and *Proteus*, as well as two extinct genera from the Miocene of Europe. All were small, perennibranchiate (gill-bearing as larvae and adults) salamanders.

Other salamandroid salamanders appeared in the Late Cretaceous. *Proamphiuma* is the first fossil amphiumid. Unlike many fossils with "pro" in their names, *Proamphiuma* is a structural precursor to *Amphiuma* (Paleocene–Recent) and the relationship actually may be that of ancestor to descendant. The amphiumids have remained a strictly North American group throughout their 60+ million year history.

Two extinct families, Batrachosauroididae and Scapherpetonidae, follow in the fossil record shortly after the amphiumids. The scapherpetonids were a group of moderate-sized salamanders living from the Late Cretaceous to the Early Eocene of North America. These salamanders are related to the present-day dicamptodontids, and *Scapherpeton* and *Piceoerpeton* share the *Dicamptodon* body form. *Lisserpeton* appears to have had an elongate body and likely reduced limbs. Interestingly, one species of *Piceoerpeton* occurred on Ellesmere Island within the present Arctic Circle.

The batrachosauroidids have a skull structure that suggests a proteid–batrachosauroidid relationship. The batrachosauroidids were a diverse group of at least five genera of paedomorphic salamanders, ranging from the Late Cretaceous to Late Pliocene of North America, and Paleocene–Eocene of Europe. They were like the modern mudpuppies (*Necturus*) in body form and size.

Both salamandrid and dicamptodontid fossils occur in Paleocene deposits of Eurasia and North America. The earliest dicamptodonids consist of five monotypic genera (e.g., *Ambystomichnus*/North American Paleocene, *Geyeriella*/European Paleocene), each based on fossil remains from a single specimen. Of these, *Ambystomichnus* is the only salamander recognized exclusively on a fossilized trackway. This trackway is similar to that of *Dicamptodon*, differing only in stride length, which indicates that the trackways were produced by a salamander nearly twice as long as the extant *Dicamptodon*. The *Ambystomichnus* trackway was associated with a redwood flora, a plant community relationship that still exists for the dicamptodontids. The ambystomatids do not appear until the Early Oligocene. All fossils represent North American *Ambystoma*.

The salamandrids have the most speciose fossil record. Their fossil record contains representatives of 18 genera and more than 50 species. Living genera, such as *Notophthalmus* and *Triturus*, extend as far back as the Miocene, *Taricha* to the Oligocene, and *Salamandra* and *Tylototriton* to the Eocene. The extinct genera derive principally from the Paleocene to Oligocene. The fossil species of the extinct and extant genera match the extant species in size and body form and probably shared the modern ones' diversity of behaviors and ecology.

Today, the plethodontids are the most speciose of the salamanders, yet they have a meager fossil record. Half a dozen genera are represented and four of these occur no earlier than the Pleistocene. A few vertebrae attributable to *Aneides* have been found in an Early Miocene deposit in Montana, and a fossil trackway from the Early Pliocene of California has been referred to *Batrachoseps*.

Of the two families of cryptobranchoid salamanders, only cryptobranchid fossils have been found. *Cryptobranchus* is represented by two occurrences, one from the Paleocene of Saskatchewan and another from the Pleistocene of Maryland. *Andrias* has a much more extensive history. The oldest *Andrias* fossils are from the European Miocene, and *Andrias* persisted there at least to the Pliocene and in North America to the Miocene. Within its present range, *Andrias* has been found only in Pleistocene deposits in Japan. The fossil forms were also giant salamanders, one estimated to be more than 2 m long (TL).

The only other salamanders to reach such lengths were some fossil sirenids. They may have reached 2 m (TL), although it is difficult to confirm such lengths because all sirenid fossils are isolated or short series of vertebrae. Sirenids first appeared in the Late Cretaceous as the giant *Habrosaurus*, which survived into the Early Paleocene. This siren looked much like its living relatives, except for specialized shovel-shaped teeth. Sirenids are absent then until the mid-Eocene when *Siren* makes its appearance. *Pseudobranchus* occurred first in Pliocene deposits of Florida. The entire history of sirenids is North American with the earliest fossils from Montana and Wyoming and later ones from the Gulf Coast.

Frogs

Frogs (Salientia) contain two major groups: the proanurans, the earliest and structurally most primitive frogs; and the anurans, all subsequent frogs. The anurans also divide into three subgroups based on specialization in anatomy. These three groups appear sequentially (archaeobatrachians, mesobatrachians, and neobatrachians) and surprisingly chronologically (primitive to advanced) in the fossil record. The first frogs appeared in the Southern Hemisphere. Thus, frogs are believed to have had a Gondwanan origin and many of modern families also appear to have arisen and radiated in the southern continents. Yet not all families are Gondwanan; a few have fossil and modern histories confined to the Northern Hemisphere, sharing Laurasian origins and radiations with the salamanders.

The first frog was *Triadobatrachus* from the Early Triassic of Madagascar. Although it had more vertebrae than later frogs and a short stumpy tail, it was clearly a frog. *Triadobatrachus* (Proanura: Protobatrachidae) is probably not the ancestor of later frogs, nonetheless it provides a glimpse at the anatomy of frogs in transition from early temnospondyls to anurans. Its small size (about 10 cm SVL) suggests that frog evolution emphasized small size, unlike many earlier amphibian lineages. A single fossil is all that has been found of the proanurans, and it may represent a juvenile of an aquatic form or a metamorphosing individual of a semiterrestrial one.

Frogs do not appear again until the Early Jurassic, more than 30 million years later. These Early Jurassic frogs (*Notobatrachus* and *Vieraella*) are representatives of the extant Leiopelmatidae (Archaeobatrachia) and derive from Patagonian Argentina. They possess a suite of primitive characteristics, such as nine presacral vertebrae, free ribs, and partially fused astragalus-calcaneum, and these are shared with *Ascaphus* and *Leiopelma*. *Vieraella* was a small frog (about 28 mm SVL); *Notobatrachus* was much larger (12–15 cm SVL), roughly three times the size of the modern leiopelmatids. These two frogs are the only leiopelmatid fossils, leaving a nearly 200 million year hiatus between their Jurassic occurrence and their modern relatives.

The next group of frogs to appear was the Discoglossidae, another group of primitive frogs that has survived into the Recent epoch. The modern discoglossids are more diverse than the leiopelmatids with five genera and more than a dozen species. Their fossil history shows a similar diversity of forms. *Eodiscoglossus* appeared in the Late Jurassic of Spain and persisted into the Early Cretaceous. Two genera appeared in the Late Cretaceous of western North America; one of them (*Scotiophryne*) survived into the Paleocene. They are absent throughout the Eocene, then in Europe one genus reappears in the Oligocene, and abruptly five genera in the European Miocene. Three of these are the modern *Alytes*, *Bombina*, and *Discoglossus*.

The paleobatrachid frogs (Mesobatrachia) are recently extinct, but they were a long-lived family. They appeared first at the Jurassic–Cretaceous boundary and went extinct in the Pleistocene. Throughout their entire history, they were confined to Europe, with one questionable Cretaceous occurrence in North America. Although apparently very abundant, they were only moderately speciose with less than two dozen species recognized throughout their 120 million year history. All paleobatrachids were strictly aquatic species and moderate to small frogs, generally less than 50 mm (SVL). They had long, robust hindlimbs and long digits on both the fore- and hindfeet. *Neusibatrachus*, the oldest paleobatrachid, occurred first in the Late Jurassic, then without a fossil record until the Miocene. *Paleobatrachus*, with 12 species, spanned the Eocene to Pliocene period. Their fossils are very abundant in a series of freshwater deposits in eastern Czechoslovakia. In this area, volcanic gases apparently poisoned the waters of streams and ponds periodically, causing massive die-offs of all aquatic animals. These gases also stimulated diatom blooms and the skeletons of the diatoms buried the frogs and even their tadpoles. Burial was rapid and imprints of soft parts remain to help paleontologists reconstruct the anatomy and life histories of the paleobatrachid frogs.

The paleobatrachids and pipids are sister groups, and all paleobatrachids resembled the modern clawed frogs (*Xenopus*). The pipids do not appear in the fossil record until the Early Cretaceous, but they are more likely the ancestors, rather than the descendants, of the paleobatrachids. The paleobatrachid's restricted distribution (Europe) throughout their history (Late Jurassic, Eocene to Pleistocene) contrasts sharply with the presence of pipids in South America and Africa since the Cretaceous. Three genera are known from the Early Cretaceous of the eastern Mediterranean, suggesting an early radiation of the African pipids. *Xenopus* is confined now to Africa; however, two species were present in

Brazil and Argentina during the Late Paleocene. *Xenopus* also occurred early in Africa, from the Late Cretaceous of Nigeria and the Oligocene of Libya. It is an amazingly adaptable frog genus and even today it is the most speciose of the pipid frogs.

Pelobatids, another "moderately primitive" group of extant frogs, appear in the Late Cretaceous of Asia and North America. The Cretaceous taxa (e.g., *Eopelobates* and *Kizylkuma*) differ sufficiently from their later-appearing relatives to be placed in a separate subfamily (Eopelobatinae). The *Eopelobates* had a long existence from the Late Cretaceous to the mid-Miocene and an equally broad geographic occurrence from western North America through temperate Asia to Europe. The eopelobatine species were generally moderate-sized (5–6 cm SVL), terrestrial frogs. They lacked the spades on their heels, so characteristic of the modern pelobatids, but presumably shared many of the latters' natural history. Of the modern pelobatids, the tropical Asian megaphryines have no fossil history. The pelobatines or spadefoots appear in the Eocene of Europe (*Pelobates*) and Early Oligocene of North America (*Scaphiopus*). The related pelodytid frogs had a brief appearance in the Eocene of central Europe and the Miocene of western North America.

The fossorial rhinophrynids made an early Tertiary appearance from the Late Paleocene into the Oligocene in western North America. They have no subsequent fossil record.

The advanced frogs (Neobatrachia) also began to appear in the early Tertiary but somewhat later than the mesobatrachians. Surprisingly, considering their present diversity, neobatrachians are neither abundant nor diverse throughout much of the Tertiary. Only in the Pliocene and Pleistocene do they appear regularly in fossil beds. Excluding fossil records from the Pliocene, only the bufonids, hylids, leptodactylids, pelodryadids, and ranids have Tertiary representatives. *Bufo* has a nearly continuous record in South America from its first occurrence in the Late Paleocene. It also was present in North America and Europe from the mid-Tertiary onward. No other bufonids are known as fossils. *Hyla* appeared in the Oligocene in North America and in the Miocene in Europe. The only other fossil hylid is *Proacris* from the Miocene of Florida. The leptodactylids have a somewhat more diverse fossil history. The ceratophryines were represented in the Miocene of Argentina by *Wawelia*, the telmatobines by two genera in the Oligocene and Miocene, and a sprinkling of *Eleutherodactylus* and *Leptodactylus* species in the Pleistocene. The Miocene *Australobatrachus* is the first fossil pelodryadid and was contemporaneous in the Late Miocene with the still extant *Litoria*. The widespread and diverse ranids are represented in the fossil record only by *Ptychadena* in the Moroccan Miocene and an assortment of nearly 50 species of *Rana* from the Oligocene and onward of Europe and Miocene through Pleistocene of North and Central America.

CHAPTER 3

Reptiles

All reptiles reproduce via shelled, amniotic eggs. This evolutionary innovation encases the embryo within a fluid-filled container and eliminates the hazardous free-living embryonic stage of amphibians. Reptiles are, thus, freed from the necessity of locating and using a moist or aquatic habitat for egg laying.

Even though the evolution of the amniotic egg marks the transition from anthracosaurs to reptiles, the amniotic embryo is not uniquely reptilian. Mammalian embryos also develop an amnion that encloses them in fluid-filled sacs and forms part of their placenta. Other features, such as internal fertilization, pulmonary respiration, a single occipital condyle, two sacral vertebrae, and epidermal scales, are used commonly to characterize reptiles, yet each also occurs in other vertebrates, although all five occur together only in reptiles.

Reptiles are a distinct evolutionary lineage. The commonly used characteristics are supported by a set of uniquely reptilian ones. These traits tend to be aspects of anatomy that are not easily observed and may seem trivial. Nonetheless these anatomical features show the unity of reptiles. For example, several aspects of the reptiles' skin occur nowhere else. The epidermal covering (scales, feathers) arises as dermal evaginations. The finished covering is heavily keratinized and contains α- and β-keratins; β-keratin is uniquely reptilian. The free nerve endings innervating the epidermal portion of the skin have large expanded tips in contrast to the tapered tips of all other vertebrates. The eyes of some reptiles are protected by large mobile nictitating membranes, and the intraocular muscles of the iris are striated, not smooth muscles. The reptilian skull has a small tabular bone separated from the opisthotic bone, a small suborbital fenestra, a large posttemporal fenestra, a horizontal ventral margin, and no anterior coronoid bone in the lower jaw. The cervical vertebrae have midventral keels, and in adults the second intercentrum fuses to the axis. These unique anatomical features and others support the monophyly of the Reptilia.

MODERN GROUPS OF REPTILES

The Reptilia consists of two ancient lineages, the anapsids and the diapsids (Table 3.1). The living representatives of the former are the turtles, and of the latter, the crocodilians, birds, tuataras, lizards, and snakes. As noted in the preface, the historical perspective of herpetology has excluded birds, so that "reptiles," as used here, equals all of the above taxa except birds. Subsequent discussions, whether anatomical, behavioral, or ecological, characterize this "reptiles" group, and statements may also apply to birds.

There seems little chance of mistaking a turtle for any other animal. All turtles, past and present, have their bodies encased in bony shells. The shell may be reduced, nonetheless portions remain to protect the back and underside. All living turtles lack teeth and have substituted horny jaw sheaths for cutting and crushing. Crocodilians are similarly unmistakable among extant vertebrates. The elongate head, body, and tail dwarf the short, strong limbs. The back is covered with abutting, rectangular epidermal scales that are matched by an armor of bony plates beneath the scales. The head appears to be all jaws filled with strong, conical teeth (Fig 3.1).

The tuataras, lizards, and snakes comprise the lepidosaurs, a group of reptiles more closely related to one another than to any other reptiles. All lepidosaurs share a transverse cloacal slit, a notched tongue tip, shedding of the skin as a unit, and other traits. The tuataras are lizardlike but represent an early divergence within the lepidosaurian lineage. The squamates or snakes and lizards display a

TABLE 3.1 ▼▲▼▲▼▲▼▲▼▲▼▲▼▲▼▲▼▲▼▲
Hierarchical Classification Derived from the Sister Relationships of the Tetrapod Vertebrates

- Tetrapoda
  - Amphibia
    - Lissamphibia
      - Gymnophiona
      - Caudata
      - Salientia
  - Anthracosauria
    - Anthracosauroideae
    - Batrachosauria
      - Seymouriamorpha
      - Cotylosauria
        - Diadectomorpha
        - Amniota
          - Synapsida
            - Mammalia
          - Reptilia
            - Anapsida
              - Captorhinidae
              - Testudines
            - Diapsida
              - Araeoscelidia
              - Sauria
                - Archosauria
                  - Pseudosuchia
                    - Crocodylia
                  - Ornithosuchia
                    - Aves
                - Lepidosauria
                  - Sphendontida
                  - Squamata

[From Gauthier et al. (1988a,b)]

Note: Cateogry titles are not assigned to the hierarchical ranks above family, and some category levels or steps are absent.

variety of body forms and sizes; all squamates share paired penes in males and a highly modified vomeronasal organ.

Turtles

Turtles are the only tetrapods with their limb girdles lying inside the rib cage. The embryonic growth of the rib rudiments and the dermal shell skeleton

FIGURE 3.1

A sampling of adult body forms (habitus) in living reptiles. *Caretta* (Cheloniidae), *Crocodylus* (Crocodylidae), *Sphenodon* (Sphenodontidae), *Varanus* (Varanidae), and *Trimeresurus* (Viperidae), clockwise from upper left. Scale: bar = 10 cm.

outward and over the girdles is a major embryonic and evolutionary alteration. The resulting armored body readily identifies members of this group, no matter how the shell has been modified subsequently.

Although not highly speciose (240+ species), turtles occur worldwide and are often locally abundant. Two groups (Table 3.2) are recognized by differences in the plane of head retraction. The pleurodires or sidenecks lay the head and neck onto their shoulder. The cryptodires or hiddennecks retract the neck vertically posteriorad into the body cavity, producing a strong S-shaped curve in the neck.

The pleurodires contain two families of freshwater turtles, confined to the Southern Hemisphere. As their name suggests, many snake-neck sidenecks (Chelidae) have extremely long and thin necks. The chelids are the only freshwater turtles of Australia and New Guinea; they and the pelomedusids are the principal freshwater turtles of South America. Most chelids are moderate-sized with shell lengths less than 25 cm, and most have oblong, domed shells. *Chelus fimbriatus*, the matamata and source of the family name, is an exception with a flattened

TABLE 3.2 Classification of Extant Turtles and Crocodilians

Reptilia
- Anapsida
  - Testudines (Testudinata)—turtles
    - Cryptodira
      - Chelonioidea
        - Family Cheloniidae
        - Family Dermochelyidae
      - Chelydroidea
        - Family Chelydridae
      - Testudinoidea
        - Family Emydidae
        - Family Testudinidae
      - Trionychoidea
        - Family Carettochelyidae
        - Family Dermatemydidae
        - Family Kinosternidae
        - Family Trionychidae
    - Pleurodira
      - Family Chelidae
      - Family Pelomedusidae
- Diapsida
  - Archosauria
    - Crocodylia—crocodilians
      - Family Alligatoridae
      - Family Crocodylidae
      - Family Gavialidae

and ridged shell reaching 45 cm. The other sideneck turtles, the Pelomedusidae, occur in South America, Africa, and Madagascar. Most pelomedusids share the size range and shell shape of the chelids, but a few are large riverine turtles (*Podocnemis*, *Peltocephalus*) with shell lengths (carapace length/CL) of 40–105 cm.

The cryptodires comprise nine families of turtles (Table 3.2), and these families segregate into three distinct groups. The seaturtles (Cheloniidae and Dermochelyidae) are large (>60 cm minimum adult CL), streamlined turtles with flipperlike forelimbs. The cheloniid seaturtles have a typical bony shell covered by large epidermal scutes. The six cheloniid species (five genera) are nearshore animals and predominantly subtropical-tropical. However, the loggerhead seaturtle (*Caretta caretta*) nests on warm-temperate beaches. The leatherback (*Dermochelys*) is a giant, reaching shell lengths greater than 1.5 m and weights of nearly a metric ton. Its dorsal shell is covered by leathery skin instead of epidermal scutes. Even though leatherbacks nest only on tropical beaches, they are pelagic turtles and feed annually in subarctic and cold-temperate seas.

The three remaining groups of turtles are mainly freshwater and terrestrial. The softshell-musk turtle group (Trionychoidae) contains very dissimilar-appearing turtles. The softshells (Trionychidae) and pig-nose turtle (Carettochelyidae) share a leathery shell without epidermal scutes and a snorkel- or piglike snout. Carettochelids are represented today only by a single species in the the rivers of northern Australia and southern New Guinea. Trionychids are much more diverse with species in North America, Africa, and Asia. All trionychids have flattened and reduced bony shells. Trionychids and carettochelyids are moderate-sized turtles with adult lengths greater than 25 cm CL. The musk turtles (Kinosternidae) seldom reach this size, and most are much smaller (<15 cm CL). They have oblong, domed shells and release a malodorous musk. They occur in streams and ponds of the Americas. They have a large riverine relative (*Dermatemys*) in northern Central America.

The three chelydroids are large-headed, aquatic turtles with long tails. The snapping turtles (Chelydridae) are freshwater turtles, principally North American, although *Chelydra* has some isolated neotropical populations. *Macroclemys*, the alligator snapping turtle, is the giant (carapace lengths >1 m) of this group. It lives in deep, slow-moving waters of the North American Gulf Coast rivers. The big-headed turtle, *Platysternon*, is a distant relative of the former two genera and lives in mountainous streams of southern China and northern Indochina.

The testudinoid turtles contain nearly two-thirds of the living turtles. All have hard, bony shells covered with epidermal scutes. The Euroamerican pond turtles (Emydidae) contain mainly aquatic and semiaquatic turtles such as the sliders and cooters of southeastern North America, but the American box turtles, *Terrapene*, are predominantly terrestrial species. The Testudinidae consists of two subfamilies, the tortoises and the Euroasian pond turtles (batagurines). The tortoises are strictly terrestrial turtles. All tortoises have elephantine hindlimbs and most have high-domed shells. They occur on all continents except Australia and Antarctica, and even on the oceanic islands of Aldabra and the Galapagos. The pond turtles have their greatest diversity in Southeast Asia with nearly a dozen species there, but they also possess one group of turtles (*Rhinoclemmys*) in Central and northern South America.

Crocodilians

Three families of crocodilians survive today (Table 3.2). All crocodilians are semiaquatic, hunting and feeding in water but basking for a few hours each day out of water. They are moderate to large reptiles; most species attain total lengths of 3 m as adults. The Alligatoridae contains the Chinese and American alligators and the caiman of Central and South America. Alligators and caimans are denizens of freshwater habitats, ranging from swamps and marshes to lakes and rivers. Only one species of gharials (Gavialidae) persists today. This fish-eating crocodilian has a long, narrow snout with the teeth protruding like a pincushion at its tip. It lives in the large rivers of northeastern India and the Indus River of Pakistan. The crocodiles (Crocodylidae) occur throughout the world in the tropics. In most regions, there will be a freshwater species in the rivers and lakes and a different species in the brackish and saltwater habitats.

Lepidosaurs

The lepidosaurs are the dominant and most abundant group of extant reptiles. They represent two ancient lineages, the sphenodontidans and the squamates (Table 3.3). The Sphenodontida are the tuataras and consist of two species of buck-toothed, lizardlike reptiles of New Zealand. In contrast, there are over 5000 species of squamates.

Squamates live everywhere but on Antarctica and a few remote, desolate oceanic islands. They display a variety of body forms unequaled by any other tetrapod group, and range in total adult length from a 12-mm gecko to an anaconda over 10 m long. Amidst the great diversity of squamates, five lineages are readily recognizable (Table 3.3): iguanians, gekkotans, autarchoglossans, amphisbaenians, and snakes. Each is a distinct group. The amphisbaenians and snakes may have arisen from early autarchoglossan lizards and can be considered specialized autarchoglossans. The iguanians include three families: Iguanidae, Agamidae, and Chamaeleonidae. They are predominantly small to moderate-sized lizards, although many iguanas (Iguaninae: *Amblyrhynchus*, *Conolophus*, *Cyclura*, *Iguana*) exceed 1.5 m in total length. The iguanids are mainly New World lizards and are the most speciose of the lizard families occurring in the Americas. Anoles, horned lizards, fringe-toed lizards, and basilisks are only a few of the many American iguanids. About a dozen species of iguanids live in Madagascar and two species in Fiji. The agamids share many of the same body forms as the iguanids and live in the full spectrum of habitats from cold deserts to wet rain forests. Agamids inhabit Africa, Eurasia, and Australia. Nowhere do the distributions of the iguanids and agamids overlap. On the other hand, the chameleons occur together with the agamids in Africa, Southwest Asia, and India. The chameleons are a bizarre group of lizards with their compressed bodies, mittenlike feet, and projectile tongues.

The gekkotans contain the geckos and pygopods. Geckos are mostly nocturnal lizards. They are small to moderate-sized, usually have a delicate skin with small scales, large eyes covered with spectacles, and often expanded digital pads. The eublepharid geckos include five genera, found primarily in the Northern Hemisphere. The terrestrial, desert-adapted *Eublepharis* of Southwest Asia and India and *Coleonyx* of southwestern North America represent the typical eubleph-

TABLE 3.3 Classification of Extant Sphenodontidans and Squamates

Reptilia
  Diapsida
    Lepidosauria
      Sphenodontida
          Family Sphenodontidae
      Squamata
        Gekkota[a]—geckos
          Family Eublepharidae
          Family Gekkonidae
        Iguania[a]
          Family Agamidae
          Family Chamaeleonidae
          Family Iguanidae
        Autarchoglossa[a]
          Anguimorpha
            Family Anguidae
            Family Helodermatidae
            Family Varanidae
            Family Xenosauridae
          Scincomorpha
            Family Cordylidae
            Family Dibamidae
            Family Gymnophthalmidae
            Family Lacertidae
            Family Scincidae
            Family Teiidae
            Family Xantusiidae
          Amphisbaenia—worm lizards
            Family Amphisbaenidae
            Family Bipedidae
            Family Rhineuridae
            Family Trogonophidae
          Serpentes

[a] These taxa comprise the reptiles commonly known as Lizards.

arids. The gekkonid geckos are a diverse group of 700± species (75+ genera), pantropical in occurrence. Not all gekkonids fit the gecko model; the pygopods of Australia are snakelike.

The amphisbaenians or worm-lizards are a peculiar group of burrowing lizards. All have the body scales arranged in annuli, reduced eyes, a short tail, and no limb (with one exception, *Bipes* has well-developed forelimbs), thus appear superficially like earthworms. All use their heads for digging. They show their greatest diversity in Africa and South America, although one family each occurs in northwestern Mexico (Bipedidae), southeastern North America (Rhineuridae), and northern Africa-Arabian Peninsula (Trogonophidae).

The autarchoglossan lizards include nearly a dozen families of lizards and possibly the ancestors of the amphisbaenians and snakes. The Autarchoglossa contains two lineages of "typical" lizards, the anguimorphs and scincomorphs. Both groups have limbless and reduced-limbed members; however, the majority have well-developed limbs. The beaded lizards (Helodermatidae) of western North America, the monitors or goannas (Varanidae) of Africa, Asia, and Australia, the xenosaurs (Xenosauridae) of northern Central America and southern China, and the glass lizards, slowworms, alligator lizards and relatives (Anguidae) of the Americas and Eurasia make up the anguimorphs. The first three families share a pebblelike skin surface, whereas anguids commonly have rectangular, abutting scales covering a subdermal bony armor. The lacertoid scincomorphs (Gymnophthalmidae, Lacertidae, Teiidae, and Xantusiidae) have mainly elongate bodies, necks, and heads and strong limbs; granular scales cover their sides and backs and large rectangular plates on their undersides. The lacertids occur throughout temperate and tropical Africa and Eurasia. In contrast, the gymnophthalmids are confined to tropical America, and the teiids and xantusiids mainly to the American subtropics and tropics. The scincoid scincomorphs (Cordylidae, Dibamidae, and Scincidae) have overlapping scales over the entire body. The cordylids occur only in Africa and usually are stout-bodied and -limbed lizards, although a few species are limbless. The dibamids of the East Indies and north-central Mexico have degenerate eyes and no limbs. Skinks are the most speciose lizards. These shiny-scaled lizards occur worldwide and include strong-limbed and limbless forms as well as every stage in between.

Snakes are another group (Table 3.4) of limbless squamates (some have hindlimb rudiments projecting as cloacal spurs). But unlike the preceding limbless squamates, all snakes (but blindsnakes) have the two halves of the lower jaw (dentaries) flexibly united. They have a few other unique characteristics, but many characters possessed by most snakes also occur in one or more of the other squamate families.

The blindsnakes (Anomalepididae, Leptotyphlopidae, and Typhlopidae) are small burrowing snakes with rudimentary eyes, blunt or depressed-pointed heads, short tails, and shiny smooth, equal-sized scales around the body. They inhabit the tropics worldwide and are moderately speciose (>300 species). They appear unlike any other snakes and likely diverged early from the main lineage of snakes. Of the typical snakes, the boas (Boidae, Bolyeriidae), pythons (Pythonidae), Loxocemidae, and Aniliidae have retained hindlimb and pelvic vestiges that appear externally as cloacal spurs. The boids and pythons contain the giants of the snake world, the anaconda (*Eunectes murinus*, to 11 m) and reticulated python (*Python reticulatus*, 8–9 m). Both families have smaller species, and some boids are less than 0.5 m as adults. Pythons are African, Asian, and Australian

TABLE 3.4
Classification of Extant Snakes

- Reptilia
  - Diapsida
    - Lepidosauria
      - Squamata
        - Serpentes (Ophidia)—snakes
          - Scolecophidia
            - Family Anomalepididae
            - Family Leptotyphlopidae
            - Family Typhlopidae
          - Alethinophidia
            - Family Acrochordidae
            - Family Aniliidae
            - Family Atractaspididae
            - Family Boidae
            - Family Bolyeriidae
            - Family Colubridae
            - Family Elapidae
            - Family Loxocemidae
            - Family Pythonidae
            - Family Uropeltidae
            - Family Viperidae
            - Family Xenopeltidae

snakes; the boids occur worldwide with the highest diversity in tropical America. The bolyeriid boas are a pair of peculiar species with a unique joint in the middle of each maxilla. They survive today only on a single island in the Indian Ocean. The Loxocemidae contains a single species (*Loxocemus bicolor*) living in the forest of Mexico. The Acrochordidae, Aniliidae, Uropeltidae, and Xenopeltidae are also primitive snakes with limited geographic ranges. The wart or elephant-trunk snakes (Acrochordidae) are aquatic snakes of Southeast Asia. Their scales are tiny keeled granules, and their skin is folded and wrinkled, appearing two sizes too large for their bodies. The uropeltids or shieldtails are burrowing snakes with short tails modified to plug the burrow. Although restricted to southern India and Sri Lanka, there are nearly 50 species of shieldtails. The aniliids and xenopeltids of tropical America and Asia have less than a dozen species between them.

The remaining families of snakes (Table 3.4) are anatomically more advanced than the preceding snakes. Many of the advancements involve the loss or extreme reduction of structures. For example, the advanced snakes have no vestige of limbs, their left lung persists as a tiny rudiment, and the skull is highly kinetic. The Colubridae contains about 70% of the total number of snake species. They occur worldwide, live in all aquatic and terrestrial habitats, and range in total length from 15 to 350 cm. Some are venomous, but most are not. All Atractaspididae, Elapidae, and Viperidae are venomous. The atractaspidids or burrowing asps of Africa and Southwest Asia are dull, innocuous-appearing snakes whose long fangs and mobile jaws permit them to hook-bite prey in narrow tunnels. The viperids also have long fangs on rotating upper jaws. The jaws can erect the fangs as the mouth opens and lay the fangs against the roof of the oral cavity when the mouth closes. The vipers include the true vipers of Africa and Eurasia and the pitvipers of the Americas and Asia. The elapids contain the American coral snakes, African and Asian cobras, Asian kraits, Australian tiger snakes, and the seasnakes of the IndoPacific seas. Almost all elapids have short, permanently erect fangs.

GENERAL ANATOMY

Skin

The skin of reptiles has the same cellular organization as in the amphibians and other vertebrates. The inner dermis consists mainly of connective tissue, and the outer epidermis has a germinative cell layer (stratum germinativum) basally and above are numerous differentiating and dying cell layers (stratum corneum). Differentiation produces an increasingly thick, keratinous cell wall and the eventual death of the cell. This basic pattern is variously modified in the different reptilian lineages and occasionally on different parts of the body of the same individual. Reptiles uniquely produce β-keratin as well as α-keratin, which they share with the other vertebrates. β-Keratin is a hard and brittle compound, α-keratin an elastic and pliable one.

On all or most of the body, the skin is divided into scales. The scales have an assortment of names (plates, scutes, shields, laminae, lamellae, scansors, tubercles) depending on the taxonomic group, size and shape of the scales, and the scale's location on the body. Some of the names are interchangeable (scutes = shields), whereas others are specific (scansors = scales beneath digits). All reptilian scales are keratinized epidermal structures, but those of the lepidosaurs are not strict homologues of crocodilian and turtle scales. Scales commonly overlap in the squamates but seldom do in crocodilians and turtles.

Two patterns of epidermal growth occur. In crocodilians and turtles, the germinative cells divide continuously throughout an individual's life, stopping only during hibernation or torpor. This pattern is shared with most other vertebrates, whether fish or mammal. A second growth pattern, where growth is discontinuous, occurs in lepidosaurs (Fig. 3.2). Upon shedding of the outer epidermal sheath (Oberhautchen), the germinative cells enter a resting phase

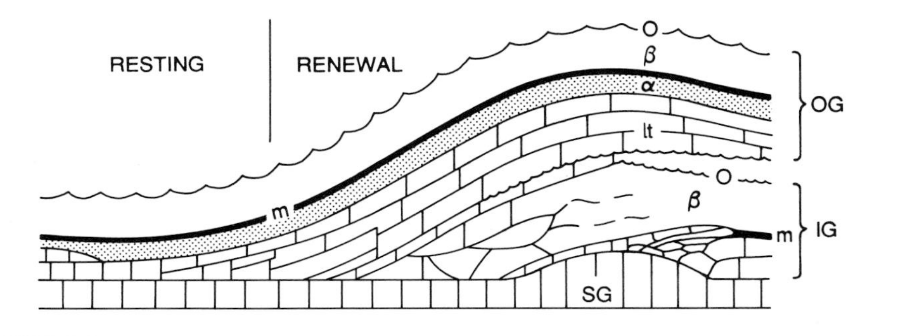

FIGURE 3.2
Diagram of the sequential cellular changes during a single shedding cycle in squamate epidermis. Abbreviations: α, α-layer; β, β-layer; IG, inner generation layers; lt, lacunar tissue; m, meso layer; O, Oberhautchen; OG, outer generation layers; SG, stratum germinativum. [Adapted from Landmann (1986).]

with no mitotic division. The renewal phase begins with the synchronous division of the germinative cells and the differentiation of the upward-moving scales into two distinct layers separated by a narrow layer of cell secretions.

The surface of each reptilian scale is entirely β-keratin, and the interscalar space or suture is α-keratin. This distribution of keratin produces a durable and protective scale surface and flexible junctures between the scales for flexibility and expansion of the skin. Although the preceding is the typical pattern, the scales on the limbs of some turtles have α-keratin surfaces, and in the softshell and the leatherback turtles, the entire shell surface is an α-keratin skin. In most of the hard-shelled turtles, the scutes and sutures contain only β-keratin. The two-layered epidermis of the lepidosaurs has an α-keratin inner layer and an β-keratin Oberhautchen.

An anomaly of special interest is the occurrence of scaleless snakes in several species of colubrids and viperids. Scales are largely although not totally absent. The labial and ventral scales are usually the only ones present; the remainder of the skin is externally a smooth sheet of soft, keratinous epidermis. Genetically scalelessness appears to be a simple Mendelian homozygous recessive.

Glands, Claws, and Other Structures

The scales of crocodilians, turtles, and some lizards (e.g., anguids, cordylids, scincids) are underlain by bony plates (osteoderms or osteoscutes) in the dermis. The bony organization of the osteoderms matches the bipartite organization of the dermis. The outer layer is spongy and porous bone; the inner layer is compact and dense bone. Usually osteoderms are confined to the back and sides of the animal and attach loosely to one another in symmetrical rows and columns to permit flexibility, yet maintain a protective bony armor. In crocodiles and a few lizards (*Heloderma*), the osteoderms fuse with the dorsal skull elements, forming a rigid skull cap. The carapace and plastron (upper and lower shell) of turtles arose from the fusion of osteoderms with vertebrae and ribs dorsally, and osteoderms and sternum ventrally.

Reptiles have a variety of skin glands. Although common and widespread over the body, the glands are small and inconspicuous. Most of them are multicellular glands. Their secretions are mainly lipid- and wax-based compounds that serve as waterproofing, surfactant, and pheromonal agents.

Aggregations of glandular tissues do occur. In turtles, musk glands are present, except in tortoises (Testudininae) and the pseudemyd turtles. The glands are usually bilaterally paired and lie within the bridge between the top and bottom shell, opening to the outside through individual ducts in the axilla and inguen or on the bridge. Male tortoises have a mental gland just behind the tip of the lower jaw. Both male and female crocodilians have paired mandibular and cloacal glands. The occurrence of large glands is more erratic in the lepidosaurs. Some geckos, iguanids, and agamids have a series of secretory pores on the underside of the thighs and pubis. Each pore arises in an enlarged scale and extrudes a waxy compound containing cell fragments. These femoral and precloacal (pubic) pores do not open until the lizards attain sexual maturity and often occur only in males. Thus they appear to be sexual scent glands, although this remains unconfirmed. Snakes and some autarchoglossan lizards have paired scent glands at the base of the tail; each gland opens at the outer edge of the cloaca opening. These saclike glands release copious amounts of semisolid, malodorous fluids. For some species, the fluid may serve for defense, in other situations for sexual recognition. Other glandular aggregations occur but are limited to a few reptiles. For example, a few Australian geckos have specialized squirting glands in their tails; an assortment of marine and desert species of turtles, crocodilians, and lepidosaurs have salt glands.

Specialized keratinous structures are common in reptiles. All limbed species with functional digits have claws, that is, the keratinous sheaths encasing the tips of the terminal phalanges. The sheaths have three layers, with the outermost of hard B-keratin. The claws form either as full keratinous cones (crocodilians and turtles) or as partial cones (lepidosaurs). The upper and lower jaw sheaths of turtles are also keratinous structures and replace the teeth as the cutting and crushing surfaces. Hatchling turtles, crocodilians, and *Sphenodon* have an egg carbuncle on the snout. This keratinous projection slices open the egg shell and allows the hatchling to emerge.

A dozen or more types of small, epidermal sense organs occur in reptiles, particularly in the lepidosaurs. Most are barely visible, appearing as tiny pits or projections. They are strictly epidermal structures and persistent ones, not being shed during the sloughing cycle. Presumably, most of these organelles are tactile; however, the recent discovery of a light-sensitive region on the tail of a seasnake suggests a broader range of receptors and sensitivities. These organs are often concentrated on the head but are also widespread on the body, limbs, and tail.

Ecdysis

Different epidermal organizations and growth patterns result in different shedding or sloughing patterns. In crocodilians and the nonshell epidermis of turtles, cell growth is continuous and the outer surface of the skin is shed continuously in flakes and small sheets. Whether the scutes of hard-shelled turtles are retained or shed seasonally is species specific. When retained, successive scutes form a flattened pyramidal stack, because an entire new scute develops beneath the older scute at the beginning of each growing season. Scute growth is not confined to the margins, although each new scute is thickest there and much compressed beneath the older scutes.

The shedding pattern in lepidosaurs is more complex and intimately tied to the unique epidermal growth pattern. In the tuataras and most lizards, the skin sheds typically in large patches. Snakes usually shed the skin as a single piece. But for all lepidosaurs, the sequence of epidermal growth and shedding (Fig. 3.2) is identical. During the resting stage, the epidermis has a basal germinative layer of cells, a narrow band of α-precursor cells, a thin mesos layer of mucus and other cell secretions, and externally the beginnings of an outer-generation layer capped by the Oberhautchen. The resting stage ends as cell proliferation and differentiation begin in the outer-generation layer. Then the germinative cells begin to divide. As each newly formed layer of cells is pushed upward and outward by cell division below them, the cells differentiate and produce the inner-generation layer. This inner-generation layer forms the precursor of the scales (outer-generation layer) for the next epidermal cycle. As the Oberhautchen nears completion, the outer-generation layer separates from the inner layer and is shed, completing the shedding or sloughing cycle. This cycle is repeated at regular intervals when food is abundant. This growth-shedding (renewal) phase requires about 14 days. The resting phase may last from a few days to many months.

Coloration

In general, reptiles have two sets of color-producing cells. Melanocytes are scattered throughout the basal layers of the epidermis. During the renewal phase of epidermal growth, the melanocytes send pseudopodia into the differentiating keratocytes and transfer melanin. The melanin-bearing keratocytes occur in the β-layer of crocodilians, iguanid lizards, and snakes, and in the α- and β-layers in many other lizards.

The second set of color-producing cells are the chromatophores. The chromatophores lie one atop another in the outer portion of the dermis. A single layer of xanthophores (i.e., lipophores and erythrophores) lies directly beneath the basal membrane of the epidermis. Beneath them are two to four layers of iridophores (guanophores and leukophores), and at the bottom are large melanophores. This organization may represent the general pattern for all color-changing reptiles, since stacked chromatophores are absent in some some static-colored species. The presence, density, and distribution of chromatophores within each layer vary within an individual and among species to produce the different colors and color patterns.

Muscles and Skeleton

The long independent evolution of each reptilian lineage is no more evident than in the musculoskeletal system. The outward appearance of crocodilians, turtles, and lepidosaurs is strikingly different, with body shape indicating the form of the underlying muscles and skeleton—the effector and the support systems.

As in amphibians, the reptilian musculoskeletal system is adapted primarily for terrestrial limbed locomotion and modified secondarily for aquatic and terrestrial limbless locomotion. Reptiles retain considerable lateral flexure of the body in their limbed gaits, and only in the archosaurs does dorsoventral flexure become an important component of locomotion.

Head and Hyoid

The reptilian skull consists of the three basic vertebrate units: dermocranium, chondrocranium, and splanchnocranium. The reptilian chondrocranium encases the brain ventrally, laterally, and posteriorly, and encapsulates the olfactory and otic structures. The anterior portion remains cartilaginous, even in adults, and consists mainly of continuous internasal and interorbital septa and a pair of nasal conchae supporting olfactory tissue. Between the eyes and ears, the chondrocranium ossifies as the basisphenoid, and further posteriorly, the basioccipital; a pair of exoccipitals, and the supraoccipital bones develop below and behind the brain (Fig. 3.3). They encircle the exit (foramen magnum) for the spinal cord. Below the foramen magnum, the exoccipitals and basioccipital jointly form a single occipital condyle, the articular surface between the first cervical vertebra (atlas) and the skull. Portions of each otic capsule remain cartilaginous, although much of the capsule becomes the epiotic, prootic, and opisthotic bones.

The columella of the middle ear is a splanchnocranial element, as are also the quadrate and the epipterygoid (small in lizards and turtles, lost in snakes and archosaurs). The quadrate is a large bone on the posterolateral margin of each side of the skull. It bears the articular surface for the lower jaw. On the mandible, the articular bone provides the opposing articular surface and is the only splanchnocranial element of the lower jaw. The reptilian hyoid arch is much reduced; a large midventral plate or body usually has three processes (arms) extending upward and posteriorly from it.

Dermal bones form the major portion of the reptilian skull and mandible.

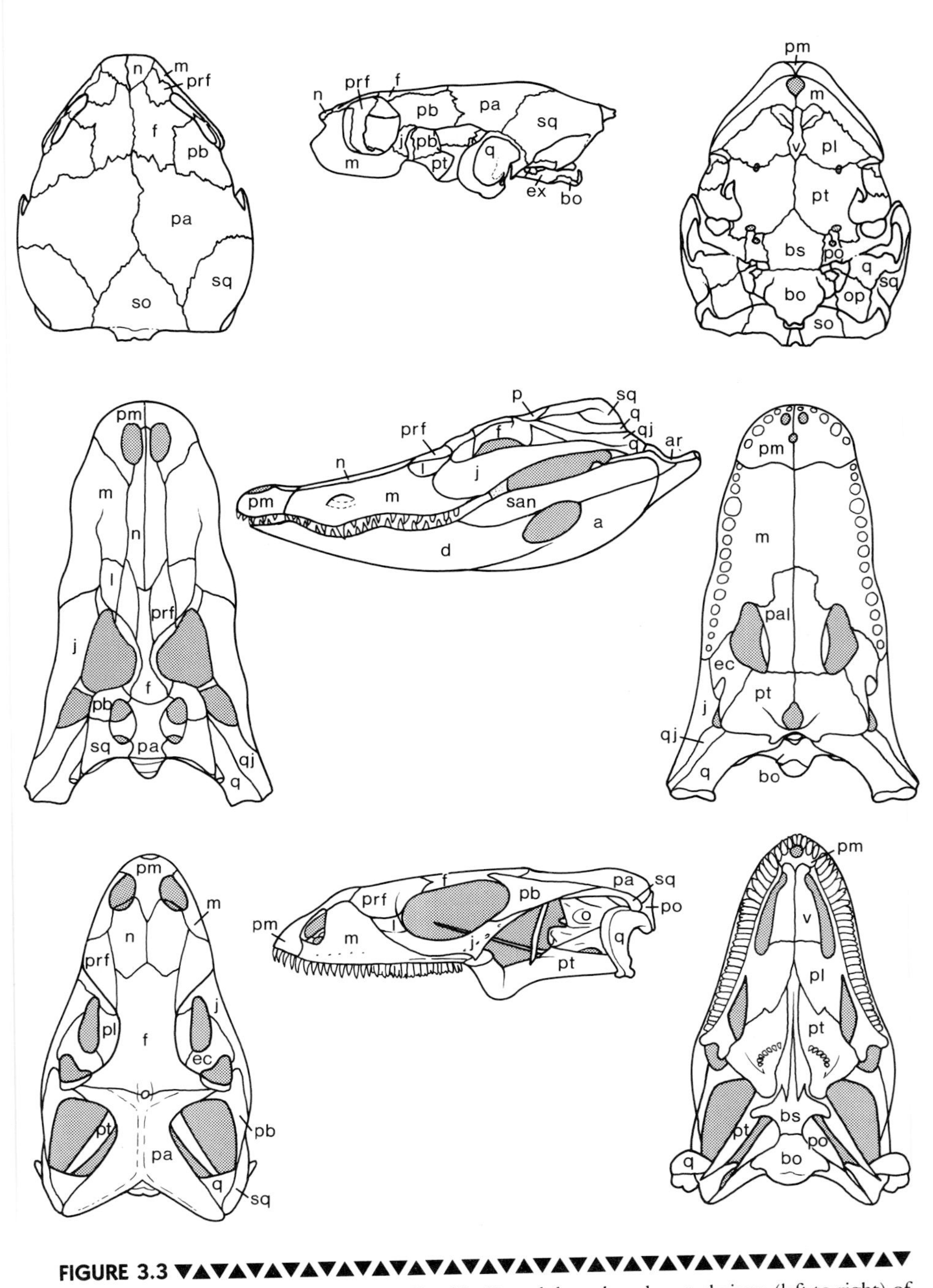

FIGURE 3.3 ▼▲▼

Cranial skeletons of an anapsid and two diapsids. Dorsal, lateral, and ventral views (left to right) of the turtle *Pseudemydura umbrina*, the crocodilian *Alligator sinensis*, and the lizard *Ctenosaura pectinata* (top to bottom). Abbreviations: a, angular; ar, articular; bo, basioccipital; bs, basisphenoid; d, dentary; ec, ectopterygoid; ex, exoccipital; f, frontal; j, jugal; l, lacrimal; m, maxillary; n, nasal;

They form over and around the endochondral bones of the two other cranial units. The roof of the dermocranium contains the nasals, prefrontals, frontals, and parietals (all paired) from anterior to posterior. The upper jaws (premaxillae and maxillae) join the roofing bones directly. The cheek and temporal areas contain a postorbital, postfrontal, jugal, quadratojugal, and squamosal bone on each side. The primary palate or roof of the mouth consists of premaxillae and maxillae anteriorly, a median vomer bordered laterally by the palatines, and posteriorly by the pterygoids and occasionally a parasphenoid. When a secondary palate forms, as in crocodilians, it derives largely from the premaxillae and maxillae. A few other dermal bones, for example, septomaxilla or lacrimal, are present in some extant reptiles. The jugal, quadratojugal, prefrontal, postfrontal, and squamosal are absent (individually or in sets) in some taxa.

The mandible (lower jaw) contains numerous dermal bones (paired): dentary, splenial, angular, surangular, coronoid, and prearticular (Fig. 3.3). Only the dentary bears teeth, and in the upper jaw, only the maxilla, premaxilla, palatine, and pterygoid bear teeth. Teeth may be absent on one or more of these teeth-bearing bones. In turtles, teeth are entirely absent; their cutting and crushing functions are performed by the keratinous jaw sheaths.

Typical reptilian teeth are cone-shaped and arranged in a single, longitudinal row. This basic shape has been variously modified, for example, laterally compressed and serrated edges in some herbivorous lizards or elongated and posteriorly curved in snakes. The teeth attach to the bone by sitting in sockets (thecodont; e.g., crocodilians), arising from a one-sided groove (pleurodont; most lepidosaurs), or attaching directly to the bone surface (acrodont; some lizards). Tooth replacement is continuous throughout life, except in most acrodont forms, which replace teeth only as juveniles.

The skulls of the two major reptilian lineages, anapsids and diapsids, are distinct (Fig. 3.3). In the anapsid (turtle) skull, the bony temporal arcade (parietal, squamosal, postorbital, jugal, and quadratojugal) is entire. This condition is the primitive reptilian state. In the diapsid (crocodilian, squamate) skull, the temporal area has two openings (fenestrae), an upper fenestra between the parietal, postorbital, and squamosal and a lower one between the squamosal, jugal, and quadratojugal. Both of these skull types have been modified in extant reptiles. Most living turtles have emarginated temporal arcades, leaving a small arch of bone behind each eye. Only a few turtles, such as the seaturtles, retain a nearly complete arcade. The crocodilians retain the basic diapsid architecture (Fig. 3.3), although the upper (superior temporal) fenestra is small. In lepido-

---

op, opisthotic; p, pa, parietal; pal, pl, palatine; pb, postorbital; pm, premaxillary; po, prootic; prf, prefrontal; pt, pterygoid; q, quadrate; qj, quadratojugal; san, surangular; so, supraoccipital; sq, squamosal; v, vomer. [Adapted from Gaffney (1979), Iordansky (1973), and Oelrich (1956), respectively.]

saurs, only *Sphenodon* retains the two fenestrae. The squamates have only one (upper) or no fenestra, losing the lower temporal arch (squamosal-quadratojugal-jugal) in the first case and losing the middle arch (squamosal-postorbital) or both lower and middle arches in the latter.

The loss of arches and fenestrae in the diapsid skull is associated with the increased flexibility (kinesis) of the skull. Kinesis derives from the presence of hinges between various sections of the skull. A hinge in the back of the skull (a metakinetic joint) between the dermal skull and the braincase (parietal-supraoccipital) is the oldest kinetic joint, occurring in the earliest reptiles and today in *Sphenodon*. Two other joints develop in the dermal roofing bones. A mesokinetic joint between the frontals and parietals occurs in many lizards, and a prokinetic joint lies between the nasals and prefrontal-frontals in many snakes. The most striking kinesis is streptostyly (quadrate rotation) of the lepidosaurs, particularly in the snakes; here each quadrate is very loosely attached to the dermocranium and has a free ventral end. The ligamentous attachments allow the quadrates to rotate and to swing forward-backward and inward-outward, thus improving the jaw's grasping ability and increasing the gape.

The complexity in the arrangement and subdivision of muscles mirrors the diversity of the bony architecture of the head. There are no facial muscles, but the diversity of jaw and tongue muscles permits a wide range of feeding and defense behaviors. The jaw's depressor and adductor muscles arise from within the temporal arcade and attach to the inside and outside of the mandible. With highly kinetic skulls, the muscles are more finely subdivided and permit a wider range of movements of the individual bones, including those of the upper jaw. The throat muscles are typically flat sheets of muscles extending onto the neck. Beneath these muscles, the hyoid muscles are thicker sheets and longer bundles attaching the hyoid body and arms to the mandible, back of the skull, and the cervical vertebrae.

Vertebral Column

The trend for increased rigidity of the vertebral column begun in early tetrapods is further elaborated in reptiles. The vertebrae form a firmly linked series with the addition and elaboration of intervertebral articular surfaces and a complex fragmentation of the intervertebral muscles. In reptiles, the rigidity combines with regional differentiation of the vertebrae. This regionalization permits different segments of the column to have different directions and degrees of movement, and is reflected in the architecture of both bones and muscles.

There is no such thing as a typical reptilian vertebra or vertebral column. However, some features are shared by most reptiles. The centra are the weight-bearing units of the vertebral column. Commonly each centrum is a solid spool-shaped bone, but in *Sphenodon* and some geckos, the notochord persists and perforates each centrum. A neural arch sits astride the spinal cord on each centrum. The legs (pedicels) of each arch fuse to the centrum or insert into notches on the centrum. The neural spine varies from short to long, wide to narrow, depending on its position within the column and the type of reptile. The intervertebral articular surfaces, the zygapophyses, arise from the top of pedicels, a pair in front and a pair in the rear of each vertebra. The anterior zygapophyses have inward- and upward-facing articular surfaces, the posterior ones with outward- and downward-facing surfaces. The angle of these articular surface determines the amount of lateral flexibility. When angled toward the horizontal, lateral flexibility between adjacent vertebrae increases, but if angled toward the vertical, rigidity increases. The pedicels also bear the articular surfaces for the ribs; for two-headed ribs, the upper surface is the transverse process or diapophysis, the lower surface the parapophysis. The ribs of living reptiles are one-headed and articulate with the transverse process with the exception of crocodiles. In many lepidosaurs, accessory articular surfaces occur at the base of the neural spine; a zygosphene projects from the front of the arch into a pocket, the zyganthrum, on the rear of the preceding vertebra. The articular surfaces between the centra are variable, but the procoelous ball-and-socket condition is widespread, occurring in all extant crocodilians and most lepidosaurs. The most variable central articular patterns occur in the cervical vertebrae, where, for example, procoelous, opisthocoelous, and biconvex centra exist in the neck of an individual turtle.

Regional differentiation of the vertebrae is well marked in crocodilians. There are nine cervical, fifteen trunk, two sacral, and numerous caudal vertebrae. The first two cervical vertebrae, the atlas and axis, are constructed of several unfused components. The atlas bears a single anterior surface for articulation with the occipital condyle of the skull. The axis and subsequent cervical vertebrae bear two-headed ribs that become progressively longer toward the trunk. The first eight or nine trunk vertebrae have ribs extending ventrally to join the sternum and to form the thoracic basket. The remaining thoracic vertebrae have progressively shorter ribs. The ribs of the sacral vertebrae anchor the vertebral column to the ilia of the pelvic girdle. The caudal or postsacral vertebrae become sequentially smaller and laterally compressed. They also progressively lose their processes.

The limbed lepidosaurs show the same regional differentiation pattern as in the crocodilians. Vertebral number is much more variable, although all share a pair of sacral vertebrae. Generally there are eight cervical vertebrae, and ribs exist only on the posterior four or five vertebrae; however, *Varanus* has nine and chamaeleonids have three to five. Trunk vertebrae are even more variable in number; 16–18 vertebrae appears to be the primitive condition, but vertebral number can be less (11 in chamaeleonids) or considerably more in elongated lizards, particularly in limbless and reduced-limbed anguids and skinks. Caudal vertebrae are similarly variable in number. In limbless squamates, differentiation is limited; atlas and axis are present, followed by a large number (50–300) of trunk (precloacal) vertebrae, several cloacal vertebrae, and a variable number (10–120) of caudal vertebrae.

In contrast, vertebral number is nearly invariable in turtles. All living turtles have eight cervical vertebrae, and cervical ribs when present are rudimentary and confined to the posteriormost vertebrae. The variable neck lengths of the different turtle species arise from elongation or shortening of vertebrae lengths. There are ten trunk or dorsal vertebrae. The first and last are attached but not fused to the carapace. The middle eight are firmly fused (co-ossified) with the neural bones of the carapace. The trunk ribs extend outward and fuse with the costal bones of the shell. The two sacral vertebrae link the pelvic girdle to the vertebral column by short, stout ribs. Caudal number is variable, but <24 in most species.

The division of the vertebral column muscles into epaxial and hypaxial bundles remains in reptiles, although the distinctiveness of the two sets is not obvious. Similarly, the segmental division also largely disappears in reptiles. Most axial muscles span two or more vertebral segments and often have attachments to several vertebrae. The complexity of the intervertebral muscles is greatest in the limbless taxa. Unlike fish, their undulatory locomotion is not a uniform wave of contraction but requires individualized contraction patterns, depending on which part of the body is pushing against the substrate. Turtles lack trunk musculature. Epaxial and hypaxial muscles, however, do extend inward from the neck and tail to attach to the carapace and dorsal vertebrae.

Girdles and Limbs

The limb and girdle skeletons of extant reptiles share many components with the appendicular skeleton of living amphibians; nonetheless, the morphology and function of the muscular and skeletal components are also different. Much less of the reptilian cartilaginous (endochondral) skeleton remains unossified. The reptilian rib or thoracic cage is linked to the pectoral girdle through the sternum. A shift in limb posture occurred with the development of a less sprawled locomotion. Lizards and salamanders share gait patterns and considerable lateral body undulation when walking or running, but the lizards have more elevated postures and a greater range of limb movements. No reptile has a musculoskeletal system so tightly linked to saltatory locomotion as that of frogs.

The early reptiles had a pectoral girdle of five dermal components (paired clavicles and cleithra, and an interclavicle) and the paired, endochondral scapulocoracoids, each with two or three ossification centers (scapula, coracoid, or anterior and posterior coracoids). A cleithrum lays on the anterolateral edge of each scapula. Cleithra disappeared early in reptilian evolution, so none exists in extant reptiles. The interclavicle is a new girdle element, lying ventromedial and superficial to the sternum. The clavicles extend medially along the base of the scapulae to articulate with the anterior ends of the interclavicles. The endochondral components lie deep to the dermal ones. The scapula is the vertical element, the coracoid the horizontal one; at their junction, they support the glenoid fossa for the articulation of the humerus. The coracoids of the left and right sides meet medially, usually narrowly separated by a cartilaginous band, which is continuous posteriorly with the broader, cartilaginous sternum. The sternum bears the attachments for the anterior thoracic ribs and often a pair of posterior processes that receive the attachments for additional ribs. Posterior to the thoracic ribs, a series of dermal ribs (the gastralia) may support the ventral abdominal wall. These ribs are superficial to and are not joined to the thoracic ribs or any sternal processes, although the connective tissue sheath of the gastralia may attach to the epipubis of the pelvic girdle.

Crocodilians, *Sphenodon*, and some lizards have gastralia (abdominal ribs). This structure and the sternum are absent in snakes and turtles. The plastron (ventral portion of the shell) of turtles is largely a bony neomorph (= novel and unique structure), only the clavicles and interclavicle appear to have become part of the plastron. Snakes have also lost all of the pectoral girdle elements and many of the limbless lizards have greatly reduced the size of the endochondral elements and occasionally lose the dermal ones. Even limbed lizards show a reduction of dermal elements; the interclavicle reduces to a thin cruciform rod of bone in most. Chameleons lack the clavicles and interclavicles. Clavicles are absent in crocodilians, but the interclavicle remains as a median rod.

The reptilian pelvic girdle contains three pairs of endochondral elements: the vertical ilia attaching to the sacral vertebrae dorsally, and the horizontal pubes (anterior) and ischia (posterior) form a ventral plate that joins the left and right sides of the girdle. An acetabulum occurs on each side at the juncture of these three bones. These elements persist in all living reptiles, except in most snakes. In all, the puboischiac plate develops a pair of fenestrae that often fuse into a single large opening encircled by the pubes and ischia. The plate becomes V-shaped, as the girdle deepens and narrows. In most reptiles, the ilia are rodlike. In a few primitive snake families, a rod-shaped pelvic bone remains on each side. Its precise homologues are unknown, but it does bear an acetabulum and usually processes that are labeled as ilial, ischial, and pubic processes. The femur is vestigial and externally covered by a keratinous spur.

The early reptiles had short, robust limb bones with numerous processes. In the modern species, the propodial elements (humerus, femur) are generally smooth, long, and columnar with a slight curve; their heads are little more than rounded ends of the bony element. Only in the turtles are the heads elevated and tilted from the shaft as distinct articular surfaces. The epipodial pair (ulna and radius, tibia and fibula) are of unequal size, with the ulna or tibia the longer, more robust weight-supporting element of the pair. With the rotation of the epipodium (forearm), the ulna developed a proximal olecranon process and sigmoid notch for articulation with the humerus. The tibia lacks an elevated process but has a broad proximal surface for femoral articulation. The mesopodial elements (carpus or tarsus) consist of numerous small blocklike bones. The arrangement, fusion, and loss of these elements are highly variable, and the wrist or ankle flexure usually lies within the mesopodium. The metapodial elements (metacarpus or metatarsus) are elongate and form the base of the digits (phalanges). The basic phalangeal formula for the reptilian forefoot (manus) is 2-3-4-5-3 and for the hindfoot (pes) 2-3-4-5-4. Most extant reptiles have lost phalanges within digits and not uncommonly entire digits.

The pectoral girdle and forelimbs attach to the axial skeleton by muscles extending from the vertebrae to the interior of the girdle or to the humerus. Similar patterns of muscular attachment exist for the pelvic girdle and hindlimb, although this girdle also attaches directly to the vertebral column via the sacral ribs-ilia buttress. Within the limbs, the single-jointed muscles serve mainly as rotators and the multiple-jointed muscles (origin and insertion separated by two or more joints) are extensors and flexors, many extending from the distal end of propodium to the manus or pes.

Nerves and Sense Organs

Nervous Systems

Just as reptiles display a multitude of body forms, they display a variety of brain sizes and morphologies. In all, the basic vertebrate plan of two regions (forebrain, hindbrain) is maintained and flexure of the brain stem is limited. The braincase is commonly larger than the brain, so that its size and shape does not accurately reflect dimensions and morphology of the brain.

The forebrain forms the cerebral hemispheres, thalamic segment, and optic tectum of adult reptiles. The hindbrain becomes the cerebellum and medulla oblongata. The cerebral hemispheres are pear-shaped with olfactory lobes projecting anteriorly and ending in distal olfactory bulbs. These lobes range from long, narrow stalks with tiny bulbs (many iguanid lizards) to short, stout stalks and bulbs (tortoises). Their sizes reflect the importance of olfaction in the animal's life. The thalamic area is a thick-walled tube compressed and hidden by the cerebral lobes and the optic tectum. The dorsal, epithalamic portion has two dorsal projections. The anteriormost projection is the parietal (parapineal) body; in many lizards and *Sphenodon*, it penetrates the skull and forms a parietal eye. The posterior projection, the epiphysis (pineal organ), is typically glandular (turtles, snakes, and most lizards), although in some lizards (Fig. 3.4) and *Sphenodon*, it is a composite with a rudimentary retinallike structure (parietal body) and glandular tissue. Crocodilians lack a parietal-pineal complex. The ventral portion of the thalamic area is the hypothalamus. In addition to its nervous function, the thalamus, hypothalamus, and adjacent pituitary gland function together as a major endocrine organ. The posterior portion of the forebrain has the optical tectum dorsally and the optic chiasma ventrally. The cerebellum and medulla are small in extant reptiles.

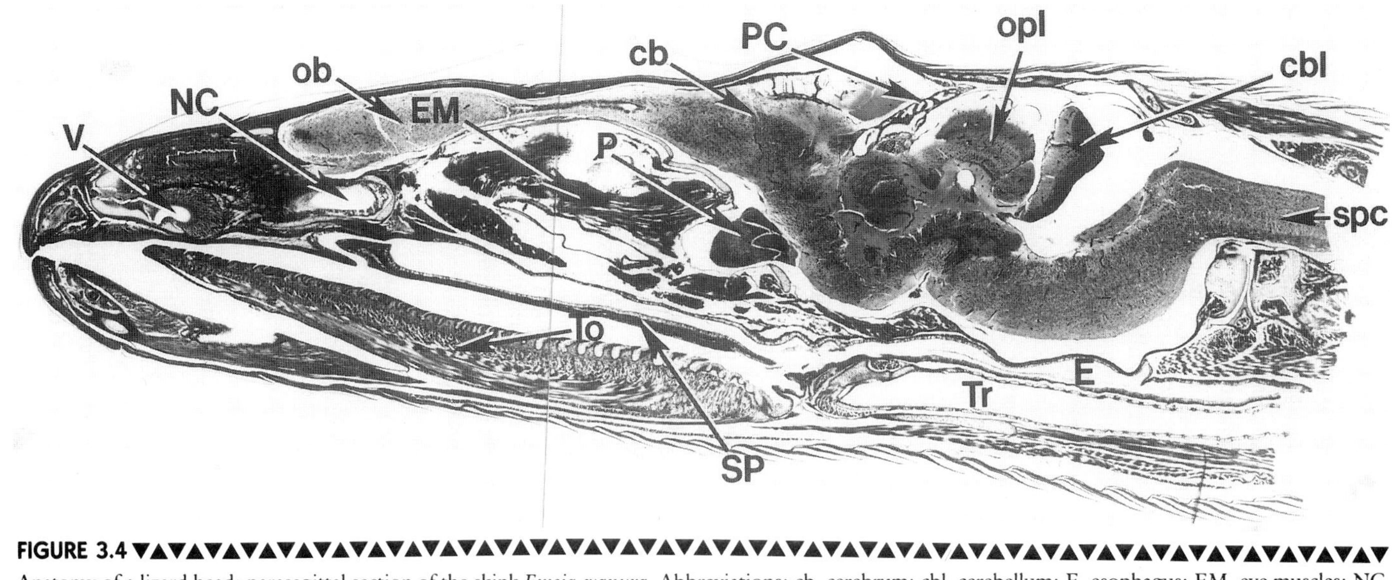

FIGURE 3.4

Anatomy of a lizard head; parasagittal section of the skink *Emoia cyanura*. Abbreviations: cb, cerebrum; cbl, cerebellum; E, esophagus; EM, eye muscles; NC, nasal cavity; ob, olfactory bulb; opl, optic lobe; P, pituitary gland; PC, pineal complex; SP, secondary palate; spc, spinal cord; To, tongue; Tr, trachea; V, vomeronasal organ. (D. Schmidt)

The spinal cord extends from the medulla posteriorly to the end of the vertebral column in all reptiles. The diameter of the cord is nearly uniform from brain to base of tail, except for slight expansion in the region of the limbs. Spinal nerves arise segmentally (a bilateral pair in association with each vertebrae) along the entire length of the cord. The organization of the spinal nerves is similar in all reptiles. The dorsal root contains somatic and visceral sensory neurons; the ventral root has somatic and visceral motor neurons.

Sense Organs

Sense organs allow an animal to perceive its internal and external environments and then to take appropriate action to escape a predator, to catch prey, to increase cutaneous circulation when overheating, to release gonatrophine to initiate the maturation of gonads, and a multitude of other actions required to live and reproduce.

Cutaneous Sense Organs

Cutaneous sense organs are especially common in reptiles and occur in a variety of forms. In addition to pain and temperature receptors, several types of intraepithelial mechanoreceptors register pressure, tension, or stretching within the skin. Free-ending mechanoreceptors with discoid endings (terminals) occur over most of the body, and branching terminal mechanoreceptors lie within the hinge between scales of lepidosaurs. There are also mechanoreceptors confined to the dermis with coiled, lanceolate, or free terminals. On the surface of the skin, tactile sense organs are abundant; they range in shapes from buttonlike to smooth and barbed bristles.

The pit organs of some boids, pythonids, and viperids are specialized epidermal-dermal structures housing radiant (infrared) heat receptors. In *Boa*, these receptors (both intraepidermal and intradermal ones) occur scattered on unmodified supra- and infralabial scales. In *Python*, a series of pits occurs in the labial scales, and the heat receptors are concentrated on the floor of the pit. In the crotaline snakes, a pit organ occurs on each side of the head between the naris and the eye. Their openings face forward and their receptor fields overlap, giving them stereoscopic infrared vision. Further, the heat receptors lie within a membrane stretched across the pit.

Ears

The ears are paired, and each consists of three components: an outer, a middle, and an inner ear. The outer ear is the cavity external to the tympanum. It occurs only in crocodilians and some lizards. A special muscle allows the crocodilians and most geckos to close the ear cavity. The tympanum may be flush with the side of the head (many lizards) or covered with scales (turtles). The middle ear consists of the tympanum and ear ossicles (columella and extracolumella) within an air cavity. The tympanum receives sounds and transmits the vibrations along the extracolumella-columella chain to the oval window of the inner ear. The middle ear cavities are large in turtles, large with left and right cavities connected in crocodilians, small and nearly continuous with the pharynx in most lizards, narrow canals in snakes, and usually absent in amphisbaenians. The columella is typically a slender columnar bone and its cartilaginous tip, the extracolumella, has three or four processes onto the tympanum. In snakes, the columella abuts against the quadrate bone for transmission of vibration.

The inner ear cavity is encased in bone and is fluid-filled. The inner ear membranous labyrinth floats therein. The three semicircular canals of reptiles share the same form as in other vertebrates (see description in Chapter 1) and are equilibrium organs. The sacculus is a large sac hanging from the utriculus. The cochlear duct with the auditory sensory area projects ventrally from the sacculus and lies adjacent to the oval window.

Eyes

The eyes of most reptiles are large and well developed. The eyes have degenerated only in a few fossorial species and groups; they disappear completely (no pigment spot visible externally) only in a few species of scolecophidians. Structurally, the eyes are similar to those of amphibians (see Chapter 1). Reptiles (except snakes) have a ring of bony plates (scleral ossicles) embedded in the sclera around the cornea. Pupils range from round to elliptical (usually vertical, occasionally horizontal). The eyeball and lens are usually spherical. Rather than moving the lens for accommodation, lens shape is changed by the contraction of radial muscles in the ciliary body encircling the lens. Crocodilians and turtles share a duplex (rods and cones) retina with the other vertebrates and possess single and double cones and one type of rod. In squamates, the retina has been differently modified in snakes and lizards. Primitive snakes have a simplex retina of only rods; advanced snakes have a duplex retina of cones and rods, although the cones are probably "transformed rods." In lizards, their simplex retina contains two or three different types of cones.

Nasal Organs

The nasal organs are bilaterally paired. Each organ consists of an external naris, vestibule, nasal cavity proper, nasopharyngeal duct, and internal naris. These structures serve mainly as air passages and are lined mainly with nonsensory epithelium. The sensory or olfactory epithelium lies principally on the roof and anterodorsal walls of the nasal cavity. These passages and cavities are variously modified in the different reptilian groups. The vestibule is a short tube in turtles and snakes, much longer and often curved in lizards. A concha covered with sensory epithelium projects into the nasal cavity from the lateral wall. *Sphenodon* has a pair of conchae, squamates and crocodilians one, and turtles none.

Jacobson's organ or the vomeronasal organ is an olfactory structure, used primarily to detect nonaerial, particulate odors. It arises embryologically from the nasal cavity, but remains connected to this cavity as well as to the oral cavity only in *Sphenodon*. In squamates, it communicates with the oral cavity by a narrow duct. Odor particles are carried to the duct by the tongue. Well developed in squamates, this organ is absent in crocodilians; in turtles, it does not occupy a separate chamber but lies in the main nasal chamber.

Internal Sense Organs

Structurally, the proprioceptors of the musculoskeletal system are similar to those of amphibians. Taste buds occur in many reptiles on the tongue and scattered in the oral epithelium. Structurally, they appear similar to those of amphibians and share the same sensory responses. In squamates, taste buds are abundant in fleshy-tongued taxa and greatly reduced or absent in taxa (e.g., snakes) with heavily keratinized tongue surfaces.

Heart and Vascular Network

Blood

Blood plasma is colorless or nearly so in most reptiles. A few skinks and crotaline snakes have green or greenish yellow blood. In addition to dissolved salts, proteins, and other physiological compounds, the blood transports three types of cells: erythrocytes, leucocytes, and thrombocytes—all are nucleated. The erythrocytes are the hemoglobin-bearing cells and the most numerous blood cells. They carry the gases to and from the body tissues. Leucocytes comprise five or six cell types; all participate in maintenance and cleaning activities. The thrombocytes assist in clotting.

Arterial and Venous Circulation

In general organization, the arterial and venous network of vessels is similar to that of adult amphibians. There are differences between reptilian groups. For example, the pattern of vessels to and from the trunk of snakes and turtles is not the same. But, the major trunk vessels leading from the heart and to the viscera, head, and limbs, and those vessels returning the blood to the heart, are more similar among species and groups than they are different.

The pulmonary arteries typically arise as a single trunk from the cavum pulmonale (right ventricle) and bifurcate into right and left branches above and in front of the heart. The systemic arteries (aortae) arise separately but side by side from the cavum venosum (left ventricle). The left systemic curves dorsally and bifurcates into a small ductus caroticus and the larger systemic branch. The right systemic bifurcates in front of the heart; the cranial branch forms the major carotid network and the systemic branch curves dorsally to join the left systemic branch. This combined aorta (dorsal aorta) extends posteriorly and its branches go to the limbs and the viscera. The major venous vessels are the jugular veins draining the head and the postcaval vein receiving vessels from the limbs and viscera. The jugular and postcaval trunks join into a common sinus venosus; it in turn empties into the right atrium.

Lymphatic Network

The lymphatic system of reptiles is an elaborate drainage network with vessels throughout the body. This network of microvessels gathers plasma (lymph) from throughout the body, and the smaller vessels merge into increasingly larger ones that in turn empty into the main lymphatic trunk vessels and their associated sinuses. The trunks and sinuses empty into veins. Major trunks collect plasma from the limbs, head, and viscera, forming a network of vessels that outline the shape of the reptile's body. The occurrence of valves is irregular and plasma flow can be bidirectional; however, the major flow in all trunks is toward the pericardiac sinus and into the venous system. There is a single pair of lymphatic hearts in the pelvic area but no lymph nodes.

Heart

There is no single model for the generalized reptilian heart. Heart size, shape, structure, and position are linked to other aspects of each reptile's anatomy and physiology. The animal's physiology is a major determinant of heart structure and function, but phylogeny and behavior play a role; in snakes, heart position correlates to arboreal, terrestrial, and aquatic habits. Among these numerous variables, three general morphologies are recognized.

The "typical" reptilian heart of turtles and squamates is three-chambered: two atria and a ventricle with three chambers or cava (left to right—arteriosum, venosum, pulmonale). The right atrium receives the venous blood (unoxgenated) from the sinus venosus and empties into the cavum venosum of the ventricle. The left atrium receives oxygenated blood from the lungs via the pulmonary veins and empties into the cavum arteriosum. Because the three ventricular cava communicate and the muscular contraction of the ventricle is single-phased, oxygenated and unoxygenated blood mixes and blood exits simultaneously through all arterial trunks. Blood in the cavum pulmonale flows into the pulmonary trunk, and blood in the cavum venosum into the aortae.

Monitor lizards (varanids) possess a higher metabolic rate than other lizards and differ also in heart structure. The differences are in the architecture of the ventricular cava. The cava communicate with one another. The cavum venosum is small, little more than a narrow channel linking the cavum pulmonale with a greatly enlarged cavum arteriosum. Ventricular contraction is two-phased so that the pumping cycle creates a functionally four-chambered heart. Although mixing of unoxygenated and oxygenated blood can occur and probably does under some circumstances, the cavum pulmonale is isolated during systole (contraction) and unoxygenated blood is pumped from the right atrium to the lungs. Within the crocodilians, the ventricle is divided into separate right and left muscle components. Uniquely, the two aortae in crocodilians arise from different ventricular chambers, the left aorta from the right chamber and right aorta from the left chamber. This arrangement provides an opportunity for unoxygenated blood to bypass the lungs under special physiological circumstances, such as during diving, by altering the pattern of ventricular contraction.

Digestive and Respiratory Structures

The buccal (oral) cavity and pharynx are shared passageways for the movement of air in and out of the respiratory passage and for movement of food and water into the digestive tube. Air enters and exits the buccal cavity through the internal nares-nasal chamber-external nares route. Food and water enter directly into the buccal cavity through the mouth.

Digestive Structures

The mouth of reptiles opens directly into the buccal cavity. The lips bordering the mouth are flexible skin folds but not movable in lepidosaurs and are nonexistent in crocodilians and turtles. Tooth rows on the upper and lower jaws form a continuous border along the internal edge of the mouth, except for turtles with their keratinous jaw sheaths. In reptiles, teeth typically serve for grasping, piercing, and fragmentation of food items. Only in a few species do teeth cut and slice (e.g., *Varanus*) or crush (*Dracaena*). A well-developed tongue usually occupies the floor of the mouth. There are numerous tongue morphologies associated with a variety of feeding behaviors, such as the projectile tongues of chameleons or the telescoping tongues of varanoid lizards and snakes. The roof of the buccal cavity is the primary palate. There are two pairs of openings anteriorly; the small Jacobson's organ openings lie immediately in front of the larger internal nares. The crocodilians have a secondary palate that creates a separate respiratory passage from the internal nares on the primary palate to the beginning of the pharynx; this passage allows air to enter and exit the respiratory system while food is held in the mouth. A few turtles and snakes (aniliids) have developed partial secondary palates.

The pharynx is a small antechamber behind the buccal cavity. A valvular glottis on its floor is the entrance to the trachea. On the rear wall of the pharynx above the glottis, a muscular sphincter controls the opening into the esophagus. The eustachian tubes, one on each side, open onto the roof of the pharynx. Each tube is continuous with the middle-ear chamber to permit the adjustment of air pressure on the tympanum. Middle ears and eustachian tubes are absent in snakes.

The esophagus is a distensible, muscular-walled tube of variable length between the buccal cavity and the stomach. In snakes and turtles, the esophagus may be a quarter to a half of the body length. It is proportionately shorter in reptiles with shorter necks. The stomach is a heavy muscular and distensible tube, usually J-shaped and largest in the curved area and rapidly narrows to a thick muscular sphincter, the pylorus or pyloric valve. This valve controls the movement of the food bolus from the stomach into the small intestine. The small intestine is a long narrow tube with little regional differentiation externally or internally; the pancreatic and hepatic ducts empty into its forepart. The transition between the small and large intestine is abrupt. The diameter of the latter is several times larger than that of the former, and often a small outpocketing, the caecum, lies adjacent to the juncture of the two intestines. The large intestine or colon may be a straight or C-shaped tube emptying into the cloaca. The large intestine is the least muscular and has the thinnest wall of the entire digestive tube.

Strictly speaking, the cloaca is part of the digestive tract since it derives from the embryonic hindgut. There is a muscular sphincter (anus) between the large intestine and the cloaca. This dorsal portion of the cloaca is the coprodaeum and is the route for the exit of feces. The urodaeum or urogenital sinus is a ventral outpocket of the cloaca and extends a short distance anterior and beneath the large intestine. Digestive, urinary, and genital products exit via the vent, a transverse slit in turtles and lepidosaurs, and a longitudinal one in crocodilians.

Digestive Glands

The oral cavity contains numerous glands. The small, multicellular mucous glands are a common component of the epithelial lining and comprise much of the tissue on the surface of the tongue. Larger aggregations of glandular tissue, both mucous and serous, form the salivary glands (labial, lingual, sublingual, palatine, and dental; see Fig. 5.4). In the venomous snakes, the venom glands are modified salivary glands. Mucous glands occur throughout the digestive tube. The stomach lining is largely glandular with several types of gastric glands. The small intestine also has many small glands within its epithelial lining. The liver and pancreas also produce secretions that assist in digestion. The liver is usually the largest single organ in the visceral cavity and occupies the anteroventral quarter of the cavity. The pancreas is a smaller more diffuse structure lying within the visceral peritoneum.

Respiratory Structures

Air exits and enters the trachea through the glottis at the rear of the pharynx. The glottis and two or three other cartilages form the larynx, a simple tubular structure in most reptiles. The larynx is the beginning of the trachea, a rigid tube of closely spaced cartilaginous rings (incomplete dorsally in squamates) within its wall. The trachea extends down the neck beneath the esophagus and forks into a pair of bronchi, each of which enters a lung.

Lungs

For most lepidosaurs, the lungs are simple saclike structures. The bronchus empties into a large central chamber. Numerous faveoli (small sacs) radiate outward in all directions, forming a porous wall around the central chamber.

The walls of the faveoli are richly supplied with blood and provide the major surface for gaseous exchange. Chameleons, agamids, and iguanids have the central chamber of each lung divided by a few large septae. These septae partition the lung into a series of smaller chambers, each of which possess porous faveolar walls. Varanids, crocodilians, and turtles also have multichambered lungs. A bronchus extends into each lung and subdivides into many bronchioles, each ending in a faveolus. In some lizards, smooth-walled tubes project from the chamber and beyond the surface of the lung. No gas exchange occurs in these air sacs, rather the sacs may permit the lizard to hold a larger volume of air. The sacs are used by some species to inflate their body to intimidate predators.

The development of air sacs is even more extensive for snakes in association with their highly modified lungs. A single functional right lung and a small, nonfunctional left lung is the common condition. A functional left lung occurs only in few primitive snakes (e.g., *Loxocemus*), and in these snakes, it is distinctly smaller than the right lung. The trachea-right bronchus extends into the lung and empties into a chamber with a faveoli-filled wall as in most lizards. Snake lungs are typically quite long, half or more of the snake's body length (Fig. 3.5), and usually the posterior third or more is an air sac.

Many snakes also possess a tracheal lung. This lung is a vascular, faveolar sac extending outward from where the tracheal rings are incomplete dorsally, and posteriorly it abuts the right lung. Breathing occurs by the expansion and contraction of the body cavity. In saurians, the thoracic cavity is enlarged (inhalation) by the contraction of the intercostal muscles drawing the ribs forward and upward. Compression of the cavity (exhalation) occurs when the muscles relax and the weight of the body wall and adjacent organs squeezes the lungs. In crocodilians, the diaphragm contracts enlarging the thoracic cavity for inhalation and abdominal muscles contract driving the liver forward for exhalation. In turtles with their rigid shell, the posterior abdominal muscles and several pectoral girdle muscles serve to expand and compress the body cavity for breathing.

Other Respiratory Surfaces

Reptiles are dependent on their lungs for aerial respiration. None has developed a successful substitute for surfacing and breathing in the aquatic environment. Long-term submergence in reptiles is possible owing to a high tolerance to anoxia, a greatly suppressed metabolism, and varying degrees of cutaneous respiration (see Gas Exchange, Chapter 10). Softshell turtles are purported to obtain more than 50% of their respiratory needs by cutaneous and buccopharyngeal respiration when submerged, but experimental results of different investigators are conflicting. The accessory cloacal bladders of turtles have also been proposed as auxiliary respiratory structures; however, their walls are smooth and lightly vascularized, thus unlikely respiratory surfaces.

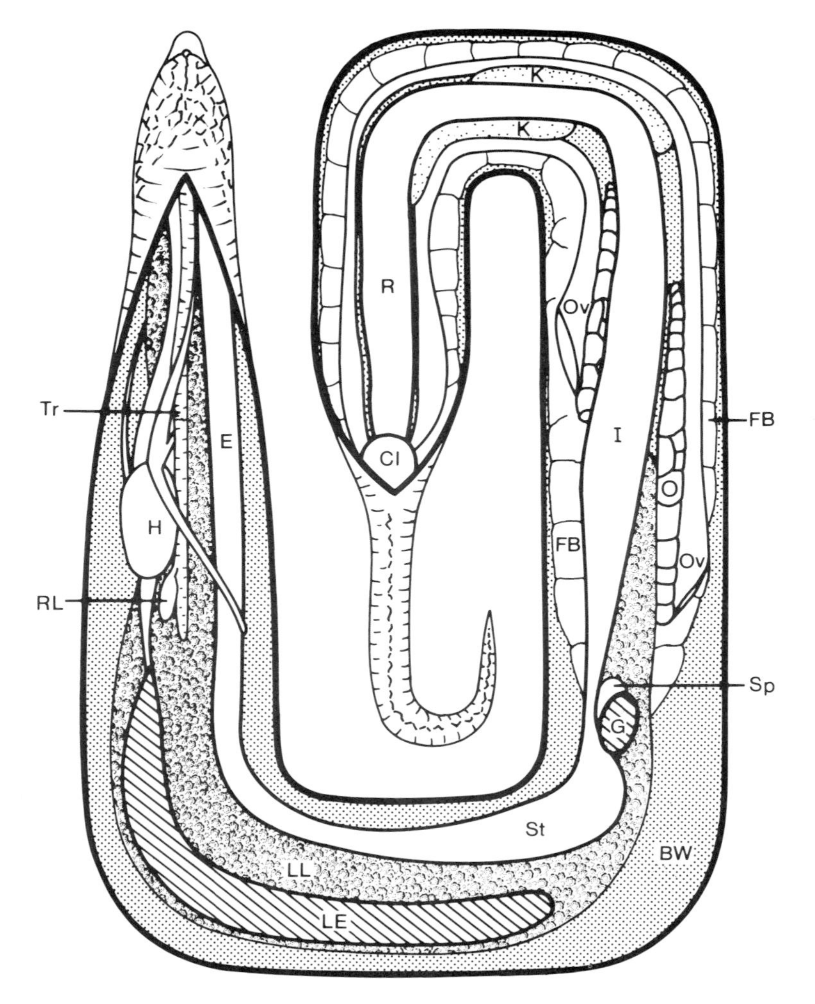

FIGURE 3.5

Visceral anatomy of a female hognose snake (*Heterodon platyrhinos*); a diagrammatic ventral view. Abbreviations: BW, body wall; Cl, cloaca; E, esophagus; FB, fat bodies; G, gallbladder; H, heart; I, intestine; K, kidney; LE, liver; LL, right lung; O, ovary; Ov, oviduct; R, rectum (colon); RL, left lung; Sp, spleen; St, stomach; Tr, trachea. [Adapted from Cope (1898).]

Urinary and Reproductive Organs

The urinary and genital organs and ducts are associated by their neighboring positions dorsoposteriorly in the body cavity and by a shared evolutionary history. Through generations of vertebrates, the male gonads have usurped the urinary ducts of primitive kidneys for the transportation of sperm. In amniotes, these ducts are strictly genital and new kidneys and urinary ducts appear.

Kidney and Urinary Ducts

The metanephric kidneys of reptiles come in an assortment of sizes and shapes—smooth, equal-sized, nearly spherical kidneys in some lizards; smooth or rugose, elongated cylinders in snakes; and lobate spheroids in crocodilians and turtles. In all forms, the kidneys lie side by side on the dorsal body wall just in front of the cloaca, and in all, a ureter drains each kidney and empties independently into the cloaca. An elastic-walled urinary bladder is present in turtles and most lizards, but absent in snakes and crocodilians. The bladder joins the cloaca through a single median duct, the urethra, through which urine enters (for storage) and exits (for evacuation).

Gonads and Genital Ducts

A pair of ovaries occupy the same location as the testes of the males, and the right ovary precedes the left in squamates. Each ovary is an aggregation of epithelial cells, connective tissue, nerves, blood vessels, and one or more germinal cell beds encased in an elastic tunic. Depending on the stage of oogenesis (see Gametogenesis and Fertilization, Chapter 7), each ovary can be a small, granular-appearing structure to a large, lobular sac filled with spherical or ellipsoidal follicles. An oviduct is adjacent to each ovary but not continuous with the ovary. The ostium (mouth) of the oviduct lies beside the anterior part of the ovary; it enlarges during ovulation to entrap the ova. The body of the oviduct has an albumin-secreting portion followed by a thicker-walled shell-secreting portion. Each oviduct opens independently into the urogenital sinus of the cloaca.

Each testis is a mass of seminiferous tubules, interstitial cells, and blood vessels encased in a connective tissue sheath. The walls of seminiferous tubules are lined with germinal tissue (see Chapter 7). The sperm produced by these tubules empties via the ductuli efferentia into ductuli epididymides of the convoluted epididymis lying on the medial face of the testis. The ductuli coalesce into the ductus epididymis that runs to the cloaca as the vas (ductus) deferens. In shape, testes vary from ovoid to spindle-shaped. The testes are usually adjacent, although the right testis lies somewhat more anterior, characteristically so in snakes and most lizards.

All living reptiles have copulatory organs. Crocodilians and turtles have a single median penis arising from the floor of the cloaca. The lepidosaurs have a pair of hemipenes, each hemipenis arising from the corner of the cloacal vent and base of the tail.

Endocrine Glands

The endocrine system is composed of a multitude of glands scattered throughout the body. Their general structure and location are similar in all tetrapods.

Like the nervous system, endocrine glands initiate and coordinate the animal's response to internal and external stimuli. As most endocrine glands transmit their messages as chemical compounds via the bloodstream, response time is slow compared to neural response. Most of us, however, can attest to the quickness of an adrenalin surge in a threatening situation.

Pituitary Gland

As in other vertebrates, the pituitary is the "master gland" of reptiles. It is a small two-part gland (see Fig. 3.4) lying beneath the thalamic portion of the brain. The neurohypophysis and the adenohypophysis are anatomically and hormonally interconnected. The former stores the neurohormones secreted by other regions of the brain and produces its own hormone, MSH (melanocyte-stimulating hormone). The adenohypophysis produces adrenocorticotropins, gonatropins, prolactin, thyrotropins, and others. The hormones of the pituitary are responsible for initiating and regulating the daily and physiological cycles.

Pineal Complex

The pineal complex of the pineal and parietal bodies projects from the roof of the thalamic portion of the brain (see Fig. 3.4). They are both photoreceptors and neuroendocrine organs. Their ability to perceive light, even when the parietal body does not penetrate the skull, is critical to their role in integrating photoperiodicity and neurohormone production. A major outcome of this integration is the regulation of the reproductive cycle; also other daily and seasonal physiological activities are dependent on the pineal complex.

Thyroid and Parathyroid Glands

The thyroid and parathyroid glands are linked because of their shared location in the throat adjacent to the larynx-trachea. They have quite dissimilar functions. The parathyroid hormones regulate the calcium levels in the blood, hence bone growth and remodeling. The thyroid is well known for its accumulation of iodine and the importance of its hormones in growth and development. In reptiles, the thyroid assumes a variety of forms. It is a single, nearly spherical organ in turtles and snakes. In crocodilians, it is H-shaped with a lobe on each side of the trachea connected to its mate by a narrow isthmus. Some lizards have this H-shape, others have a lobe on each side but no isthmus, and in others it is a single median gland. In *Sphenodon*, the gland is transversely elongated. The parathyroid appears as one or two pair of granular glands usually at the base of the throat adjacent to the carotid arteries.

Pancreas

The reptilian pancreas is a compact organ lying in the mesentery adjacent to the duodenal segment of the intestine. It possesses both exocrine and endocrine (islets of Langerhans or pancreatic islets) tissues. The islet tissue, scattered in clusters throughout the pancreas, produces four hormonal products: insulin,

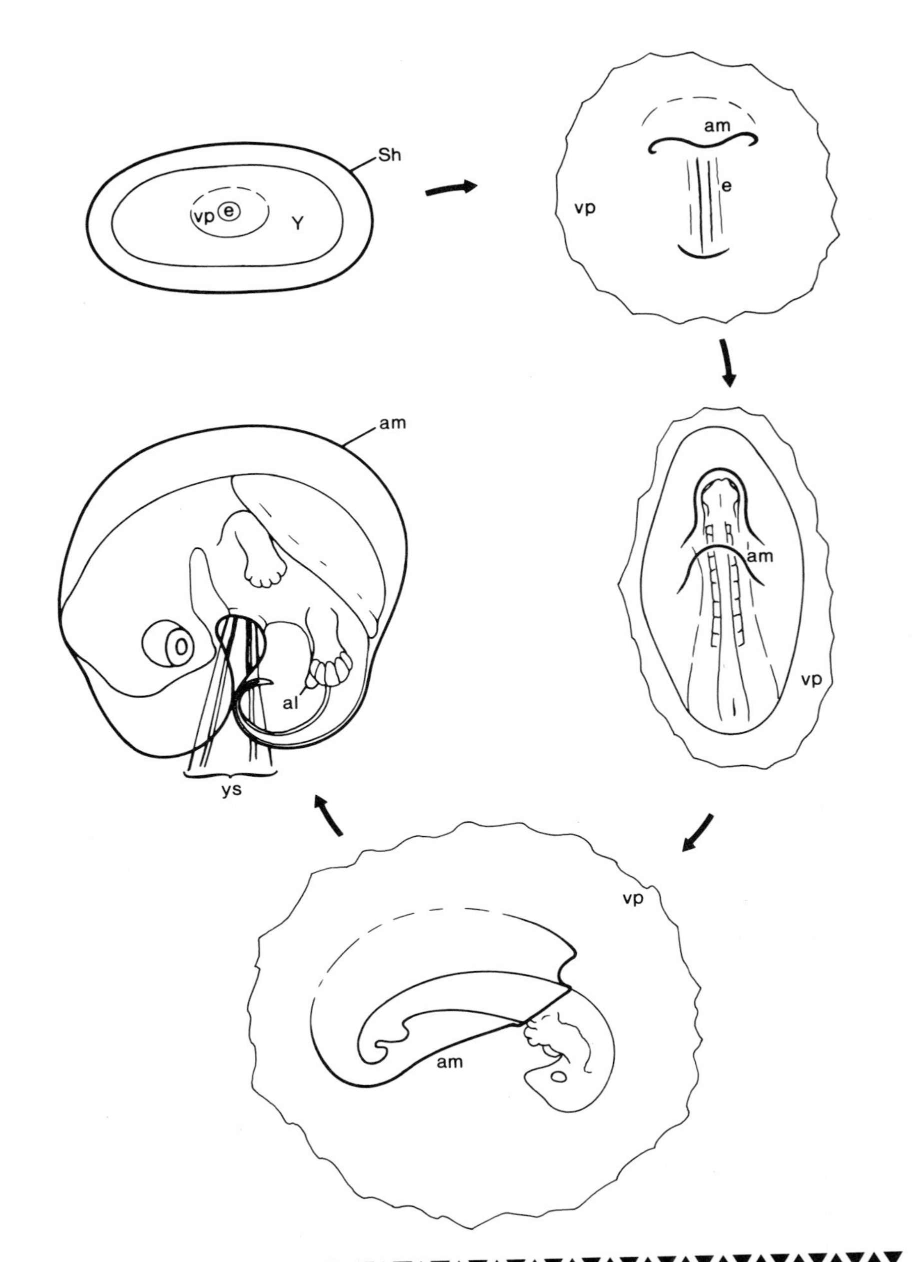

FIGURE 3.6

Selected developmental stages of a turtle embryo showing the formation of extra-embryonic membranes. From top, clockwise: shelled egg with early embryogenesis; embryonic disc during neural tube formation and initiation of amnionic fold; embryo during somite formation and developing brain cloaked by amnionic hood; embryo in early organogenesis with early outpocketing of the allantois; near term embyro encased in amnion and attachment to yolk sac. Abbreviations: al, allantois; am, amnion; e, embryo; Sh, shell; vp, vitelline plexus or extra-embryonic disc; Y, yolk; ys, yolk stalk. [Adapted from Agassiz (1858).]

glucagon, somatostatin, and pancreatic peptides. All of these hormones regulate carbohydrate metabolism. The exocrine pancreatic tissue produces digestive enzymes that empty via a duct into the duodenum.

Gonads

Aside from producing gametes, the gonads also produce sex hormones. The maturation and production of gametes is closely regulated by the brain and the pituitary gland, and the hormonal response of these organs is in turn influenced by the hormones secreted by the gonads. In the testis, two cell types produce estrogens. The Sertoli cells embedded within the germinal epithelium are largely responsible for nourishing the spermatozoa, but they may also release small amounts of androgens. The principal source of androgens is the interstitial cells between the seminiferous tubules. Similarly the supportive cells (follicle cells) of the ovary are the major producers of estrogens.

Adrenal Glands

The adrenal glands are bilaterally paired glands lying anterior to the kidneys. Each gland contains two secretory tissues usually separated into a core of interrenal tissue surrounded by adrenal tissue, although there is often some mixing of the two. Adrenalin and noradrenalin are produced by the adrenalin cells and corticosteroid hormones by the interrenal cells. Only in turtles do the glands attach to the kidneys. In lepidosaurs, the glands lie in the mesentery beside the epididymides, and in crocodilians above the testes outside of the peritoneum.

AMNIOTE EGG

The amniotic egg—really the embryo—creates its own aquatic habitat early in development. When the embryo is little more than an elongating sheet of cells on top of the yolk, the outer edges of the embryonic shield curl upward and inward, fusing above and enclosing the embryo proper (Fig. 3.6). The inner (amnion) and outer (chorion) layers form a protective sheath around the embryo. These membranes also retard evaporation and serve as a respiratory surface. If the female deposits the eggs externally, the eggs are encased in calcareous shells. The shell serves as a mechanical barrier against abrasion and penetration of the developing embryo. Later in development, the embryo forms a third external membrane, the allantois, as an outpocketing of the hindgut. The allantois becomes the major respiratory surface while serving as a receptacle for waste storage.

These extraembryonic membranes (amnion, chorion, allantois) and the yolk sac serve the embryo identically whether they are enclosed in a shell or held in the oviducts. They also provided the structural templates from which the various placentas evolved (see Live-Bearing, Chapter 7).

CHAPTER 4

Origin and Evolution of Reptiles

The earliest tetrapods were amphibious. Even those capable of living on land as adults returned to water in order to reproduce. One lineage, the amphibians, retained this amphibious life-style with occasional evolutionary ventures toward full terrestriality. Though many of these ventures were successful in terms of high abundance or diversity and geologic longevity, the amphibians remain moisture addicts. This addiction to water (to a greater or lesser extent) is not maladapted or a "lowly evolutionary" state, because terrestriality is not a higher or more successful state. Each state is simply a different adaptive complex that allows amphibious and terrestrial organisms different options for living and reproducing.

While the amphibians took the amphibious path,

the anthracosaurs and their descendants became increasingly terrestrial in all phases of their life. This lineage quickly became fully terrestrial through the development of a cleidoic (shelled) and amniotic egg. The first tetrapod is known from the Late Devonian, the first amphibian from the mid-Mississippian, and the first amniotes from the mid-Pennsylvanian (Fig. 4.1). These first amniotes were *Archaeothyris* (a pelycosaur), *Hylonomus* (a romeriid), and *Paleothyris* (a romeriid). Already the divergence of the mammalian and reptilian stocks was evident.

In deciphering the relationships among the Paleozoic amniotes, the earliest appearing reptiles are not the most primitive reptiles; later appearing taxa such as the captorhinids have more of the characteristics of transitional forms. Our understanding of the transition from early tetrapod to amniote depends on the anthracosaurs, seymouriamorphs, and diadectomorphs. This series of taxa shows increasing specialization for terrestrial existence. Because they retain some amphibious adaptations and other traits of early tetrapods, their classification has

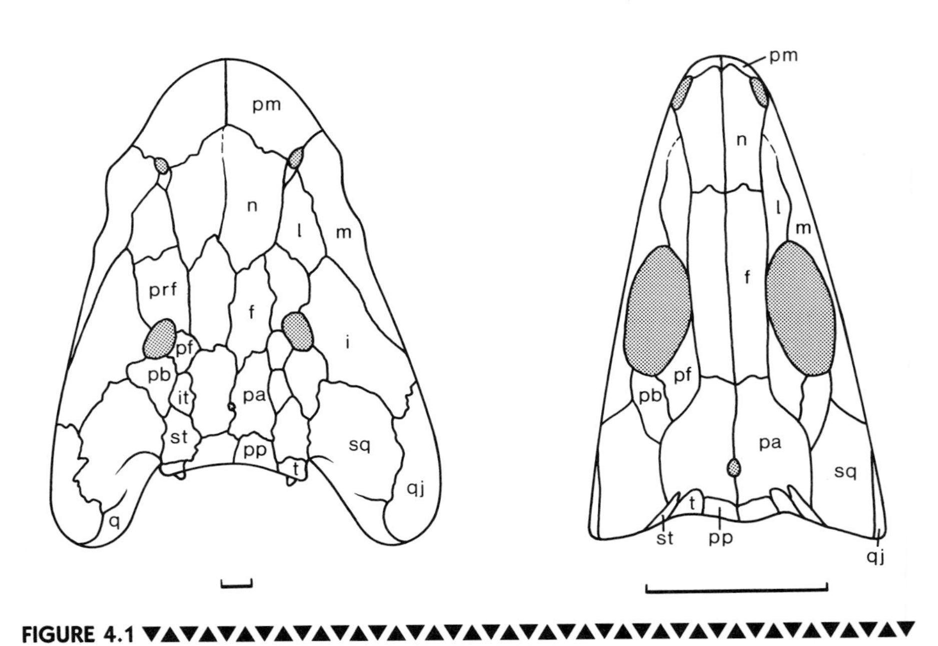

FIGURE 4.1 Comparison of the skulls of an early amphibian *Edops* (left) and an early reptile *Paleothyris* (right). Scale: bar = 1 cm. Abbreviations: f, frontal; i, jugal; it, intertemporal; l, lacrimal; m, maxillary; n, nasal; pa, parietal; pb, postorbital; pf, postfrontal; pm, premaxillary; pp, postparietal; prf, prefrontal; q, quadrate; qj, quadratojugal; sq, squamosal; st, supratemporal; t, tabular. [Adapted from Piveteau (1955) and Carroll and Baird (1972), respectively. Reproduced, with permission, from Museum of Comparative Zoology, Harvard University.]

been changeable. Anthracosaurs have most often been considered to be amphibians, the seymouriamorphs sometimes amphibians and at other times reptiles, and the diadectomorphs as basal reptiles. These groups are on the lineage to reptiles (amniotes) or direct offshoots, but they are not reptiles as defined in Chapter 3.

Many of the anthracosaurs were aquatic tetrapods (see Chapter 2), although some were definitely terrestrial, for example, *Proterogyrinus*. Anthracosaurs lived from the Late Devonian into the Early Permian. The ancestors of the amniotes, thus, diverged very early in the history of anthracosaurs, and the evolution from basal tetrapod to amniote was swift and left no record.

AMPHIBIOUS TO TERRESTRIAL TETRAPOD

Anthracosaurs are recognized as the ancestral stock to the amniotes because they possess features present in amniotes but not in Paleozoic or later amphibians. Anthracosaurs and amniotes share such features as a multipartite atlas-axis complex in which the pleurocentral element provides the major support. Both also possess five-toed forefeet with a phalangeal formula of 2-3-4-5-3 and a single, large pleurocentrum for each vertebra. These traits are also present in the seymouriamorphs and diadectomorphs.

The diadectomorphs shared a number of specialized (derived) features with early reptiles, traits that are not present in their predecessors. For example, both groups have lost temporal notches in their skulls, have a fully differentiated atlas-axis complex with fusion of the two centra in adults, and a pair of sacral vertebrae. They share a large, platelike supraoccipital bone and a number of small cranial bones—supratemporals, tabulars, and postparietals—that are lost in advanced reptiles. The stapes of both were stout bones with large foot plates, and apparently eardrums (tympana) were absent. These latter features do not suggest deafness, but that their hearing was confined to low frequencies, probably less than 1000 Hz, much like modern-day snakes and other reptiles without eardrums.

Throughout the evolution of early tetrapods, the skull became more compact and tightly linked. The various units of the rhipidistian skull (cheek plates, skull roof, braincase, and jaws) were loosely attached to one another and often weakly linked within each unit. The buoying force of their watery home reduced the need for a rigid, supportive framework, a requirement imposed by gravity on terrestrial animals. A major adaptive trend was the reduction of the otic capsule in early tetrapods, without the concurrent development of structural struts, thus the skull roof and braincase became weakly linked. Different strengthening mechanisms have appeared in the various tetrapod lineages. The diadectomorphs and reptiles shared the unique development of a large supraoccipital bone to link the braincase and skull roof. The cheek to braincase solidification occurred in three general patterns within the amniotes. The anapsids developed a strong attachment of the parietal (skull roof) to the squamosal (cheek) along with a broad and rigid supraoccipital attachment. In the diapsids, the opisthotic extended laterally to link the braincase to the cheek. A lateral expansion of the opisthotic also occurred in the synapsids, but in a different manner.

The robust stapes with its broad foot plate was also a critical strut in the strengthening of the skull. This role as a supportive strut precluded its function as an impedance matching system (see Vocalization, Chapter 8). Later, the opisthotic became the supportive unit, and the stapes (columella) became smaller and took on its auditory role. This change occurred independently in several reptilian lineages; although the results are the same, the evolutionary route to the turtle middle ear differed from that of the archosaurs and lepidosaurs. The synapsids followed an entirely different route and evolved the unique three-element middle ear seen today in mammals.

The postcranial skeleton of the early anthracosaurs was little different from that of the contemporaneous amphibians. The vertebral column consisted of robust vertebrae with stout, cylindrical centra fused to tall, stout neural arches. Cervical and trunk vertebrae supported strong ribs. The atlas-axis complex contained three centra, which permitted considerable head movement. The tail contained numerous vertebrae and was somewhat longer than the body. The girdles were robust structures of large flat surfaces, probably supporting bulky limb muscles. They had a sprawling limb posture and a gait that used the flexures/undulations of body movement to increase gait length and the propulsive force. The feet had no single joint plane, bending at many articular surfaces. Each foot functionally rolled on and off the substrate, in a distinctly digitigrade manner. The limbs of the early amniotes and diadectomorphs were less massive and better jointed. The limbs were robust and short in the anapsids and pelycosaurs, whereas the limbs were gracile and elongated in the diapsids, particularly the metapodia and phalanges of the fore- and hindfoot. Long and gracile limbs suggest more active and mobile animals with less plodding or lumbering gaits.

These skeletal changes are traceable in the fossil record, even though many of the traits identifying an amniote as an amniote do not fossilize, particularly the extra-embryonic membranes of the amniotic egg. These membranes provided a mechanism for the development of a thoroughly terrestrial egg—an egg that could be laid in a dry, although not dehydrating, environment. When and how these membranes appeared remains unknown. Numerous independent origins of direct development occurred in modern amphibians by the elimination of the larval stage. Many of these adaptations also involve terrestrial eggs, but in none of them is the embryo encased in a sac of extra-embryonic tissue or able to survive in a dry nest site.

Because both synapsids and reptiles share an identical pattern in the development of the extra-embryonic membranes, it is most parsimonious to assume that they inherited this pattern from a common ancestor. Beyond this conclusion, only speculation is possible, although it seems probable that the diadectomorphs had direct development, perhaps with traces of the amniote membranes. What is less speculative is that initially the amniote eggs were probably shell-less and required moist nest sites, just as do the direct-developing eggs of modern amphibians. Shell-less, externally deposited, amniote eggs are unknown in living amniotes, and evidence suggests that they were of brief occurrence in the early amniotes. Parsimony once again requires the conclusion that egg shells appeared prior to the divergence of synapsids and reptiles, because monotreme mammals possess egg shells like those of reptiles. The first egg shell probably appeared as a protective coat against attack by invertebrate predators and did little to alter the diffusion of water into and out of the egg. Such diffusion remains an important aspect of development in many living reptiles.

Acceptance of the appearance of a cleidoic, amniotic egg in the earliest amniotes leads also to the conclusion that amniotes had internal fertilization. Internal fertilization is not a prerequisite for direct-developing eggs, because only a few anurans with direct development possess internal fertilization. However, when an egg is encased in a shell, the encasing process must be done inside the female's reproductive tract, and if sperm is to reach the egg/ovum surface, the sperm must be placed within the female's reproductive tract as well. Furthermore, sperm delivery and fertilization must precede egg shell deposition. Internal fertilization does not require a copulatory organ; most avian archosaurs lack them yet have effective internal fertilization. However, a copulatory organ likely arose early in amniotes because mammals, turtles, and crocodilians share similar histology and erection of their penes.

Terrestrial life required changes in skin structure, and these are similarly invisible in the fossil record. The condition in amphibians suggests that the initial adaptive step was to increase skin thickness by adding more cell layers and to keratinize the externalmost layer(s). This step effectively reduces frictional damage and the penetration of foreign objects but appears to be ineffectual in reducing water loss. The amniotes further increased the thickness of the skin and the thickness of keratinized layers, thereby making the skin less permeable. Presumably, these modifications were also well developed in the earliest amniotes.

As the skin became increasingly resistant to water loss, cutaneous respiration would have become less effective and pulmonary respiration more so. Lungs probably changed in several ways. The first modifications were likely an increase in size and internal partitioning. The latter is commonly associated with increased vascularization. Once again, it seems likely that these modifications preceded the divergence of the synapsids and diapsids. When and where they occurred may be partially identified by examining rib structure and the appearance of a complete rib cage. A rib cage (thoracic basket) signals the use of a thoracic respiratory pump for the ventilation of the lungs. The rib cage appears incomplete in most anthracosaurians and seymouriamorphs, so they probably were still largely dependent on the buccal-force pump. The rib cage of diadectomorphs extends further ventrally; although it still appears incomplete, this condition may mark the transition from buccal to thoracic ventilation.

As noted earlier, the early amniotes lacked otic notches and eardrums. While not deaf, they were certainly insensitive to high-frequency sounds. It is doubtful that their olfactory sense was as limited. Well-developed nasal passages in the fossils and the presence of highly developed olfactory organs in living reptiles indicate that this sense was also well developed in the earliest amniotes. Eyes were also likely well developed at this stage, for vision is extremely important in foraging and avoiding predators in an aerial environment.

HISTORY OF EXTINCT REPTILES

Radiation of Early Amniotes

The earliest known amniotes are several contemporaneous taxa of reptiles and synapsids from a buried forest of the mid-Pennsylvanian period in Nova Scotia, Canada. These amniotes apparently lived in hollowed, upright trunks of dead trees and were buried when the forest was periodically flooded. Even though many of the Paleozoic amniotes were quite large, particularly in comparison to most living reptiles, these earliest amniotes (*Archaeothyris*, pelycosaur; *Hylonomus* and *Paleothyris*, romeriids) were small, approximately 15 cm long (SVL).

The diadectomorphs and the captorhinids, structurally more primitive than the romeriids and pelycosaurs, did not appear until the Late Pennsylvanian and Early Permian, respectively. The explosive radiation of the reptiles was still millions of years away in the Mesozoic. Nonetheless the amniotes, particularly the pelycosaurs, began to assume a dominant role in the terrestrial vertebrate communities of the Permian.

Reptilelike Mammals—The Synapsids

The pelycosaurs are the basal stock of the evolutionary line that became the mammals. The pelycosaurs were enormously successful, diversifying into two dozen genera and numerous species in six families (Table 4.1). They became the major tetrapods of the Early Permian in both abundance and number of different species. The earliest pelycosaurs were small (ca. 30 cm SVL) and lizardlike. They had large heads with big, widely spaced teeth, suggesting that they were effective carnivores of large prey. This basal stock radiated into several groups of medium to large carnivores and at least two groups of herbivores. Two different carnivore stocks, *Ophiacodon* and *Dimetrodon*, had elongated neural spines on the trunk vertebrae, forming a large sail or fin on their backs; presumably this sail was used for thermoregulation. Both of these creatures were over 3 m long. *Ophiacodon* had a narrow snout and likely fed on fish, whereas *Dimetrodon* was a broad-jawed,

TABLE 4.1

Hierarchical Classification of Early Amniotes[a]

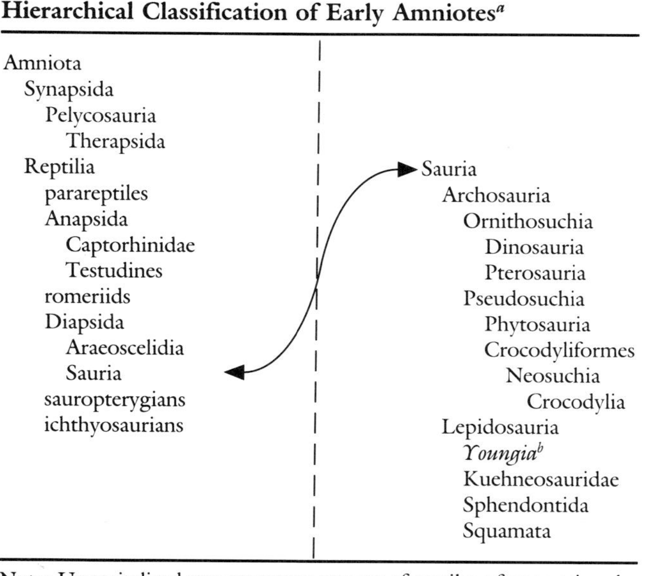

- Amniota
  - Synapsida
    - Pelycosauria
      - Therapsida
  - Reptilia
    - parareptiles
    - Anapsida
      - Captorhinidae
      - Testudines
    - romeriids
    - Diapsida
      - Araeoscelidia
      - Sauria
    - sauropterygians
    - ichthyosaurians

- Sauria
  - Archosauria
    - Ornithosuchia
      - Dinosauria
      - Pterosauria
    - Pseudosuchia
      - Phytosauria
      - Crocodyliformes
        - Neosuchia
          - Crocodylia
  - Lepidosauria
    - *Youngia*[b]
    - Kuehneosauridae
    - Sphendontida
    - Squamata

Note: Uncapitalized taxa represent groups of reptiles of uncertain relationships or polyphyletic groups. Some category levels or steps are absent.

[a] See also Table 3.1.

[b] *Youngia* represent a separate and early lepidosaurian lineage, but this lineage has not been assigned formally to a family or higher taxonomic cateogry.

terrestrial predator. *Edaphosaurus* was another large, sail-finned pelycosaur, and its closely spaced, peglike teeth indicate a herbivorous diet. There were a number of varanidlike pelycosaurs that probably shared the varanid's agility and carnivorous diet. These early pelycosaurs disappeared by the middle of the Late Permian. Their decline may have been brought about by the success of an early offshoot that gave rise to the therapsid radiation of the Late Permian.

Paleozoic Reptiles

Of the two anapsid lineages, only the captorhinids have a fossil presence in the Late Paleozoic. Although the turtle lineage must have been present, their fossils are either not recognized as the ancestors of turtles or their life-style prevented their fossilization (Fig. 4.2).

The captorhinids were medium-sized, lizardlike reptiles, although the broad-jowled head was proportionately larger than that of lizards, between a quarter and a third of snout–vent length. The teeth showed regional differentiation with

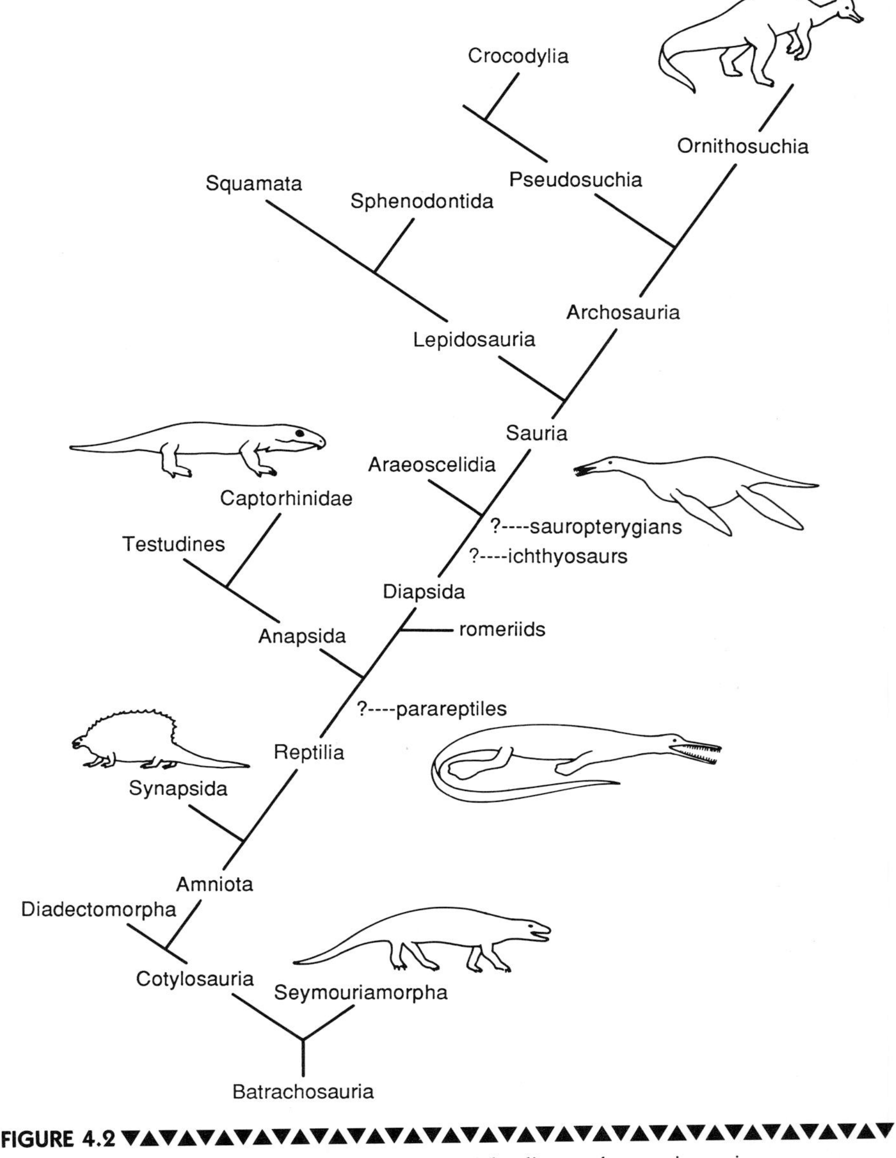

FIGURE 4.2 ▼▲▼▲▼▲▼▲▼▲▼▲▼▲▼▲▼▲▼▲▼▲▼▲▼▲▼▲▼▲▼▲▼▲▼▲▼

A branching diagram of the evolution of amniotes. The diagram has no time axis.

large, pointed incisors in front and double to triple rows of short, cone-shaped teeth in the rear. Their bodies were slender and limbs moderately long, suggesting them to be agile carnivores.

The romeriids are pre-diapsid reptiles. Although structurally similar to the captorhinids, they generally had smaller bodies and narrower heads. Undoubtedly they were small agile carnivores as well. *Petrolacosaurus* is the earliest diapsid (Late Pennsylvanian) and one of several genera of the Araeoscelida (Fig. 4.3). Araeoscelidans were all lizardlike in body and head proportions. The limbs were gracile and elongate with fore- and hindlimbs of near equal length. Their dentition was simple and indicates a general carnivorous diet. There was nothing special in their appearance to suggest that they or their similarly appearing precursors would provide the gene pool for the phenomenal radiation of the lepidosaurs and archosaurs. From their disappearance in the early Middle Permian, diapsids are absent from the reptilian communities until the Late Permian. The Late Permian diapsids represent a variety of lineages, some short-lived, others members of groups that came to prominence in the Mesozoic. The diapsid radiation began in the mid-Triassic and produced the numerous saurian (or neodiapsid) lineages of archosaurs and lepidosaurs, as well a variety of other unusual lineages.

Several groups of Paleozoic reptiles have proved exceedingly difficult to classify. For lack of a better name, the mesosaurs, millerosaurs, pareiasaurs, and procolophonoids are called the parareptiles. All four groups appeared in the mid-Permian, and only the procolophonoids continued into the Mesozoic to the end of the Triassic.

The parareptiles do not fit the characteristics of either the anapsids or the diapsids and probably did not diverge from either of these lines. Their mix of primitive and specialized traits also does not indicate that they are closely related to one another. They are reptiles, for which independent origins from the basal romeriid lineages is a reasonable, if uncertain, hypothesis. Two other groups, the sauropterygians and ichthyosaurs, share this uncertain origin. The radiation of both occurred in the Mesozoic and will be described in the following section.

The mesosaurs (Early Permian) were miniature (ca. 1 m TL), marine, gharial-like reptiles. They had long, narrow-snouted skulls, and the long, thin teeth of the upper and lower jaw projected outward and overlapped when the jaw was closed. Such jaws are excellent for catching fish with a sideward sweep of the head. The body and tail were similarly elongated and the tail laterally compressed for undulatory swimming. Nonetheless, the limbs were well developed, and the hindlimbs and feet were large, probably webbed and used as rudders.

The three other groups of parareptiles were terrestrial animals. The millerosaurs (Late Permian) were small lizardlike reptiles. Their small heads and simple conical teeth match the appearance of many agamids or iguanids living today, and they probably shared a diet of insects. The millerosaurs and mesosaurs appeared briefly, each with only a few species or genera before disappearing.

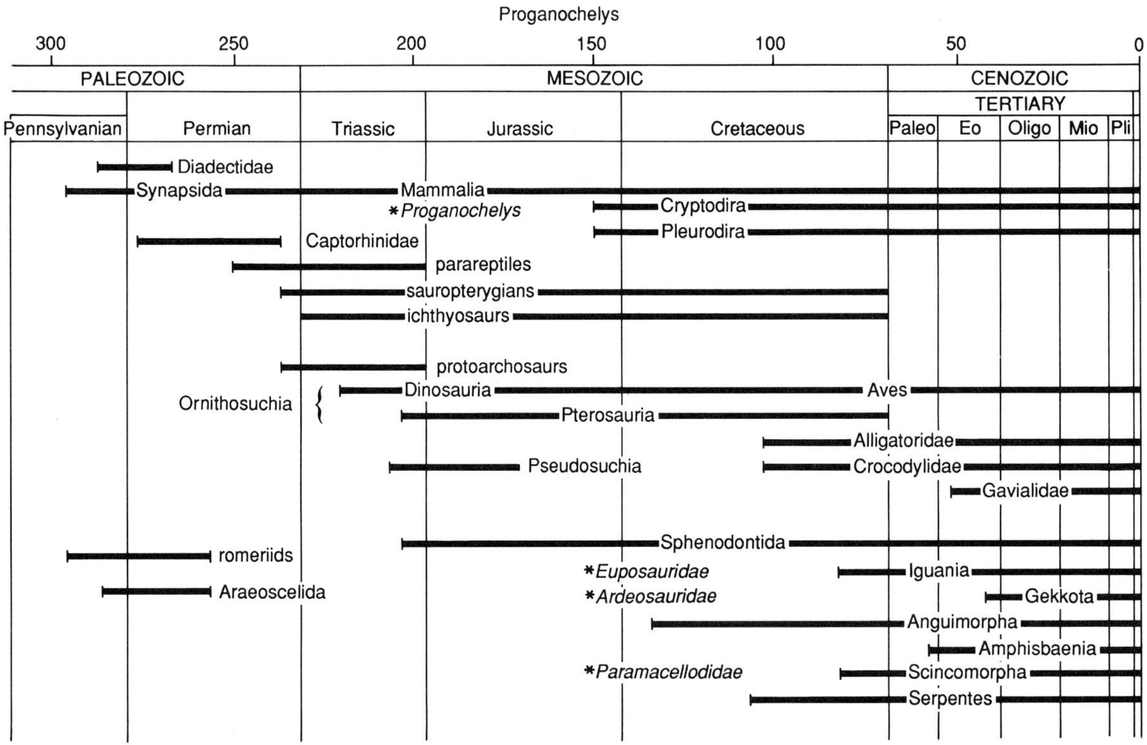

FIGURE 4.3

Geological occurrence of extinct and living reptiles. Abbreviations for Cenozoic epochs: Paleo, Paleocene; Eo, Eocene; Oligo, Oligocene; Mio, Miocene; Pli, Pliocene; Pleistocene is the narrow, unlabeled epoch at the top of the chart.

In contrast, the pareiasaurs (mid to Late Permian) and the procolophonoids (Late Permian through Late Triassic) persisted longer and showed a much greater diversity. The pareiasaurs were the giants of the parareptiles with some forms to 3 m (TL). They had large barrel-shaped bodies, elephantine limbs, and proportionately small, broad-jawed heads capped with thick bone and numerous projections. The teeth were closely spaced with laterally compressed leaf-shaped crowns. By all indications, the pareiasaurs were slow, lumbering herbivores.

The procolophonoids were small- to medium-sized lizardlike reptiles. Their stocky bodies, short limbs, and broad-jowled heads gave them the appearance of modern *Uromastyx* or *Sauromalus*, and they may have shared the herbivorous habits of these extant lizards. Unlike the pareiasaurs, their widely spaced, thick, bulbous-crowned teeth were probably used for crushing rather than mincing. Procolophonoids may have contained the ancestors of turtles. This suggestion had been discredited, but a recent discovery of numerous complete skeletons of *Owenetta* shows that this diminutive and earliest procolophonoid (Late Permian) shares many features with the oldest known turtle, *Proganochelys*.

Eunotosaurus represents another unclassified mid-Permian reptile. This small (20 cm SVL) lizardlike creature was also considered the link between the basal reptiles and turtles, because it possessed eight pairs of broadly expanded ribs on the trunk. However, the pectoral girdle lay external to the ribs, and the skull was strongly divergent from the morphology of any early turtles.

Radiation of Mesozoic Reptiles

The Mesozoic is often called the Age of the Reptiles. The diversity of reptiles that lived during this period remains unmatched by any other tetrapod group. They were the dominant terrestrial and aerial animals, and although not the dominant marine ones, some were major marine predators. The following summaries touch only briefly on this diversity.

Marine Reptiles

The sauropterygians were immensely successful aquatic reptiles that appeared early in the Triassic and remained abundant until the end of the Cretaceous. The sauropterygians consisted of two distinct but related groups, the nothosaurs and the pleisosaurs (Fig. 4.4). Both share a body form unlike that of any other aquatic tetrapod. Although streamlined, the body was large and stocky with a long, flexible neck and large flipperlike limbs. The nothosaurs (Triassic) were small- to moderate-sized (0.2–4 m TL) diapsids with the tail being a third to nearly half of the total length. This body form suggests that they swam by undulatory movements of the tail and posterior half of the body, using the limbs as rudders. The pleisosaurs arose from nothosaurian stock, probably in the mid-Triassic, but became abundant only in the Jurassic through the mid-Cretaceous. They were generally large creatures, 10–13 m in total length. The

body was barrel-shaped with a short tail, less than body length, and very large flipperlike limbs. In one group, the neck was very long, ending in a tiny head, and in another group, the neck was shorter with a large, elongated head. They swam with their limbs, rather than through body and tail undulation. How they swam is uncertain. The two most likely possibilities are aquatic flight like penguins and seaturtles (limbs move in a figure-8 stroke as in flying birds) or with the more paddlelike stroke of seals. No matter how they swam, they were undoubtedly excellent and fast swimmers.

The ichthyosaurs were also a successful group of marine reptiles, although they declined greatly in abundance in the Early Cretaceous and disappeared by the mid-Cretaceous. As their name implies, the ichthyosaurs were fishlike reptiles (see Fig. 4.4), with body forms matching those of today's faster swimmers—mackerel and tunas. They ranged in size from about 1.5 to 15 m. Their fishlike form and the presence of fetuses within the body cavity of some individuals attest to their viviparity (live-bearing). This reproductive mode was unlikely for most other Mesozoic marine reptiles, which probably had to struggle ashore to deposit their eggs like modern seaturtles.

Among the early crocodyliforms, several groups became highly aquatic and perhaps totally so (Fig. 4.5). The most specialized group was the metriorhynchids (mid-Jurassic to Early Cretaceous). At least 15 species have been recognized. All were about 3 m long with heavy, streamlined bodies and tails. The tail had a sharklike downward bend at its tip (heterocercal), and the limbs were flippers. The head was long-snouted and streamlined. From appearance, they were excellent swimmers and successful fish predators.

A lineage of aquatic lizards split early from the evolutionary line leading to the extant varanoid groups. The aigialosaurs were small (1–2.5 m TL) lizards, monitorlike in general appearance although they had shorter necks, reduced limbs but not structurally reorganized, and a laterally compressed, heterocercal tail. They occurred in the Late Jurassic to mid-Cretaceous seas and likely were the ancestral stock of the Middle to Late Cretaceous dolichosaurs and Late Cretaceous mosasaurs (see Fig. 4.5). The dolichosaurs were long-necked pleisosaurlike lizards and displayed little diversity. By contrast, the mosasaurs had a moderate adaptive radiation that produced a variety of different sizes and feeding morphologies (e.g., at least sixteen different body forms are recognized). The general body form remained lizardlike, even though the mosasaurs were highly aquatic animals. The head was elongate and narrow, joined by a short neck to an elongate trunk and tail. The limbs were modified into flippers by a shortening

FIGURE 4.4

Some Mesozic marine reptiles: *Protostega* (cryptodiran seaturtle/Late Cretaceous), *Dolichorhynchops* (pliosaurid pleisosaur/L. Cret.), *Stenopterygius* (ichthyosaur/L. Jurassic); from upper left, clockwise. [From *Into The Swim Again*, courtesy of the Natl. Museum of Natural History/Smithsonian.]

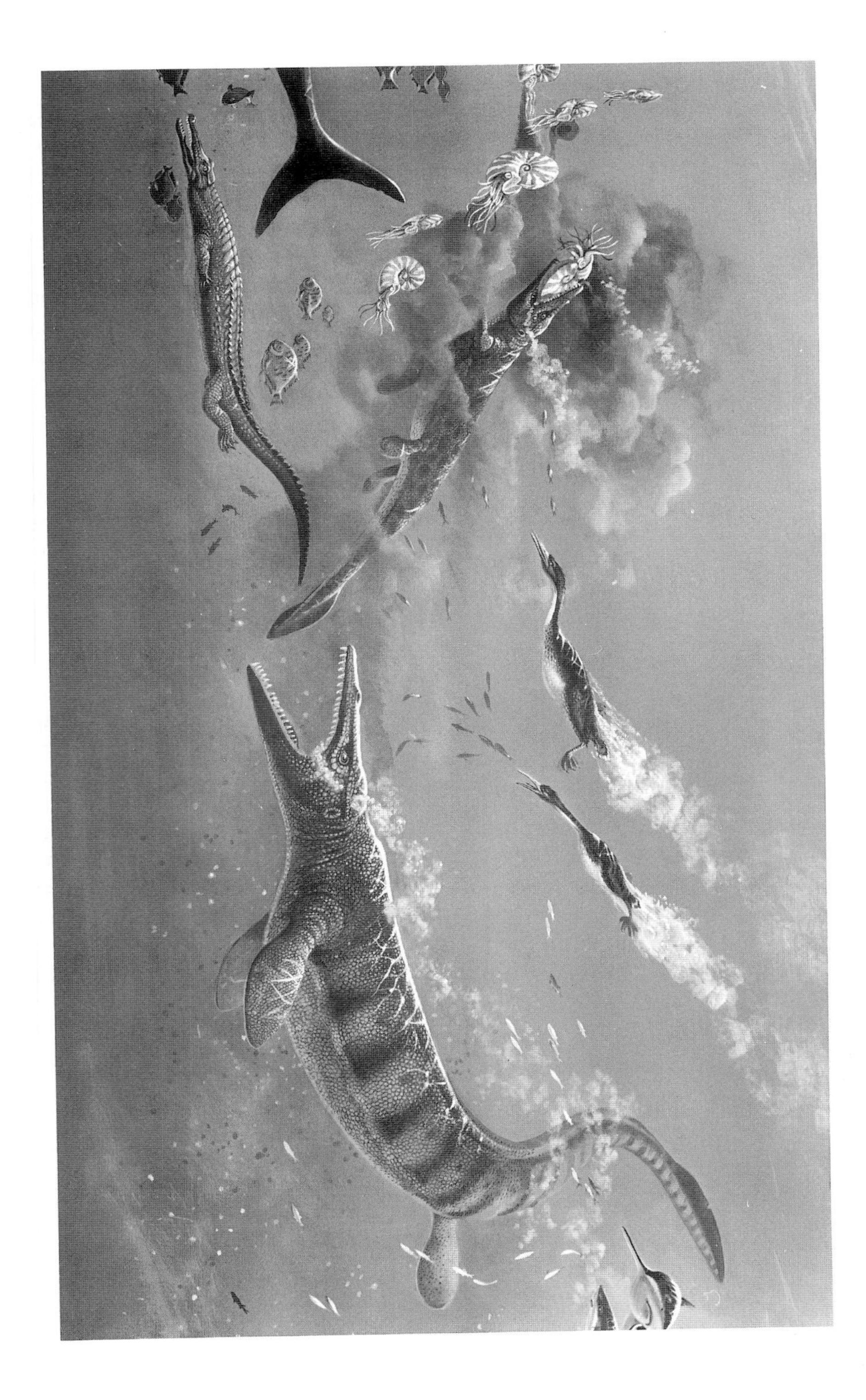

of the pro- and epipodial elements and an elongation of the meso-, metapodial elements and phalanges (hyperphalangy). The sinuous body and tail were both used in undulatory swimming, with flippers serving as rudders. Terrestrial locomotion would have been most difficult; nonetheless, they were likely oviparous so the females would have had to come ashore to lay eggs. This difficult task was compounded further by their size, as the smallest genus was 2.5 m (TL) and the largest reached nearly 12.5 m. Some mosasaurs were surface creatures, and others probably dove regularly to depths of several hundred meters for food. All were carnivorous predators.

Gliders and Fliers

Most airborne animals develop flight surfaces by modifying anterior appendages or by stretching membranes between anterior and posterior appendages. Several groups of diapsid reptiles independently modified their ribs and associated muscles to form an airfoil. This rib-cage adaptation is unique to diapsids and exists today in *Draco*, a group of Indomalaysian agamids. The thoracic ribs are greatly elongated, and for more than half of their length, they are free of the body cavity and attached to each other by a thin web of skin. Limbs are well developed, and *Draco* can run nimbly up and down tree trunks, with the elongated ribs folded tightly against the body. When pursued, they jump into the air; the elongated ribs unfold like a fan and create an airfoil that allows them to glide long distances at a gentle angle of descent.

The rib-cage airfoil appeared first in the Late Permian *Coelurosauravus*, a moderately large reptile with a airfoil span of nearly 30 cm. Although highly specialized as a glider, it possesses many primitive diapsid features but is not a member of any of the other diapsid groups. The Late Triassic kuehneosaurids also had the rib-cage airfoil and were an early divergent lineage of the lepidosaurs.

The typical animal airfoil (wings, modified appendages) was used for flight by two groups of ornithosuchian archosaurs—pterosaurs and birds. Both of these aerial reptiles modified their forelimbs as airfoils and were capable of self-propulsive, "flapping" flight. (Some proponents still argue for gliding flight only in pterosaurs.)

The pterosaurs developed a membranous wing that stretched from the posterior edge of the forelimb to the body. The proximal skeletal elements were shortened and robust for the attachment of flight muscles. Most of the wing's span attached to a greatly elongated fourth digit (i.e., elongation of metacarpal

FIGURE 4.5

Some Mesozoic marine reptiles: *Clidastes* (mosasaur/L. Cret.), *Hesperornis* (avian ornithosuchian/L. Cret.), *Tylosaurus* (mosasaur/L. Cret.), *Steneosaurus* (teleosaurid crocodyliform/L. Jur.); from upper left, clockwise. [From *Into The Swim Again*, courtesy of the Natl. Museum of Natural History/ Smithsonian.]

IV and especially the phalanges, each of which was longer than the humerus). The birds modified their specialized scales (feathers) to produce an airfoil surface. The forelimb provided the support for the feathers and the anterior edge of the airfoil. In birds, the humerus is short; the radius, ulna, metacarpals, and phalanges of the third digit are elongated.

The pterosaurs appeared in the Late Triassic as fully winged fliers and persisted as a group throughout the remainder of the Mesozoic. Nearly a hundred species of pterosaurs are recognized from small (15-cm wingspan) to the aerial giants, *Pteranodon* (7-m wing span) and *Quetzalcoatlus* (11- to 12-m wing span). The variety in size was matched by a variety of shapes and feeding habits (scavengers, insectivores, piscivores, carnivores, even filter feeders). Their cousins, the birds, did not appear until the Late Jurassic (*Archaeopteryx*), and their diversity remained low throughout the remainder of the Mesozoic or perhaps only a few kinds were fossilized.

Archosaurs

The archosaurs, "the Ruling Reptiles of the Mesozoic," are a monophyletic group with a major presence today in the form of crocodilians and birds. Indeed these two groups are survivors of the two major lineages of archosaurs. The pseudosuchians, containing the crocodyliforms, and the ornithosuchians, containing the dinosaurs, birds, and pterosaurs, represent a lineage split that is evident in the mid-Triassic and may have occurred in the Late Permian.

The proterosuchids, erythrosuchids, and proterochampsids were early offshoots of the diapsid lineage that would give rise to the archosaurs. They show a sequential alteration of the skeleton toward the archosaurian mode and a trend toward increasing size. Proterosuchids (Late Permian to Early Triassic) were moderate-sized, varanidlike reptiles with a sprawling gait. The erythrosuchids (Early to mid-Triassic) were large (ca. 5 m TL), heavy reptiles with the beginnings of a more erect limb posture and of the archosaurian triradiate pelvic girdle. The crocodilelike proterochampsids (mid to Late Triassic) also possessed an erect limb posture.

Euparkeria (Early Triassic) is the most primitive of the archosaurs. This medium-sized (ca. 50 cm SVL) diapsid appeared much like a short-necked varanid lizard. All archosaurs subsequent to *Euparkeria* are members of either the pseudosuchian or ornithosuchian lineage. These two lineages have strikingly different ankle anatomies. Both lineages radiated broadly, beginning in the mid-Triassic. The major ornithosuchian groups are the pterosaurs, ornithischian dinosaurs, and saurischian dinosaurs. Dinosaurs came in all sizes (1–25 m TL) and had an enormous variety of shapes. They had equally varied diets and occupied a wide range of habitats. Birds are ornithischian dinosaurs and likely derive from the small coelurosaurs of the mid-Jurassic.

The pseudosuchians (Crocodylotarsi) include a large number of families, most of which possessed a general crocodilian body form that was variously modified. Their diversity does not match that of the archosaurs; nonetheless, nearly two dozen families and roughly a dozen unassigned species are known from the Mesozoic. Until recently, our classification emphasized levels (grades) of specialization or divergence from the basic pseudosuchian stock. These grades, such as protosuchian or mesosuchian, contained multiple lineages and that classification is now being replaced by monophyletic groups. The new cladistic classification is not firmly established, in part because the fragmentary nature of some extinct taxa does not permit reliable determination of relationships.

The phytosaurs (Late Triassic) are the most primitive of the pseudosuchians and an early offshoot of the main crocodilian lineage. They were 2–4 m TL, ghariallike diapsids; however, their teeth were small and remained inside the mouth when closed, and the nostrils were on a raised bony mound at the base of the long, narrow snout. The aetosaurs (Late Triassic) are another early evolutionary side branch. They had a small, piglike head on a heavily armor-plated crocodilian body and tail. Their small, leaf-shaped teeth indicate a herbivorous diet, thus they have the honor of being the earliest herbivorous archosaurs.

Several other divergent lineages appeared and disappeared in the Triassic. The main crocodilian group, Crocodyliformes, had a few families (e.g., teleosaurids Fig. 4.5) in the Early Jurassic, but their diversity did not increase until the Late Jurassic and Early Cretaceous. The low Jurassic diversity results from the presence of only a few terrestrial and freshwater fossil deposits, the habitats in which crocodyliforms were radiating. The teleosaurids (Early Jurassic to Early Cretaceous, Fig. 4.5) were highly aquatic, ghariallike crocodyliforms (1–9.5 m TL) of estuarine and nearshore habitats. Their forelimbs were greatly reduced probably owing to swimming through undulatory movements of the body and tail. The hindlimbs remained large and likely served as rudders.

The teleosaurids and the metriorhynchids arose from the crocodilian stock leading to the modern crocodilians (neosuchians). The neosuchians consist of much more than the sole surviving Crocodylia and include neosuchians such as *Bernissartia*, a small alligatorlike, mollusivorous form; the monstrous, semiaquatic *Sarcosuchus* (an Early Cretaceous pholidosaurid) reaching a total length of over 11 m; and the marine *Stomatosuchus*, dyrosaurs, and pholidosaurs. The Crocodylia appeared in the Early Cretaceous and have been the prominent semiaquatic crocodilians since then. A few species were terrestrial, and the pristichampsines had hooflike feet.

Lepidosaurs

The lepidosaurs are the second major diapsid lineage. A group of Late Permian lepidosaurs (*Youngina*, *Acerosodontosaurus*, and their tangsaurid relatives) represents the first lepidosauromorph radiations. *Youngina* was a slender diapsid that would have been easily mistaken for many modern lizards and was likely an agile, terrestrial insectivore. The tangsaurids appeared similar but had laterally compressed tails and probably an aquatic life-style.

The lepidosaurs are largely absent from the fossil record until the Late Triassic when the Sphenodontida and the *Draco*-like kuehneosaurids appeared. The sphenodontidans have never been an exceptionally diverse group, and most appeared much like the living tuataras, *Sphenodon*. The exception is a small group of aquatic genera, the pleurosaurines, with greatly elongated bodies and tails, and usually a barracudalike head. The sphenodontidan miniradiation occurred from the Late Triassic to the Late Jurassic, and they were moderately abundant during this interval. Thereafter, their fossil presence declined through the Cretaceous, and no Tertiary forms have been found.

The first squamates appeared late in the Late Jurassic. The term "eolacertilians" has been used for a group of Upper Permian lepidosaurs, which include such unrelated forms as *Kuehneosaurus*, *Paleaeagama*, and *Paliguana*. Although they are lepidosauromorphs, they are early offshoots of the lineage that divided into the Sphenodontida and Squamata.

The Jurassic squamates represent at least five families that appeared briefly and apparently were extinct by the Early Cretaceous. Two of these families are considered gekkotans. The Ardeosauridae contain three genera from Palearctic deposits—*Ardeosaurus*, *Eichstaettisaurus*, and *Yabeinosaurus*. The first two genera are represented by nearly complete specimens and they are definitely gekkotans; however, *Yabeinosaurus* is a fragmentary fossil and less convincingly a gekkotan. The Bavarisauridae contains two genera, *Bavarisaurus* and *Palaeolacerta*, and shares at least two derived characters with extant geckos.

Though the Jurassic euposaurids resemble agamids and are regularly placed with the Iguania, other evidence suggests that they are actually sphenodontidans. The paramacellodids are similarly associated with the cordylids by incomplete data. The evidence for the association of the Jurassic dorsetisaurids is somewhat better, and this group may link xenosaurids and anguids. The Cretaceous marine lizards (aigialosaurids, mosasaurids, and others) are strikingly similar to the varanoids; however, this similarity may be convergent. A better case can be made for the relationship of the necrosaurids (Late Cretaceous to Oligocene), because they possess some uniquely varanoid traits and may even be the ancestor of the helodermatids.

Fossil representatives of the extant lizard families did not appear until the Cretaceous. These forms are discussed next.

HISTORY OF EXTANT REPTILES

Turtles

Turtles have a good fossil record. Their bony shell provides a durable structure for them in life and in death. Their history extends back nearly 200 million years (Late Triassic) to the first known turtle, *Proganochelys*. *Proganochelys* is unquestionably a turtle. It had its dermal (osteoderms) and axial skeletons modified into a true shell, that is, the fusion of the ribs and vertebrae to dermal bones to form a carapace and fusion of pectoral girdle elements and dermal bones to form a plastron. It also possessed a number of early amniote characteristics that were lost in later turtles. Teeth were present on the palatines but absent from the upper and lower jaws. *Proganochelys* was a large (ca. 90 cm carapace length/CL), semiaquatic turtle, well protected by its bony shell and bony neck spines (Fig. 4.6).

A pleurodire, *Proterochersis*, was nearly contemporaneous with *Proganochelys*. It was a somewhat smaller turtle (ca. 50 cm CL) and likely terrestrial. Its pelvic girdle was fused to the plastron, showing it to be the earliest pleurodire and indicating that the divergence of cryptodires and pleurodires was past. *Proganochelys* represents a split of the testudine lineage prior to the divergence of the two major lineages of turtles.

After *Proterochersis*, pleurodires are absent until the brief appearance of *Platychelys* in the Late Jurassic. Pleurodires do not occur again until the Late Cretaceous, but from then to the present, they are present in the fossil faunas, particularly those of the Southern Hemisphere. Although now confined to the southern continents, a few pleurodires occurred in the Northern Hemisphere at

FIGURE 4.6
The oldest known turtle *Proganochelys quenstedti* (L. Triassic; ca. 15 cm CL). [From Gaffney (1990).]

the least through the Miocene. Some of the Tertiary turtles were marine or estuarine and reached the size of modern seaturtles, although they did not develop the form and locomotor mode of the cryptodiran seaturtles. Chelids do not appear until the Oligocene or Miocene and only in South America and Australia.

The oldest turtle in North America and the first cryptodire is *Kayentachelys aprix* from the mid-Jurassic (185 MA/million annum before current era) of western North America. It was a moderate-sized (30 cm CL), semiterrestrial turtle. Structurally, *Kayentachelys* was a cryptodire, although it possessed a number of features not seen in modern turtles, such as small teeth on the roof of the mouth.

Cryptodires were absent also in the mid-Jurassic, but the Pleisochelyidae and Pleurosternidae appeared in the lower Late Jurassic, and cryptodires remained part of the reptilian fauna thereafter. Both families contained moderate-sized, aquatic turtles, and neither is related to any of the later-appearing turtle groups. The modern groups of turtles began to appear in the Cretaceous. The chelonioids appeared in earliest Cretaceous, represented by the desmatochelyids. These seaturtles were large (>1.2 m CL) turtles with reduced carapace and plastron. They were an early branch of the leatherback lineage, but probably had keratinous scutes on their shells. The leatherbacks (Dermochelyidae) did not appear until the Eocene and thereafter experienced a modest radiation of several genera and a dozen species. The cheloniids and their relatives, the osteopygids, appeared near the end of the Cretaceous and were even more diverse (see Fig. 4.4).

The baenoids were a group of heavy-shelled, moderate-sized turtles that appeared in the mid-Cretaceous and survived into the mid-Tertiary. They were strictly North American and probably aquatic to semiaquatic. They have no close relationships to any modern groups and likely arose from the early cryptodiran stock contemporaneous with the pleurosternids. The meiolaniids, the horned tortoises, were another early divergent lineage; yet they do not occur in the fossil record (Australia and South America) until the Late Eocene and probably survived into prehistoric times. Most were large (1 m CL), high-domed shelled species. They had large heads with a bizarre arrangement of horns or spines projecting from the posterior margin of the skull.

The modern groups (chelydroid, testudinoids, trionychoids, and their extant families) have Late Cretaceous representatives, but none represents extant genera. The modern genera began to appear in the Miocene, concurrently with the disappearance of the Early Tertiary genera, although a few of the latter remained into the Pliocene.

Crocodilians

The modern crocodilians (Crocodylia) appeared first in the mid-Cretaceous and underwent a major radiation in the early Tertiary. The body forms observed in extant species were widespread among the early Crocodylia and occurred in the earliest of the crocodyliform archosaurs. Such a long history for and the continual reappearance of these body types show how suitable the crocodilian habitus is for aquatic predators.

Both alligatorids and crocodylids occurred in the Cretaceous, but the first gavialids did not appear until the Early Eocene. As noted previously, first occurrences are not an accurate indicator of time of origin, because gharials apparently diverged from a prealligator-crocodile stock, perhaps in the Early Cretaceous. (While fossil occurrence is hard evidence of a taxon's presence, absence is equivocal evidence, because a taxon may have existed but no remains were fossilized.) Extinct gharials lived in South America (Oligocene to Pliocene), perhaps northeastern Africa (Miocene), and southern Asia (Miocene to Recent). All possessed the long, narrow snout associated with a specialized diet of fish (piscivorous). Most extinct gharial species equaled the size of the living species, but a Miocene species from India apparently reached lengths of 15–18 m and is a contender for the biggest-crocodilian-ever record.

Numerous generic and specific names have been proposed for the alligatorid and crocodylid fossils. As many names are based on a tooth or a skull fragment, the existence of dozens of species is questionable, and generic diversity is likely less than the number of names implies. Nonetheless, crocodilian diversity during the Tertiary was greater than today. Though two crocodiles are known from the Cretaceous, four genera and more than 25 species supposedly occurred in the Eocene. Most of these are species of *Crocodylus*. Only a few of the extant species are known from fossils and all are no older than the Pleistocene. The alligatorids now have a greater diversity than the crocodiles, and this diversity was also evident in the past. In addition to the several types of caimans, there were several different alligators, but the variety of body forms seems to have been no greater than today. There were also different crocodilians, such as the hoofed crocodiles (pristichampsines). Pristichampsines first appeared in the Eocene of Europe and survived into the Pleistocene in Australia.

Lepidosaurs

The history of the extant squamates begins in the Cretaceous. Teiids appeared early in the Late Cretaceous, and the agamids, iguanids, scincids, xenosaurids, anguids, helodermatids, and varanids emerged soon thereafter. Snakes are questionably reported from the Early Cretaceous, but were present by the Late Cretaceous (aniliids, madtsoiine boids, lapparentophids, and simoliophids).

All Cretaceous lizards represent extinct genera and species, and even subfamilies, with the possible exception of the anguid *Gerrhonotus*. The Cretaceous *Gerrhonotus* records seem unlikely, but fossil skull fragments match closely those of living *Gerrhonotus*. The Cretaceous teiids were the extinct polyglyphanodontines, structurally very similar to extant *Dicrodon* and *Teius*. They occurred in western North America and Mongolia. Four genera of teiine teiids were also present in the Late Cretaceous; two genera are related to the modern *Ameiva–Cnemidophorus* group, one to *Teius*, and one to *Calliopistes*. The single Cretaceous iguanid, *Pristiguana*, is a questionable iguanid, since it has several traits that associate it with the teiids. The Cretaceous scincid *Contogenys* is also an uncertain familial assignment. Its skull fragments show similarities to cordylids. In contrast, the Cretaceous *Mimeosaurus* is certainly an agamid, although it possessed a tubercular sculpturing of the skull unlike any other agamids.

All four extant families of anguimorphs occurred in the Cretaceous. Excluding the uncertain *Gerrhonotus*, the anguids are known from a single species of glyptosaurine lizards in the Late Cretaceous. The glyptosaurs (Cretaceous–Oligocene) were heavy-bodied, broad-headed lizards with an armor of tubercular-sculptured osteoderms over the head and body. Glyptosaurs remained a common group through the early Tertiary, disappearing in the mid-Miocene. The Cretaceous *Paraderma* may represent the first venomous lizard; its teeth appear to have shallow venom grooves. These grooves and other cranial traits suggest that *Paraderma* was an early helodermatid. *Palaeosaniwa*, a Cretaceous varanid, was a large lizard, probably exceeding 2 m (TL), and with its large recurved teeth, it was likely a major predator. The two Cretaceous xenosaurid genera derive from western North America and are unquestionably xenosaurids. One of them, *Exostinus*, is similar to the modern *Xenosaurus*.

In the early Tertiary, the lizard faunas were a mix of extinct and extant taxa. The Paleocene to Oligocene faunas contained mainly extinct genera of extant families, although two extinct families, Arretosauridae and Necrosauridae, were present during this time. The Eocene fauna of western North America contained iguanids, agamids, anguids, helodermatids, necrosaurids, varanids, xenosaurids, scincids, and xantusiids. Beginning in the Miocene, extant genera became prominent and were the only ones present by the mid-Pliocene. The Miocene fauna of Europe contained *Chamaeleo*, *Gerandogekko*, *Phyllodactylus*, *Lacerta*, *Iberovaranus*, and *Varanus*. An extraordinary exception to the exclusive presence of living genera in the late Tertiary or the Quaternary is the Australian varanid *Megalania*, a huge goanna. Its average size was about 1.5–1.6 m SVL, but some individuals reached total lengths of nearly 7 m (4–4.5 m SVL). These giants, probably weighing more than 600 kg, must have been formidable predators and equivalent to lions or tigers.

The first amphisbaenian (*Oligodontosaurus*) appeared in the Late Paleocene of western North America. *Oligodontosaurus* was not a generalized amphisbaenian, but a shovel-headed form. Although similar to rhineurids, which appeared first in the Early Eocene of the American West, *Oligodontosaurus* had a distinct jaw structure and is placed in its own family. The rhineurids are abundant in the Oligocene of the American West and are remarkably similar to the single species surviving today in Florida. *Hyporhina*, another Oligocene shovel-nosed amphisbaenian from the West, represents another family. It is probable that these shovel-headed families comprise a single monophyletic group. The only other fossil amphisbaenians are three amphisbaenids, one from the Late Eocene-Early Oligocene of France and two from the Early Miocene of East-Central Africa.

The oldest snake is *Lapparentophis* from the mid-Cretaceous. Typical of fossil snakes, it is known only from vertebrae (three trunk vertebrae). *Lapparentophis* is an alethinophian and presumed to be a terrestrial snake. Two other snakes, *Simoliophis* and *Pouitella*, are nearly of equal antiquity, but apparently not closely related, other than being primitive snakes. These three snake genera do not seem to be related to any of the living families of snakes, and they or descendants do not occur later in time. However, determination of relationships and habits on the basis of a few vertebrae is risky business.

Representatives of two extant groups (booids and aniliids) existed in the Late Cretaceous. Two large boids, *Gigantophis* and *Madtsoia* (Madtsoiinae), equaled the size of the largest extant boids. Known only from the Cretaceous, the unique *Dinilysia* (Dinilysiidae) was also a large snake, roughly equal in size and appearance to *Boa constrictor*. It is one of the rare fossil snake finds, consisting of a nearly complete skull and part of a vertebral column. The madtsoiine boids reappeared in the Eocene, but no records occur thereafter. One, perhaps two, species of aniliids occurred contemporaneously with the preceding booids in both geological periods.

Other booids appeared in the early Tertiary. Most of these were genera and species that disappeared by the end of the Miocene. A few were related to modern species. *Lichanura brevispondylus* from the mid-Eocene of Wyoming, for example, is a sister species of *Lichanura trivirgata*. *Eunectes* from the mid-Miocene of Colombia and *Boa constrictor* from the Early Miocene of Florida are ancestral to living species. *Wonambi*, a Pleistocene boid of southern Australia, derives from a continent that is otherwise known only for pythons, and it is further peculiar because it may be related to *Boa*.

The scolecophidians have an extremely poor fossil history. Only two fossils have been found that can be assigned to the typhlopids, and this assignment is tentative. An acrochordid appeared first in the mid-Miocene, but two earlier fossils (Paleocene and Eocene) are of a related, but extinct, family.

Colubrid snakes occurred first in the Oligocene and become increasingly common in Miocene and later fossil deposits. These Oligocene snakes represent two species of *Coluber*, a *Natrix*, and *Texasophis*. Other Miocene and Pliocene colubrids are closely related, if not identical to modern taxa, but because they are fossils, they have been given names such as *Paleonatrix*, *Paraoxybelis*, and *Protropidonotus*. The elapids have few fossil representatives and appeared first in the mid-Miocene. The viperids are similarly poorly represented by fossils and also appeared first in the Miocene.

PART II

AS PREDATOR AND AS PREY

There are very few animals that are not either predator or prey. Eating and avoiding being eaten are clearly of central importance to Darwinian fitness.

P. A. Abrams, 1986

Predator–prey relationships are often compared to an arms race, with the predators striving to improve their recognition and capture abilities, and the prey striving to become less visible and either more difficult to capture or less edible. The "striving" is the selective pressure imposed by each side on the other, but neither is "striving" for perfection. Evolution does not operate in that manner. Each organism is exposed to many different selective pressures during its life, and each system within an organism fulfills multiple functions. Evolutionary changes must meet the demands of many competing needs, thus successful adaptations improve one set of biological activities without greatly altering the performance of other activities.

Although simplistic, evolutionary constraints act on predators like a double-edged sword; gains may be only temporary and potentially detrimental to survival of the predator. By improving its rate of prey capture, a predator imposes a strong selective pressure on prey. Those individuals of the prey species best able to avoid capture survive, and their offspring dominate the next generation. Predatory pressure drives the evolution of the prey population to improve its individuals' abilities to resist or avoid capture. If the prey is successful, it is less or not suitable as a food item or more difficult to capture, and the predator has reduced its resource base. In another direction, predator species may become more successful at capturing a particular prey species and decimate the prey population, perhaps driving it to extinction. Success results now in famine and a population crash of the predator, possibly its extinction as well. A predator may become more numerous owing to its successful exploitation of prey and its higher abundance may increase its own attractiveness as prey to other predators.

Predator–prey interactions are often viewed as the interplay between a pair of species. No prey has a single predator, and few predators are dependent on a single prey species. Even the most specialized predators consume a variety of prey species, although the prey may be a particular life-history stage (e.g., eggs) or a taxonomic group (e.g., termites). Predator–prey interactions are more complex than portrayed, because of the difficulty of identifying and studying all the associations throughout the lives of prey and predators.

CHAPTER 5

Diet and Feeding

To survive, all animals must eat. This seems obvious, yet many aspects of eating remain unexplored in amphibians and reptiles. Eating is used in a broad sense and encompasses two facets, types of food eaten (diet) and procurement and processing of food items (feeding). These aspects are not unrelated but provide a useful dichotomy for examining feeding behavior.

FOOD PREFERENCES

Amphibians, with the exception of most frog larvae, are carnivores. Reptiles, with the exception of a few turtles and lizards, are also carnivores. The exceptions are herbivores. As for most generalizations, the preceding statements are correct only in a coarse-grained sense. Amphibian and reptilian diets grade from totally flesh-eating (carnivorous) to totally plant-eating (herbivorous), and within each dietary sphere or end point, there are specialists eating only a limited group of food items and generalists eating a wide range of food types. No matter what the food items, eating is the means to gain energy, nutrients, and water for body maintenance, growth, and performance of all other biological activities, for example, courting a mate or regenerating a tail.

Herbivores

Among amphibians, herbivory is limited to anuran tadpoles. This limitation is due to the difficulties of digesting fiber (see Digestion and Assimilation, this chapter). Tadpoles avoid the herbivory conundrum by consuming mainly the algal and bacterial scum (aufwachs) covering all objects immersed in the water. Herbivory in tadpoles likely occurs in many species but is poorly verified owing to few studies on the gut contents of tadpoles that examine which cells are digested and which are voided whole. Tadpoles gather their food from all levels of the water column—bottom sediments, filtering midwater phytoplankton, and skimming the surface scum. Most species specialize on a particular portion of water column and style of harvesting. For the postmetamorphic stage, there are no totally herbivorous amphibians. Very few include some plant material in their diet (see Omnivores, this chapter).

In contrast to the amphibians, hatchling and juvenile reptiles are predominantly carnivores (preying mainly on invertebrates), except those species with herbivorous adults. Herbivores occur only among turtles and lizards (Table 5.1). Most are generalist herbivores and include a mix of fruits, flowers, and foliage in their diet.

The green seaturtle (*Chelonia mydas*) is one of the few specialist herbivores among the reptiles. Individuals eat either algae or the foliage seagrasses, but not both owing to the need for different gut microfloras to break down these structurally dissimilar plants. Most of the reptilian herbivores preferentially eat young, growing foliage, which is more digestible with less fiber and higher protein content. *Chelonia* grazes on previously grazed plots and continually crops the new growth. When introduced into a ungrazed area, an individual will bite the seagrass close to the base, allowing the older, higher fragment to float away; it continues to feed in this manner until it has established a large grazed pasture that it can harvest daily without enlarging the area or foraging elsewhere.

Many testudinids (tortoises) appear to be herbivores or nearly so; their diet is a mix of locally available plants and tends to be high in foliage. An assortment of emydids are herbivores with plants comprising more than 75% of their diet. These herbivorous turtles include both large and small species as well as semiterrestrial and aquatic forms (e.g., *Cuora*, *Hardella*, *Ocadia*, *Pseudemys* — usually not all species of polytypic genera); other turtles, such as *Chrysemys*, are

TABLE 5.1 ▼▲▼▲▼▲▼▲▼▲▼▲▼▲▼▲▼▲▼▲▼▲▼▲▼
Some Reptilian Herbivores Whose Diets are Predominantly Plant Matter as Adults

Taxon	Food items
Batagur baska	Foliage, fruit, animal
Chelonia mydas	Seagrasses, algae
Melanochelys trijuga	Foliage, animal
Pseudemys nelsoni	Foliage, animal
Most testudines	Foliage, fruit, flowers
Geochelone carbonaria	Fruit, flowers, foliage, animal
Geochelone gigantea	Foliage
Gopherus polyphemus	Foliage, fruit
Angolosaurus skoogii	Foliage, animal
Aporosaura anchietae	Seeds, animal
Corucia zebrata	Foliage, fruit, flowers
Dicrodon guttulatum	Fruits
Hoplodactylus pacificus	Nectar, fruit, animal
Lepidophyma smithii	Fruit, animal
All iguanines	Foliage, fruit, flowers
Cyclura carinata	Foliage, fruit, flowers, animal
Dipsosaurus dorsalis	Flowers, foliage, animal
Iguana iguana	Foliage, fruit, flowers
Sauromalus hispidus	Foliage, flowers, fruit

Sources: Turtles—Bb, Moll, 1980; Cm, Bjorndal, 1980; Mt, Wirot, 1979; Pn, mt, Ernst and Barbour, 1989; Gc, Moskovits and M Bjorndal, 1990; Gg, Hamilton and Coe, 1982; Gp, MacDonald and Mushinsky, 1988. Lizards—As, Steyn, 1963; Aa, Robinson and Cunningham, 1978; Cz, Parker *in* Greer, 1976; Dg, Holmberg, 1957; Hp, Whitaker, 1968; Ls, Mautz and Lopez-Forment, 1978; ai, Iverson, 1982; Cc, Auffenberg, 1982; Dd, Mautz and Nagy, 1987; Ii, Rand et al., 1990; Sh, Sylber, 1988.

Note: Some possess a cellulolytic microflora in the digestive tract and/or colic modifications of the hindgut. The list does not include all well-documented cases of herbivory. The order of plants represents the importance (volume) in the taxon's diet, at least at one locality.

facultative herbivores and can survive and grow on a vegetarian diet when animal prey resources are unavailable. Most herbivores are opportunistic and will eat animal matter occasionally.

Lizard herbivores, the iguanines, a few agamids (e.g., *Uromastyx*), and a few skinks (*Corucia*), are generalist herbivores (Fig. 5.1), although at one location or season their diet may consist of a single species or particular plant parts. The

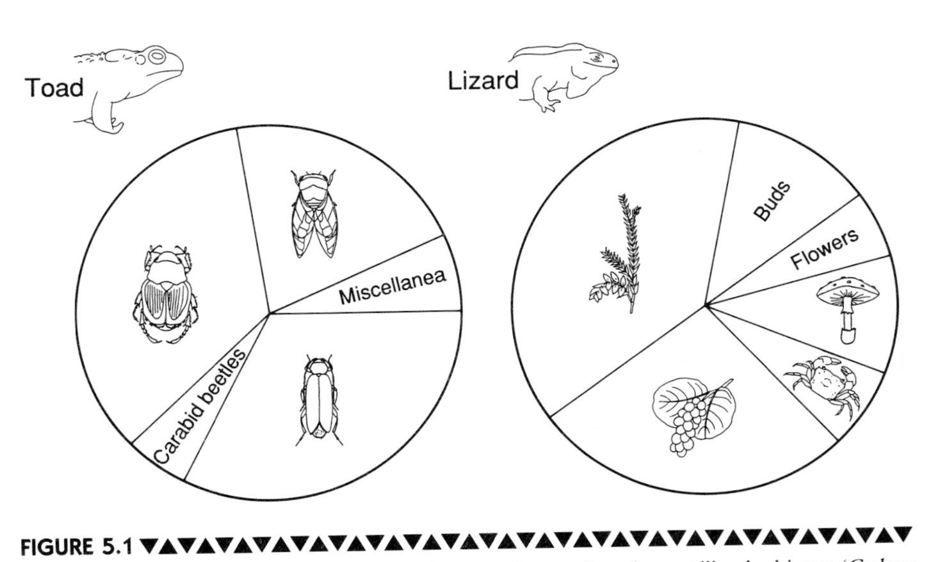

FIGURE 5.1 ▼▲▼

Diets of an amphibian insectivore (*Bufo marinus*, marine toad) and a reptilian herbivore (*Cyclura carinata*, rock iguana). The areas of the wedges reflect the proportional biomass of the various food items in the diet of these two species. [Data from Zug and Zug (1979) and Auffenberg (1982), respectively.]

association of large body size and herbivory prompted the hypothesis that large body size both permitted and required herbivory in the insectivorous taxa of lizards, because large (>300 g) lizards require more energy than foraging on invertebrates would permit. Simply put, the energy expended to capture numerous small packets of energy (insects and other invertebrates) is greater than the total energy gained from the animal prey. Feeding on plant material is much less energy expensive; although a greater bulk is required, it is easier to obtain. When the generalization was proposed in the early 1970s, the dietary and energetic information on lizards supported the transition from insectivory to herbivory in the 100- to 300-g class of lizards. Contrary to the predictions of this hypothesis, recent studies show that most young iguanines are herbivorous from the beginning of life. Also, as the dietary habits of more smaller species (iguanids) are documented, the size association disappears, although the cost/benefits of being herbivorous in a low animal-prey environment remains.

Carnivores and Omnivores

Carnivores

Postmetamorphic anurans and all lifestages of caecilians and urodelans are usually opportunistic carnivores, preying on any animal small enough to be captured and swallowed (Fig. 5.1). Because most amphibian species are small, their diets include a preponderance of arthropods, but small mollusks, worms, and vertebrates are also eaten. Other amphibians, including conspecifics (cannibalism), are prey in all lifestages (egg to adult), and in temporary pond-breeders, salamander larvae and aquatic adults may be the major predator of salamander and anuran eggs and larvae.

All crocodilians and snakes are strict carnivores, as are also the tuataras. Plant matter may be accidentally ingested, but when it is, it passes undigested through the digestive tract. Crocodilians are generalists and eat a broad variety of prey from insects and snails to turtles, fish, birds, and mammals. They eat progressively larger prey as they grow larger, although the lower acceptable size limit does not shift upward as rapidly as the upper limit. Many snakes are generalists as well, although their diets are not as catholic as that of the crocodilians. Many others, however, can be considered specialist carnivores and confine their selection to a narrow range of prey. For example, *Crotalus* usually eats rodents; *Thelotornis* and *Dispholidus* eat birds; *Oxybelis* and *Salvadora* eat lizards; *Dipsas* and *Pareas* eat slugs and snails; *Dasypeltis* eats birds eggs and *Aipysurus eydouxii* eats fish eggs; *Fordonia* eats crabs; and other snake taxa specialize on earthworms, insect larvae, orthopterans, salamanders, or centipedes.

The majority of lizards are small-bodied (<30 cm TL, <100 g), and this limits the size of their prey, hence insectivory (i.e., all invertebrates including insects) dominates. A few lizards or lizard groups are specialists: *Moloch* and *Phrynosoma* on ants; *Dracaena* on snails; *Lialis* on skinks. Turtles tend toward opportunism and many are best considered omnivores, although chelydrids are usually restricted to living or dead animals and seaturtles to groups of invertebrates (e.g., *Eretmochelys* eats sponges, sea anemones, and other reef-encrusting animals). The ultimate specialist among the turtles is *Dermochelys* with its diet of jellyfish medusae, a prey consisting of >90% water.

Omnivores

Omnivores are dietary generalists, occupying the vast middle ground between strict carnivory and strict herbivory. Omnivory is not widespread in amphibians, or perhaps not well documented. The known amphibian omnivores are anuran tadpoles, with a few exceptions. Two frogs, *Bufo marinus* and *Hyla truncata*, eat vegetable matter occasionally. A single population of marine toads at a site of low arthropod density fed regularly on pieces of foliage, but all individuals were emaciated and apparently unsuccessful at extracting much energy from the plant matter. *Hyla truncata* occurs regularly with a species of coca (*Erythroxylum ovalifolium*), eats its fruits, and may be a major seed dispersal agent for this tree. *Siren* is the only salamander known to regularly eat plants (*Elodea*) and invertebrates.

Omnivory is widespread in turtles and lizards, and as more lizards are examined carefully, omnivory has become even more widespread than formerly recognized. Several taxa have been identified as herbivores [e.g., *Angolosaurus* (cordylid), *Aporosaura* (lacertid), *Lepidophyma* (xantusiid)], and while one population or species of these three taxa may regularly and predominantly consume plant matter, others are strongly insectivorous. Omnivores appear to be present in all families of lizards. The Philippian butaan (*Varanus olivaceus*) is primarily a mollusivore but also regularly eats fruits. *Xantusia riversiana* eats animals and plants in roughly equal proportions. The full spectrum from insectivory to near-pure herbivory occurs within the iguanid *Leiolaemus*, a diverse group of small southern South American lizards. A few New Zealand geckos feed extensively on nectar; this nectivory is, however, temporally and geographically restricted. Other lizard populations may use nectar as a temporary or regular resource, such as the Madeira wall lizard *Podarcis dugesii*. Many of the pleurodiran and cryptodiran turtles include both animals and plants in their diets; usually eating animal matter, they shift to fruits or foliage when animal resources are low or when high-energy plant matter is abundant or easily gathered. Omnivory is dietary opportunism and takes advantage of readily available food resources.

PREY CAPTURE AND INGESTION

Stages of Feeding Behavior

Feeding begins when hunger stimulates the "desire" for food. The first phase of feeding is the search for food. The search may be a motile search for the preferred food items or may be a heightened awareness for potential food items while remaining stationary (ambush). Once a potential food item is located, the animal must determine whether it is appropriate (i.e., an acceptable dietary item) and capable of being captured. The item should not be too small, or obtaining and processing it will require more energy than the item provides. The item cannot be too large or its capture may not be possible and may result in injury or death of the predator. If acceptable, the animal must obtain/capture the item and process it for ingestion. The final and invisible stage is digestion (mechanical and chemical breakdown of food) and absorption (transport of digestive products into the bloodstream).

Forage Behavior

Forage behaviors for carnivores and omnivores are commonly grouped as either sit and wait (ambush) or motile-search (cruise, widely foraging) behaviors. The sit and wait predator waits for prey to pass within a capture zone and then strikes or moves to capture the prey. The motile-search predator hunts for its prey and discovers it by searching likely areas within a foraging or activity zone. The first behavior is an energetically conservative one but usually results in low prey captures; the second is energetically expensive and is rewarded usually by higher rate of prey capture. These two category labels provide a ready visual image of forage behavior, even though they do not convey the complexities of forage behaviors and may even obscure an observer's awareness of the actual mechanics and energetics of foraging.

Time analysis studies of forage behaviors of other vertebrates have revealed that animals are not either sit and wait or motile-search predators but that activity profiles grade from one category to another. All forage behaviors consist of stop and go activity (saltatory patterns) and the relative lengths of stop and go intervals differ among species and age classes. In the typical motile-search pattern, stops are brief ($<5\%$) and movements long; in contrast, stops are long ($<95\%$) and movements to capture prey short in the typical sit and wait pattern. Teiid and iguanid lizards or dendrobatid and hylid frogs, respectively, are common examples of the two extremes in forage behavior; the intermediates between the two extremes have not been carefully studied in amphibians and reptiles.

The forage behavior and preferred food items are strongly associated with a broad repertoire of other life processes such as thermal biology, home range/activity area size, daily activity profiles, aerobic scope, escape/predator avoidance behavior, intraspecific interactions, body form and feeding-apparatus morphology, and habitat preferences. A few general examples can illustrate this association but not the complexities of the association. The sit and wait *Sceloporus* (iguanid) has a smaller home range, long periods of daily activity, reduced aerobic scope (i.e., tires rapidly), and moderate hindlimb lengths, uses immobility and/or short dash to escape predators, and actively defends a home range from conspecifics. The motile-search *Cnemidophorus* (teiid) characteristically has a large activity area, shorter activity periods, high aerobic scope, long hindlimbs, moves rapidly and over long distances to avoid predators, and may or may not defend its activity area.

Such associations of form and function make lizards and salamanders excellent organisms to test hypotheses developed from foraging theory (= optimal foraging or feeding theory). Foraging theory examines the cost/benefit ratio among the choices available to an animal searching for, capturing, and processing food. Costs and benefits apply equally well to carnivores, omnivores, and herbivores and address all aspects of prey selection—why animals eat what they do. Tests of the hypotheses examine the actual metabolic cost of procuring and processing food to the actual energy gained from the food (digestion and assimilation yields).

Mechanisms for Location and Identification of Prey

Food location, identification, and procurement depend on all the senses of an animal, although usually one or two senses are the dominant participants in foraging. The following sections are brief introductions into the form and use of the various senses. The general anatomy of the sense organs is discussed in Chapters 1 and 3.

Vision

Many amphibians and reptiles are visual predators, generally relying on the movement of prey to alert them to its presence. Neurophysiological studies of the anuran eye show that prey recognition derives from four aspects of a visual image: perception of sharp edges, movement of the edges, dimming of images, and curvature of the edges of dark images. Perception is greatest when the object image is smaller than the visual field; then anurans can determine the speed, direction of movement, and relative distance of the prey. Success in capture by visual predators also appears dependent on binocular perception, because most align their head or entire body axis with the prey before beginning capture behavior. The importance of binocular aiming of the head/capture apparatus is demonstrated by chameleon search and capture behavior. The eyes of chameleons are independently movable, yet when one eye sights a prey, the head turns to allow both eyes to focus on the prey prior to releasing its projectile tongue.

The importance of vision in prey capture is also apparent from the number of species, both diurnal and nocturnal predators, with large, well-developed eyes. The noctural predators must locate prey under low light conditions and require maximum light entry into the eye for perception. The elliptical pupil, horizontal or vertical, is an adaptation that allows the maximum movement of the iris and the greatest dilation of the pupil.

Boids and viperids are able to "see" endothermic prey by the prey's infrared radiation. This long-wavelength light is sensed by trigeminal-innervated bare nerve endings in the skin of the head. Thermal receptors are localized into special organs, a pair of bilateral loreal or pit organs in crotalines (Fig. 5.2) and numerous bilaterally paired labial pits in some boids and pythonids. The pits open (face) anteriorly and provide a binocular perception field at least in the crotalines. The pits appear to be mainly short-range detectors and important in aiming the snake's strike during nocturnal foraging. Thermal receptors also occur in viperines and some boids and allow these snakes to orient toward heat targets, although the receptors are not organized in specialized structures.

Chemosensory Detection

Amphibians and reptiles possess three chemical senses: olfaction, vomerolfaction, and taste (gustation). The first two are used in prey location and identification, olfaction via airborne odors and vomerolfaction via surface odors (vomodors). The olfactory epithelium in the nasal chamber is sensitive to volatile compounds carried by the air and inspired with respiratory air or "sniffing" by rapid buccal or gular pumping. The vomeronasal (Jacobson's) organ is especially sensitive to high-molecular-weight compounds and usually requires the snout or tongue to touch a surface for their transport into the oral or nasal cavity. Olfaction acts mainly in long-distance detection, for example, the presence of food and its general location, and vomerolfaction acts as a short-range identifer, for example, specific identity of the food and its potential edibility.

FIGURE 5.2 ▼▲▼
The head of the pit viper *Crotalus molossus* (black-tailed rattlesnake) showing the opening to the infrared detector. (G. R. Zug)

Olfaction and vomerolfaction acts have long been recognized as feeding senses in salamanders, noniguanian lizards, and snakes, and used often in conjunction with vision. Motile-search predators, such as skinks, use vision as they move across open areas but depend on olfaction to locate prey in dark crevices or buried in the soil. Likewise, many salamanders probably alternate between visual and vomerolfactory searching depending on the availability of light and crypsis of the prey. Iguanian lizards and most anurans are highly visual predators and most lack well-developed olfactory-vomerolfaction structures. Historically, turtles and crocodilians were considered to be visual foragers; however, both groups produce scent for individual and species recognition and would seem capable of locating prey via odor/vomerodor. Recent experiments have shown that the American alligator can locate visually hidden food both in the water and on land.

Taste is a chemosensory sense but not used to locate prey. Taste combined with the tactile sense organs of the oral epithelium serve to identify food items once in the mouth and permit rapid acceptance or rejection. Items may be rejected because of taste or mechanical (spines) stimulation of the tactile sense.

Hearing

Using airborne sound to locate prey may occur widely in amphibians and reptiles, but it remains largely undocumented. The observations are mostly anecdotal, such as *Bufo marinus* orienting and moving toward a calling *Physalaemus pustulosus*. A recent field experiment showed that the gecko *Hemidactylus turcicus* locates calling male crickets and preys on female crickets coming to the male.

For some amphibians and reptiles, sensitivity to substrate vibrations or seismic sounds is likely a major prey detection mechanism. Seismic sensitivity may be particularly important for fossorial (burrowing) species or those with fossorial ancestors, for both the avoidance of predators and the location of prey. Neither snakes, salamanders, nor caecilians have external ears, so they probably possess a high sensitivity to seismic vibration, although actual tests are lacking for most species. Uniquely, both frogs and salamanders possess a special pathway (opercularis system) for the transmission of vibrations from the substrate to the inner ear, and the limited data indicate that salamanders are twice as sensitive as frogs. The opercularis system links the forelimb to the inner ear through the opercularis muscle that extends from the scapula to the opercular bone lying in the fenestra ovalis of the otic capsule. The muscle acts like a lever arm; vibrations received by the forelimb rock the tensed muscle thereby pushing/pulling the operculum and creating fluid movement in the otic capsule. These seismic vibrations are of low frequency, typically less than 200 Hz, and stimulate the neuroreceptors in the sacculus and lagena rather than those of either the basilar or amphibian papillae, although the latter may be stimulated by frequencies as low as 100 Hz. These low frequencies are made by such activities as the digging of insect prey or mammalian predators. In snakes, seismic vibrations appear to be transmitted via the lower jaw through the quadrate-columella to the inner ear. Snakes also detect seismic vibrations through mechanoreceptors in the skin, although not with the same fine-scaled resolution as with the ear. Other fossorial groups (e.g., caecilians and amphisbaenians) likely use mechanoreceptors for detection of seismic vibrations.

Tactile

The role of the tactile sense in amphibians and reptiles is poorly studied. Once located, prey can be identified by touch (mechanoreceptors in the skin). Aquatic amphibians use the lateral line, a string of mechanoreceptors, to sense changes in water pressure reflecting from stationary or motile objects in the near vicinity of the predator, hence permitting the identification and location of prey. This recognition would certainly be enhanced by a weak electric field (see discussion of lateral line in Nerves and Sense Organs, Chapter 1). Preliminary evidence from aquatic salamanders indicates that prey identification and size determination occur solely by the lateral line system.

Mechanisms of Prey Capture and Ingestion

Once a food item has been located and identified, the animal must procure it. For herbivores, this is simple because plant matter is stationary and the animal approaches the leaf or fruit and bites it. For animal prey, the eater must first catch the prey. Numerous behavioral and morphological adaptations are associated with catching and subduing prey. In catching motile prey, motor and sensory units are finely coordinated to intercept the moving prey, and usually the strike/catch mechanism aims at the center of mass/gravity of the prey. The center of gravity is the most stable part of the target and has the least amount of movement.

Bite and Grasp

Biting and grasping is the main capture mechanism for amphibians and reptiles. It is the simplest of mechanisms, requiring only bringing the head near the food item and a gape large enough to encompass part of the item. The head reaches the food either by movement of the entire body to the prey or by a strike, that is, rapid movement of head and neck from a stationary body. Long, flexible-necked animals (turtles, varanids) and elongate, limbless ones (amphiumas, pygopods, snakes) can and regularly do use the strike mechanism, often from ambush but also following a slow stalk of the prey. In both strikes and bites, the mouth commonly does not open until the head moves toward contact with the prey, and the bite/strike is an integrated behavior of motor and sensory units. When the open mouth encompasses the food, the tactile/pressure on teeth and oral epithelium triggers the rapid closure of the mouth. A few animals use lures to attract the prey to the predator, for example, caudal luring in some juvenile viperids, pedal luring in a few *Ceratophrys* frogs, and lingual-appendage luring in alligator snapping turtles (*Macroclemys*).

There is minimal food processing in the mouth. Teeth may crush or perforate food items, which are commonly swallowed whole. Fragmentation of food is limited to the herbivores that bite off pieces of foliage, and large lizards, turtles, and crocodilians that may use a combination of sharp jaw sheaths or teeth and limb/body movements to break up large items. Turtles have ever-growing keratinous sheaths on upper and lower jaws; each sheath provides a continuous bladelike labial surface to cut food. Tooth structure in amphibians and diapsid reptiles is highly variable, ranging from simple conelike teeth to molarlike teeth or bladelike teeth with serrated edges. Specialized diets usually are associated with specialized teeth: broad and sturdy for crushing mollusks; bladelike for cutting vegetation or fragmenting large prey; long recurved teeth for feathered prey; hinged teeth for capturing skinks.

Once captured, the food must be moved through the oral cavity into the esophagus. Several "swallowing" mechanisms are recognized in amphibians and reptiles. Inertial swallowing is mechanically the simplest and widespread in reptiles. It relies on inertia and moves the head/body over the food. The food is held stationary, the bite-grasp is quickly and slightly opened, and the head is thrust forward thereby shifting the head forward over the food. Snakes swallow by moving the head over the prey and use a rachetlike mechanism. The left and right sides of the head alternately pull the prey inward by the movement of the palatoquadrate-mandibular skeletal complex. The prey is held secure by this complex on one side of the head, while the bite-grasp of the opposite side relaxes, shifts forward, and then contracts. The alternate forward movement of the left and right sides moves the head and body over the prey.

Tongue and hyoid manipulation appears to be the principal swallowing mechanisms in amphibians. Some salamanders use hyoid-tongue retraction to swallow prey. Prior to opening the mouth, the tongue presses the prey tightly against the roof and vomerine-palatine teeth, the mouth opens quickly, and the tongue still firmly holding the prey retracts and draws the prey rearward as the mouth slowly closes. This cycle is repeated until the prey exits the buccopharyngeal cavity. Swallowing in frogs also involves tongue-hyoid movement, although the actual mechanics are known in less detail. Frogs have voluminous oral cavities and captured prey are usually completely engulfed. Apparently a similar hyoid-tongue retraction cycle without opening the mouth moves the prey rearward. This movement is assisted by the compression of the palate, visible externally by the retraction of the eyes.

Constriction

Constriction is a specialized bite and grasp technique used by numerous snakes to subdue prey. A constricting snake strikes its prey, and if its bite-grip is secure, a loop of the body is thrown on and around the prey. Then additional loops (coils) of the body encircle the prey with continual adjustment to reduce overlapping loops. As the prey struggles, it relaxes a portion of its body, only to have the encircling loop tighten in that area. The tightening continues and suffocation occurs rapidly in endothermic prey, less rapidly in ectothermic ones. When struggling ceases and the prey is dead or unconscious, the snake relaxes its coils, locates the head of the prey and begins to swallow it. Some other limbless amphibians and reptiles may use constriction to subdue prey; there are such anecdotal reports for *Amphiuma*.

Bite-and-Venom Injection

Are fangs and venom glands structures for defense or feeding? Though effective in defense, most evidence suggests that their origin and development was for prey capture. The independent origins of venom injection systems in several distinct lineages of snakes and the high specificity of venom for prey type emphasize their origins for feeding. All members of the Helodermatidae, Elapidae, and Viperidae are venomous as are several groups of colubrids. Venom subdues the prey either by anesthetizing or killing it. A nonstruggling prey is much safer and less energy demanding to capture and swallow than a struggling one. Also a predator can eat larger prey if they don't resist capture and consumption. Many of the viperids add a third benefit to the injection of venom by including proteolytic components, which speed digestion (at least in colder areas) by initiating tissue lysis internally in the prey.

A venom delivery system contains three items: a gland to produce the venom; a duct to transport venom from the gland to the injection system; and fangs (modified teeth with open or closed canals) to inject the venom into the prey. The fangs of helodermatids and most venomous colubrids bear a single groove on one face of each enlarged tooth, whereas the fangs of elapids and viperids have closed canals. Venom is produced continuously in the venom gland and stored in a venom-gland chamber. When the predator bites a prey, the muscles over or around the gland contract and squeeze a portion of the venom down the venom duct and into the fang canal. Snakes regulate the venom dose depending upon the size of the prey and possibly how much venom is available. Viperids and some elapids strike, bite, inject venom, and release the prey; whereas most elapids and colubrids, and *Heloderma,* maintain their bite-grasp and chew the wound to ensure the deep penetration of the venom.

The venom of each species is a composite of several compounds (Table 5.2) that work synergistically to subdue the prey. Typically, the venom of a species emphasizes either tissue destruction or neurological collapse. The tissue-destruction venoms subdue the prey because the prey goes into shock. Neurological-collapse prevents nerve impulse transmission and interrupts all motor activity including respiration. The immobile prey can then be eaten safely.

Projectile Tongues

Tongues are small and usually have no or limited mobility in aquatic animals. Tongues became important in terrestrial animals when water was no longer present to carry food through the oral cavity into the esophagus (see preceding Bite-and-Grasp section). A protrusible tongue for sampling the environment and gathering food probably evolved early in terrestrial tetrapods, because protrusion is widespread in amphibians and reptiles. Many bite and grasp feeders (herbivores and carnivores) use their tongues to retrieve small items. The tongue is simply extended through the mouth and the item is touched by the tip or dorsal surface of the tongue. The item is held by sticky saliva and the tongue is retracted. The most dramatic tongue protrusions are the projectile tongues, and they have evolved independently six or more times in amphibians and reptiles.

Most frogs use a flick-tongue mechanism. The tongue is attached at the front of mouth, directly to the cartilage symphysis joining the right and left sides of the mandible. When prey is identified, the frog orients itself toward the prey. The mouth opens as the lower jaw drops downward, and simultaneously the submentalis muscle (linking left and right mandibles beneath the middle of the tongue) contracts and yanks the symphyseal cartilage downward. This movement

TABLE 5.2 ▼▲▼

Major Components of Reptilian Venoms with Some Examples of Each Component

Enzymes

All venoms contain several different enzymes; more than 25 enzymes occur in reptilian venoms.

Proteolytic enzymes	digest tissue protein and peptides causing hemorrhagic necrosis and muscle lysis; also known as endopeptidases. [common in crotalines, less in viperines, absent in elapids]
Thrombinlike enzymes	interfere with normal blood clotting, by acting as either an anticoagulant or procoagulant. [common in viperids, rare in elapids]
Hyaluronidase	breaks down mucopolysaccharide links in connecting tissue and enhances diffusion of venom. [in all venomous snakes]
Phospholipase A	modifies muscle contractability and makes structural changes in the central nervous system by interfering with prey's motor functions. [common in colubrids, elapids, viperids]
Acetylcholinase	interrupts ganglionic and neuromuscular transmission and eventually affects cardiac function and respiration. [common in elapids, absent in viperids]

Polypeptides

The polypeptides are toxic nonenzymatic proteins of venoms. These toxins commonly act at or near the synaptic junctions and retard, modify, or stop nerve impulse transmission.

Crotactin	produces paralysis and respiratory distress. [in rattlesnakes, crotalines]
Cobrotoxin	acts directly on heart muscle to cause paralysis. [in cobras, *Naja*]
Viperatoxin	acts on medullary center in brain, resulting in vasodilation and cardiac failure. [in *Vipera*]

Miscellanea

Various ions and compounds in venoms but as yet with no recognizable prey-type or taxonomic group association.

Inorganic ions	sodium, calcium, potassium, iron, zinc, and others; some enhance the activity of specific enzymes.
Glycoproteins	anticomplementary reactions that suppress normal immunological tissue response.
Amino acids and biogenic amines	

Note: Reptilian venoms are an admixture, consisting mainly of enzymatic and nonenzymatic proteins.

pulls the anterior end of the tongue downward and momentum imparted to the tongue flicks the posterior end outward. The weight of the tongue's posterior half (now the projectile tip) stretches the tongue to twice it length, and as the upper surface of the tongue hits the prey, the tip wraps over the prey. A quick retraction with the prey glued to the tip brings the prey into the mouth (Fig. 5.3).

Projectile mechanisms in salamanders derive from modifications of the hyoid apparatus. Projectile tongues appear to have evolved three times in the salamanders. The general mechanism is the presence of a pedestallike tongue tip that is driven out of the mouth by the movement of a median hyoid rod. The posterior, bilaterally paired hyoid arms lie in the floor of mouth like a partially opened fan with the hinge-tip pointed anteriorly. When the hyoid muscles contract, the fan closes and drives the tip outward. The movement is rapid and the momentum, as in frogs, assists in stretching the tongue as much as 40–80% of the salamander's body length. The projectile tongue of chameleons also shoots forward by a hyoid mechanism.

Filter Feeding

The diets of most tadpoles consist mainly of algae and bacteria, hence tadpoles are microphagous ("small eating"). Microphagy requires a filter or straining mechanism to capture tiny items and direct them into the gut. Tadpoles use the movement of water in through the mouth, buccopharyngeal cavity, and out through the gills (branchial arches) for both respiration and food entrapment. The internal gill slits serve as a mechanical filter and prevent the passage of items larger than 5–10 μm. Strings or a web of mucus captures the smaller particles (2–5 μm).

The buccopharyngeal cavity of tadpoles is large, more than half the volume of the head in most tadpoles. The upward and downward movement of the buccal floor in association with opening and closing of the mouth and gill filter valves (vela) moves water through this large cavity. As the mouth opens, the floor drops and draws water into the cavity; the vela prevent a major backflow through the gill openings. The mouth then closes and the floor rises, forcing the water outward through the gill slits. The flow of water brings the food particles to the rear of the cavity and in contact with the gill filter surfaces. The large particles cannot pass through the filter and water flow bounces them rearward toward the esophagus. The smaller particles touching the surface are snared by strings of mucus. A combination of water movement and ciliary activity moves the strings and trapped food rearward. The strings aggregate into larger clumps before passing into the esophagus with the larger food particles. The rate of filtration is regulated by the volume of food entering the mouth cavity. When particle suspension density is high, the buccal pump works slower to prevent the gill filters and mucous traps from clogging, and conversely if particles are sparse.

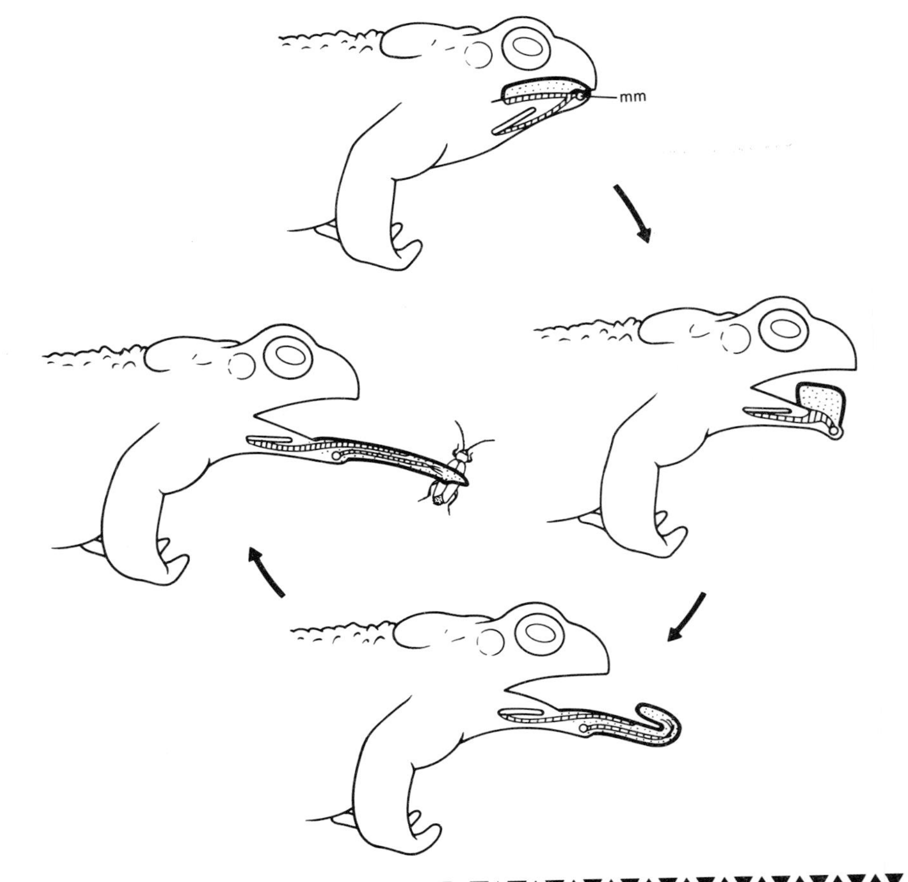

FIGURE 5.3 ▼▲▼

The anatomical mechanics of an anuran projectile tongue (*Bufo marinus*, marine toad). The four schematic stages show the projection sequence from tongue at rest on the floor of the oral cavity (top) to its full extension and capture of an insect (left). Five anatomical features are highlighted: soft tissue of tongue (stippled); two muscles (barred), genioglossus from hyoid to base of tongue and hyoglossus from mentomeckelian element (mm) to base of tongue; two skeletal elements (white), hyoid horn lying below tongue and mm at tip of jaw. Projection begins (right) with the mouth opening, the mm snaps downward by the contraction of a transverse mandibular muscle (not shown), and the genioglossus contracts to stiffen the tongue. The tongue flips forward (bottom) from the momentum generated by the downward snap of the mm and the genioglossus contraction; the two tongue muscles now relax and are stretched. The tongue is fully extended and turned upside down (left) and the dorsal surface of tip encircles the prey. The genioglossus and hyoglossus muscles contract drawing the tongue with the adhering insect back through the mouth as it closes. [Adapted from Gans and Gorniak (1982). Reprinted with permission, from Wiley-Liss. A Division of John Wiley and Sons, Inc.]

Suction Feeding

Aquatic salamanders, pipid frogs, and some turtles capture prey by opening their mouths simultaneously with the enlargement of their buccal (mouth) cavity. The prey is literally sucked into the mouth by the rush of water flowing into the reduced-pressure cavity created by the enlarged buccal cavity. The matamata turtle, *Chelus fimbriatus*, offers the most vivid demonstration of suction (gape and suck) feeding. Either from ambush or by slow stalking the matamata moves its head so that it is aligned with the prey, often a fish or tadpole. The head shoots forward, simultaneously the hyoid musculature contracts, dropping the floor of the buccal cavity. With the valvular nostrils closed, a tremendous suction results from the three to four times increase in volume of the buccal cavity. Just prior to reaching the prey, the mouth opens and prey and water surge into the buccal cavity. The mouth is shut, but not tightly; the buccal floor rises, expelling the excess water without losing the prey. The success of this prey capture technique depends on accurate alignment of the head to the prey, good timing, and rapid enlargement of the buccal cavity.

The hellbender, *Cryptobranchus alleganiensis*, can capture prey alongside its head and not just straight ahead. This primitive salamander is capable of asymmetrical movements of its lower jaw and hyoid apparatus, which allow it to open its mouth on only one side. The key feature is the ligamentous attachment of the left and right dentaries at the front of the mouth. The flexible attachment permits one side of the jaw to remain in place while the opposite side swings downward accompanied by a unilateral depression of the hyoid apparatus, yielding asymmetrical suction.

Digestion and Assimilation

General Digestive Physiology and Structure

The digestive tract from mouth to anus is a tube with variously modified walls for processing and transporting the ingested food (digesta). The buccal cavity with teeth and tongue serves mainly for capture, manipulation, and initial retention of prey (see preceding section). Fragmentation or perforation by the teeth and coating with mucus and oral gland secretions are minor but beginning steps in the breakdown of the food items. Oral glands are present in nonaquatic amphibians and reptiles; in amphibians, they are mainly mucous glands within the oral epithelium. These glands also occur in reptiles; however, glandular tissue has become concentrated in salivary and venom glands (Fig. 5.4). Reptilian salivary glands secrete principally mucus and minor amounts of enzymes (systematic enzymatic surveys are largely lacking for amphibians and reptiles). Venom and Duvernoy's glands are modified salivary glands and produce larger quantities of proteolytic enzymes and other compounds.

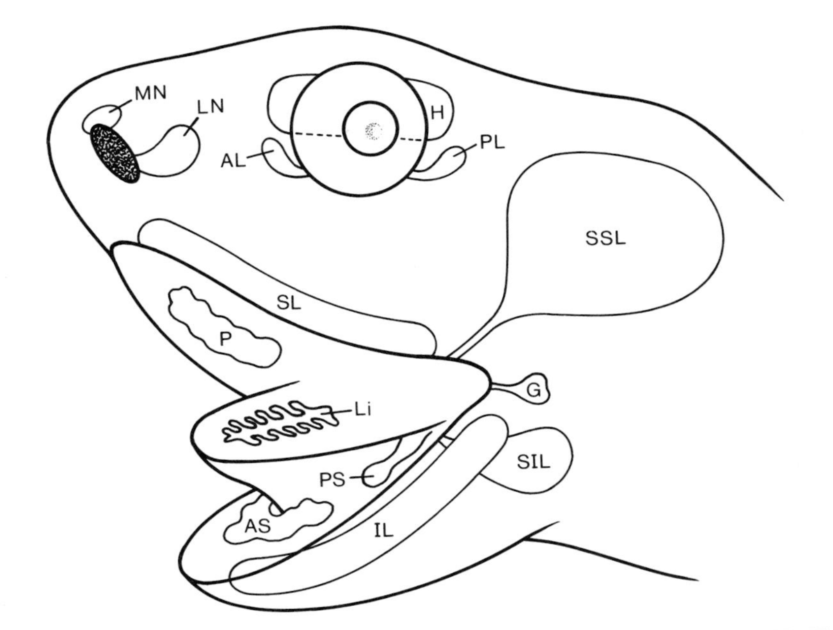

FIGURE 5.4 Common cephalic glands of reptiles; schematic reptilian head showing relative position of cephalic glands. No reptile possesses all the glands illustrated. Abbreviations: Nasal glands—LN, lateral nasal; MN, medial nasal. Orbital glands—AL, anterior lachrymal; H, Harderian; PL, posterior lachrymal. Salivary/oral glands—AS, anterior sublingual; G, glandula anguli oris (posterior); IL, infralabial; Li, lingual; P, palatine; PS, posterior sublingual; SIL, specialized infralabial (mandibular); SL, supralabial; SSL, specialized supralabial (Duvernoy's or venom glands). [Adapted from Kochva (1978) and Saint Girons (1988).]

The esophagus is little more than a passage from mouth cavity to stomach; any digestive activity therein is minor and infrequent. It is of various lengths and can serve as a temporary storage area for food items too large or too numerous to fit in the stomach. It is highly extensible as are most subsequent portions of the digestive tube. In some species, the esophagus moves food by peristalsis; in others, a combination of gravity and extrinsic body movement move food rearward. The stomach is the first digestive area serving principally for the mechanical reduction of food items and the initial cellular and chemical disruption of the digesta. The stomach's muscular walls are lined with glandular epithelia, anteriorly secreting hydrochloric acid and pepsinogen and posteriorly secreting mucus (also chitinase in some insectivorous taxa). The HCl and pepsinogen begin the mechanical breakdown of the digesta as well as killing prey swallowed alive and arresting putrefaction. Contraction of the muscular stomach wall aids the mechanical fragmentation of the prey. Crocodilians commonly have hard objects (e.g., stones, pine knots) in their stomach; these objects probably assist in the fragmentation of digesta in a manner analogous to the gizzard stones of feathered reptiles. The pyloric valve between the stomach and intestine of amphibians and reptiles is variously developed but does appear to regulate the movement of digesta rearward, permitting the passage of fluids and small fragments and retaining large fragments in the stomach for further processing.

Most digestion (alteration of chemical structure) and absorption occur in the intestine. The small intestine has numerous glands in the wall and receives secretions via ducts from the liver-gallbladder and pancreas. Bile and pancreatic secretions play the major role in digestion. Bile contains phospholipids of critical importance in the digestion and absorption of fats by the conversion of triglycerides into fatty acids and monoglycerides, which can be metabolized directly. Pancreatic secretions contain a wide spectrum of enzymes capable of breaking down proteins, carbohydrates, and fatty acids into compounds that are readily absorbed and assimilated. Most of these enzymatic activities and absorption of the useable compounds occur within the small intestine; however, the multitude of amphibian and reptilian diets produces numerous differences in general digestive area (stomach-intestine) morphology and regional specializations. For example, carnivores typically have long small intestines and short large intestines; herbivores possess the converse morphology. In herbivores, the inner walls of the intestine are highly folded to increase surface area and enhance absorption. To further slow the passage of the digesta and assist absorption of fermentation products, the colon of iguanine herbivores has a series of baffles. Some reptilian herbivores and omnivores have a caecum at the junction of the small and large intestine, where digesta is retained to increase its digestion time.

Digestive physiology and biochemistry have been examined for only a few species of amphibians and reptiles, for the alligator (*Alligator mississippiensis*) most intensely. Likely, many features of the alligator digestive physiology will be applicable to many other carnivorous reptiles. Crocodilians are strict carnivores and are totally unable to digest plant proteins and all carbohydrates. Experiments with a full spectrum of vegetable proteins and polysaccharides found that these proteins passed through the digestive tract undigested; of the carbohydrates, only glucose was absorbed. Although fat digestion has not been tested, feces are nearly fat-free even when high-fat foods are eaten. Bone and animal protein are also readily reduced and absorbed.

Digestive efficiency varies with the nature of the food item, and usually a taxon's preferred food will be the type processed most efficiently. Other factors also influence digestive efficiency in amphibians and reptiles, most notably rate of consumption, which is temperature dependent. Digestive efficiency (digestion coefficient, DC) measures the number of calories removed from the food relative to the number of calories consumed, that is, DC = (consumed − defecated) / (consumed). At low temperatures, ectothermic amphibians have very low digestive efficiencies; in general, efficiency increases with rising temperature and is maximal at or above the taxon's preferred temperature range. Passage or clearance time (interval for passage of digesta from consumption to defecation) also varies directly with temperature; however, the association between passage and efficiency need not be positive. A higher rate of food consumption usually causes a shorter passage time, and the passage rate can exceed the time necessary for maximum digestion and absorption. Temperature also affects the assimilation efficiency (energy absorbed and converted to animal tissue relative to the energy consumed) of ectotherms. Assimilation efficiency increases with temperature to a point where the animal's metabolism and activity use more energy than is absorbed.

Amphibians and reptiles are noted for their ability to survive for long periods without eating. Their low energy requirements relative to endotherms are unquestionably a major factor in their resistance to starvation, as is their ability to reduce their metabolism even lower than normal. The actual physiological mechanics of this resistance have been examined in only a few species (e.g., *Alligator*, *Amphiuma*, *Phrynops*), although it is likely similar in all amphibians and reptiles. When starved, these animals show a strong reduction in the synthesis of metabolic reserves and a decrease in insulin levels in the blood. This response, similar to the mammalian one, effectively increases the glycogen stores in the liver and muscle, and begins as soon as an individual experiences a drop in energy intake.

Digestion of Plant Materials

Herbivory poses a digestive problem for vertebrates. The major component of all vascular plants is the cell-wall fiber, and no vertebrate produces cellulase to break down cellulose. Thus, vertebrate herbivores must depend on the presence of a gut microflora of cellulolytic bacteria. Without such a microflora, it is doubtful that a reptile could eat and process enough plant matter to survive on a strictly herbivorous diet.

To maintain an efficient gut microflora, a constant and elevated body temperature appears necessary. Other requirements are a constant food supply, slow passage of food items to permit adequate time for bacterial degradation, an anaerobic gut environment, regulation of gut pH, and removal of fermentation waste by-products. Lowland tropical reptiles feed year-round and maintain fairly high and constant body temperatures; for such reptiles, once a cellulolytic microflora is obtained, it is improbable that the microflora would need to be renewed. Such microflora stability is less certain for temperate-zone reptiles owing to the low temperatures of the body core and possible absence of a food bolus during dormancy. Low temperature and/or the usual purging of the digestive tract prior to hibernation or estivation might well eliminate a specialized microflora. Only a single temperate species, the gopher tortoise (*Gopherus polyphemus*), has been closely examined and it efficiently digests a high-fiber diet and effectively absorbs the nutrients generated by the bacterial fermentation in the hindgut. It either retains a microflora bolus or restores its microflora each spring.

The importance of the gut microflora is emphasized by its absence in the herbivorous Aldabran tortoises (Fig. 5.5) and their low digestive efficiency (30%) in contrast to its presence and about 65% DC for the redfooted tortoises and 85% DC for the green iguana. How and when gut microflora is acquired are unknown except for *Iguana iguana*. There are suppositions for other reptilian herbivores but no direct observations. *Gopherus polyphemus* defecates within its burrows and presumably eats some of its feces prior to emerging in the spring. But where do the juvenile gopher tortoises obtain their fiber-digesting microfauna and for that matter the young of all other reptilian herbivores? In mammalian herbivores, gut microflora acquisition poses no problem, because the young and the parents are closely associated from birth through weaning. The mammalian mother regularly licks the young and the young feeds from the mother's mammary glands, so young mammals acquire the microflora early from the ingestion of the mother's saliva or fecal material. This close association of mother and offspring does not exist for any reptilian herbivore. For *Iguana iguana*, a complex behavioral mechanism has evolved to ensure the acquisition of plant-digesting microbes. The hatchlings eat soil before emerging from the nest cavity and continue to do so after emergence as they begin to feed regularly on plants. After a few days, the young iguanas move from the low shrubbery around the

FIGURE 5.5 Juvenile Aldabran tortoises (*Geochelone gigantea*) eating a leaf from their shade tree. (G. R. Zug)

nesting area upward into the canopy and join the older juveniles and/or adults; here they consume the feces of their seniors, and this inoculate ensures the presence of the correct microflora in their guts.

ENERGETICS

The total energy demands of most amphibians and reptiles are low, although not insignificant (i.e, based on the few examples studied in sufficient detail to estimate an annual energy budget). Standard or basal metabolic rate is easily measured by confining an animal in a chamber and recording its oxygen consumption (O_2 ml g^{-1} hr^{-1}; volume of oxygen per unit of body mass per unit of time) or carbon dioxide expired. Respiration serves as the measure of the body's energy utilization because all life processes run on cellular oxidation. Standard metabolic rates do not exceed 0.5 O_2 ml g^{-1} hr^{-1} for amphibians and reptiles and are usually less than 0.25 O_2 ml g^{-1} hr^{-1}. These rates are 10–20% of those of similar-sized endothermic vertebrates. Because amphibians and reptiles are ectothermic and body temperatures fluctuate depending on an individual's activities and ambient temperature, the rate also fluctuates directly with body temperature. The rates are also a function of body mass; smaller animals usually have higher rates per unit body mass than larger ones. This relationship is expressed by the power or allometric equation $MR = am^b$; MR, standard metabolic rate (ml O_2 hr^{-1}); *a*, slope of the regression line; *m*, body mass (g); *b*, exponent (0.60–0.85, varies with temperature and taxonomic group).

The standard metabolic rate represents only the energy expended by an animal to maintain its baseline physiological activities at rest. Any activity immediately elevates the metabolic rate, and an all-out effort to escape a predator or subdue a prey may increase the metabolic rate five to eight times. The difference between the resting or standard rate and maximum O_2 consumption at peak sustainable exertion is an animal's aerobic metabolic scope and represents a measure of an animal's ability to perform work (its activities) without metabolically borrowing from cellular energy reserves (anaerobic metabolism), which results in the accumulation of metabolic end products, the depletion of energy reserves, and fatigue. Aerobic scope is quite variable among species and is intimately associated with a species' locomotor and behavioral characteristics. Sit and wait predators have lower aerobic scopes than motile-search foragers (iguanid versus teiid lizards) and strong jumping frogs tire quickly compared to hopping species (ranids versus bufonids).

This aspect of energetics provides one level of insight into an animal's or species' energy needs, although only for a moment in time. A broader perspective appears from the development of an individual's or species' energy budget; that is, how many calories are consumed and how are they used (measured energy/work units, joules)? Energy budgets (Table 5.3) identify the energy resource allotments for an entire season, year, or an animal's entire life, and are expressed by the descriptive equation $C = P + R + U + F$, where C is the food (energy) consumed; P and R are the amounts of energy assimilated and used in production (P) and maintenance (R, respiration); and U (excreta) and F (egesta) are the portions of the food processed but without extraction of energy and discharged by the urinary and digestive system. The P and R categories overlap in a living organism, but for the purposes of measurement, they are mutually exclusive categories with growth, gamete production, and energy storage as production energy utilization, and resting and activity metabolism for maintenance. While conceptually simple, energy budgets are still difficult to construct because they require a knowledge of the full life history and physiology of a species and the actual consumption and expenditures of an individual for its entire life or even an entire season. Thus, there are few energy budgets for amphibians and reptiles, though due to a growing recognition of their importance more are being compiled. The example for a single individual turtle (Table 5.3) offers a glimpse at the division of and total energy utilization during the turtle's life. Annual energy use increases as the animal grows larger and requires more energy to maintain its increasing body mass, although the percentage of energy used for maintenance remains nearly constant. On the production side, growth dominates energy expenditure until the turtle (female) reaches sexual maturity; then egg production consumes a significant portion. If the turtle had been a male, the reproductive cost (production of gametes) would not have been as great, but the activity-maintenance expenditure would have likely been greater, on account of the male's

TABLE 5.3 ▼▲▼
Example of an Annual Energy Budget for the Painted Turtle (*Chrysemys picta*)

		Percent utilization				
		Maintenance		Production		
Year of life	Energy used	Rest	Active	Growth	Eggs	Storage
1	96.2	49	44	6	0	1
3	452.6	48	43	8	0	1
5	818.3	51	45	3	0	1
7	1177.9	51	46	2	0	1
8	1591.3	43	38	2	15	2
10	1823.3	43	40	1	14	2
12	1908.9	44	41	<1	14	0
14	1996.9	45	40	<1	14	0

[Modified from Congdon et al. (1982).]

Note: Energy used (kJ, kilojoules) is the total amount metabolized, not the total amount consumed; table condensed by deleting alternate years.

search for females and other similar reproductive costs that cannot be measured directly (as can be the caloric content of the gametes).

The importance of energy budgets lies in examining how and why individuals divide and use their energy resources. This information allows us to address the evolutionary implications of behavior, physiology, ecology, and other life processes by viewing the potential energy trade-offs or cost/benefits of one adaptive "choice" or another. Foraging theory uses the energetic data to study why and how animals select their food items by using a cost/benefit analysis of feeding. The cost is the energy expended to locate, gather, and process food items, and the benefits are the energy acquired to grow, mature, and reproduce. The "decision" that must be made at the beginning of each feeding bout is whether the energy gained will exceed the energy expended. The decisions are not conscious ones, but ones honed through time by evolution, where the individuals making better choices more frequently than poor ones have a higher survival rate and produce more offspring.

Ectotherms seem better able to conserve or to shift energy utilization within maintenance and production categories than can endotherms. When food resources are low, a juvenile reptile or amphibian can suspend growth, use available energy for maintenance, and later reinitiate growth without structural impairment, and the individual can do this many times without detrimental results. Low energy demands permit ectotherms to rapidly store excess energy (fat reserves) during peak food abundance; these reserves enhance survival during nonfeeding periods and increase reproductive output.

CHAPTER 6

Defense and Escape

To reproduce, an animal must survive a gauntlet of life-threatening situations from birth to maturity. Many, such as starvation or drowning, are unpredictable owing to the vagaries of weather and other abiotic events. Predation is a constant threat, whether an animal is at rest or active, and has been the selective force driving the evolution of many adaptations. The prey that successfully avoid predation are the ones capable of reproducing and placing their adaptation in the next generation.

FOES AND RUSES

In the preceding chapter, successful foraging behavior was recognized as four sequential actions: searching; location and identification; capturing and swallowing;

digestion and assimilation. The goal of prey organisms is to avoid the third phase, and selection has favored those adaptations that remove an individual prey early from the forage sequence, particularly prior to contact by a predator. Once again, energetics are important. The best defense costs the least energy; struggling with a predator only to escape with an injury is less effective than avoiding the predator. In an evolutionary sense, each animal strives to maximize its energy intake without increasing its likelihood of detection and capture by a predator. Within most species, a set of antipredator devices operate at different stages of a predator's feeding sequence and against multiple predators with different feeding behaviors and sensory detection mechanisms.

At the first level of a predator's search, the prey seeks to escape detection. The best solution is complete avoidance by absence from the predator's search path. Many predators develop search images for common prey items. Prey can avoid this type of predator by having low densities or appearing to have low densities. Appearance of rarity results from altering activity cycles (daily and seasonal) to avoid predators, using different parts of the habitat, or possessing multiple appearances (polymorphism). Early detection of the predator is an effective means of avoiding predators; larval amphibians are alerted to a predator by the release of an "alarm" substance from a captured larva. Other devices depend on disrupting or confusing the predator's detection senses. Immobility is highly effective against predators that detect prey by movement. Crypsis and confusion also reduce the probability of detection. In the first, a prey may look, smell, or sound like some unedible item; in the second, detection of individual prey is reduced by distortion of observation, for example, erratic movements or schooling. These distortions also decrease the probability of capture.

Both crypsis and confusion also operate at the second level of defense—to avoid identification once detected. A predator may see a *Bufo typhonius*, but if the predator identifies the toad as a leaf, the defense was successful. Similarly, an experienced predator seeing an edible *Plethodon cinereus* and mistaking it for a toxic *Notophthalmus viridescens* has been fooled by a mimetic resemblance. The brightly colored and highly visible *Notophthalmus* efts advertise their presence and toxicity. Such advertisement (aposematism) of toxicity or distastefulness requires predators to learn the association of a color, color combination, behavior, or other sensory traits with the unsuitability of a potential prey as food.

Once recognized and identified as a potential food item, prey still have antipredator defenses to avoid capture. The most obvious is to escape before the predator is close enough to be caught. This action requires the prey to recognize that it has been selected as the predator's next meal. It is energetically wasteful for an individual to leave a feeding station if it is not "prey"; thus, it must balance escape reactions against the likelihood of capture. Such a balance occurs because many lizards modify their flight distance relative to their body temperature. At normal activity body temperatures, ectothermic prey allow predators to come closer before flight than when body temperatures are lower and their movements slower. Flight reactions are various, and all aim to remove the prey from the predator's perception. Flight may be short to a burrow or any other location inaccessible to the predator. Flight can be long, placing the prey at a location far from the predator, or it can be a short, erratic flight ending in immobility so that the predator loses track of the prey. Some prey allow predators to touch them or nearly so and depend on a bright flash of color or peculiar behavior to startle the predator and create a brief interval for escape.

Should the precontact defenses fail, prey still have options to avoid being eaten. They may be too strong to be restrained and cause predators to stop their attacks because the effort is too energy demanding and/or the struggle may injury the predator. Spines, mucus, or other surface defenses may prevent the predator from tightly holding or consuming the prey. Finally, if eaten, the prey may be emetic causing it to be vomited free or poisonous causing the predator either to die or become sick and remember to avoid that particular prey type in future encounters.

Defense options are numerous, and most species employ several at different stages in the predator's feeding cycle and as defense against multiple predators. Some of these defenses are highlighted in the following sections. The mechanisms are artificially divided into categories for description, but each is part of an integrated adaptive set involving gross and microscopic structures, behavior, physiology, and ecology.

Predators

Most amphibians and reptiles are small, hence are suitable prey to a broad assortment of invertebrate and vertebrate omnivores and carnivores. The variety of predators is so large that only a sampling of types is possible (Fig. 6.1).

The early lifestages of amphibians and reptiles are particularly susceptible to invertebrate predators. Tadpoles are regular prey for dragonfly larvae, diving beetles, water bugs, and other predaceous aquatic arthropods and their larvae. Salamander larvae experience the same intense pressure from these arthropod predators. Leeches and a few arthropods attack amphibian eggs, although fungus and bacteria may cause higher mortality. Reptile eggs also suffer fungal and bacterial attack, and as in amphibians, one or two rotten eggs in a clutch may serve as an infestation source to kill the entire clutch. Small terrestrial amphibians and reptiles are often captured by spiders, scorpions, centipedes, and other predaceous arthropods, especially those invertebrate predators that are as large as or larger than their vertebrate prey.

In permanent water habitats, fish are major predators of amphibian larvae, and field experiments demonstrate that the presence of fish changes the larvae's time and microhabitat of feeding. The changes in larval behavior are sufficient to affect their growth rates, hence size and age at metamorphosis. Adult amphibi-

FIGURE 6.1 ▼▲▼
Herpetological predation. A *Leptodeira* eating a *Hyla*. (K. Miyata)

ans and small aquatic reptiles are also captured by large predaceous fishes. Juvenile and adult frogs, salamanders, snakes, and turtles have been found in the stomachs of the largemouth bass, *Micropterus salmoides*.

Amphibians and reptiles are major predators of amphibians and reptiles. Salamander larvae are common predators on anuran and other salamander larvae. In eastern North America, *Ambystoma* and *Notophthalmus* larvae prey on the other amphibian larvae with which they share ponds. Their presence and densities affect the structure of the amphibian larval communities, greatly altering species densities. Only a few larval anurans are strict carnivores and even fewer prey on other tadpoles, for example, *Anotheca*, some *Ceratophrys*, *Lepidobatrachus*, and *Pyxicephalus*. From the same clutch of *Scaphiopus bombifrons* eggs, a few individuals may become megatadpoles with modified beaks and carnivorous habits; perhaps this morph is an adaptation to permit a few spadefoot tadpoles to metamorphose from rapidly drying pools.

Many adult amphibians prey heavily on other amphibians as well as on reptiles, birds, and mammals. Most are generalists, and when large enough, they eat anything smaller than themselves that passes within reach. *Amphiuma*, *Dicamptodon*, *Pyxicephalus adspersus*, *Rana catesbeiana*, and *R. tigerina* are a few of the large gluttonous amphibians. Within the reptiles, many species of snakes prey heavily or entirely on amphibians: for example, *Farancia abacura* on amphiumas; *Diadophis punctatus* on plethodonid salamanders; *Xenodon* and *Heterodon* on *Bufo*; and *Causus*, *Leptodeira*, and *Notechis* on a variety of anurans. Some species of snakes are specialists on lizards and others on snakes. Only a few lizards preferentially eat amphibians or reptiles, although many larger species (*Varanus*) capture frogs, snakes, and other lizards opportunistically. Crocodilians and carnivorous turtles such as *Chelydra* also regularly include amphibians and reptiles in their diet.

Birds prey heavily on amphibians and reptiles. Wading birds consume many adult and larval frogs, watersnakes, and small turtles during their shallow-water feeding forays. Raptors catch large numbers of lizards and snakes, and many passerines (both specialized carnivores such as shrikes and nonspecialists like blackbirds, the icterids) eat small amphibians and reptiles. Roadrunners, secretary birds, and a few other species specialize on lizards and snakes. Mammals as a group are not major amphibian and reptilian predators, although small felids, mustelids, viverrids, procyonids, and others likely attempt to capture and eat any amphibian or reptile encountered. Clearly, small mammals can have a devastating effect on amphibian and reptile communities as has been repeatedly demonstrated by the extinction or near extinction of insular lizard and snake populations where the mongoose and/or housecats have been introduced. Apparently, even the omnivorous armadillo has reduced the abundance of surface and subfossorial amphibians and reptiles in Florida since its introduction there in the mid-twentieth century.

Parasites

Parasites are predators in the sense that they feed on their prey/host and that their attacks may be fatal. Death of the host may be attributed directly to the parasite's destruction of the host's tissues and physiological functions, or the parasites may weaken the host so that it becomes easy prey for another predator. This latter aspect may be "intentional," since many parasites require multiple hosts and must pass from an intermediate host (an amphibian or reptile) to the main host by consumption. Even where death does not result from a parasitic infection, an individual's reproductive potential can be reduced. As a result, antiparasite defenses have evolved to avoid parasitic attacks or to reduce the parasites' survival and debilatory effect once it has gained entry to the host.

Few antiparasite mechanisms have been examined carefully in amphibians and reptiles. These animals share many of the features of the immunological system of mammals, and those mechanisms serve to resist or modulate some parasite attacks. Another mechanism for combating bacterial infection is elevation of body temperature, because high temperature inactivates or kills bacteria. When infected, some lizards behaviorally select and maintain body temperatures significantly above their normal activity temperatures. This behavioral-fever mechanism appears to reduce the infection and improve the lizard's resistance. Another antiparasite mechanism that has received little attention is the amphibians' epidermal poison glands. All amphibians possess these glands, and their toxicity to many invertebrate and vertebrate predators appears to be low to nonexistent. Perhaps these glands appeared early in amphibian evolution as a protection against bacterial and fungal infections of the moist skin and still serve that function. Magainins, isolated from the skin of *Xenopus*, have exceptional antibiotic and antifungal properties. They or related compounds likely exist in other amphibians.

The effects of parasite infections are largely unknown for amphibians and reptiles. There are lists of what parasite is present in what species, but there are few data on parasite life histories, how infestation affects an individual amphibian's or reptile's health, growth, or reproductive output, or what is the parasite incidence and duration within populations and the effect on population structure and dynamics. That individuals and populations are affected by parasites is acknowledged, at least, when mass die-offs occur, such as the microsporidian epidemic of English *Bufo bufo* in the early 1960s, the septicemialike decimation of *Rana pipiens* populations across northern North America in the early 1970s, or the current high incidence of possibly viral-induced papillomas in Floridan populations of *Chelonia mydas*.

The best studied parasite–host interaction system within amphibians and reptiles is lizard malaria. Malaria occurs in most lizard families and is essentially worldwide. A study of *Sceloporus occidentalis* populations in northern California showed that about 40% of the populations had malaria. Within these populations, less than a third of the population was infected, and males were more commonly infected than females. Infected lizards were adversely affected by their malaria (Table 6.1), although there were no apparent differences in structure and dynamics between infected and noninfected populations. In Panamanian populations of *Anolis limifrons*, adult males also had the highest incidence of infections in all seasons; however, there was no evidence of differences in general health, feeding, or reproductive behavior between noninfected and infected males. Elimination of a parasitic infection is possible, although the mechanisms are unknown. Nearly all male spadefoots, *Scaphiopus couchii*, leave the breeding aggregation with a monogenean trematode infection, yet 50% have shed the parasites prior to hibernating.

Amphibians and reptiles suffer all the usual vertebrate parasites: internally by bacteria, protozoans, and various groups of parasitic "worms" and externally by helminths and arthropods. All individuals probably have endoparasites of one kind or another as well as one or more ectoparasites. The level of virulence is unknown, but in most populations, individual frogs or snakes appear healthy, so many parasites must be benign and/or the host resistant and durable.

TABLE 6.1
Effect of Malaria on the Performance of Western Fence Lizards (*Sceloporus occidentalis*)

Criterion	Performance (%)
Hemoglobin concentration	76
Metabolic rate, active	85
Burst running speed	89
Running stamina (2 min)	83
Fat stored, female	75
Clutch size	86
Growth rate	96
Mortality	114

[Modified from Schall (1983).]

Note: The values are the level of performance of a sample of malaria-infected lizards compared to the performance of noninfected lizards.

CRYPSIS AND CONFUSION

Avoiding predators is the safest and usually the least energy demanding antipredator defense. As noted earlier, avoidance obtains either by absence from the predator's forage area or by presence without detection. Absence is not always possible, and many animals have evolved mechanisms that prevent their detection. This crypsis or biological deception operates in all sensory spheres, although the visual sphere is most studied and the one emphasized here. Crypsis works by an organism assuming another identity (mimicking) of a nonprey item (model) and misdirecting (duping) the predators' attention elsewhere. Mimicry can be to specific items or to a general background. Further, this resemblance can match items or backgrounds of no interest to the predator, so the predator is neither attracted or repelled but simply ignores the prey (camouflage). Conversely, the resemblance may match an item known to the predator but usually avoided (homotypic mimicry and aposematism).

Camouflage

Amphibians and reptiles often appear boldly colored or patterned when seen out of their normal habitats, but in their habitats, these colors and patterns blend with their background and make the animal difficult to detect. Many wear shades of green, brown, and gray and live on matching surfaces. The presence of spots, blotches, crossbars, and stripes further adds to the camouflage by disrupting the animal's outline. Immobility and postures with limbs appressed to the body further decrease the chance of detection. The numerous green frogs (e.g., *Centrolene*, *Hyla*, *Litoria*, *Rana*) or blotched-patterned snakes (e.g., *Boa*, *Python*, *Nerodia*, *Crotalus*) attest to the effectiveness of these colors and patterns. Where bright colors are used for courtship or warning displays, the bright colors remain hidden until the animal moves or intentionally displays the hidden surface. The inguinal or groin patches of many anurans appear and disappear quickly as the frogs move and may startle or distract a predator. The bright ventrolateral belly patches of many male lizards (*Sceloporus*) or dewlaps (throat-fans of *Anolis*) are courtship advertisements, but can be quickly hidden to avoid detection by a predator. Bright colors (red, orange, yellow, white) are also effective in camouflage. Small, bright spots, lines, and other irregular shapes contribute to the disruption of body shape and assist in matching background colors. Even when bright colors cover most of the body surface, they may match background color, for example, orange *Agkistrodon contortrix* among fallen leaves. The bright contrasting bands of the coral-snake patterns (black and white or yellow; black, yellow, and red) effectively conceal snakes in forest-floor litter (Fig. 6.2).

Modifications of body shape also enhance the effects of color camouflage. *Pipa*, *Phrynosoma*, trionychid turtles, viperid snakes, and other amphibians and reptiles are dorsoventrally compressed, and this flattening of the body eliminates the edge effect when these animals rest on a flat substrate. Adding spines and other appendages to body edges furthers the disruption of body shape and prevents a match with a predator's search image, for example, *Chelus fimbriatus*,

FIGURE 6.2 The Mexican colubrid *Scaphiodontophis zeteki* combines coral snake mimicry with camouflage. (R. W. Van Devender)

Phrynosoma, and *Ceratobatrachus*. By combining cryptic coloration and background matching, detection by a predator is often avoided. In addition to matching the background to the camouflage color and pattern, the animal must also match light level and direction for the resulting pattern of light and shadows to be successful.

Color and pattern are also effective in misdirecting a predator's attack. Boldly colored tails cause a predator to strike at a disposable body part or one that can withstand repeated bites without serious injury (e.g., blue tails in juvenile skinks). Stripes and bands also mislead predators when the prey is moving. The uniformity of longitudinal stripes on elongate animals gives the impression of immobility when instead the striped prey is escaping. Movement of a banded animal results in a flickering of light and dark. This flickering may confuse the predator, give the impression of movement in the opposite direction, or, if prey speed is great enough, create the illusion of a unicolored prey that visually disappears when it stops and again becomes banded.

While prey avoid predators by camouflage, predators are also served by camouflage. A predator hidden from its predators is also hidden from its prey, allowing it to ambush or approach closely without detection before attacking its prey.

Aposematism and Mimicry

Bright colors and contrasting patterns may warn a potential predator that the bearer is dangerous. This warning advertisement (aposematism) usually means that the bearer is unpalatable, injurious, or lethal. Species with one or more of these traits occur in amphibians and reptiles, and these traits would seem to be the ultimate in antipredator defense. Against some predators, they are successful, but in the continuous evolutionary seesaw between defense and conquest, some predators capture and consume even the most toxic prey.

Salamanders, frogs, and snakes have aposematic species. All salamandrids are aposematic, and all possess highly toxic secretions of the granular glands (the nature of poisonous secretions is described in the Structural and Chemical Deterrents section). *Salamandra salamandra* and efts of *Notophthalmus viridescens* display their toxicity with bright colors (red or yellow markings on black background and brillant orangish red, respectively). Other salamandrids (e.g., *Taricha*, *Cynops*) are cryptic above, but when threatened expose a brightly colored underside of red or orange. Indeed, most aposematic amphibians and reptiles combine their warning coloration with an unusual behavioral display. *Cynops*, *Notophthalmus*, and *Taricha* have the unken-reflex, where the salamander is immobilized with the head and tail bowed over the back exposing the colorful venter. *Salamandra* arches its back and stands on extended limbs to appear larger than it is. Other salamandrids have similar postures displaying the warning colors and/or placing the surface with concentrations of granular glands toward the attacking predator. Both the bright color and the stereotypic display provide the predator with a memory tool to reinforce its learning to avoid this species in the future, particularly if it pursues the attack and is poisoned but not killed. Most other terrestrial salamanders (Ambystomatidae, Hynobiidae, and Plethodontidae) have noxious skin secretions and stereotyped behaviors to ensure that an attacking predator comes in contact with the secretions (see Fig. 6.5). Some secretions, such as in *Plethodon glutinosus*, substitute stickiness for noxiousness, and the compound is capable of immobilizing the predator or gluing its mouth shut.

Of the frogs, the poison-dart frogs (*Dendrobates*, *Epipedobates*, *Minyobates*, *Phyllobates*) are the boldest advertisers of their toxicity. Many have vivid blue, green, red, orange, or yellow markings on a black background or nearly unicolored red to orange. They are diurnal species and openly forage on the forest floor. Few other anurans are aposematic; they are camouflaged but when attacked expose bright-colored surfaces with special displays, for example, unken-reflex in *Bombina* or inguinal ocella in *Physalaemus*.

The American coral snakes (*Micrurus*, *Micruroides*) have bold banded patterns (Fig. 6.3) and when disturbed will often lift and wave their tails. Their

FIGURE 6.3 ▼▲▼

The boldly banded coral snake (*Micrurus fulvius*) can be inconspicuous on the forest floor. (R. W. Van Devender)

aposematism is usually associated with their venomous bites, but anecdotal evidence also suggests that their flesh is distasteful. Both traits, singly or in combination, would serve as antipredator defenses. Bold banded patterns are present in other venomous snakes (Asian *Bungarus*, Australian *Vermicella*) and likely are aposematic. The saw-toothed viper, *Echis carinatus*, may use the rasping-sound produced by rubbing the body scales together as an auditory aposematic warning, and the high-pitched buzz of the rattlesnakes (*Crotalus*) may also represent auditory aposematism.

Palatable and harmless species can take advantage of predators' avoidance of aposematic species by mimicking the latter in form, color, and behavior. The red-backed salamander (*Plethodon cinereus*) mimicks the *Notophthalmus viridescens* eft stage. The palatable *Desmognathus imitator* and *D. ochrophaeus* have individuals matching the geographically variable patterns of distasteful *Plethodon jordani*. The mimicry of a harmful or distasteful model by a harmless or palatable mimic is called Batesian mimicry. In Müllerian mimicry, a strongly toxic or noxious model is mimicked by a mildly toxic or noxious mimic. This double-level reinforcement teaches the duped predators more rapidly to avoid the models and mimics. The eft *Notophthalmus* and *Pseudotriton* of eastern North America form a Müllerian complex.

The situation with American coral snakes and their colubrid mimics is less easy to categorize. The matches in color and pattern of *Lampropeltis*, *Pliocercus*, and species of several other genera with sympatric species of *Micrurus* prove mimicry. These resemblances are striking, because the *Micrurus* patterns vary geographically and the mimics' patterns also change geographically matching those of the model. In some instances, the mimicry may be Batesian with harmless mimics, and in other cases, Müllerian with mildly venomous mimics.

Mimicry duping the other senses of predators is likely, although unconfirmed. One possibility is auditory mimicry of the gecko *Teratoscincus scincus* to the saw-toothed viper *Echis carinatus*. The rasping sound would seem effective against nocturnal mammalian predators.

Anatomy of Color

Skin color and pattern derive from the presence, density (vertical and horizontal), arrangement, and type of chromatophores in the skin (for the anatomy of chromatophores, see Chapters 1 and 3). For example, light reflecting from the melanophores and scattered as it passes through the semitransparent iridophores yields blue. This blue becomes green when the light then passes through the lipidophores. Other combinations of chromatophores produce other shades and colors; the relative abundance of the chromatophores and the density and arrangement of the pigment bodies in these cells also govern the resulting colors and their intensity. Colors can darken or be obscured by the melanin in the melanophores of the epidermis. The movement of the melanin particles (melanosomes) in the dendritic appendages of the dermal melanophores produces the abrupt lightening or darkening observed in chameleons, other lizards, and anurans. The skin lightens when the melanin is concentrated in the bodies of the cells and darkens when dispersed into the appendages. The melanin is dispersed in response to MSH (melanophore-stimulating hormone) secreted by the pituitary gland. Some snakes change color by the expansion of skin to expose a differently colored interscalar skin.

Many factors determine an individual's and a species' coloration. The predator–prey arms race, with the prey improving its match to its background and the predator improving its discriminatory ability, is a major factor. Physiological regulation is another one. Ectotherms are dependent on the reflection and absorption of long-wavelength light (infrared) to regulate heat gain and loss. This role of color is most evident in desert lizards that lighten and darken according to their heat load, but it is also evident in ontogenetic color change and permanent habitats of different light/heat intensities. The striped and black morphs of the red-backed salamander (*Plethodon cinereus*) have different heat tolerances, and their relative abundances correlate with annual habitat temperatures.

DEFENSE AND ESCAPE BEHAVIOR

If crypsis and escape fail, the prey's next defense is to repulse the predator either by bluff, actual aggression, body-surface defenses, or a combination of these. If successful, these back-to-the-wall defenses stop the predator's attack with no or easily repaired injuries. Many amphibians and reptiles repel a predator by bluffing aggressive actions. Widemouth hissing and striking by snakes are a visible threat and frighten many but the most determined or experienced predators. These threat displays are often associated with special postures and/or body structures that make the prey appear larger and more threatening. Another line of defense is to modify the body surfaces to make the animal less appetizing, difficult or dangerous to handle, or difficult to penetrate.

Structural and Chemical Deterrents

Amphibians and reptiles display a wide range of structural and chemical antipredator devices. The amphibians with their ever-present granular skin glands emphasize chemical defenses. These defenses range from irritating and mildly distasteful to emetic and lethal. The granular glands may be evenly spread across the dorsal surface as in dendrobatids or concentrated in large glandular masses as the parotoid glands and warts of bufonids. Glandular masses are evident on many salamanders and frogs, and their locations complement their use in defense behaviors. *Bufo* and *Ambystoma* meet their predators head-on and have large parotoid glands (Fig. 6.4). Tail-lashing salamanders (*Bolitoglossa*, *Eurycea*) have heavy concentrations on the tail. The predator cannot approach and grab the prey without being exposed to gland secretions. Even in species with less striking defense behaviors and glandular concentrations (e.g., *Hyla*, *Rana*), the predator receives a dose of the secretion as soon as it takes the prey into its mouth, and irritating secretions usually are sufficient to have the prey ejected.

The tabulation of the noxious and toxic components of amphibian chemical defenses has only just begun. The known compounds fall into four groups: biogenic amines, peptides, bufodienolides, and alkaloids. The biogenic amines include serotonin, epinephrine, and dopamine; all affect the normal function of the vascular and nervous systems. The peptides comprise compounds such as bradykinin that modify cardiac function. The bufodienolides and alkaloids are similarly disruptive of normal cellular transport and metabolism and are often highly toxic.

With the exception of snake venoms, the chemical defenses of reptiles are more disagreeable than harmful. Turtles have musk (Rathke) glands opening on

FIGURE 6.4 Protection by glandular secretions. A marine toad (*Bufo marinus*) in a defensive posture with poisonous secretion oozing from its parotoid gland. (G. R. Zug)

the bridge of their shells; musk secretions have not been demonstrated as defensive, but to the human nose, the odor of the kinosternid and chelid turtles is pungent. Snakes have paired cloacal glands that are aimed and emptied on predators. Geckos also have cloacal glands that may or may not be used in defense; however, the squirting tail glands of the Australian *Diplodactylus spinigerus* are clearly defensive against invertebrates. There is no reason to assume that the glands of any amphibian or reptile have a single role. Their secretions, even the most poisonous one, probably also serve in individual and species recognition for reproductive and territorial behaviors. Similarly some may serve as lubricants, waterproofing agents, or antibacterial or -fungicidal protectants. Aside from glandular secretions, many amphibians and reptiles defecate and/or urinate when grasped by a predator.

Phrynosoma, the horned lizards, are notorious for their blood-squirting-from-eye defense. Although unconfirmed, the blood does not appear to be toxic or noxious. Blood-squirting is largely reserved for canid predators and is elicited only after the lizard is mouthed. Squirting may stimulate the canid's prey-tossing behavior and give the lizard an opportunity to escape when tossed out of the canid's visual field. *Phrynosoma* are excellent examples of multiple levels of predator defense. They are well camouflaged by coloration and body form, bear caps of cranial horns used in defense during predator attacks and making them difficult to swallow, inflate their bodies to create an illusion of larger size and to absorb the predator's blows with less internal damage, and squirt blood from orbital sinuses.

The softer, more permeable skin of amphibians has fewer structural modifications to increase its resistance to predator attacks. Aside from the assorted bony or keratinous spines that occur on the limbs and trunks of some frogs (most are associated with reproduction or digging), only the fusion of the skin to the dorsal skull roof may be defensive. This fusion provides strength to both skin and skull and, for a few species, has been suggested that they use the top of their head to block entry to their rest retreats. The more heavily keratinized skin of reptiles provides a more durable body armor, and many modifications have evolved to give it even greater strength. The turtle shell with thick epidermal plates is a most obvious defense structure. Crocodilians and some lizards (most anguimorphs) have the epidermal scutes or scales underlain by bony osteoderms; this combined barrier makes penetration by a predator's teeth difficult, and both groups use a spinning, thrashing movement to escape from the jaws of predators. Enlarged and spiny scales make a biting grip painful and difficult to hold. The horniest horned lizards (*Phrynosoma*) occur in the areas of highest predator densities. The spiny tails of *Ctenosaura*, *Uromastyx*, and many other lizards strike a painful blow. There are other, subtler structural modifications protecting the bearer. The tiny chameleons, *Bradypodion*, have the transverse processes of the vertebrae curved dorsally over the neural arches to form a shield over the spinal cord. When touched, *Bradypodion* freezes, releases its grip, and falls to the ground;

during the fall, it rights itself so that it always lands with its vertebral shield side up, and birds treat it as an inedible object.

Specialized Behaviors

Stereotypic postures and movements are integral parts of all preceding defense mechanisms, particularly those associated with direct predator–prey interaction when the prey has been discovered and cannot escape. Stereotypy serves in several ways: to startle the predator and allow time for escape; to orient the prey's body to improve its defensive position; and to provide an unambiguous memory record for a disagreeable and possibly unsuccessful capture experience. Startling behaviors are classed as protean defenses (i.e., sudden, erratic change in appearance or movement) and deimatic defenses (i.e., frightening display, posture, or noise). The unken-reflex (Fig. 6.5) is an example of the first and sudden spreading of a hood and hissing of a cobra for the second. These behaviors simultaneously place the prey in a better position to defend itself, either passively by presenting a toxic surface or actively by placing its biting, scratching, thrashing defenses toward the predator. If a predator is thwarted in a disagreeable manner, it will learn to avoid prey with similar behaviors in the future.

Thanatosis (death feigning, letisimulation, catalepsy) is a peculiar defense behavior and unsatisfactorily explained for edible prey. It occurs sporadically in frogs, salamanders, lizards, and snakes and in these either as rigid immobility or a limp relaxed form. For the noxious or toxic salamanders, a predator will mouth and reject the prey, but for palatable taxa, thanatosis would seem to provide the predator with an unresisting meal. Its occurrence in unrelated taxa (e.g., *Eleutherodactylus*, *Echinosaura*, European *Natrix*) suggests that it is effective against some predators, those that would reject carrion or motionless prey, or because motionlessness may cause inattentiveness and allow the prey to slip away.

Shedding Body Parts

A loss of a body part may be catastrophic, but survival is better than death. Many animals have evolved structural and behavioral mechanisms to maintain the predator's attention to an expendable body part while the animal escapes. This escape adaptation is effective, judging by the numerous independent origins within the animal kingdom. Amphibians and reptiles are capable of losing limbs or suffering large wounds and recovering, but only salamanders are able to regenerate their digits or portions of a lost limb, and only salamanders and some lizards are capable of regenerating their tails.

Caudal Autotomy

Autotomy (*autos*, self; *tomos*, cut) is the self-induced loss of a body part. As such, it contains neurological and behavioral aspects as well as structural ones.

The structural modification for autotomy is a plane of weakness in the middle of each vertebra (intravertebral) or between vertebrae (intervertebral). Autotomy occurs in some salamanders (mainly in plethodontids) and widely in the lizards (absent entirely only in the varanoid and chamaeleonids), tuataras, and some snakes.

Caudal autotomy has been best studied in the lizards. Losing the tail is a last-resort defense, and the tail is not shed casually. Its loss has significant cost for the individual. One cost is the loss of energy (as fat or protein) stored in the tail and the cost of replacing the tail. Quick replacement is necessary because locomotor performance is usually reduced by tail loss, so reduced-tailed lizards have more difficulty escaping from predators. The cost of replacement stops general growth in juveniles and suspends the production of gametes in mature lizards. There are also social costs in loss of social dominance or defense of a large territory. Despite these high costs, survival outweighs them.

Caudal autotomy works because it draws the predator's attention to a potentially disposable body part. A predator's attack is directed to the tail by its contrasting color and/or extra movements. The tail is not shed until firmly grasped by the predator, and when shed, it becomes more active. This thrashing and waving maintains the predator's attention while the lizard escapes.

Fragile Skin

A few skinks (*Ctenotus*, *Lerista*) and many geckos (e.g., *Ailuronyx*, *Aristelliger*, *Gehyra*, *Teratoscincus*) have fragile skin that can be shed when grasped by a predator. Losing a portion of the skin is not self-induced and occurs only when the lizard is seized. It is nearly impossible to capture these lizards without damaging their skin. It is so striking that one gecko received its scientific name *Gehyra mutilata*, mutilated gecko, from this adaptation.

Usually, the lizard loses its skin from the upper surface of its trunk, neck, and posterior part of the head (Fig. 6.6). The entire epidermis and most of the dermis is shed. Unlike the typical squamate skin, sheddable skin has a stratum compactum that is divided into upper and lower layers rather than a single thick stratum compactum. The two layers are joined loosely by a thin collagenous network, allowing the two surfaces to slide pass one another, and when necessary, the outer layer slips sufficiently far to detach. There are further zones of weakness within the outer layer to limit the detachment to the area where the lizard is grasped. These zones lie between the large tubercles and are crossed only by

FIGURE 6.5 ▼▲▼

Some amphibian defense postures. From top left, clockwise: unken-reflex, *Bombina orientalis* (Chinese fire-bellied toad); unken-reflex, *Notophthalmus viridescens* (red-spotted newt); head hiding and tail lashing, *Pseudotriton montanus* (mud salamander); tail lashing, *Eurycea lucifuga* (cave salamander). (frog, R. W. Van Devender; salamanders, C. K. Dodd, Jr.)

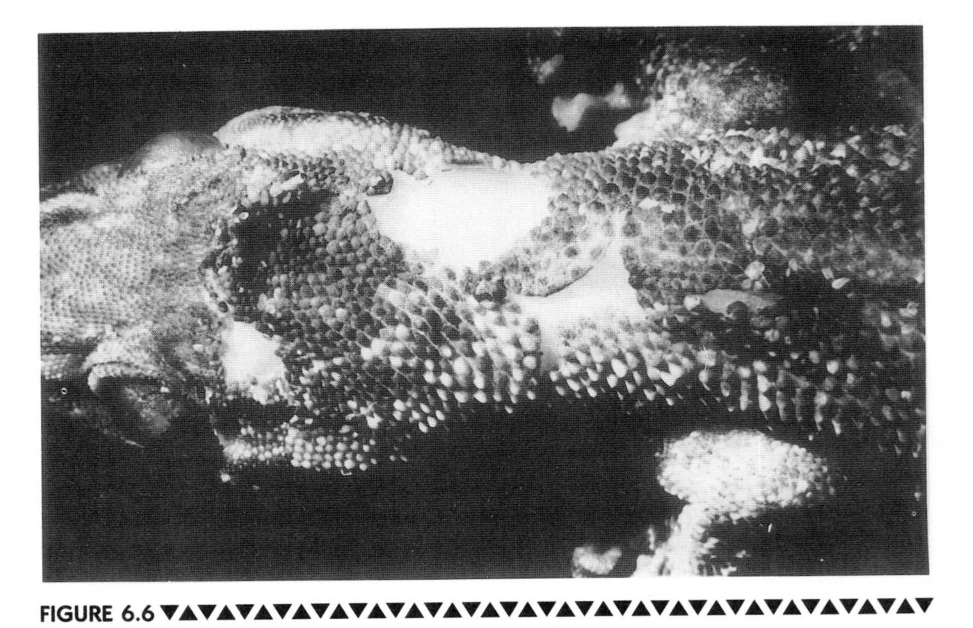

FIGURE 6.6 ▼▲▼
The fragile and torn skin on the neck and shoulders of a gecko (*Ailuronyx seychellensis*). (A. M. Bauer)

small collagenous fibers. An additional structural modification in fragile-skinned lizards is the firm attachment of the stratum compactum to the underlying connective tissue sheath of the muscles. This firm attachment anchors the lower layer and permits the outer layer to tear away. In most lizards, the dermis has a looser attachment to the sheath and is able to slide over the muscles and avoid tearing. (See Figs. 1.2 and 3.2 for illustration of skin structure.)

Safety by Association

Schooling

Of the aquatic amphibians and reptiles, only tadpoles form social swimming aggregates. Such aggregates or schools have been reported for less than three dozen species (e.g., *Bufo americanus*, *Scaphiopus bombifrons*, *Xenopus laevis*). Schooling behavior is presumed to be a predator defense mechanism but may also enhance feeding. As a defense, it can reduce predation in several ways: when prey are aggregated, many predators have difficulty focusing on a single individual and make more false strikes; a captured prey releases an alarm substance that alerts all members of the school and the school leaves the area of predation; individual predators may become satiated (see later); and predators learn more quickly to avoid prey with chemical defenses. The latter advantage accrues to *Bufo* tadpole schools. *Pyxicephalus adspersus* and *Leptodactylus ocellatus* females shepherd their tadpoles and provide an extra level of predator defense.

Predator Saturation

As in schooling, an individual benefits (i.e., gains higher probability of survival and reproducing) by being present in an aggregation even if the predators can easily recognize and capture prey. Frog and salamander breeding aggregations and seaturtle arribada (simultaneous mass nesting) provide multiple predators with an abundance of prey (adults in the first, eggs and hatchlings in the second instance) but on the average allow more individual prey to survive because the predators become glutted and predation stops. Whereas if the prey did not aggregate and were available at the breeding area for a long time, the predators would not stop eating prey and the overall total of prey killed would be higher. Predator saturation can evolve only in prey species with extremely high reproductive output and/or high density of breeding adults. Careful field observations have demonstrated that predation is less on larger choruses of *Physalaemus pustulosus* than on smaller choruses, and the reduced risk per individual does not reduce each male's chance of mating.

Living with Predators

Many amphibians and reptiles eat ants and termites as a major or exclusive food, but only a few have been able to develop a commensal relationship that permits them to live in ant or termite colonies. The soldiers of ants and termites are formidable defenders of the colony; thus if an animal can be accepted by the colony, it will receive the same protection as the colony. A few frogs have achieved this acceptance. *Pseudophryne nichollsi* occasionally occurs in the nest of the Australian bulldog ant, a very ferocious species that boils out of its nest at the least disturbance and attacks the source of the disturbance. A small South American frog (*Lithodytes lineatus*) also lives in the nests of an aggressive ant (*Atta*). Another Australian frog, *Myobatrachus gouldi*, feeds exclusively on termites and lives in their mounds. Other frogs and lizards use termite nests at least occasionally as residences, and some lizards, notably Australian varanids, regularly deposit their eggs in termite mounds. Two North American snakes, *Storeria* and *Diadophis*, hibernate, often in small aggregations, in ant nests.

The mechanism for living with ants or termites is unknown for all but *Leptotyphlops humilis*. This tiny blind snake lives with several species of ants and termites, and when attacked, its writhing defense coats itself completely with its cloacal sac secretions. This coating repels ants and termites and allows the blind snake to stay in the ants' nest or move along their pheromone trails. This coating also repels ophiophagous snakes.

Many spiders prey on small amphibians and reptiles. Yet several anurans

(*Chiasmocleis ventrimaculata*, *Gastrophyrne olivacea*, *Physalaemus pustulosus*) live in the burrows of American tarantulas. Field experiments with *Chiasmocleis* show that its tarantula host not only tolerates its guest but recognizes them; the tarantula will attack small predators of the frogs and eats other species of frogs. The tarantula seems to recognize the frogs by the "taste" of their skin secretions.

PART III

LIFE CYCLE: REPRODUCTION, DEVELOPMENT, AND GROWTH

Evolution places heavy emphasis on reproduction, for the way an organism reproduces affects profoundly its contribution to future generations.

S. C. Stearns, 1976

All aspects of life are aimed toward reproduction and the perpetuation of the genetic self. Thus, reproductive biology is central to much of biological research, ranging from studies on the molecular events at fertilization through developmental anatomy and the endocrinology of circadian cycles to measures of reproductive potential in evolutionary biology. Amphibians and reptiles have and continue to be major contributors to studies in developmental and reproductive biology. The large and robust eggs of amphibians have been a mainstay in embyrological research, providing the eggs for the first nuclear transplants, tractable cells and tissues for tracing molecular events in differentiation, and "simple" illustrative materials for beginning students.

Amphibians and reptiles are the paragons of reproductive experimentalists, among both extinct and extant taxa. A dozen or more amphibian lineages evolved direct development and terrestrial eggs. Similarly, dozens of reptilian lineages independently evolved internal development; several amphibian groups have gone this way also. Some populations of amphibians and reptiles have dispensed with males and rely on parthenogenetic reproduction. These reproductive adaptations and others equally strange hold much fascination for biologists and offer many clues and perhaps answers to questions in evolution.

CHAPTER 7

Modes of Reproduction and Development

Reproductive behaviors are categorized in numerous ways. The basic division for amphibians and reptiles is the location of fertilization. Does fertilization occur externally or internally, that is, outside or within the reproductive tract of the female? Each site delimits the habitats and other options available to courting individuals and to developing embryos. External fertilization limits the developmental environment to moist-aquatic habitats or to special nonreproductive tract structures for carrying the developing embryos internally. Internal fertilization permits external or internal development and establishes a different set of options for the evolution of fertilization and development.

GAMETOGENESIS AND FERTILIZATION

Preparation for reproduction begins at the cellular level. Hormones trigger sex (germ) cell division and growth within the gonads. This gametogenesis, in turn, produces hormones, and the interplay of gonadal hormones with those of brain, pituitary, and other organs prepares the individual physiologically, structurally, and behaviorally for reproduction.

Gamete Structure and Production

Male and female gametes (sex or germ cells) are strikingly different. The male gamete, spermatozoon (= sperm), consists of a small head, middle piece, and tail. These parts come in various shapes and sizes in different species but always retain a spearlike form. The head is mainly the cell nucleus capped by an acrosome, whose secretions enable the sperm to penetrate the ovum's cell membrane. The middle piece carries the mitochondria (energy source) and basal contractile rods for the tail or flagellum (two or more flagella in primitive frogs). Vibrations of the tail propel the sperm to the ovum. The ovum is many times larger than the sperm and contains cytoplasm and yolk. The ovum becomes the embryo upon fertilization.

Gametogenesis is the growth and developmental sequence of undifferentiated germ cells to mature sex cells, either ova or spermatozoa. Male germ cells (spermatogonia) line the inside of the seminiferous tubules (see Gonad and Genital Ducts in Chapters 1 and 3). Sperm production (spermatogenesis) begins by the proliferation (mitosis) of the spermatogonia (Fig. 7.1). Some of these grow and mature into primary (1°) spermatocytes; others remain as undifferentiated spermatogonia for future sperm production. Each 1° spermatocyte divides (meiosis I) to form a pair of secondary (2°) spermatocytes, each of which in turn divides (meiosis II) into a pair of spermatids. Each spermatid transforms (spermiogenesis) into a spermatozoon by the condensation and structural reorganization of cellular organelles. As germ cells move through the spermatogenic stages, they are pushed inward toward the lumen of the seminiferous tubule by the peripheral production of spermatocytes. The spermatozoa then drop into the lumen and move through the tubules to be temporarily stored in the ducts of the epididymides, awaiting ejaculation.

Oogenesis also follows a similar cell division cycle of mitosis and meiosis, but with a very different final product. Oogonia arise from the primordial germ cells in the wall of the ovary. Each is surrounded initially by a single layer of epithelial (follicle) cells forming a primary follicle. As an oogonium enlarges into a 1° (primary) oocyte, the follicle cells multiply and form a multiple-layered secondary follicle (Fig. 7.2). The 1° oocyte divides into a 2° (secondary) oocyte retaining most of the cellular material and a tiny polar body, a cell with little more than a nucleus. For most vertebrates, the 2° oocyte begins the second

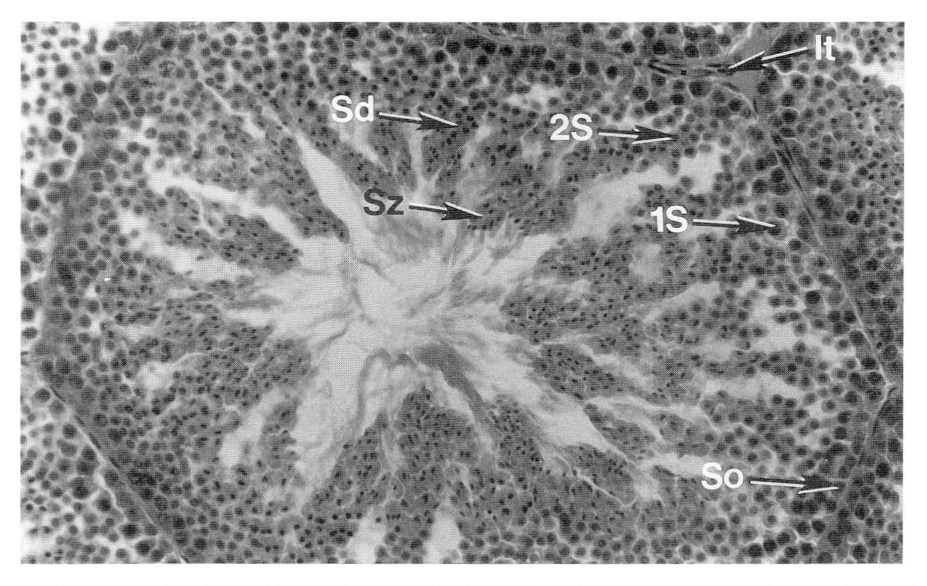

FIGURE 7.1 ▼▲▼▲▼▲▼▲▼▲▼▲▼▲▼▲▼▲▼▲▼▲▼▲▼▲▼▲▼▲▼▲▼▲▼▲▼

Spermatogenesis. Cross section through a seminiferous tubule of the skink *Carlia bicarinata*. Abbreviations: 1S, primary spermatocyte; 2S, secondary spermatocytes; It, interstitial tissue; Sz, spermatozoon; Sd, spermatid; So, spermatogonium. (M. Barber)

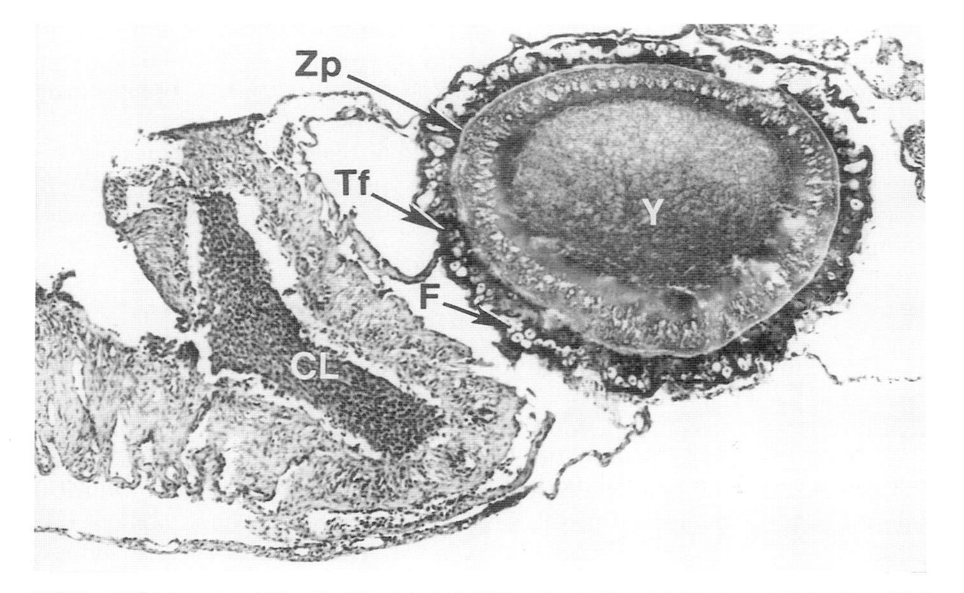

FIGURE 7.2 ▼▲▼▲▼▲▼▲▼▲▼▲▼▲▼▲▼▲▼▲▼▲▼▲▼▲▼▲▼▲▼▲▼▲▼▲▼

Oogenesis. Cross section through an ovary of the skink *Carlia bicarinata*, showing a corpus luteum (left) and a maturing follicle (right) with its ovum. Abbreviations: CL, corpus luteum; F, follicular cells; Tf, theca folliculi; Y, yolk; Zp, zona pellucida. (D. Schmidt) meiotic division and stops at metaphase; division is completed when sperm enters the ovum. The 2° oocytes then begin a period of maturation and growth through the accumulation of yolk. In this vitellogenic phase, lipids move from fat stores throughout the body via the bloodstream to the liver, where they are converted to vitellogenin. The oocytes selectively absorb (pinocytosis) the vitellogenin and enzymatically convert it into the yolk platelet proteins (lipovitelline, phosvitin). The first phase of vitellogenesis is usually slow, requiring several months to attain 0.5–0.6 the size of the mature ovum. Vitellogenesis may pause and then rapidly (2–4 weeks) be completed just before ovulation. The mature ovum is 10- to 100-fold its previtellogenic size. Ovulation occurs with the rupture of the follicular and ovarian walls; the ovum is shed into the body cavity, although the infundibulum of the oviduct is nearby. The remains of the follicle reorganize and form the corpus luteum (Fig. 7.2).

The ovum becomes an egg as it passes through the oviduct and obtains protective membranes. For amphibians, these membranes consist of several layers of mucoproteins and mucopolysaccharides produced by the walls of the oviduct. The number of layers range from two to six or seven, most commonly three to four layers. All amphibian eggs have a vitelline membrane next to the ovum; this is the fertilization membrane that lifts from the surface of the ovum upon fertilization. The other membranes are more gelatinous and of variable thickness. Depending on the species, the eggs may be expelled singly in their gelatinous coats, as gelatinous strings, clusters, flat sheets, or in other aggregate shapes. This covering provides some protection from desiccation and predators. Reptilian ova are encased in a more durable and resistant shell (cleidoic eggs). The reptilian ovum is sequentially coated by albumen in the uppermost part of the oviduct, then by several thin layers of protein fibers, and finally the fiber layer is impregnated with calcium carbonate (calcite crystals in crocodilians and squamates, aragonite crystals in turtles). Shell formation and degree of calcification vary among reptilian genera.

Fertilization: Transfer and Fusion of Gametes

The term "fertilization" is used for two related but different activities. At the organismal level, fertilization (= mating) concerns the behaviors associated with the male depositing sperm on or near the ova or where the female can collect them (courtship behaviors are discussed in Chapter 8). Fertilization also refers to the actual penetration of sperm and fusion of female and male pronuclei. External fertilization applies to this fusion of ovum and sperm outside of the female's body; internal fertilization occurs within the female's body, almost always in the oviducts.

At the cellular level, spermatozoa must attach to and penetrate the cell membrane of the ova. For each ovum, many sperm reach the egg surface, but only a few adhere to the egg membrane. Then, the cell membranes of the sperm head and ovum fuse, and the sperm pronucleus moves into the ovum cytoplasm. The ovum reacts with the separation and elevation of the vitelline membrane and the completion of the meiotic division of the ovum nucleus. The elevation of the vitelline membrane removes the other sperm from the egg surface, thereby preventing polyspermy and subsequent abnormal development. (Salamanders have polyspermic fertilization, although all but one sperm pronucleus degenerates.) The sperm pronucleus fuses with the ovum pronucleus, restoring the diploid condition and mitosis (somatic cell division) begins.

The simplest fertilization is the synchronous shedding of eggs and sperm into water. No tetrapods rely on this chancy type of external fertilization. Instead, they use courtship to bring females and males together for the delivery of sperm to eggs. Courtship ensures that both sexes have mature gametes and that the gametes are brought as closely together as possible. In external fertilization (most frogs, cryptobranchid and hynobiid salamanders), the male sheds his sperm on the eggs as they emerge from the female's cloaca. Male frogs grasp (Fig. 7.3) in front of the hindlimbs (inguinal amplexus) or behind the forelimbs (axillary amplexus) so that the cloacal openings of the two sexes are adjacent. In salamanders, mating behavior may place the female's and male's cloacae adjacent, or the

FIGURE 7.3
Axial amplexus in *Proceratophrys cristiceps* (horned frog). (A. Cardoso)

male follows the female and deposits sperm on the egg mass during or after deposition.

Although these behaviors promote close association of eggs and sperm, placement of sperm within the female's reproductive tract increases the probability of sperm and egg contact. A few frogs (*Ascaphus*, two *Eleutherodactylus*/leptodactylids, four *Nectophrynoides*/bufonids), all salamandroid salamanders, all caecilians, and all reptiles have internal fertilization. A diverse array of behaviors and structures accomplish this. Frogs and *Sphenodon* use cloacal apposition. Males of the other reptiles, *Ascaphus*, and caecilians have intromittent organs for depositing sperm deep in the cloaca adjacent to the oviduct openings. The phallodeum of the caecilians is an eversible pouch of the cloacal wall that is everted into the female's cloaca through a combination of muscular contractions and vascular hydraulic pressure, and withdrawn by a retractor muscle. The organ in *Ascaphus* is a combination of modified cloaca and residual tail. In turtles and crocodilians, a penis of spongy connective tissue erects and retracts by vascular pressure; it is structurally similar and probably homologous with the mammalian penis. In lepidosaurians, the penis was lost and later replaced by hemipenes, eversible pouches, one (hemipenis) at each posterior corner of the cloaca. Only one is everted (vascular pressure) during copulation and withdrawn by a retractor muscle.

Salamanders with internal fertilization produce spermatophores in the cloaca and deposit them externally. The spermatophore consists of a proteinaceous pedestal capped by a sperm packet. Male salamanders have elaborate courtships that stimulate females to move over the spermatophores and pick up the sperm packets with their cloacal lips.

Fertilization occurs in the upper portion of the oviducts (all reptiles and caecilians) or in the cloaca (salamanders). This location is essential for reptiles, because fertilization must occur prior to egg shell deposition. Sperm can penetrate the gelatinous capsules of amphibian eggs. Fertilization may occur immediately after copulation or commonly be delayed (salamanders, turtles, snakes). The delay may be only a few hours to years after copulation. Sperm storage is essential for delayed fertilization, and sperm storage structures occur in salamanders, turtles, and squamates. In salamanders, the roof of the cloaca bears a cluster of tubules, the spermatheca, opening independently or by a common duct into the main cloacal chamber. Sperm are held there and expelled by muscular contraction as the eggs enter from the oviducts. Sperm storage tubules do not form a single aggregate in reptiles, although they are confined to the upper-middle portion of the oviducts between the infundibulum and the shell-secreting area in turtles and to the base of infundibulum and lower end of the shell-secreting area in squamates. Because of the latter position, their function for long-term storage has been questioned. The mechanism for expelling sperm from the tubules is unknown.

Sperm storage and mating with multiple males suggest the possibility of multiple paternity among the resulting offspring with the potential of sperm competition. Although multiple paternity has been demonstrated in only a few species of salamanders and snakes, it is likely more widespread. A selective advantage accures to any male that can restrict access to his mated female, thereby ensuring that his sperm are used exclusively in fertilizing her ova. Like a predator–prey interaction, the male's goal may conflict with the female's. Her evolutionary success might be better served by mixing her genome with several males, thereby increasing the adaptive diversity and the survival of her offspring.

The cloacal plug mechanism in *Thamnophis sirtalis* is a male adaptation to eliminate multiple mating. At the conclusion of copulation, a portion of the male's ejaculate solidifies in the female's cloaca preventing immediate access by another male. This "ounce" of prevention is reinforced by the presence of a pheromone in the plug that repels other males. Further, a male in copula releases an inhibitory pheromone that repels competing males and may temporarily sterilize a male that lingers too long.

Reproduction without Fertilization

Some amphibian and reptilian populations (species) are unisexual, consisting entirely of females. The eggs of these females are special; they develop without the fusion of ovum and spermatozoan pronuclei (Fig. 7.4). Two distinct cellular mechanisms permit this unusual initiation of vertebrate development. In gynogenesis, females of the unisexual species parasitize males of the bisexual species. The males' spermatozoa are required to initiate development, but the sperm's pronucleus does not join with the female's and soon degenerates. In parthenogenesis, males are not required; the female's eggs are capable of self-activation/fertilization. The precise mechanism for self-activation of development remains unknown.

All known unisexual species of amphibians and reptiles arose through the hybridization of two bisexual species or of a bisexual and a unisexual species. Thus, each unisexual species has one set of chromosomes from each parent species if diploid, or one or more sets from the parental species if polyploid. The genetic complement of each individual within each unisexual species is identical. This clonal inheritance results from aberrant gametogenetic mechanisms that prevent recombination and produce gametes that are genetically identical to the maternal somatic cells (Table 7.1).

In other vertebrates, hybridogenesis maintains a third type of unisexual population. Like gynogenetic species, hybridogenetic ones depend on males of a bisexual species; however, the male pronucleus joins the female's and recreates hybrid offspring with each generation. Unexpectedly, one group of European frogs (*Rana* klepton *esculenta*) is hybridogenetic but bisexual (a few localized unisexual populations do occur; see later). The unusual origin and maintenance

TABLE 7.1
Genera of Unisexual Amphibians and Reptiles

Genus	Number of species	Mode of reproduction	Representative species
Ambystomatidae			
Ambystoma	3+	G&H	*platineum*
Ranidae			
Rana	5	H&P	*esculenta*
Agamidae			
Leiolepis	1	P	*triploida*
Chamaeleonidae			
Brookesia	1	P	*affinis*
Gekkonidae			
Hemidactylus	3+	P	*garnotii*
Hemiphyllodactylus	1	P	*typus*
Heteronotia	4+	P	*binoei*
Lepidodactylus	1+	P	*lugubris*
Nactus	2?	P	*pelagicus*
Gymnophthalmidae			
Gymnophthalmus	2+	P	*underwoodi*
Leposoma	1	P	*pericarinatum*
Lacertidae			
Lacerta	5	P	*unisexualis*
Teiidae			
Cnemidophorus	12+	P	*uniparens*
Kentropyx	1	P	*borckianus*
Xantusiidae			
Lepidophyma	2	P	*reticulatum*
Typhlopidae			
Rhamphotyphlops	1	P	*braminus*

[In part from Vrijenhoek et al. (1989).]

Abbreviations: G, gynogenetic reproduction; H, hybridogenetic; P, parthenogenetic. The actual mode of reproduction is unconfirmed for most of these unisexual species.

of hybridogenetic, gynogenetic, and parthenogenetic populations has resulted in a nomenclatural debate. Should the populations or clones be recognized as distinct species and be formally named or should their name reflect their parentage? The issue remains unresolved. Here parthenogens are recognized as distinct species, because each does represent a unique and cohesive genetic unit reproductively linked to past populations. The klepton (from Greek *klept* to steal) concept is used for gynogens and hybridogens.

Hybridogenesis

The *Rana* klepton *esculenta* plexus is a complex reproductive and genomic melange. *Rana* kl. *esculenta* probably arose several times by hybridization of *R. lessonae* and *R. ridibunda* in different localities to produce a diploid (RL) and

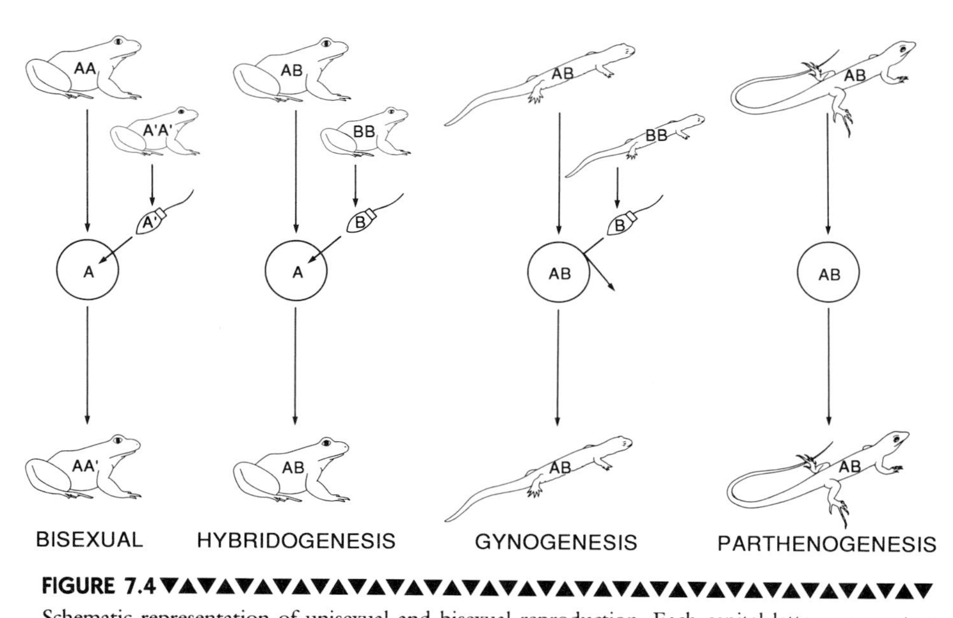

FIGURE 7.4 Schematic representation of unisexual and bisexual reproduction. Each capital letter represents a haploid complement of chromosomes.

the two triploid (RLL, RRL) biotypes; regionally different cytogenetic structures and behaviors indicate additional origins (Table 7.2). Hybridization between *R. ridibunda* and the Spanish *R. perezi* and an unnamed Italian species have produced two other kleptons in the waterfrog plexus.

Rana kl. *esculenta* is widespread in Europe (from France to central Russia), and everywhere it is sympatric with *R. lessonae*, and both are sympatric with *R. ridibunda* in central Europe. Diploid populations of *esculenta* are the rule; triploid populations are rare. Most *esculenta* populations persist not by continual hybridization of *lessonae* and *ridibunda* but by female *esculenta* parasitizing male *lessonae* (by using their sperm). Male *esculenta* mate with both *esculenta* and *lessonae* females, but only the offspring of the latter pairing survive through metamorphosis. *Rana* kl. *esculenta* persists because *esculenta* produces haploid gametes with a *ridibunda* genome (Table 7.2). These gametes arise by a premeiotic shedding of the *lessonae* genome and then a duplication of the remaining *ridibunda* genome followed by normal meiotic division. Males exist in these hybridogenetic populations because of XX-XY sex determination in waterfrogs and the presence of X and Y gametes from the *lessonae* males.

Gynogenesis

In mid-northeastern North America, many breeding aggregations of mole salamanders, the *Ambystoma laterale–jeffersonianum* complex, are composed of

TABLE 7.2 ▼▲▼▲▼▲▼▲▼▲▼▲▼▲▼▲▼▲▼▲▼▲▼▲▼▲
Genome Compositions of Offspring from Matings within The *Rana* klepton *esculenta* Nexus

		Females		
		LL (*lessonae*)	RR (*ridibunda*)	RL (*esculenta*)
Males	LL	LL	RL	RL, RRL, RLL
	RR	RL	RR	RE (RR)
	RL	RL	RE	I

[In part from Graf and Pelaz (1989).]

Note: *Rana* kl. *esculenta* is usually diploid and produces haploid gametes by a premeiotic exclusion of the *lessonae* genome and reduplication of the *ridibunda* genome so that normal meiosis yields haploid *ridibunda* genomes. Male *esculenta* are usually sterile.

Abbreviations: I, inviable; LL, *lessonae* genome; RE, variable exclusion of *lessonae* genome; RR, *ridibunda*; RL, *esculenta*; RRL, RLL, *esculenta* triploid genomes.

diploid males and females and polyploid individuals, usually females. When the composite nature of these breeding populations was first recognized, it was assumed that both the diploid individuals [*A. laterale* (genome LL) and *A. jeffersonianum* (JJ)] and the polyploid females [*A.* klepton *tremblayi* (LLJ), *A.* kl. *platineum* (LJJ)] were genetically distinct and reproductively isolated species, and the unisexual polyploids were maintained by gynogenesis. This explanation was accepted until the reproductive behavior and the genomic composition of numerous individuals in these breeding aggregations were examined, then hybridogenesis was proposed. The issue remains unresolved, because the frequency and composition of the diploid, triploid, and tetraploid genomes vary greatly among breeding populations and diploid males occur in low frequency or are absent. Multiple explanations, including gynogenesis, hybridogenesis, retrogression, and introgression, are likely necessary.

The genomic composition of this mole salamander plexus is now known to contain the following hybrids: diploids — LJ, LT (T = *texanum*); triploids — LLJ, LJJ, LLT, LTT, LJT, LTTi (Ti = *tigrinum*); tetraploids — LLLJ, LJJJ, LLJJ, LLLT, LTTT, LLTT, LTTTi, LJJTi, LJJT. This diversity can only have arisen by hybridization. Some breeding aggregations have unisexual populations of the hybrids. For example, an aggregation in Illinois contained bisexual populations of *texanum* and *maculatum*, and LJJ and LJJT females, thus gynogenesis or parthenogenesis is required to maintain the hybrid unisexual populations.

Parthenogenesis

Parthenogenetic reproduction has been confirmed only for a few unisexual populations of reptiles but is assumed to be the exclusive mechanism for all of them (Table 7.1). Inheritance within a parthenogen is strictly clonal; the initial hybrid female and all her subsequent offspring share the same genome. In some unknown fashion, hybridization appears to have altered normal mitotic-meiotic divisions during gamete formation. The actual cytogenetic events have been traced only in *Cnemidophorus uniparens*. Here premeiotic mitosis lacks nuclear and cytoplasmic division, yielding a tetraploid oogonium; normal meiosis produces a diploid gamete, identical to the maternal somatic genome.

Although the cytogenetic mechanism initiating development is unknown, pseudocopulation occurs in some parthenogens. Comparison of hormone levels in the courted female and the courting female (*Cnemidophorus tesselatus*, *uniparens*, and *velox*) shows that the courted female is preovulatory and the courting female postovulatory or oogenetically inactive. The courtship and pseudocopulation stimulate ovulation, thus one female is acting as a surrogate male. Pseudocopulation also occurs regularly in *Lepidodactylus lugubris* and may be a common (and necessary?) reproductive behavior of lizard parthenogens.

ASPECTS OF PARENTAL INVESTMENT

Parental investment is defined as "any investment by the parent in an individual offspring that increases the offspring's chance of surviving at the cost of the parent's ability to invest in other offspring" (Trivers, 1972). Investment in this sense for amphibians and reptiles is any energy expended by a parent over and above that required by the embryo to grow and develop sufficiently to hatch and to begin to obtain food. For those amphibians and reptiles laying and abandoning their eggs, the major investment is the production of yolk in excess of that needed for hatching and an initial feeding bout. Most species invest in the survival of their offspring by producing an excess of yolk with which the hatchling can meet its energy needs prior to the attainment of an effective feeding behavior and to survive unusual vagaries of the environment. Examination of lipid content in eggs has shown that hatchling turtles (e.g., *Chelydra serpentina*, *Chrysemys picta*) receive 50–60% more lipid than required for hatching and emergence, and *Alligator mississippiensis* ova have in excess of 70% of their hatching needs.

When and how parents make an investment to enhance the survival of offspring may be functionally and developmentally divided into preovulatory and postovulatory contributions. An example of preovulatory investment is the addition of extra yolk. Preovulatory investments are energy expenditures during gametogenesis that later improve subsequent vigor and survival of gametes, developing zygotes, and hatchlings or newborns. Postovulatory contributions begin at ovulation and the gamete's transfer into the genital ducts and end when the parent leaves the eggs or offspring. Because the nature of the contributions differs strikingly between when the gametes are within the genital ducts and when they are expelled, it is convenient to discuss these parental investments as predepositional and postdepositional; also fecundity (number, size, and frequency of the clutch/litter) and other reproductive parameters are not exclusive measures of parental investment but are also measures of an individual's and its population's allocation of resources (see Life-History Patterns, Chapter 11). The difficulty (and unnaturalness) of separating expenditures into theoretical categories is illustrated by attempting to divide the energy invested in an egg's covering into what is and is not necessary for the embryo's survival or by partitioning a male's energy expenditures into what portion enhances his reproductive potential and survival versus what portion provides a suitable developmental habitat and food for the hatchlings.

Selection of egg deposition sites is the most basic or elementary step of postovulatory-postdepositional parental investment and is a universal among amphibians and reptiles. There is little adaptive value to the production of a "healthy" egg or zygote if it is deposited in a lethal environment. Selectivity, however, can appear most casual, for example, toads (*Bufo*) laying strings of eggs helter-skelter in shallow, temporary pools or musk turtles (*Sternotherus*) seemingly dropping their eggs beneath leaf litter. While these examples represent a low level of selectivity, the eggs are still deposited in sites capable of supporting development and eventual hatching of the eggs. Most other amphibians are more selective, for example, bullfrogs (*Rana catesbeiana*) laying eggs only in areas of dense aquatic vegetation or a green seaturtle (*Chelonia mydas*) digging nests only to abandon them whenever site conditions are unsuitable.

Selection and preparation of egg-deposition sites are often energetically costly. Foam nest construction has evolved independently in several lineages of frogs (e.g., *Physalaemus* /leptodactylid, *Rhacophorus*/rhacophorid). Foam nests require the combined efforts of the amplexing male and female to beat the matrix of the emerging eggs into a frothy foam. The foam hardens in the air and provides a shelter from predators, high temperatures, and dehydration. When the tadpoles hatch, they may remain in or near the foam for its shade and to eat it. Digging an egg or brooding cavity requires a high energy commitment. Most turtles require a minimum of an hour to dig the egg chamber and even more time may be spent searching for a suitable nest site, exposing the female to a high risk of predation.

Egg survival is enhanced by brooding. Brooding behavior spans activities from a parent's presence in the vicinity of eggs through guarding to incubation (Fig. 7.5). For many amphibians and reptiles, the actual behavior of a brooding parent (either female or male) remains undocumented, and the protective or selective value of the parent has been tested for only a few species (Table 7.3). Nonetheless, the selective value of brooding is likely high owing to its widespread

FIGURE 7.5 ▼▲▼▲▼▲▼▲▼▲▼▲▼▲▼▲▼▲▼▲▼▲▼▲▼▲▼▲▼▲▼▲▼▲▼

The male glass frog *Hyalinobatrachium valerioi* guarding two egg masses on the underside of a leaf. (R. W. McDiarmid)

occurrence. Brooding behavior occurs in caecilians, salamanders (e.g., *Cryptobranchus*, *Amphiuma*, many plethodontids), frogs (*Leiopelma*, *Alytes*, *Dendrobates*), crocodilians (all species), and squamates (*Eumeces*, *Farancia*, *Ophiophagus*, pythons).

Posthatching or postparturition care is rare in amphibians and reptiles. Crocodilians are well known for helping the hatchlings exit from eggs and nest, transport to a nursery area, and posthatching "baby-sitting." Some frogs have similar parental behaviors; *Pyxicephalus adspersus* herds its tadpoles, and *Hemisus* digs tunnels from nest chambers to ponds.

Live-Bearing

Brooding provides the developing embryos with a guardian who can repel some predators, reduce desiccation, and generally increase survivorship of the clutch. Brooding has a drawback in that the parent and embryos are site-fixed. Predators may consume both the parent and its clutch, or abiotic events (e.g., floods) may kill both. Carrying the embryos allows the parent to escape predators and lethal environments. However, it increases the cost of parental investment owing to the inability to reproduce until the transported embryos depart. Preda-

TABLE 7.3 ▼▲▼▲▼▲▼▲▼▲▼▲▼▲▼▲▼▲▼▲▼▲▼
Examples of Brooding Amphibians and Reptiles

Amphibians	Reptiles
Ambystoma opacum	*Caiman crocodilus*
Amphiuma tridactylum	*Crocodilus niloticus*
Cryptobranchus alleganiensis	
Desmognathus ochrophaeus	
	Cyclura carinata
	Eumeces fasciatus
Hyalinobatrachium colymbiphyllum	*Ophisaurus attenuatus*
Eleutherodactylus coqui	
Hyla rosenbergi	
Leiopelma archeyi	*Ophiophagus hannah*
Leptodactylus ocellatus	*Ptyas mucosus*
	Python molurus
	Sistrurus catenatus

Sources: Salamanders—Ao, Bishop, 1941; At, Baker, 1945; Ca, Smith, 1907; Do, Forester, 1981. Frogs—Cc, McDiarmid, 1978; Ec, Townsend et al., 1984; Hr, Kluge, 1981; La, Bell, 1985; Lo, Vaz-Ferreira and Gehrau, 1972. Crocodilians—Cc, Ouboter and Nanhoe, 1987; Cn, Cott, 1961. Lizards—Cc, Iverson, 1979; Ef, Vitt and Cooper, 1989; Oa, Fitch, 1989. Snakes—Oh, Oliver, 1956; Ptm, Daniel, 1983; Pym, Vinegar et al., 1970; Sc, Reinert and Kodrich 1982.

Note: Parental behavior in the taxa has been verified by laboratory or field experiments or by extensive field observation.

tor risk is potentially greater for the parent because the embyros' weight may hinder escape maneuvers.

The widespread occurrence of embryo transport (live-bearing) in most groups of invertebrates and vertebrates shows its effectiveness for increased survivorship of parents and offspring. Of the extant amphibian and reptile groups, live-bearing is absent only in turtles, crocodilians, and birds. Live-bearing spans a continuum from a few days to months, ending with the completion of development and "birth." This continuum encompasses the full span from ovoviviparity (embryo transport without nutrition beyond the original yolk supply) to viviparity (embryo transport with continuous maternal energy supplements). Its converse is oviparity, the transport of unfertilized eggs or eggs fertilized internally and development arrested until eggs are laid.

External fertilization excludes urogenital adaptations for live-bearing species but does not prevent live-bearing. Anurans have a wide spectrum of ovoviviparous adaptations from short-term transport to retention of embryos through metamorphosis. *Alytes obstetricans* males carry the fertilized egg strings until hatching and then move to pools for release of the tadpoles. Dendrobatids have terrestrial eggs that produce tadpoles, which "climb" onto the male's back and are carried to water for the completion of development. Each female *Gastrotheca* (hylid) has a pouch of skin on her back; the male pushes the fertilized eggs into this pouch with his hindlimbs. Depending on the species, *Gastrotheca* embryos may be carried full-term or have a free-living tadpole stage. Female *Rheobatrachus silus* swallow the fertilized eggs and turn off the digestive system until the froglets are regurgitated. The preceding is only a sampling of the extra- or nonoviducal ovoviviparity in anurans; no frog has evolved nonoviducal viviparity. Among the few species with internal fertilization, embryos of *Eleutherodactylus jasperi* and two species of *Nectophrynoides* develop through metamorphosis in the oviducts, nourished by yolk. *Nectophrynoides liberiensis* and *N. occidentalis* are viviparous; their embryos feed on oviducal secretions.

All caecilians have internal fertilization, but only a few (typhlonectids and some caeciliaids) are viviparous. Similarly most salamanders have internal fertilization, yet only two or three salamandrids are ovoviviparous and two (*Salamandra atra*, *Mertensiella l. anatalyana*) are viviparous. The viviparous embryos of both groups feed on oviducal secretions and possibly lysed ova.

Ovoviviparity and viviparity occur in about 15% of the world's lizards and snakes. With internal fertilization, embryo transport is oviducal. Many species retain the embryos in shelled eggs, and at least a dozen species retain the eggs until the embryo is near term and lay the eggs only a few days prior to hatching, for example, *Opheodrys vernalis* and *Lacerta agilis*. Others (Boinae) have eliminated the shell, although the embryos rely entirely on original yolk for nourishment. Viviparous squamates obtain nourishment via placentae. Placentae are modifications of the extra-embryonic membranes to form a thin-walled vascular plexus adjacent to a similar plexus of the maternal oviducal wall. The nearness of the two vascular beds permits the transfer of nutrients to the embryo and metabolic waste products to the female. Two general types of placentae occur in squamates. The most common one is the chorioallantoic placenta derived from membranes of the same name, and the yolk-sac placentae of three different forms but all sharing the yolk sac as a major fetal membrane component. Ovoviviparous species often have modifications to enhance gaseous exchange between mother and embryos, and both ovoviviparous and viviparous species retain corpora lutea to secrete luteal hormone (LH) to maintain gestation.

There are numerous hypotheses for the evolution of viviparity in squamate reptiles. Most invoke a climatic factor such that ovoviviparity and viviparity are adaptations to permit the survival of embryos in habitats where embryos in eggs could not complete development (e.g., the microenvironment is too cold or too short to allow normal development). Egg retention allows the female to seek microclimates matching the developmental requirements of her embryos. Although this is a suitable explanation for some live-bearing squamates, this hypothesis is totally inadequate for amphibians, because embryos and adults function at the same temperatures. Live-bearing arose multiple times and likely under multiple selection regimes, in all instances as a means to increase embryo survivorship.

DEVELOPMENT

Fertilization initiates cleavage, the division of the large, single-celled ovum into a ball of normal-sized cells (amphibians) or a sheet of cells on one pole of the yolk mass (reptiles). The restriction of embryonic cells to one pole of the egg results from the reptiles' greater energy investment (yolk) in each ovum. The amphibian zygote typically receives enough yolk to fuel its development through organogenesis but not beyond. The embryo must then feed itself (i.e., become free-living) to fuel the completion of its development and metamorphosis to the juvenile stage. The energy provisions of the reptilian zygote are sufficient to fuel its entire development and often the first weeks of juvenile life. These two contrasting sequences are indirect and direct development, respectively. The recently proposed terms "exotrophic" and "endotrophic" development provide a more explicit label for these two modes by stressing the source of energy (Greek *trophe* for food; ex-, outside; endo-, inside). All reptiles are endotrophic; the energy for their development derives totally from the female. Most amphibians are exotrophic with the female providing only part of the developmental energy needs; a few amphibians, such as *Plethodon* salamanders and *Eleutherodactylus* frogs, are endotrophs. With few exceptions, amphibian endotrophs lay terrestrial eggs and a parent broods them.

Exotrophic zygotes display rapid differentiation (tissue and organ formation) with little or no growth. The embryo hatches as a larva capable of feeding itself immediately or soon thereafter. Larval development emphasizes growth (increase in size) and eating is necessary to obtain the energy and materials for this growth. Differentiation continues but at a slower pace until the latter stages of larval development when the larva rapidly transforms (metamorphoses) from its larval body form to the adult body form. Metamorphosis is especially demanding in frogs with their transformation from an aquatic body form to a terrestrial one.

Endotrophic zygotes combine differentiation and growth. The yolk permits an uninterrupted maturation of tissues and organs. When differentiation is complete, the endotroph emerges from the egg in a minature adult body form.

Because the parental investment per egg is significantly different between exotrophs and endotrophs, clutch size is strikingly different. The former produces tens to thousands of small (<5 mm) eggs; the latter commonly produces less than ten large (>10 mm) eggs, although larger endotrophs may produce over a hundred eggs per clutch, for example, seaturtles. The average duration of development (6–10 weeks) is much the same for most exo- and endotrophs and in both is highly dependent on temperature (Table 7.4).

Amphibian Modes

Categorization of reproductive modes builds on the type of fertilization and development and yields four major divisions: external fertilization (EF) and external development (ED); external fertilization and internal development (ID); internal fertilization (IF) and external development; and internal fertilization and internal development . Each of these modes can be subdivided further by using developmental habitats and parental and larval morphology and behaviors. For example, 29 modes have been proposed for the anurans; these may be useful for tracing the evolution of specific reproductive adaptations. The major divisions permit the identification of general evolutionary trends, some of which were described in the preceding sections.

All major modes occur in amphibians. EF–ED is the primitive condition, absent only in caecilians and present in about 20% of the salamanders and 98% of the frogs. EF–ED salamanders have aquatic reproduction and free-living larvae (i.e., indirect development). Many EF–ED frogs also have aquatic, free-living larvae; however, direct development is also common, having arisen independently one or more times in the leptodactylids, microhylids, myobatrachids, ranids, rhacophorids, and sooglossids. These frogs lay large-yolked terrestrial eggs in which the embryo skips the tadpole stage and development (direct) proceeds until the young hatches as a miniature adult. Typically, a parent broods direct-developing eggs. Other frogs, but many fewer, have nonfeeding larvae. The EF–ID mode is confined to frogs, and the internal development is nonoviducal, that is, in vocal sacs, back-pouches, and others as mentioned earlier. Both indirect (larval) and direct development occur in this mode. Internal fertilization occurs in most salamanders and all caecilians; their development may be indirect or direct. IF–ED is uncommon in frogs, and IF–ID (ovoviviparity and viviparity) is rare in both salamanders and frogs but common in caecilians.

Modes show some associations with climate and habitat. Only ED modes, mainly with larval stages, occur in cold temperate habitats. ED and direct development become increasingly common from cool temperate to tropical climates; indirect development (frogs) occurs in all climates. Direct development is strictly nonaquatic although moist microhabitats or embyro-parental transport is required. Mode of fertilization has no climatic association but does relate to some aspects of clutch size.

Within a mode and a lineage, clutch size is usually directly proportional to body size. Egg size is inversely proportional to clutch size. Within mode and lineages, there is no apparent relationship between egg/clutch size and climate;

TABLE 7.4
Selected Reproductive Parameters for Some Amphibians and Reptiles

Taxon	Clutch and litter size	Egg and fetus size	Duration	Female size	Mode
Ambystoma talpoideum	226–401	4	80+	46	IF–ED
Cryptobranchus alleganiensis	319–450	6	1080+	330	EF–ED
Necturus maculosus	32–91/54	5	65[a]	127	IF–ED
Ascaphus truei	50–85/68	3	36	45	IF–ED
Bufo japonicus	6000–14,000	2	120±	84	EF–ED
Eleutherodactylus coqui	16–41/28	4	21	35	EF–ED
Hyla rosenbergi	1780–3050/2350	2	40	87	EF–ED
Chrysemys picta	4–6/5	32	76	80	IF–ED
Caretta caretta	64–198/126	41	55	850	IF–ED
Alligator mississippiensis	2–58/39	74	65	1830[b]	IF–ED
Cyclura carinata	2–9/4	31	90	225	IF–ED
Gymnophthalmus underwoodi	1–4/2	9	50	40	IF–ED
Varanus komodoensis	1–30/18	85	240	745	IF–ED
Diadophis punctatus	1–10/4	25	60	280	IF–ED
Crotalus viridis	2–8/5	267[c]	110	760	IF–ID
Pituophis melanoleucus	4–15/8	50	55	790	IF–ED

Sources: At, Shoop, 1960; Ca, Bishop, 1941; Peterson et al., 1983; Nml, Shoop, 1965; At, Metter, 1964; Bj, Maeda and Matsui, 1989; Hr, Kluge, 1981; Cp, Ernst, 1971; Cc, Caldwell, 1959; Am, Joanen, 1969; Cc, Iverson, 1979; Gu, Hardy et al., 1989; Vk, Auffenberg, 1981; Dp, Fitch, 1975; Cvo, Macartney and Gregory, 1988; Pmd, Parker and Brown, 1980.

Note: Values for clutch and litter size are range/mean; egg and fetus size, mean of longest dimension (mm); duration, mean number of days to metamorphosis, hatching, or birth; female size at sexual maturity, snout–vent length except carapace length in turtles (mm); mode, see definitions and abbreviations in text.

[a] Duration to hatching, external evidence of metamorphosis indistinct.

[b] Total length.

[c] Neonate snout–vent length.

however, large clutches of small eggs are invariably aquatic and indirect. IF–ID species seldom have clutches greater than 10 embryos and often 1–3. IF–ED clutches may exceed 50 eggs but most are much smaller. Large clutches are possible only with EF–ED, although some EF–ED species may lay only 1–2 eggs, usually with direct development.

Hatching and Metamorphosis

In amphibians, hatching may occur early in embryogenesis (indirect development) or at the conclusion of embryogenesis (direct development). Although these two hatchling types are very different creatures, both face the same obstacle, the protective gelatinous egg capsules. The hatching mechanism is known only for a few species, but all share "hatching" glands on the snout and head of the embryo. These secrete proteolytic enzymes that weaken or dissolve the capsules and permit escape from the capsules. Only *Eleutherodactylus* has an egg tooth that slices the capsules; enzymatic weakening of the capsule is a likely precursor.

Metamorphosis is the shift from an aquatic embryonic stage to a terrestrial growth and maturation stage. The shift is nearly imperceptible in caecilians and salamanders but dramatic in frogs. All anurans change skin structure, lose the lateral lines system, resorb gills and develop lungs, modify feeding mechanism and behavior, convert from axial to appendicular (limbed) locomotion (with major musculoskeletal modifications in anurans), and alter their physiology.

During much of the larval stage, growth is emphasized over differentiation, but both occur throughout. Roughly in the last quarter of larval life, metamorphosis begins. Its timing is variable within and among species. Within a species, growth and differentiation (tissue and organ development) respond differently to temperature, for example, with decreasing temperatures, growth continues after differentiation stops. These temperature effects and others, such as crowding, affect the timing of and size at metamorphosis. Once begun, metamorphosis proceeds rapidly to its climax. Thyroxin is the stimulant for this final transformation. The thyroid is present early in larval life, but it is inactive, possibly suppressed by the absence of TSH (thyroid-stimulating hormone) from the pituitary, and the body tissue is insensitive to thyroxin. Metamorphosis begins with the production of TSH and body tissue stimulation by thyroxin.

Paedomorphosis

Many salamanders are paedomorphic, retaining larval traits as adults. This phenomenon was first discovered when axolotls, *Ambystoma mexicanum*, were brought into A. Duméril's laboratory, and some reproduced without metamorphosis. Even more astounding (for the 1860s) was that others from this laboratory colony did metamorphose into typical terrestrial salamanders. We have since come to recognize that there are many grades of paedomorphosis from reproducing larvae to normal-appearing adults that harbor some larval characteristics. Also paedomorphosis arises either from slow development of somatic tissue relative to the gonads (neoteny) or from precocial maturity of gonads truncating somatic development (progenesis).

Distinguishing between these two paedomorphic origins is often difficult; however, most overtly paedomorphic salamanders (e.g., *Cryptobranchus*, *Proteus*, *Siren*, *Ambystoma tigrinum*, *Haideotriton*) are neotenic (Fig. 7.6). Neotenic sala-

FIGURE 7.6 ▼▲▼

The neotenic salamander *Siren intermedia* (lesser siren). (R. W. Barbour)

manders may be facultative (capable of metamorphosis) or obligate neotenic (incapable of metamorphosis); experimentally, these two types are recognized by their response to thyroxin; the former complete metamorphosis at low concentrations, the latter show no or limited response to high thyroxin concentrations. The latter are fixed genetically to their larval form; the faculative neotenes are not. However, studies of species (e.g., *Ambystoma talpoideum*, *Notophthalmus viridescens*) with neotenic and nonneotenic populations indicate that the neotenic response does have a genetic base and may appear relatively quickly under intense selection. Paedomorphosis is known in only one frog species, the Brazilian hylid *Sphaenorhynchus bromelicola*, in which gonad maturation begins prior to metamorphosis, and over 50% of a cohort has mature gonads at the end of metamorphosis.

Reptilian Modes

The major modes for reptiles are two: internal fertilization and external development (IF–ED), and internal fertilization and development (IF–ID = ovovivi- or viviparity); development is direct in all reptiles. IF–ED is the primitive reptilian condition and remains the common mode for extant reptiles. It is the exclusive mode for turtles, crocodilians, and *Sphenodon*; it is the dominant mode in squamates. IF–ID has arisen independently in numerous squamate lineages, and in many cases, viviparity occurs in those species at the highest latitudes or elevations (= cold habitats).

As in amphibians, clutch size may be directly proportional to body size (e.g., snakes, turtles), although many lizards have fixed clutch sizes (commonly two eggs, e.g., many gecko genera and the skink *Emoia*). Likely because of internal development, egg size is directly proportional to body size when compared within lineages.

Developmental Environment

All reptiles lay terrestrial eggs. Although reptilian eggs/embryos drown in waterlogged nests, moisture must be available to the developing embryos for proper growth and differentiation. The requirements are variable depending on the species and the nature of its eggshell. Generally leathery eggs with little calcium in the shell require nests of higher humidity or soil moisture than the hard (heavily calcified) eggs. At least in turtles, the amount of moisture available will determine the size of the hatchling; smaller hatchlings come from low-moisture nests.

Reptilian development is also temperature dependent and faster at high temperatures. For most species, temperatures below 18°C are lethal, and for many <24°C reduces hatching success. Thus, nest site selection is critical to ensure high temperatures and adequate moisture for the duration of embryogenesis. Many crocodilians build nests of plant material, presumably relying on the heat generated by decomposition to incubate the eggs. Some varanids and a few other lizards and snakes use termitaria with their regulated temperatures as nest sites; but for most species, nest sites are in open areas and insolation is the heat source.

An amazing discovery in 1971 revealed that incubation temperatures influence the sex of hatchling turtles (*Testudo graeca*, *Emys orbicularis*). This datum was surprising because sex was assumed to be genetically controlled in vertebrates. Subsequent studies have shown that this phenomenon of temperature-dependent sex determination (TDSD) is widespread in reptiles. Typically vertebrates display genotypic sex determination (GSD) with an individual's sex irreversibly fixed by its genotype. The most common GSDs in reptiles (amphibians also) are male/female heterogamety of the XY/XX (male heteromorphic), ZZ/ZW (female heteromorphic), and homomorphic sex chromosomes (sex chromosomes undifferentiated, but sex determination as in heterogametic forms). To date, TDSD appears ubiquitous in crocodilians, widespread in turtles, and limited in lizards (only in geckos and lacertids).

The general mechanism of TDSD is that sex determination usually occurs in the second trimester of development and the "average" temperature during that period regulates gonad differentiation. Within the threshold temperature range, the gonads may become either ovaries or testes, and above or below predominantly one type or the other. In most crocodilians and lizards, males result from high temperatures, females from low ones; in turtles, females develop at high temperatures and males at low ones; in a few crocodilians, turtles, and lizards, females develop at high and low temperatures, males at intermediate ones. The adaptive significance of TDSD remains an enigma and explanatory hypotheses are controversial. The main feature would seem to be the head start in growth advantage given to the high-temperature sex (i.e., larger hatchlings).

Hatching in Reptiles

Several mechanisms have arisen in reptiles for emergence from the egg. In turtles, a keratinous protuberance on the tip of the snout (egg-carbuncle) slices through the various encasing layers; since turtle embryos extract calcium from the eggshell during their embryogenesis, the eggshell is easier to penetrate. Crocodilians and *Sphenodon* also possess an egg-carbuncle on the snout. Squamates have a special egg tooth that develops on the premaxillary and projects forward. Emergence is assisted by a parent only in crocodilians.

Hatching is often an extended affair, requiring several hours to a day for complete emergence. A few turtles have delayed emergence, hatching in autumn but not emerging from the nest until spring.

GROWTH

Amphibians and reptiles display two pulses of growth, embryonic and juvenile growth. Embryonic growth is greater than juvenile growth, because embryos have an abundant, high-quality energy resource (yolk) that requires little energy for processing. Juveniles (also free-living amphibian larvae) face a variable food supply with lower energy yield and must expend more energy obtaining the food and avoiding predation. From hatching/birth to sexual maturity, a reptile or amphibian will increase 3- to 20-fold in length and in some species over 100-fold in mass. Growth may or may not continue indefinitely throughout life; data in most cases are inadequate to state whether a species has definite or indefinite growth.

Mechanics of Growth

Growth is the addition of new tissue in excess of that required for the replacement of worn-out or damaged tissue. As a cellular process, growth rate in ectotherms depends on temperature, slowing and ceasing as temperatures decline; excessive temperatures also slow or halt growth because maintenance and metabolic costs exceed energy procurement. Growth also relies on the availability and quality of food. In this respect, ectotherms have an advantage over endotherms, by ceasing to grow during food shortages and renewing growth when food becomes available.

All tissues grow during juvenile life, although their rates vary; however, our attention is usually fixed on changes in overall size and most often on measures of length. Mass is more variable owing to numerous factors (hydration, gut contents, reproductive state, etc.) changing an animal's weight yet not changing its overall length. Skeletal growth is the ultimate determiner of size as the skeleton is the supportive framework of the animal. The skeletal elements of amphibians and reptiles usually lack epiphyses and grow by apposition (one layer on top of another). These attributes offer the potential of continual growth and are one reason for the assumption of indeterminate growth in these animals. Other reasons for assuming indeterminate growth are the large sizes of some species and the continuation of growth long after sexual maturity (contrasting sharply with growth stopping soon after maturity in endotherms).

Both indeterminate (attenuated) and determinate (asymptotic) growth (Fig. 7.7) exist in amphibians and reptiles, but the evidence for one or the other is lacking for most species. Indeterminate growth is likely less frequent than commonly assumed. Adult size for most species lies within a narrow range, suggesting that growth stops. Some reptilian species with epiphyses show fusion of the epiphyses in older adults. The difficulty of distinguishing between the two growth patterns in natural populations is that a narrow adult size range may only indicate that high mortality truncates the growth/size potential of the species or population.

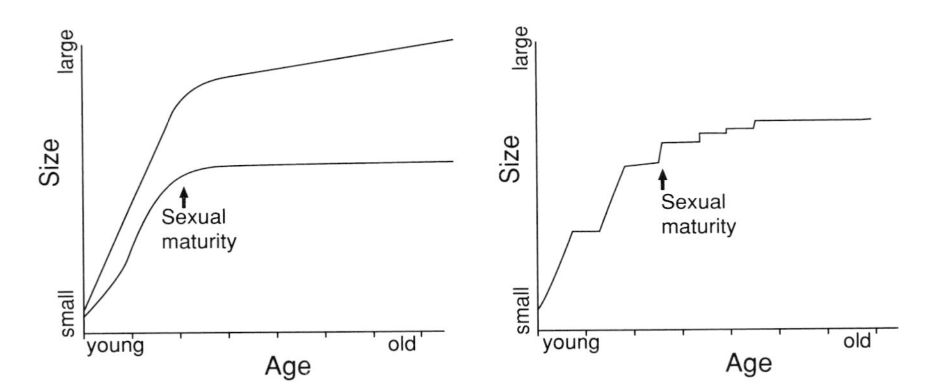

FIGURE 7.7 

General growth pattern trends for amphibians and reptiles. Left graph: top growth curve shows constant growth rate as juvenile, and upon attaining sexual maturity, growth slows but continues (indeterminate growth); bottom curve shows a sigmoid growth pattern during juvenile stage, and at sexual maturity, growth slows and soon stops (asymptotic or determinate growth). Right graph: hypothetical growth for an ectotherm in a seasonal environment follows a pattern of rapid growth during equable seasons and no growth during adverse seasons.

Whatever the end point, juvenile growth is rapid and slows as sexual maturity is approached. Most juvenile growth fits one of two curvilinear patterns: parabolic growth beginning rapidly and remaining so for most of juvenile life, or sigmoid, slow initial growth becoming rapid and then slowing again. Both patterns show a plateauing effect associated with maturity, because of the reallocation of energy resources to reproduction. Individual growth curves are not smooth (Fig. 7.7), particularly so in ectotherms; growth proceeds fast or slow depending on the abundance of food, halts for months at a time in species from seasonal environments (including tropical wet and dry seasons), and proceeds rachetlike for the first few years of sexual maturity (because energy is alternately allocated to reproduction and growth).

The ultimate size of an individual depends on its genetic potential, size at hatching, abundance and quality of food during juvenile growth, and its sex. Heredity sets the potential range of growth rates and size or age at sexual maturity. Beginning with hatchlings of the same size, faster growth or longer juvenile life yields larger adults, unless dealing with different species or different sexes (i.e., sexual dimorphism). These factors and others yield the variations in adult size within and between species.

Age

The importance of age is not how long an individual lives but the time required to reach major life-history events. Hatching/birth, sexual maturity, and reproductive senility are the principal events; reproductive periodicity or the time interval between production of offspring, although not an event, is another critical age-related aspect of an individual's life history. In biphasic amphibians, two intervals are critical: embryogenesis within the egg and larval period to metamorphosis. All of these events are regularly subjected to selection in the life of a species.

Age at sexual maturity ranges from 4 to 6 months (*Arthroleptis poecilonotus*/ranid) to 7 years (*Cryptobranchus*/hellbender) for amphibians and 2–4 months (*Anolis poecilopus*/iguanid) to 40–50 years (*Chelonia mydas*/green seaturtle) for reptiles. These marked extremes reflect differences in adult size only in part, because not all small species mature so quickly or large ones so slowly (Table 7.5). Age of maturity is a compromise among many selective-pressure variables with the "goal" of maximizing an individual's contribution to the next generation. Maturing and reproducing quickly is one method to achieve this goal; however, small body size reduces the number and/or size of offspring and also smaller adults tend to experience higher predation. Maturing later and at a larger body size permits the production of more and/or larger offspring but increases the probability of death prior to reproducing and may yield a smaller total lifetime output of offspring. The resulting diversity in size and age at sexual maturity, number and size of offspring, and the frequency of reproduction illustrate the options adopted for attaining reproductive success.

TABLE 7.5
Natural Longevity of Select Amphibians and Reptiles

Taxon	Adult size	Age at maturity	Maximum age
Cryptobranchus alleganiensis	330	84	300
Desmognathus quadramaculatus	73	84	124
Eurycea wilderae	34	48	96
Bufo americanus	72	36	60
Rana catesbeiana	116	36	96
Chrysemys picta	119	72	360
Geochelone gigantea	400	132	840±
Trachyemys scripta	195	50	288
Sphenodon punctatus	180	132	420+
Cnemidophorus tigris	80	21	94
Galloti stehlini	120	48	132+
Uta stansburiana	42	9	58
Diadophis punctatus	235	32	180+
Pituophis melanoleucus	790	34	180+

Sources: Salamanders—Ca, Peterson et al., 1983; Dq, Bruce, 1988; Organ, 1961; Ew, Bruce, 1988. Frogs—Ba, Kalb and Zug, 1990; Rc, Howard, 1978. Turtles—Cp, Wilbur, 1975; Gg, Bourne and Coe, 1978; Grubb, 1971; Ts, Frazer et al., 1990. Tuataras—Sp, Castanet et al., 1988. Lizards—Ct, Turner et al., 1969; Medica and Turner, 1984; Gs, Castanet and Baez, 1991; Us, Tinkle, 1967; Medica and Turner, 1984. Snakes—Dp, Fitch, 1971; Pmd, Parker and Brown, 1980.

Note: Body size is for female at sexual maturity (mm, snout–vent length except carapace length for turtles); age of maturity for female (months); maximum age (mo).

Longevity is usually linked conceptually with long life, when the actual focus is length of life, either brief or long, and the reproductive life span of an individual or species. The reproductive life span of some species is a single reproductive season and a total longevity of one year (e.g., *Uta stansburiana*), whereas for others the reproductive life span is a decade or longer (*Geochelone gigantea*). Annual or biennial species have little time for growth so are usually small species; the opposite is not true for the long-lived species where a cryptic or secretive life-style may serve as well as large size.

CHAPTER 8

Dynamics of Reproduction

Mating is the final and least energy-demanding act of reproduction. Preparations for mating embrace the production and maturation of gametes and the physiological, morphological, and behavioral alterations to attract and retain mates. Although these preparations are not apparent until mating nears, they demand a major portion of an individual's energy resources and the synchrony of reproductive readiness.

TIMING AND RHYTHMS

Synchrony of reproductive readiness is achieved either by continuous reproduction within a population or by having members of the population mate nearly simultaneously. Continuous (acyclic or aseasonal) and cyclic (seasonal or discontinuous) reproduction arise not from selection on the reproducing adults but mainly through selection on hatchlings/neonates and incubating eggs. These two stages are fragile; their appearance must coincide with the most hospitable conditions permitting maximum survivorship, that is, minimum physiological stress, few predators, abundant food supply. Thus, natural selection molds each population's reproductive physiology to local conditions. Species in seasonal environments have cyclic reproduction to avoid harsh climatic extremes; species in aseasonal environments may have continuous reproduction.

Continuous reproduction is less widespread in amphibians and reptiles, even in the tropical species, than might be expected; in part because much of the tropics and subtropics are seasonal (wet/dry, wet/wetter). This seasonality as in temperate climates results in fluctuations in food supply and predator densities and less favorable climatic conditions. Species with continuous reproduction show two general patterns: individuals capable of reproducing year-round and individuals reproducing cyclically but asynchronously so the population is continuously reproductive although individuals are not. Females are commonly cyclic in continuous reproductive populations, because of the high energy demands of oogenesis/vitellogenesis. In cyclic reproduction, each individual alternates between reproductive readiness and quiescence; the duration of each state is variable, and often the sexual cycle of the entire population is synchronized (Fig. 8.1).

These three general patterns are not discontinuous, and various combinations generate the diverse reproductive repertoire of amphibian and reptilian reproductive behaviors. One striking feature of cyclic reproduction is the contrast between female and male gametogenetic cycles within the same species. Gametogenesis of the two sexes may be synchronized; sperm and ova mature simultaneously, mating and fertilization occur concurrently or nearly so with gamete maturation. One variant of this pattern is delayed fertilization or development.

Gametogenesis of the two sexes is often asynchronous. Typically, spermatogenesis finishes well before oogenesis; nonetheless, the female is receptive and mating occurs at the conclusion of spermiogenesis and the female stores the sperm until ovulation. In variants, mating begins at the conclusion of spermiogenesis and continues until ovulation, or the male stores sperm until the female is receptive at the conclusion of her oogenesis. Overlaying these patterns with a climatic or seasonal periodicity increases the number of possible variants, for example, spermiogenesis complete in late summer or early fall with mating in fall, spring, or fall and spring.

A Multitude of Patterns

Temperate and most tropical amphibians are cyclic. Their requirements for an aquatic environment for their eggs and larvae demand a synchronization of reproduction with periods of high rainfall. Most cyclic amphibians are also

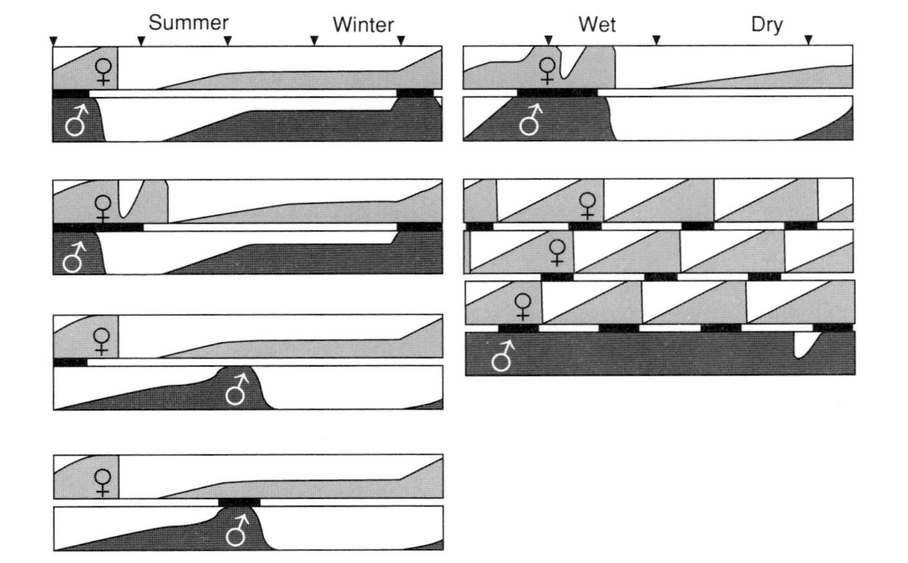

FIGURE 8.1 ▼▲▼

Major reproductive patterns for amphibians and reptiles. Left diagrams, temperate-zone cyclic patterns; right diagrams, tropical patterns. The height to the shaded areas denotes the level of gametogenic activity and maturity of gametes; the dark bar between female and male diagrams marks the mating season. Left: synchronous gametogenesis, female monestrous (top); synchronous, female polyestrous; asynchronous gametogenesis, male sperm storage; asynchronous, female sperm storage (bottom). Right: cyclic synchronous gametogenesis, female polyestrous (top); acyclic reproduction of population, continuous spermatogenesis, cyclic oogenesis (bottom).

monestrous with females producing one set of ova during each reproductive cycle. In cool temperate areas, most frogs (*Bufo*, *Hyla*, *Rana*) are spring breeders. Upon mating (potentially multiple times for males), the gonads regress and are quiescent for varying durations (weeks to months). Gametogenesis begins anew and is usually completed prior to hibernation, particularly in males and the very early spring breeders (Fig. 8.1). Summer breeders (e.g., *Hyla chrysoscelis*, *Rana clamitans*) often have a more extended breeding cycle with some females polyestrous (producing two or more egg clutches in a single breeding season). In warmer climates, anuran reproduction remains cyclic but breeding periods are often longer, and multiple clutches are more likely and less tied to spring-early summer breeding. Some species (*Rana berlandieri*) have fall and spring breeding, others may breed throughout spring and summer (*Limnaoedus ocularis*), and others are winter breeders (*Pseudacris ornata*). The time of breeding remains dependent on the availability of suitable larval habitats.

Larval survivorship also controls the breeding behaviors of tropical anurans.

In habitats with distinct wet and dry seasons, tropical anurans are commonly cyclic and often have short breeding periods that may begin just prior to the onset of the wet season, or early or at the end of the wet season. Some species (*Physalaemus pustulosus*, *Bufo marinus*) are continuous breeders, and in equable conditions, reproduction occurs in every month. In seasonal areas, reproduction occurs opportunistically. During dry intervals, the number of gravid female marine toads gradually increases, because none have the opportunity to lay their eggs. Thus, the number of gravid females breeding with the first adequate rain (females cyclic, males acyclic) is directly proportional to the length of the preceding dry period. A similar phenomenon occurs in the túngara frog *P. pustulosus* (females and males acyclic) with the added benefit of larger clutch size upon renewed breeding. Female túngaras have five to six gamete generations developing within their ovaries and upon the maturation of one a new one starts; hence, when breeding is delayed, the number of mature ova gradually increases as successive generations mature.

Among the salamanders, tropical plethodontids are potentially the only acyclic group; however, although males show continuous spermatogenesis and mating is possible year-round, females likely have cyclic oogenesis because they often brood their egg clutches. Temperate salamanders are cyclic with two common patterns: winter-spring mating and egg deposition (most ambystomatids, hynobiids, and salamandrids) or late summer-fall mating and spring egg deposition (*Necturus*, *Salamandra*, and many plethodontids). In the former, male and female gametogenetic cycles are synchronized and fertilization occurs immediately (salamanders with external fertilization) or shortly after mating. Typically, these synchronous species reproduce annually. The asynchronous species are often biennial; the energy demands of brooding the eggs or of live-bearing require a year free of parenting for the female to produce a new clutch of ova. The tropical caecilians probably share the acyclic male–cyclic female pattern with the tropical salamanders.

Reproduction in turtles is cyclic; in all but a few species, gametogenesis is annual for both sexes but asynchronous. In temperate species, males complete spermiogenesis in late summer and sperm are stored in specialized receptacles in the male's genital tract; the female completes oogenesis in the spring and then mating, fertilization, and egg-laying occur. Some individuals in some populations (*Trachemys scripta*) may be polyestrous and produce multiple clutches in a single reproductive season, likely using sperm from that spring's mating. Seaturtles usually reproduce on a multiple-year (2–5) cycle and are polyestrous (3–7) during the year of reproduction with about a two-week interval between each egg-laying bout. Mating may occur only at the beginning of the egg-laying season or regularly following each egg-laying. The multiple-year cycle for the female is assumed to be related to the high energetic demands associated with egg production and reproductive migrations. The males may be on an annual or multiple-year cycle, though this remains unknown. Crocodilians are also cyclic, annual breeders with synchronous gametogenesis.

Squamates have a variety of patterns: cyclic and acyclic, synchronous and asynchronous gametogenesis, monestrous and polyestrous, annual to multiple-year cycles. Only a few examples can be presented here. Oviparous lizards from temperate areas usually display a synchronous pattern with spring gametogenesis and early summer mating and egg-laying (*Eumeces fasciatus*, *Xantusia vigilis*) or an asynchronous pattern with fall spermatogenesis and either early winter or spring mating (*Sceloporus*). Tropical oviparous lizards may show the same general cyclic patterns in strongly seasonal habitats or an acyclic pattern in climatically more uniform habitats; however, in the latter case, the males are often acyclic and females cyclic but some females are gravid (bearing mature ova) year-round or nearly so (*Carlia bicarinata*, *Chamaeleo hohneli*). Anoles are tropical lizards and have a unique pattern of continuous egg production for several months although not circumannual. Ovum maturation alternates between left and right ovaries (allochronous); both ovaries are in active oogenesis but out of phase, so when ovulation (a single ovum) occurs on one side, maturation of an ovum in the opposite ovary begins. Males show continuous spermatogenesis in both testes. Snakes show the same diversity of reproductive patterns as lizards. Temperate species are cyclic with asynchronous gametogenesis; spermatogenesis begins in late spring and is complete by autumn, and mating is usually in the spring coinciding with ova maturation. The majority of tropical species are also cyclic with synchronous gametogenesis usually of long duration.

Timing and Mechanisms

Among the dozens of different reproductive patterns in amphibians and reptiles, the majority are cyclic, and the acyclic ones also have some level of periodicity (i.e., population acyclic but not all members). The costs of reproduction require periodic recharging of energy and nutrient stores—especially in females, if not in males, because of the higher energy expenditure for oogenesis. In temperate areas, periodicity could result simply by falling temperatures halting body functions and rising ones turning them on. While temperature does stop and start physiological activities in ectotherms, the control and coordination of reproductive rhythms spring from the interplay among endogenous rhythms and exogenous ones (light/day length, temperature, rainfall, food); however, few endogenous rhythms have been identified, even within single species. The known ones are principally refractory rather than inherent, free-running cycles. In these instances, gonadal tissue does not respond to gonadotrophins or external cues at the end of the gametogenic cycle.

Gametogenesis, steroidogenesis, and the fat body cycle are intimately linked processes. Since the fat bodies provide the energy resources for gametogenesis (especially, for vitellogenesis), their size increases prior to gamete formation and declines as mature gametes are produced. The production of sex hormones (androgens, estrogens) is also in phase with gametogenesis and involves gonadal tissue (Sertoli cells and interstitial tissue in testes, follicle cells and corpora lutea in ovaries) (Fig. 8.2). Testosterone is the principal androgen in amphibians and reptiles, progesterone and estradiol the main estrogens. Commonly, although not universally, estradiol levels are greatest at the beginning of vitellogenesis and progesterone at ovulation; testosterone peaks with spermiogenesis and mating.

The gametogenic and steroidogenic activities are regulated by gonadotrophins (FSH/follicle-stimulating hormone and LH/luteinizing hormone) secreted by the pituitary gland. The stimulatory and inhibitory actions of the amphibian and reptilian gonadotrophins are different (not less well developed) from their action in mammals. Unquestionably, some of the difficulties in deciphering gonadotrophin function in these ectotherms have resulted from experimenters using hormone extracts from mammals rather than from amphibians or reptiles. Amphibians, crocodilians, and turtles have FSH- and LH-like gonadotrophins, and squamates a single FSH/LH-like gonadotrophin. In both amphibians and reptiles, the gonadotrophins have little direct influence on growth, gameto-

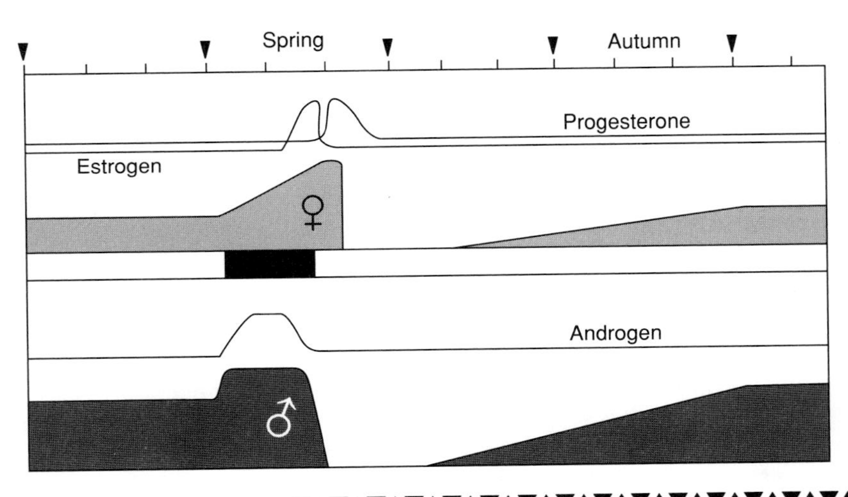

FIGURE 8.2 ▼▲▼

Schematic diagrams of sex steroid production in relation to gametogenic cycle of a temperate-zone reptile. Steroid levels match the peaks of gametogenesis; androgen production begins simultaneously with spermiogenesis and continues until the testis regresses; estrogen production occurs during final maturation of ovarian follicles, stopping at their maturation and ovulation. Corpora lutea produce progesterone, which continues while ova remain in the oviducts; production declines and corpora lutea degenerate with egg-laying, but in viviparous taxa, progesterone produced throughout pregnancy.

genesis, and estrogen production, whereas FSH may stimulate ovarian progesterone and testicular testosterone production.

External cues are necessary to ensure the synchrony of an individual's reproductive preparations with those of other members of its population. Because early experiments emphasized birds, photoperiod was assumed to be the principal timing cue for amphibians and reptiles as well. Instead, temperature is the major cue for temperate-zone amphibians and reptiles and its effects may or may not be modulated by photoperiod. The situation for tropical species is less clear; both temperature and photoperiod fluctuations are small, and it is uncertain whether together or individually they are sufficiently distinct to serve as a timing cue. Major temperature changes (although minor in total degrees) occur with the change in wet/dry seasons in the tropics, and rainfall and humidity or temperature or both may be the timing cues. Rainfall/humidity is critically important for reproduction in temperate-zone amphibians and many reptiles as well; however, the availability of adequate moisture serves as a permissive factor rather than a timing cue. Food availability is also a permissive factor, modifying reproductive output rather than timing its occurrence. Sexual/social interactions often serve as the ultimate cue to initiate mating, stimulate ovulation, or begin egg-laying; such interactions, particularly in male–male interactions, may be inhibitory.

Physical and Behavioral Indicators

Special reproductive colors or structures and special behaviors announce the reproductive readiness of males and females. These secondary sexual dimorphic features are most common in males. Females invest their energies in the production of gametes, usually proclaimed by abdomens swollen with eggs. Otherwise, females commonly retain juvenile coloration, whereas males announce their sex with new colors or patterns and often with new or extravagantly modified structures. The male's modifications serve to attract and retain mates, and often to defend mates, territory, or calling site.

Dimorphic color patterns occur sporadically within the amphibians. Among the salamanders, only male *Triturus* (salamandrids) commonly have brighter colors and bolder patterns. Similarly, only a few anurans are color dimorphic, for example, some *Bufo* males lose their juvenile cryptic patterns and become more uniformly colored and more spinose. Male crocodilians, snakes (most), and turtles (most) match the females' coloration. The Asian river turtles *Batagur* and *Callagur* are exceptions; in the former, the skin of males darkens to black contrasting sharply with the lighter shell. Sexually dimorphic color patterns are common among many lizards. In some species (*Sceloporus*, *Anolis*), males assume their adult colors at maturity (colors brightening during the breeding season) or males have a similar pattern to the females' that is transformed by bright colors when sexually active (*Agama*). The contrast between sexually dimorphic and

Dimorphism in tympana size in the adult green frog *Rana clamitans* (female, left; male, right). (R. W. Van Devender)

monomorphic taxa usually emphasizes the contrast between those relying on vision in their reproductive behaviors versus those depending on auditory, chemical, and tactile signals.

Anurans are noted for their vocalizations and are dimorphic in the occurrence of vocal sacs (almost exclusively in males). Of the half-dozen types of vocal sacs, all are partially regressed until the breeding season when they enlarge and often become pigmented (black/gray in bufonids, yellow/tan in hylids). A less well documented but likely phenomenon is when the larynx and associated vocal structures undergo morphological alterations. Dimorphism in the auditory apparatus of other vocalizing amphibians and reptiles (Table 8.1) is not known.

Multiplication or enlargement of glandular structures is common among taxa using chemical communication during courtship and mating. The secretions in most cases appear to identify the bearer as the correct species and have a stimulatory role during courtship with conspecifics. Male salamanders characteristically have swollen cloacal lips when breeding, resulting from the enlargement of the cloacal glands. Courtship (hedonic) glands are evident on the chins of many plethodontids, and although not as evident, glands on the back and base of the tail become active. Many male anurans also develop skin glands on the chin, chest, belly, or upper limbs. The function in most species is not known, but those of the rotund microhylids (e.g., *Breviceps*) "glue" the small male to the

TABLE 8.1 ▼▲▼▲▼▲▼▲▼▲▼▲▼▲▼▲▼▲▼▲▼▲▼▲▼▲▼▲▼▲▼▲▼▲▼▲▼
Vocalizing Taxa of Amphibians and Reptiles, Exclusive of Anurans

Taxon	Frequency	Taxon	Frequency
Ambystomatidae	+	*Sphenodontidae	+ + +
Ambystoma maculatum			
Amphiumidae	+	Agamidae	+
Cryptobranchidae	+ +	*Brachysaura minor*	
Andrias davidianus		Anguidae	+
Dicamptodontidae	+ +	*Ophisaurus apodus*	
Dicamptodon ensatus		Chamaeleonidae	+
Plethodontidae	+ +	*Chamaeleo goetzei*	
Aneides lugubris		Cordylidae	+
Salamandridae	+	*Cordylus cordylus*	
Triturus alpestris		*Eublepharidae	+ + +
*Sirenidae	+ +	*Coleonyx variegatus*	
Siren intermedia		*Gekkonidae	+ + +
		Gekko gecko	
		Lialis burtonis	
*Testudinidae	+ +		
Geochelone gigantea		*Iguanidae	+
		Anolis grahami	
*Alligatoridae	+ + +	*Lacertidae	+ +
*Crocodylidae	+ + +	*Gallotia stehlini*	
Gavialidae	+ + +	Scincidae	+
		Mabuya affinis	
		Teiidae	+
		Cnemidophorus gularis	

Sources: Salamanders through *Triturus,* Maslin, 1950; *Siren,* Gehlbach and Walker, 1970. Turtles—Gans and Maderson, 1973; *Geochelone,* Frazier and Peters, 1981. Crocodilians—Garrick et al., 1978; gharial, Whitaker and Basu, 1983. Tuatara—Gans et al., 1984. Lizards—*Anolis,* Milton and Jenssen, 1979; Lizards, Böhme et al., 1985.

Note: Families marked with an asterisk have one or more species presumably using vocalization for intraspecific communication. The frequency of vocalization within a family or higher group is subjectively estimated: + + +, more than 50% of species; + +, moderate; +, rare, one or few species in a speciose group. Some examples of voiced species are included.

much larger female during amplexus. Most reptiles possess cloacal glands whose secretory activity is greatest during the breeding period. Males (infrequently females) of many lizards bear a series of precloacal and/or femoral pores, the functions of which have remained uncertain even after considerable experimentation; nonetheless their activity is greatest in sexually active individuals. The skin lipids of some female snakes, although not secreted by distinct glands, elicit courtship behavior in males.

The development of keratinous spines and pads is a regular feature of male amphibians that amplex their mates, particularly in an aquatic environment or where other males attempt to dislodge an amplexing male. Keratinous pads on "thumbs" is widespread in anurans (e.g., bufonids, discoglossids, hylids, leptodactylids, ranids). Keratinous excrescences appear in various shapes and sizes on the inner surface of the limbs and chests of salamanders (usually salamandrids) and irregularly among anurans. Keratinous structures are not exclusively sexual structures in reptiles as they are in amphibians, yet the cranial and dorsal spines and crest tend to be proportionally larger in mature males than in females. In a few emydine turtles (*Pseudemys*, *Trachemys*), the forefoot claws elongate, enhancing the titillating courtship.

Structural enlargement or better development of bisexually shared features occurs frequently. Premaxillary teeth or nasolabial groove termini (cirri) enlarge in some male plethodontids (*Eurycea*). The tusks on the lower jaw of some carnivorous frogs (e.g., *Ceratobatrachus*, *Pyxicephalus*) enlarge in males. The forearms of many male anurans become more muscular and robust. The dewlaps of male anoles are significantly larger than the females' dewlaps. The epiplastral lip of the bottom shell in many male tortoises is thicker and more projecting. These and other enlargement occur in combative species, particularly where actual physical contact between males is a regular event during the breeding season. Cloacal spurs in booid snakes and geckos are larger in males than in females and assist in copulation.

Size differences between females and males are common. Females are typically the larger sex in salamanders, frogs, turtles, and snakes, and males are larger in lizards and crocodilians. Large size offers a greater body volume for eggs and potentially results in more or larger eggs. The evolution of larger females is the selective trend where combat between males and/or female preference does not select for larger males (i.e., absence of sexual selection).

Migration from feeding areas to mating-nesting areas is a major behavioral modification in salamanders, many frogs, some turtles, and a few squamates. The distances may be small (5–50 m) or great (100–2000 km), but all are associated with the necessity of locating suitable egg-laying and developmental environments. Another behavioral modification, defense of home areas (territoriality), occurs in many amphibians and reptiles and under different guises, some of which are examined in the next section.

MATE ATTRACTION AND SELECTION

Once reproductively ready, an individual must find a mate. Mate location is not a problem for most amphibians and reptiles, because of the synchrony of readiness and the tendency to aggregate. Mate procurement is less certain for the male. Even with equal numbers of females and males, not every male will mate; all females will. For a male to attract females, he must demonstrate his superiority to the females. Perhaps he has a stronger voice, a higher-quality territory, or a more aggressive display; the qualities sought by the female are not always evident to the human eye.

Selection and Defense of Breeding Sites

Reproductive activities occur either within the males' home ranges or at special breeding areas. Energetically, it is more efficient for courtship and mating to occur within the normal home ranges of females and males; migration to special breeding areas occurs only where home ranges do not provide suitable developmental conditions for the eggs and offspring. Thus, migration is a common feature in the life cycle of many terrestrial amphibians and some aquatic reptiles, the former requiring water for their eggs and larvae and the latter avoiding water for their eggs. Pond-breeding amphibians move from their terrestrial or arboreal homes to temporary or permanent ponds. Commonly males precede the females and arrive hours or days before the females. Male frog choruses can attract and guide the females to the breeding areas; female salamanders find the breeding ponds by other clues, principally olfactory (see Orientation and Navigation in Chapter 9). Migrating reptiles are mainly turtles (chelonioids, some testudinoids, and some pelomedusids), a few lizards (iguanines), and laticaudine seasnakes. Courtship and mating in these reptiles may precede the migration, occur during the migration, or occur at the breeding site. The timing and location of mating appear to be populational rather than species-specific phenomena. Green seaturtles (*Chelonia mydas*) mate at the nesting beaches in many populations, but in at least one Australian population, females copulate with males as they pass through the males' home ranges on their way to the nesting beaches.

For many migrant amphibians, spacing at the breeding site is nonrandom. Often different species share the same area, and each species prefers different locations for their breeding activities. Salamanders tend to partition the site temporally, arriving sequentially rather than simultaneously. In eastern North America, three *Ambystoma* may use the same temporary woodland pond; *A. opacum* arrives in late autumn, lays eggs, and larvae emerge when submerged by winter rains, *A. tigrinum* breeds in mid-winter, and *A. maculatum* in early spring. Frogs also partition sites temporally, but each species also has preferred calling microhabitats and commonly three or four species (often more in the tropics) may breed on the same night (Fig. 8.3). When call-site microhabitats are limited, intraspecific competition can arise for the better call-sites. Each calling male defends a territory and excludes other calling males within this zone. Defense is mainly vocal with one male dominating intruding males by the "strength" of his call, usually involving lower-frequency calls; occasionally another male cannot be ousted by voice, and forceful physical repulsion is necessary. The result is a

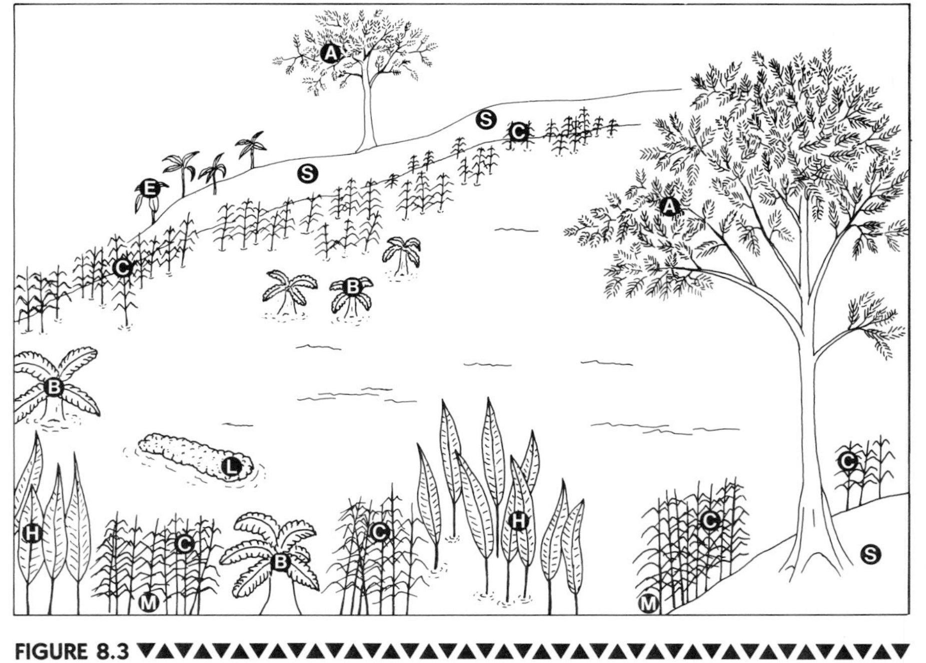

FIGURE 8.3 ▼▲▼▲▼▲▼▲▼▲▼▲▼▲▼▲▼▲▼▲▼▲▼▲▼▲▼▲▼▲▼

Microhabitat partitioning of anuran calling sites. A small pond in lowland Costa Rica hosts the breeding of ten species of hylid frogs, two *Leptodactylus*, *Bufo marinus*, and *Rana vaillanti*. This scenograph depicts site selection of the hylid breeding choruses on a single evening. The different letters designate different hylid species. [Adapted from Duellman (1967).]

chorus of regularly spaced calling males and silent males in the interstices or edges of the acoustic territories.

Most nonmigrant amphibians appear to breed in defended home ranges, and many such territories are commonly held throughout the seasons of activity, although defended most vigorously during the breeding season. All species of *Plethodon* have sharply delimited territories, marked by scents/pheromones; each male mates with females with overlapping or adjacent territories. *Dicamptodon* and *Cryptobranchus* hold aquatic territories that persist by physical eviction of male competitors. Terrestrial anurans (dendrobatids, leptodactylids, ranids) define and defend their territories by vocalization and when that fails by head butting, wrestling, or biting.

Territoriality, although widely publicized, appears proportionately less common in reptiles than in amphibians. Territories are a common feature of crocodilian and iguanian (Agamidae, Chamaeleonidae, Iguanidae) behaviors but are less evident in other reptiles, which may defend a specific feeding or resting site but not an entire home range. Iguanians are visual animals, and males are highly visible in the defense of their territories. Males select exposed perches and proclaim their occupancy by bright colors and bobbing heads (*Agama*, *Sceloporus*), flashing dewlaps (*Anolis*, *Draco*), and other overt displays. These behaviors usually are sufficient to repel competing males and attract females. Nonterritorial reptiles frequently compete for females with aggressive behaviors that commonly occur in the presence of a sexually receptive female. In snakes, many species have stylized, not stereotypic, combat dances where one male asserts his dominance over the other. In turtles and lizards, aggression begins with body posturing and threat displays and often ends with biting.

Attraction and Recognition of Opposite Sex

Courtship encompasses all reproductive activities preceding mating (i.e., transfer/release of gametes). Communication is its key aspect with three facets: attraction, assessment, and primer/releaser. All senses (vision, auditory, chemosensory, and tactile) are potentially involved, singly or in concert. The first signals serve to bring females and males together. Once together, signals assist in the assessment of the sex, species identity, and "genetic quality" of the potential mate. If the potential mate passes the assessment review, signals may increase the partner's reproductive readiness as well as elicit its cooperation in the transfer of gametes. A signal may serve a single role or multiple roles; a sexual pheromone can identify the sex and species of the producer, perhaps leave a trail for a potential mate, and even hint at the quality of the producer or promote mating. These facets are important to females and males; however, females have larger investments in gametes and in some species are more discriminating than the males. Since courtship is a stepwise communicative sequence with the signal of one partner stimulating a response from the other one, an incorrect signal may cause either partner to break off before consummation.

As noted previously, only in exceptional circumstances would a female fail to find a male and mate. A male's reproductive success is much less assured. Evolutionarily, the "goal" of each male is to mate with as many females as possible to promote the presence of his genome in later generations. Thus, competition for access to females is intense. The intensity is compounded because of the female's selectivity; her heavy investment in gametes obligates her to select the most fit male to ensure her genetic representation in subsequent generations. These two aspects of mate choice drive the evolution of courtship behaviors and especially many of the striking features of male reproductive anatomy and behavior. Sexual selection or female choice produces males with bright colors, ornate structures, or behaviors to attract and retain females. Male–male competition tends to promote larger or more robust males who can hold a territory or a female against all challengers. Further, this selectivity results in most amphibian–reptilian mating systems being polygynous.

Courtship Behaviors

Salamander behaviors span a spectrum from no apparent courtship (cryptobranchids, some hynobiids) to long, mutual-stimulatory (ritualized) courtships (ambystomatids, plethodontids). In the former, the male ignores the female until oviposition begins, then he sheds his sperm over the eggs. While not visible, the female may be attracted to the male's territory and be stimulated by pheromones; similarly the male responds to the sight and perhaps the chemicals of the eggs. Most other salamanders use tactile, chemical, and visual signals during courtship. Even in brief encounters (some rapelike), males stroke the female on the head, back, and especially around the cloaca, and her acceptance of a spermatophore depends on the proper signals. The ritualized courtships consist of a long sequence of interactive signals, and the recognition of a signal by one sex leads to it providing the next signal and so on until the male releases a spermatophore that is picked up by the female (Fig. 8.4). Vision seems to play a minor role in these courtship behaviors; tactile and chemosensory signals are more important.

Caecilian courtships are largely unknown, but likely include tactile and chemical signals. Auditory signals dominate frog courtship (see next section); however, not all frogs are vocal, and even in the vocal ones, tactile, visual, and possibly chemical signals serve in the final approach and amplexus. In calling species, females select and locate males by their calls; once in the vicinity of the male, other signals continue the courtship. Most *Bufo* males stop calling and grasp any female that comes near them; most *Hyla* males amplex females only after the females have touched them; other frogs (e.g., *Polypedates leucomystax*) require the female to posture in front of the male. *Dendrobates* are diurnal frogs with territorial males; females find males by sight and initiate courtship by nudging or jumping on the male. After a brief bout of touching one another, the female deposits her eggs near the male (without amplexus) and departs. For most frogs, the tactile role of amplexus must be emphasized; it stimulates ovulation in some frogs and probably oviposition in all.

Reptiles use various combinations of visual, chemical, and tactile signals in courtship. Aquatic and terrestrial turtles apparently locate one another by vision, although tortoises, kinosternids, and sidenecks regularly release scents that may also serve as sex attractants. Males rely on tactile signals, ranging from gentle titillations to hard bites and knocks, to obtain the females' cooperation. Tortoises (testudines) have the most ritualized courtships with courting males using head-bobs, nudges, and chin-gland rubs.

Crocodilians are highly territorial, and courtship occurs within the male's permanent territory or in one temporarily established in the aggregating species. Territories are established and defended by male–male contact and threats, although *Alligator mississippiensis* vocalization may be a territorial declaration. Females typically initiate courtship by entering the male's territory using a submissive profile. Courtship activities include snout contact, body and head rubbing, sound production (vocal, exhalation, water bubbling), and assorted

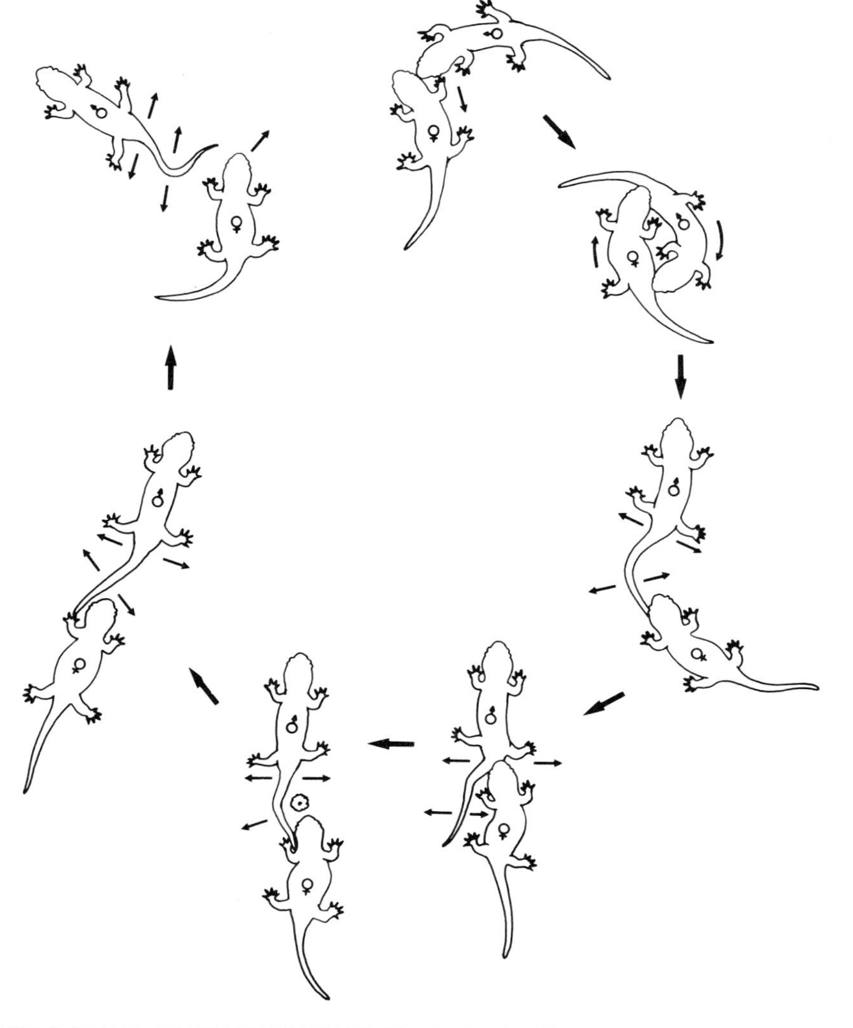

FIGURE 8.4 ▼▲▼

Courtship sequence of the mole salamander (*Ambystoma talpoideum*). Sequence begins at top center with male nosing female and proceeds clockwise; female nudges male's cloaca (bottom center), stimulating him to deposit spermatophore; she briefly examines spermatophore, then moves over it, picks off the sperm packet with her cloaca, and departs. [Adapted from Shoop (1960).]

body posture and movement displays. The contact and rubbing likely involve chemosensory as well as tactile signaling.

Squamates show a wide variety of courtships, some emphasizing visual signals, others tactile and chemosensory. Most iguanians are territorial and visual animals. Male territories are commonly larger than those of females and encompass or overlap the territories of several females. Defense of the territory gives the male nearly exclusive access to females within his space. His territorial displays also serve as courtship displays and are stereotyped, species-specific, and highlighted by bright or contrasting colors. When examined closely, the head-bob and push-up patterns of iguanids and agamids possess consistent sequence, timing, and elevation of movements within populations (Fig. 8.5). The high specificity of the male's display repels conspecific males and attracts conspecific females, informing them of the male's specific identity and his "genetic quality." At least in one *Anolis* species, the male display is a primer, as females exposed to displaying males were more receptive and fecund than females isolated from displaying males. Coloration in the visual species serves mainly to identify the sex of an individual at a distance. Evidence for this function comes from experimental studies with females masqueraded as males and vice versa; territorial males attack the masqueraded females (halting attacks where chemical cues identified her at contact) and ignored the masqueraded males. Coloration or lack thereof may inform a male that the individual is a subadult and not an adult female (*Lacerta*) or a nonreceptive female (*Holbrookia propinqua*). Among the nonterritorial lizards (many autarchoglossans), tactile and chemical cues dominate courtship. Chemical cues assist males in locating females, and then licking, nudging, and rubbing about the head and body are common precopulatory behaviors. Male biting of the neck and forebody is also a regular component of courtship and copulatory

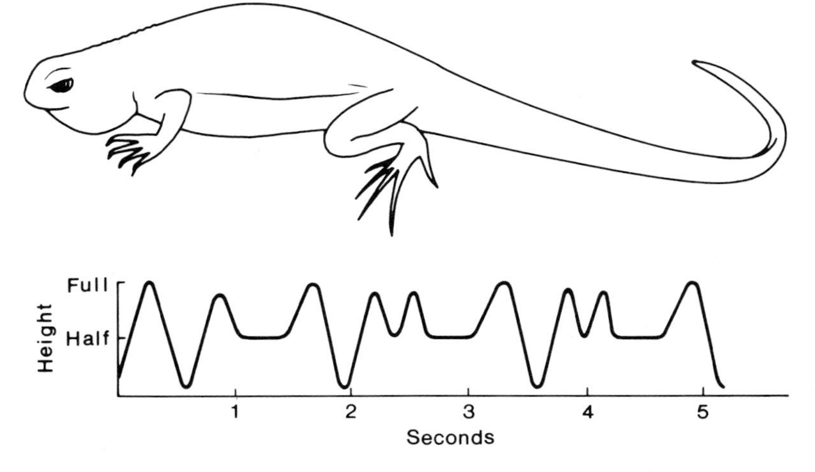

FIGURE 8.5 ▼▲▼▲▼▲▼▲▼▲▼▲▼▲▼▲▼▲▼▲▼▲▼▲▼▲▼▲▼▲▼▲▼
Display posture and movement-sequence diagram for a male desert iguana (*Dipsosaurus dorsalis*). The line in the diagram denotes the relative height of the head during a push-up defensive display sequence. [Adapted from Carpenter (1961).]

behaviors. Tactile and chemical signals are also the main cues among snakes. Snakes find mates by sight and smell (tracking), then confirm species identity by vomerolfaction. Courtship is strongly tactile with the male crawling over the female or throwing body loops across her; overlapping or intertwined body contact is maintained throughout copulation.

Vocalization

Among amphibians and reptiles, vocalization is nearly the exclusive courtship domain of frogs (see Table 8.1 for exceptions). Most frog species have voices. In contrast, a few salamanders squeak under duress, many crocodilians bellow territorial declarations and call softly in courtship, many geckos click and chirp to note personal space, some tortoises grunt during sexual encounters, and some snakes hiss warnings. Frog calls communicate different messages but mainly reproductive ones. Advertisement calls are the prominently, widely heard vocalizations (by herpetologists and all other listeners around a breeding chorus). These calls are addressed to the females to tell her the species identity, the location, and the genetic quality of the caller. Courtship and aggression calls are usually modified advertisement calls; the first aimed at a potential mate to confirm the male's mating readiness (in some species also made by females) and the second directed at an encroaching male to defend the calling site. Males and females give release calls to remove amplexing males. Predator attacks produce distress calls, which may startle the predator and permit the caller to escape.

No two species produce identical advertisement calls; yet many calls are similar, although rarely produced by sympatric species and often by totally unrelated and geographically distant species. Calls typically contain three or fewer notes and last less than two seconds, although the call may be repeated rapidly in sets (call groups) (Fig. 8.6). Most calls are between 40 and 4000 Hz; this frequency range is a good broadcast spectrum, avoiding acoustic interference from other sounds, for example, insect songs (4–7 kHz) or wind (100–200 Hz). Most calls span a broad frequency spectrum of 1–2 kHz, but most of the energy is concentrated in narrower frequency bands, usually with two or three dominant frequencies, which closely match the receptor sensitivities of that particular species. For relatively small animals, frog calls are exceptionally loud with sound pressures of 90–120 decibels at 50 cm; loudness is not size related since some of the smallest species produce the loudest calls and vice versa.

These characteristics provide anurans with an effective signaling device, one that is finely tuned to the sensor capabilities of the intended receiver, yet can be adjusted to improve its transmission or message. Two such adjustments are changes in intensity and modulation to produce courtship and aggression calls. Defense of call space is critical, but a caller must not spend all his time on defense because aggression calls have little attractiveness for females and are energetically expensive. Minor modulation (pulse) shifts and slight elevation of call rate toward a full aggressive call alerts the intruding males that his presence is known; if the

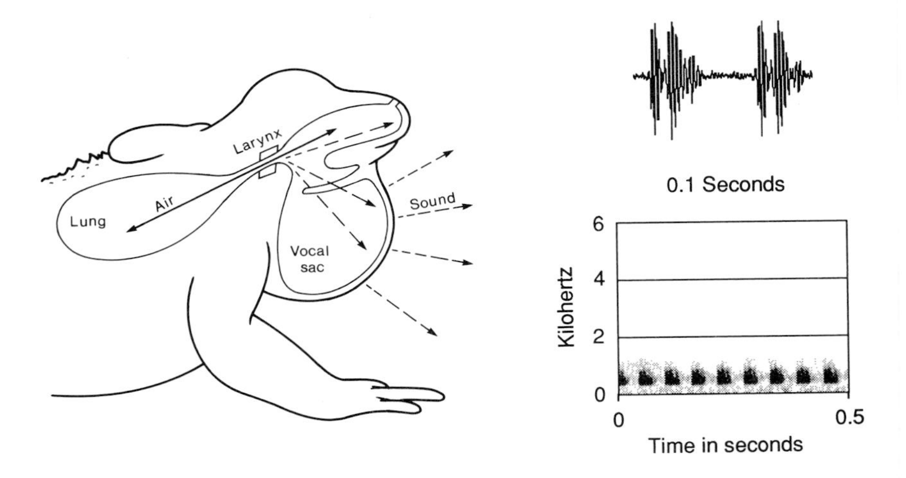

FIGURE 8.6 ▼▲▼

Sound production and call structure of *Bufo marinus* (marine toad). Sound production (left) uses aspects of the respiratory ventilation cycle without releasing air to the outside. Before calling begins, the buccopharyngeal force-pump inflates the lungs and vocal sacs. Then with the nostrils closed, the body muscles contract, pushing a pulse of air through the larynx, vibrating the vocal cords. Sound radiates outward and vocal sacs resonate it. The call of *B. marinus* is a deep, long trill of many continuous pulses (>50) and lasts for several seconds. The waveform (right top) and audiospectrograph (right bottom) show the energy envelope and pulse structure of brief segments of a call. Each pulse lasts about 0.03 sec with 18–20 notes sec^{-1}; the dominant frequency is 500–1000 Hz. [Morphology adapted from Martin and Gans (1972); Reprinted, with permission, of Wiley-Liss, a division of John Wiley and Sons, Inc. © 1972. Call analysis courtesy of W. R. Heyer.]

intruder does not depart, the call is adjusted further. Graded adjustment allows the male to continue to advertise although not at the most effective level. Similarly in dense choruses, males shift the timing of their calls to avoid overlap with their nearest neighbors and increase the call rate and number of call notes to improve their attractiveness. Since the producers are ectotherms, call frequency varies with temperature; frequency rises or falls with temperature. The auditory receptor's sensitivity changes similarly, thus remaining attuned to the species-specific call frequency. Call frequency is also size dependent (intraspecifically) with larger individuals producing lower-frequency calls. The body-size-to-call-frequency relationship is not applicable to interspecific comparisons.

In frog choruses, often only a small proportion of the males successfully attract females. The evidence for determining the whys of success are just being gathered, and not surprisingly the call features for success differ among the species examined. In most instances, females are attracted to the lower-frequency calls, that is, the callers are the larger members of a chorus. In others, the males calling most regularly and longer get the females, and in species with extended breeding seasons, the males calling on the greatest number of nights have the highest success rates. These aspects suggest that female choice favors the larger, stronger, and more durable males.

If females can recognize the "better" males, other males can also. Indeed, satellite males are found sitting quietly near calling males in many frog species. The goals of satellite males likely vary for different species, depending on the species-specific preamplectic behaviors. If females do not initiate amplexus, a satellite male can intercept females attracted by the calling male. However, this sexual parasitism may be less successful than it appears, because females in many anuran species can dislodge an unwanted suitor. Where females initiate amplexus, the amplexing male is carried away from his calling site by the female, and a satellite male may temporarily use the site. This has a dual advantage of obtaining a high-quality calling site and of avoiding the attention of predators until a primary site is obtained. Likely most satellite males are younger and smaller than the calling ones and highly unlikely to obtain females by calling, thus by adopting the satellite behavior they increase their chance of mating.

PART IV

INDIVIDUALS AND THE ENVIRONMENT

. . . organisms are functionally indivisible and cannot be split into the conventional compartments of morphology, physiology, behaviour and genetics. Each of these is only one aspect of the organism as a whole . . . it is the organism which deals with physical environment.

G. A. Bartholomew, 1964

The world's climates provide a smorgasbord of choice for organisms. Some organisms have chosen only a few, amphibians and reptiles have selected broadly. They now live almost everywhere except in areas of intense cold and the depths of the oceans. Each area selected imposes its own set of conditions for individual survival. The individual's survival requires its own actions and reflects a long history of selection for the mechanisms to make the correct responses.

Individuals respond physiologically and behaviorally to maintain stable internal environments (homeostasis) for normal cellular and biochemical functions. Appropriate responses may be instantaneous or require hours or days, may be cyclic and require daily or seasonal adjustments. Heat, water, and O_2 and CO_2 content of air — too much or too little — initiate the individual's response. The diversity of climates requires a diversity of responses; amphibians and reptiles have met the challenges.

CHAPTER 9

Spacing, Movements, and Orientation

Animal spacing is neither random nor uniform. Both distributions require homogeneous environments with uniform dispersion of resources and, for a random pattern, noninteractive individuals spaced either side by side or distant. Environments are not homogeneous, resources are of variable occurrence in space and time, and individuals interact intra- and interspecifically.

LOCAL DISTRIBUTION OF INDIVIDUALS

Distribution of organisms occurs at two levels, the population and the individual. Populational distribution involves habitat selection as defined by the availability of resources required for the survival of individuals of a particular species, for example, moisture and temperature regimes within a species' physiological tolerances, availability of food for both growth and reproduction, and shelter for protection from predators and weather. These and other aspects of species' distributions are examined in Chapter 12. Spacing addresses the local distribution of individuals within environmentally favorable habitats and is concerned primarily with the placement of an individual relative to conspecifics (i.e., individuals of the same species) for access to and use of resources. The main resources are food, water, shelter, and mates; their distributions directly influence spacing.

Conceptually, every individual has a home range, that is, the area occupied by an animal in pursuit of its normal activities. For sedentary amphibians and reptiles, individuals spend most of their lives — eating, sleeping, and mating — within a specific area, perhaps 10 m^2 or 1000 m^2 (Table 9.1), and never venture beyond self-defined borders. Such home ranges are easily recognized and mea-

TABLE 9.1 ▼▲▼

Home Range and Resource Defense in Select Adult Amphibians and Reptiles

Taxon	Area	Female area	Male size	Defense: Terr	Defense: S-S	Habits
Batrachoseps pacificus	3.6	?	42	?	?	Terrestrial
Desmognathus fuscus	1.4	?	45	?	?	Semiaquatic
Salamandra salamandra	10	>	82	?	?	Terrestrial
Atelopus varius	<20	=	25	+		Terrestrial
Rana clamitans	65	=	60	+		Semiaquatic
Syrrhophus marnocki	328	=	20	+		Terrestrial
Terrapene c. triungis	52,000	=	115	–	–	Terrestrial
Trachemys scripta	397,500	<	200[a]	–	±	Aquatic
Crocodylus niloticus	7990	<	2100[b]	+		Aquatic
Sceloporus merriami	535	<	45	+		Terrestrial
Varanus olivaceus	20,500	<	450	±	+	Arboreal
Xantusia riversiana	17	=	65	–	+	Terrestrial
Acrochordus arafurae	15,000	?	900	–	?	Aquatic
Carphophis amoenus	253	?	215	?	?	Semifossorial
Natrix natrix	99,000	>	700	?	?	Terrestrial

Sources: Salamanders—Bp, Cunningham, 1960; Df, Ashton, 1975; Ss, Joly, 1968. Frogs—Av, Crump, 1986; Rc, Martof, 1953; Sm, Jameson, 1955. Turtles—Tct, Schwartz et al., 1984; Ts, Schubauer et al., 1990. Crocodilians—Cn, Hutton, 1989. Lizards—Sm, Ruby, and Dunham, 1987; Vo, Auffenberg, 1988; Xr, Fellers and Drost, 1991. Snakes—Aa, Shine and Lambeck, 1985; Ca, Barbour et al., 1969; Nn, Madsen, 1984.

Note: Area, mean home range size of adult males (m^2); female area, smaller, same, or larger than males; size, snout–vent length of adult males at sexual maturity (mm; carapace length in turtles); defense by male, territorial or site/resource-specific, + yes, – no, ? unknown.

[a] Plastron length.

[b] Total length.

sured by biologists. Other species have less regular, more diffuse areas and patterns of activity, and these activity areas are less easily categorized as home ranges. Some of this diversity in amphibian and reptilian occupancy and movement pattern will illustrate the functional difficulties of applying the home range concept to every species. First, the concept has no implication of exclusiveness of occupancy (one animal or a herd can occupy the same area). Second, a home range's boundaries can expand or contract, or an individual can establish a new home range in a new location; however, an individual occupies only one home range at a time.

During the nonbreeding season, many terrestrial amphibians (e.g., *Ambystoma maculatum*, *Plethodon cinereus*, *Bufo marinus*, *Rana temporaria*) have small- to moderate-sized home ranges away from water. Within these areas, an individual may have one or more resting and feeding sites (activity centers) and will use one site for a day, a week, or longer before shifting to another site. Not all sites are visited every day or every month, nonetheless the periodic occurrence at sites and the persistent occupancy of the total area adjacent to these sites delimit an individual's home range. Such home ranges are largely two dimensional, although semifossorial salamanders and frogs add a third dimension by retreating downward during dry weather and, at least for some salamanders, individuals may remain active, although subterranean movements are likely confined to less than 0.5 m^2. For some of the terrestrial species (*Plethodon*, *Dendrobates*), courtship, mating, and egg-laying occur within the home range, and indeed the individual may spend its entire adult life in one area. For pond breeders (*Ambystoma*, *Bufo*), the forest-floor home range is abandoned during the breeding season, and most individuals appear to return to the same home range after breeding.

Shape and dimensions of home range relate to a species' habits. Semiaquatic species (*Desmognathus monticola*, *Rana macrodon*) have narrow linear home ranges along a stream edge or lake shore. Semiterrestrial and arboreal species (*Bolitoglossa*, *Centrolene*) have three-dimensional home ranges; terrestrial species have two-dimensional ones. Home range size is partially influenced by body size, although not all large species have large home ranges or vice versa (Table 9.1).

Recognition of home ranges among reptiles has the same spectrum of difficulties as encountered with amphibians. Home ranges are easily delineated in most crocodilians, semiaquatic and terrestrial turtles, and sit-and-wait lizards. Most aquatic turtles likely have a large foraging area (= home range) through which they move on a daily to seasonal basis; length of tenure in one small area depends on adequacy of resources. Temporary absence from the home range occurs only during egg-laying excursions ashore. Surprisingly one of the larger aquatic turtles, *Chelonia mydas*, may have one of the smallest home ranges owing to its habit of creating a submarine pasture and continually grazing therein. In contrast, *Dermochelys* appears to be constantly on the move following the seasonal blooms of its jellyfish prey. Motile-search lizards tend to have moderately large home ranges through which they search on a regular basis. Many snakes likely share the home range pattern of the motile-search lizards, but at least two other patterns occur among sit-and-wait species, and these patterns do not easily fit the home range concept since there is a regular shift of foraging sites by long-distance movements (Fig. 9.1). In the prairie rattlesnake (*Crotalus viridis*), individuals wander until they find an area of high prey density, staying and moving about in this locality until density drops and prey capture becomes infrequent. The bushmaster (*Lachesis muta*) similarly moves until a potential prey area is located and remains there only long enough (usually 2–4 weeks) to capture and digest its prey before moving again; it is unknown whether the bushmaster moves within a large, circumscribed area or randomly along a nonrepeating track.

Home range size is a direct function of the resources required for an animal's growth, maintenance, and reproduction, the relative availability of resources, amount of intra- and interspecific competition for resources, and foraging behavior. These factors and requirements further differ for the age, sex, and species of an individual, thus explaining some of the variability in home range size and type among amphibians and reptiles.

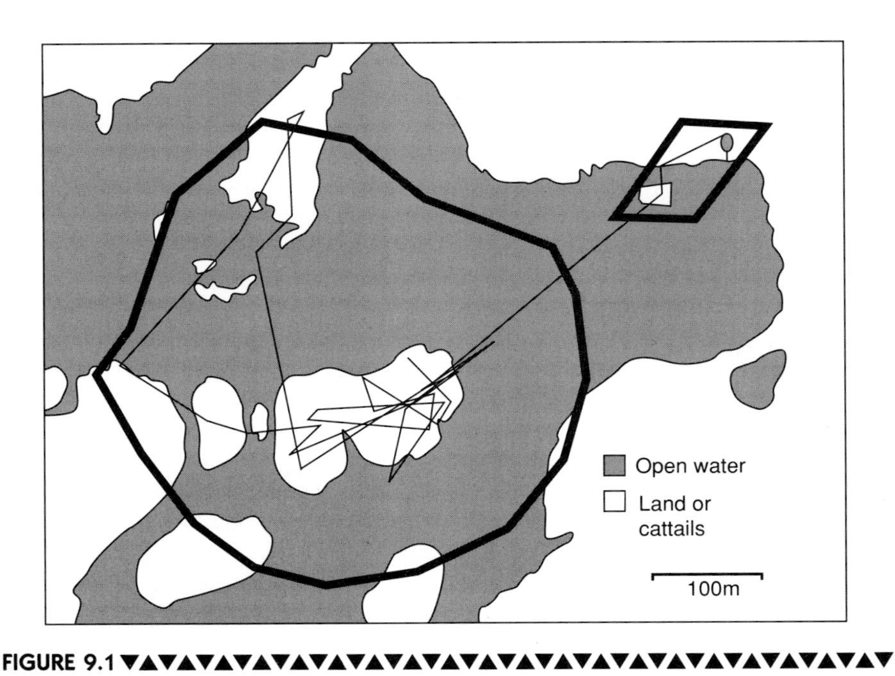

FIGURE 9.1 ▼▲▼▲▼▲▼▲▼▲▼▲▼▲▼▲▼▲▼▲▼▲▼▲▼▲▼▲▼▲▼▲▼▲▼

Movements and home range of a female *Nerodia sipedon* (northern water snake) during a single summer. Duration and frequency of occurrence indicate a midseason shift in home ranges (outlined in bold line). [Adapted from Tiebout and Gary (1987).]

Another aspect of space utilization is the defense of resources. Where one or more resources are insufficient for all individuals, individuals may gain an evolutionary advantage by defending the resource(s); however, a scarce resource, such as space, may be shared by multiple individuals. Among amphibians and reptiles, there are three levels of defense: territorial, site specific, and none. Because territoriality is defense of the home range or a major portion thereof, this defense mode is the most evident (to human observers) although not the most widespread. Territoriality occurs among terrestrial frogs (*Atelopus*, *Dendrobates*, *Eleutherodactylus*), terrestrial salamanders (*Plethodon*), crocodilians, and iguanian lizards. Site-specific defense is much more widespread but less well documented for most amphibians and reptiles; this defense mode involves only a point resource such as a resting site, basking rock, or food item. It seems doubtful that any individual or species displays total nondefense of all of its resources; rather it is nondefensive of some resources and site specific for others.

Territories exist for the defense of different resources. The territories of *Atelopus* are principally reproductive areas defended by males to increase their reproductive success. In *Plethodon*, the territories protect food and shelter; males and females maintain independent territories with little or no overlap, and only during the breeding season does aggression between sexes decline to permit reproduction. The territories of iguanid and agamid lizards protect food, shelter, and mates; both sexes defend territories, but male territories are larger and usually overlap those of one to several females. Adult females and males tolerate one another, although not others of the same sex, except juveniles.

Maintenance of territories is energetically costly and potentially injurious through increased aggression with trespassers and exposure to predators. Territorial species use a variety of devices to reduce the cost. Occupancy can be signaled by vocal, visual, or odor devices, usually in combination or sequentially. Territorial frogs use vocalization, color, and postural displays. *Plethodon* relies mainly on pheromone markers and displays a "dear-enemy" behavior by being less aggressive to neighbors than to strangers. Lizards use postural displays reinforced with special structural and color traits in males, occasionally in females.

The same defensive devices are used in site/resource-specific defense. Vocalization of male frogs declares their occupancy and willingness to defend their calling site. Geckos and other vocalizing lizards call to claim a resting or feeding site. Monitors and other carnivorous lizards use threat displays (vocal and visual) to guard food. Male snakes use ritualized combat bouts (tactile) and/or pheromones to gain or protect access to females.

All defensive behaviors result in individuals being spread broadly throughout a habitat with the dominant individuals clustering in the better resource areas and subordinates in marginal areas. A spatially limited resource results in aggregations. Breeding congregations of frogs or salamanders in temporary ponds are the most obvious amphibian aggregations and female nesting aggregations (e.g., seaturtles) in reptiles; however, even within aggregations, individuals of most species still display site-specific defense (i.e., excluding others from a narrow zone around itself). Feeding aggregations may temporarily form where food is locally abundant. Usually a dominance hierarchy forms with the dominant individuals feeding first or in the area of greatest food abundance. Nondefensive aggregations (Table 9.2) are typically climatic and/or space related, such as sleeping aggregations in marine iguanas. In climatic aggregations, defensive behaviors are suppressed. The aggregations may result from atypical or regular seasonal fluctuations. Drought causes dense aggregations of aquatic animals in the remaining and shrinking pools; such aggregations are annual events in the Venezuelan llanos where caiman, pleurodiran turtles, and other aquatic vertebrates pack densely together each dry season. Winter or dry-season hibernacula are other annual aggregations to avoid lethal climatic extremes.

TABLE 9.2 ▾▴▾▴▾▴▾▴▾▴▾▴▾▴▾▴▾▴▾▴▾▴▾▴▾▴▾▴
Examples of Social, Nonreproductive Aggregations of Amphibians and Reptiles

Taxon	Purpose
Salamanders, mixed seven species	Hibernation
Plethodon glutinosus	Estivation
Salamandra salamandra	Hibernation
Bufo tadpoles	Schooling
Hyla meridionalis, Pelodytes, Triturus, and *Podarcis*	Hibernation
Limnodynastes juveniles	Conserve water
Xenopus laevis tadpoles	Schooling
Terrapene ornata	Hibernation
Terrapene ornata and *Kinosternon flavescens*	Hibernation
Crocodilian hatchlings	Reduce predation
Alligator mississippiensis	Feeding
Amblyrhynchus cristatus	Sleeping
Diadophis punctatus	?Conserve water
Pelamis platurus	Feeding
Storeria dekayi	?Conserve water and hibernation
Thamnophis (3 species), 3 other snake genera, *Ambystoma,* and *Pseudacris*	Hibernation
Typhlops richardi	?conserve water

Sources: Amphibians—s, Bell, 1955; Pg, Humphries, 1956; Ss, Lescure, 1986; B, Wassersug, 1973; Hm, Van den Elzen, 1975; L, Johnson, 1969; Xl, Wassersug, 1973. Reptiles—To, Carpenter, 1957; c and Am, Lang, 1989; Ac, Boersma, 1982; Dp, Dundee, and Miller, 1968; Pp, Kropach, 1971; Sd, Noble, and Clausen, 1936; T, Carpenter, 1953; Tr, Thomas, 1965.

HOMING AND MIGRATION

Amphibians and reptiles are predominantly sedentary creatures, spending their entire life within a few hectares. Even most species that seasonally must migrate to breeding areas seldom travel more than a kilometer and usually much less. However, a few amphibian species are capable of prodigious feats of homing, and a few reptiles regularly migrate hundreds of kilometers.

Daily Movements

Long-distance and en masse movements have long attracted the attention of naturalists. The less spectacular daily movements have been largely ignored until recently, because the cryptic habits of most species make individuals difficult to track. Knowledge of movement and activity patterns are critical for assessing an animal's costs and benefits in life-history and evolutionary studies. The costs/benefits and goals of daily (intramural) movements differ markedly from long-distance (extramural) movements. Daily movements center on feeding, short-term physiological regulation, predator avoidance, and periodic mating activities. These local movements are within the home range or its neighborhood, energetically less expensive, and safer owing to familiarity with the area and nearness of refuges (Fig. 9.2).

Distance and duration of intramural movements correlate mainly with age, sex, body size, foraging behavior, territoriality, quality of habitat, and climatic conditions. Interactions among these factors are complex, and each generalization has numerous exceptions. Nonetheless, daily movements are generally short and brief, being principally concerned with food finding. Sit-and-wait predators move from refuge to feeding station, feed when prey passes, and return to the refuge. For example, *Hyla chrysoscelis*, a forest treefrog, uses water-filled, knothole cavities in trees as daily resting sites. At dusk, they emerge and sit near their cavity to catch passing prey; occasionally, they will climb higher in the tree to feed in the canopy. Before sunrise, they return to the cavity vacated the evening before. Territorial species, for example, *Plethodon cinereus* or *Anolis limifrons*, have similar routines combined with an occasional territorial-odor marking sallies in *Plethodon* or a defensive display or combat in *Anolis*. Motile-search species move from a refuge and alternate between bouts of searching and resting (and/or thermoregulation). Searching entails higher energy costs, but motile searchers usually find and capture proportionally more prey (energy) than sit-and-wait predators.

Seasonal Movements

Extramural movements encompass three general patterns or causes. The first are cyclic or seasonal movements where the entire population or specific age classes migrate from their regular daily activity area to a new area, usually for

FIGURE 9.2
A basking aggregation of *Amblyrhynchus cristatus* (marine iguana). (K. Miyata)

breeding, egg-laying, or hibernation. The second pattern is migration to escape an environmental disaster, typically aquatic species moving from a drying pond. The third pattern is dispersal. The first two patterns are two-way movements with the individuals returning to their original area if they survive. Dispersal is a one-way movement with individuals leaving their "home" area, usually never to return.

Seasonal movements are widespread in amphibians. All species with indirect development living away from water must return to water to breed. The list of frogs and salamanders making annual pilgrimages to breeding congregations is enormous. These breeding migrations are commonly short-distance movements (<0.5 km), although actual distances have been confirmed for only a few species and the distances are assumed from the local occurrence of individuals during nonbreeding seasons. The seasonal, reproductive movements in reptiles are in the reverse direction, from water to land. In most crocodilians, the females nest in or adjacent to their home ranges, but in some populations, nesting sites are not available in the feeding area and males and females migrate to nesting areas, for example, *Crocodylus niloticus* moves several kilometers from home range to nesting areas.

Female aquatic turtles move ashore to lay their eggs. The distances traveled are variable within a species and depend on the nearness of suitable nesting sites. The overland distances are typically <100 m, but may be >1 km if suitable nesting areas are not available nearby. Females, sometimes accompanied by males, may make long-distance aquatic migrations to reach nesting beaches, 5–10 km for aquatic species (*Apalone*, *Pelomedusa*) and as much as as 2000 km for some seaturtles (from Brazil to Ascension Island for *Chelonia mydas*). Some turtles (*Kinosternon subrubrum*, *Clemmys guttata*) annually leave their aquatic homes for a terrestrial sojourn or aestivation.

Seasonal migrations are much rarer in lizards or at least little observed. A notable example is a female *Iguana iguana* nesting aggregation on an islet in the Panama Canal. Females assemble from the forest of an adjacent larger island to dig deep egg chambers. The selection, defense, and digging of a nesting site requires 4–7 days before the females leave their nests and return to the forest canopy. In other localities, the female iguanas nest individually or in smaller groups. An Australian agamid, *Amphibolurus ornatus*, shows a complementary adult–juvenile migration; each year reproductive adults move into granite outcrops, and as the males establish their territories, the juveniles are driven out.

Annual movements are much more widespread in snakes, particularly in denning species that share a common hibernation site (hibernaculum). The most studied examples are of two North American species, *Thamnophis sirtalis* and *Crotalus viridis*; both are live-bearers and have populations in areas with long frigid winters. In each species, individuals from a wide surrounding area share a hibernaculum. Late each summer, individuals move independently from outlying areas into the hibernaculum and in spring return to feeding areas. The movement pattern differs for the two species. The garter snakes move to marshy habitats and establish home ranges; the rattlesnakes move in a circuit, stopping for stays of various length in areas of high prey density. In some areas, laticaudine seasnakes will aggregate in sea caves for egg-laying. Anecdotal evidence suggests that a wide variety of temperate-zone colubrids make autumn and spring movements to and from hibernation sites.

Dispersal and Transient Movements

Animal dispersal is the movement outward from a home area and conceptually often implies colonization of new areas, either intentionally by the animal's own movement or accidentally by a predator, flood, or other transport. Indeed, colonization is one aspect of dispersal, but several movement patterns fit the general definition of dispersal. These patterns are difficult to categorize because they form a continuum from brief, short-distance movements to permanent, short- or long-distance movements. For amphibians and reptiles, this continuum can be divided roughly into three categories: populational renewal; transient dispersal; and permanent dispersal.

The dispersal of metamorphosing amphibian larvae from their natal pond into the adjacent habitats of their parents is a seasonal or annual event and a necessary one for the maintenance of local populations. The distances are usually minor, and the individuals become established in peripheral or vacant sites within the area of the local population and, when mature, will return to their natal pond to breed. (Return to natal areas is a common assumption but is confirmed only in a few species, e.g., *Bufo fowleri*.) Similar dispersal occurs in reptiles and direct-developing amphibians, but is often later in life as larger juveniles in territorial and/or parental-care species. The juveniles stay within the home range of a parent until they are no longer tolerated and are driven out. Among the dispersing hatchlings or juveniles, a few may make long-distance, extramural movements and become residents elsewhere.

Identifying and tracking individuals making extramural movements is difficult. Extramural movements are known to occur because new individuals appear in studied populations and colonize vacant habitats. Most of the known examples derive from long-term studies of turtle populations and show the continuum from transient to permanent dispersal. In the slider (*Trachemys scripta*), adults occasionally make short excursions (<0.5 km) from their home range; the excursion may last a day or a month, may be overland or within the lake, and then the individual returns to its home range and long-term residency. Longer-distance and long-term movements also occur; some represent permanent dispersal and others are transient movements, but the two classes are difficult to discern because some individuals have been absent from their original home pond for several years only to reappear. Both male and female sliders disperse, but the frequency for males is higher and more adult females and males make extramural movements than juveniles. These movements are independent of the individual's reproductive condition, although they are important owing to the possibility of increasing the dispersing individual's reproductive success. Further they demonstrate a regular interpopulational gene flow. These examples indicate that similar intramural and extramural dispersal occurs in other amphibian and reptilian populations but such events can be recognized only in populations of marked individuals that are regularly censused over many years.

Orientation and Navigation

Directed movements to and from various sites within an individual's home range require the ability to recognize landmarks or cues and to choose the desired direction from these cues. This ability (piloting) is the simplest form of orientation and is possessed by all amphibians and reptiles. Piloting depends exclusively on local cues. Compass orientation operates independent of local cues and operates like a pointer, giving the user a preferred direction of movement. The compass direction may or may not point toward the individual's home. True navigation is the ability to orient and move to a goal (home range, breeding site, etc.) without use of local cues or landmarks. Navigation requires compass orientation and an internal map; the navigator must be able to locate her/his position and the goal-site on the map, select the compass direction, and proceed toward the goal, making appropriate bearing correction to compensate for course drift. Navigation remains unconfirmed for amphibians and probably exists for only a few reptiles (*Alligator*, seaturtles).

Homing is used broadly for goal-specific orientation and refers to any of the foregoing mechanisms. Homing experiments use spatial and temporal displacements to identify the external cues and internal mechanisms involved in orientation and navigation. All sensory receptors have been implicated in orientation. Vision and olfaction are the major ones for amphibians and reptiles. Even though one sensory modality may dominate in a species orientation, it is not used in exclusion of the other senses and use is likely hierarchical.

Acoustic cues do not appear to be a major orientation device for any reptile. Frog choruses, often mentioned as directional cues, remain speculative for male orientation. At least one male must find the breeding site without the acoustic cue; other cues must be available to all males for none can rely on the successful orientation of another male and his advertisement. Females can rely on male vocalization to lead them to the breeding chorus and then to a preferred male. Other cues are necessary for the return home.

Visual cues serve for short- and long-distance movements. Diurnal species use visual landmarks for movement within their home ranges and such landmarks, at least in the form of silhouettes, are also available to nocturnal forms. However, visual cues extend beyond the recognition of local landmarks and include celestial cues (sun, moon, stars, and/or skylight polarization patterns). Celestial cues serve for intra- and extramural movements and require an internal clock for the accurate selection of compass direction. The simplest mechanism is Y-axis orientation (Fig. 9.3), used in a variety of frogs, salamanders, and turtles (e.g., *Acris crepitans*, *Taricha granulosa*, *Chrysemys picta*). This mechanism derives its name from the animal setting its orientation axis ("Y") perpendicular to a fixed feature ("X axis") of its home range, such as a shoreline. After a sudden escape reaction, the animal uses the Y axis to return home. Since the orientation axis is fixed but the cue (sun, moon, linear-polarized light) is not, animals using Y-axis orientation have a circadian clock to compensate for the changing, but predictable, location of the cue. Y-axis orientation is a variant of compass orientation; in the latter, the animal (e.g., *Notophthalmus viridescens*) has a preferred compass direction that is not fixed on a specific feature such as a shoreline. This celestial-compass mechanism may operate in the long-distance migrants like seaturtles, but its use for

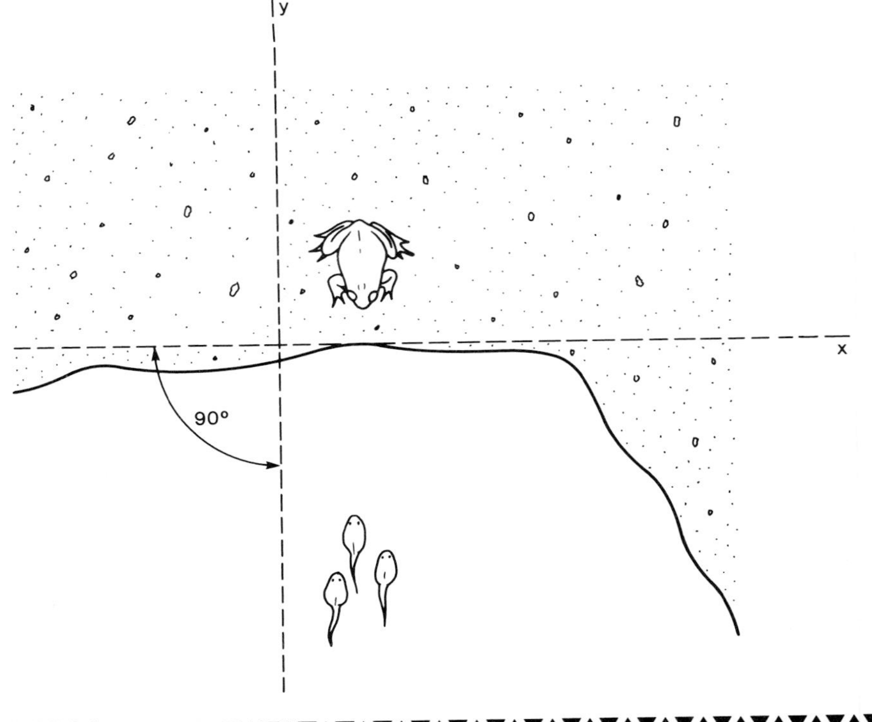

FIGURE 9.3

Y-axis orientation is a type of celestial orientation. The animal establishes a homing axis (Y) perpendicular to its home (e.g., shoreline, the X axis). Normal escape reaction is into pond for the frog or to shallow water for the tadpoles; return follows the compass direction of the Y axis. [Adapted from Adler (1970).]

them is speculative. Hatchling seaturtles use a simpler light cue to direct them to the sea. They hatch at night, scan the horizons, and select the brightest area; before the proliferation of street and house lights, the brightest horizon was always over the ocean.

A fascinating aspect of celestial orientation is that orientation occurs even in individuals without eyes but not in all species. Orientation is possible because some species possess extraoptic photoreceptors and the celestial cue can be perceived by this nonoptic receptor. Experimentation has identified the brain's pineal complex as the receptor involved in this form of orientation. Whether for blind or sighted individuals, celestial cues are the only visual cues that offer an opportunity for true navigation. Such navigation is not yet confirmed for any amphibian or reptile. Seaturtles are the most likely candidates owing to their long-distance migration. The argument that poor sight excludes visual navigation by seaturtles is negated by the ability of other species to detect celestial cues without eyes.

Odor is a major orientation cue for salamanders and some anurans. Indeed, the two homing records for amphibians (>12 km for *Taricha rivularis*, 3 km for *Bufo bufo*) appear to be dependent mainly on olfaction. Exactly how odor is used and what specific odors are used in these long-distance migrations remain unknown. Downwind orientation for ground-hugging amphibians seems an unlikely possibility. Yet? Also, does a bog or pond retain a characteristic odor signature year after year? On a shorter-distance scale, *Desmognathus ochrophaeus* and other egg-guarding plethodontids use odor to locate and recognize their eggs when returning from feeding or escape-reaction excursions. Of the reptiles, snakes are the most olfactory-dependent group with odors as dominant social signals. As yet, there is no evidence for odors as orientation cues in adults other than tracking or following a conspecific. Because snakes touch only one side of upright frictional surfaces when they crawl, another snake can determine the direction of a movement from an odor trail by checking which side has the odor and moving in that direction. This trailing mechanism allows newborn prairie rattlesnakes (*Crotalus viridis*) of one subarctic population to follow adult trails to the winter hibernaculum. Olfaction has been suggested for seaturtle orientation, and although unconfirmed, choice experiments indicate that juvenile *Lepidochelys* prefer water from the vicinity of their natal beach.

Orientation using the earth's magnetic fields has a long history of speculation but only a short history of evidence and that only for a few species. Experiments have shown that a variety of amphibians (*Notophthalmus*, *Bufo*) and reptiles (*Alligator*, *Chelonia*) can detect magnetic fields; however, only one field investigation (juvenile *Alligator mississippiensis*) has provided strong evidence of its use for orientation or navigation. The inclination of the lines of force in the geomagnetic field provide a source for the development of a map sense. Geotaxis (via gravity) has been suggested for salamanders and frogs but is unlikely except as a minor cue in short-distance homing. Seismic and water current cues are also likely minor cues; however, the recent recognition that hatchling seaturtles use ocean swells for orientation cautions the rejection of any potential cue without thorough testing in a variety of taxa.

Aside from the location of a goal-site, two other aspects of homing behavior are critical: site fidelity and return to natal area. Site fidelity has been tested widely in mark–recapture studies of adult amphibians and reptiles and repeatedly confirmed. Individual salamanders or frogs return to the same breeding pond, frequently by the same route, and female seaturtles return to the same nesting beach and even to within a few meters of their previous nests after 3- to 4-year absences. The lack of site fidelity is more striking than its converse, but it does occur. Toads will transfer their loyalties to a different pond if the original one disappears or is unsuitable, and occasionally a seaturtle will lay one clutch out of a season's four or five at a different beach 5–50 km from its regular beach. Such behavioral plasticity is necessary. Individuals cannot be genetically hardwired to a single breeding or hibernation site; survival of the individual and its reproductive success require some flexibility because no site is immune to degradation or destruction. The benefit of fidelity is its time and energy efficiency; a known site can be reached directly and quickly, and with a high probability that it is suitable for breeding and contains potential mates.

Offspring's return to their natal site is less easily tested. Its occurrence among short-distance migrants is highly probable owing to juveniles dispersing within and peripheral to their parents' habitat. For long-distance migrants, natal-site return is less certain, however, the recent acquisition of molecular-genetic techniques permits such tests. *Bufo fowleri* tadpoles (1 week old) with a unique genetic marker were transplanted to new ponds, and of the few that survived to adulthood, 73% returned to their new natal pond and 24% to immediately adjacent ponds. Comparing the nucleotide sequences in mitochondrial DNA demonstrated that four green seaturtle (*Chelonia mydas*) populations from the Atlantic and Caribbean are nearly identical, but the Ascension Island population uniquely possesses one divergent sequence site, strongly indicating the repeated use of Ascension by a single maternal lineage of turtles.

ACTIVITY CYCLES

An activity cycle divides the life of an animal into periods of rest/dormancy and movement/alertness and focuses on the latter period. Hence, daily cycles are labeled diurnal, crepuscular, and nocturnal (day, sunrise and sunset, and night, respectively), and seasonal cycles are vernal, aestival, and autumnal (spring, summer, and fall), targeting the periods of an animal's greatest energy expenditures. The periodicity and active phase of these cycles are phylogenetically constrained and environmentally regulated for most species. Individuals of a diurnal species remain diurnal throughout their lives, but they can adjust their activity to avoid or reduce climatic extremes.

Phylogenetic constraints reflect past adaptations that established specific physiological capabilities and mechanisms. Geckos (Eublepharidae, Gekkonidae) illustrate both the existence of phylogenetic constraints and the potential for their evolutionary modification. Geckos are widespread, speciose, and abundant lizards. Their name is nearly synonymous with nocturnal lizards and immediately conveys a mental image of a particular behavior and body form. Yet, geckos are not exclusively nocturnal; the Indian Ocean day-geckos (*Phelsuma*), American *Gonatodes*, and some New Zealand geckos are diurnal (as are the divergent pygopods). Did the shift from a nocturnal to diurnal cycle arise because of the absence of other diurnal lizards or low predation on diurnal lizards?

In addition to geckos, the type of daily cycle has a strong taxonomic association in amphibians and reptiles. Amphibians are predominantly nocturnal animals, beginning their activities at sunset and retiring at or before sunrise. Nocturnal behavior benefits these animals that have little control of evaporative water loss because humidity increases at night. The diurnal exceptions are aquatic and semiaquatic species (although many of these are nocturnal as well) and humid forest frogs (Dendrobatidae, *Atelopus*). In contrast, reptiles are mainly diurnal. Turtles are largely diurnal, although many species shift to a nocturnal schedule for egg-laying, possibly to avoid predation or heat overload. Crocodilians are highly visible daytime baskers, yet they are largely nocturnal in feeding and breeding activities. Of all the reptiles, crocodilians are the most opportunistic and least tied to a diurnal or nocturnal schedule. With the exception of geckos and the uncertain status of many fossorial species, lizards are diurnal creatures; the night-lizards (Xantusiidae) are misnamed, being strictly diurnal. Snakes contain both diurnal and nocturnal taxa. With minor exceptions, viperids are crepuscular and nocturnal. The majority of the elapids are diurnal; kraits and several other elapid genera, but not American coral snakes (*Micrurus*), are nocturnal. The colubrids contain a vast array of both nocturnal and diurnal species. The boids and pythons are mainly nocturnal.

The preceding generalities do not convey the variability of activities within nocturnal and diurnal daily cycles. As noted earlier, movement patterns vary widely, thus an animal may be active, that is, sensory alert and physiologically aroused, yet in a resting posture for most or all of its active phase. A bushmaster or rattlesnake moves a short distance from its resting area to a prey-scent trail and will wait, ready for its prey. Similarly, a slimy salamander will wait at the mouth of its burrow for passing prey. This posed-for-action but inactive behavior is characteristic of many sit-and-wait foraging amphibians and reptiles. The motile-search foragers spend much, but certainly not all, of their active phase in motion. While movement defines the active phase of an activity cycle, an animal — ectotherm or endotherm — moves only if the potential benefits of movement exceed the potential cost. The benefits and costs are not evaluated before each movement but are part of an animal's innate foraging behavior and, like predator escape and courtship, are honed by evolution.

The level and character of daily activity varies within a species (Fig. 9.4). Differences exist because of age, sex, reproductive state, quality of habitat, and climatic conditions. During the breeding season, adult male *Sceloporus virgatus* (diurnal iguanid) are active longer and move more each day than adult females and also are active almost everyday. In nonbreeding periods, females are active more days than males, although the daily movements of active females and males are the same. At this time, young males are nearly twice as active as older males. Patterns of this sort occur in many amphibians and reptiles; breeding male frogs or territorial male salamanders expend more energy in their reproductive behaviors than do females. Another common feature of amphibian and reptilian daily activity patterns is the irregularity of daily appearances; an individual is simply not active everyday. In the marine toad (*Bufo marinus*), individuals appeared at their feeding stations on the average only every third or fourth night; some individuals appear almost every evening, others much more irregularly and as infrequently as once every 8–10 nights even during good toad-weather. This

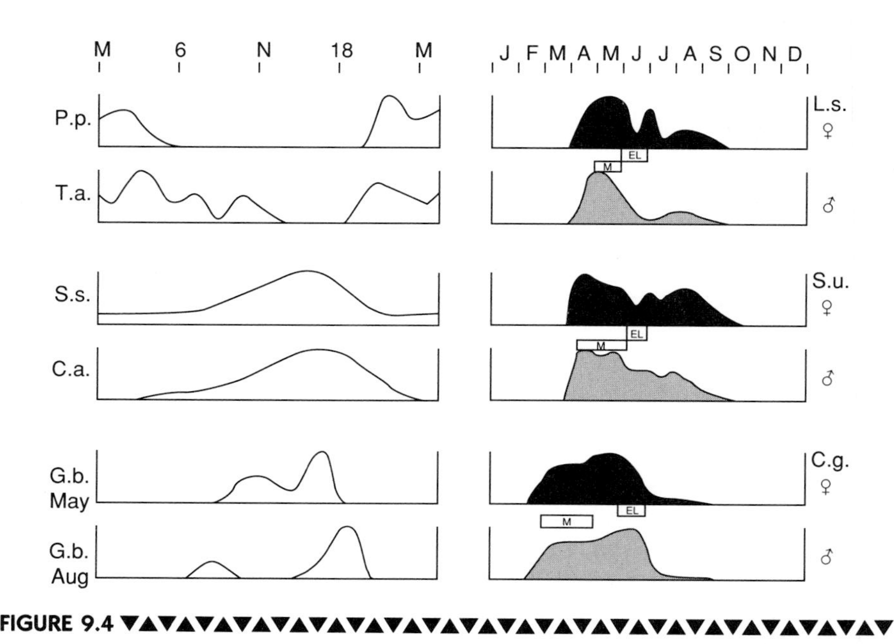

FIGURE 9.4 Examples of daily and seasonality activity patterns in amphibians and reptiles. Diel patterns (left) begin at midnight (M) and map percentage of population active hourly for a 26-hr period. Annual patterns (right) begin in January and map percentage of population active monthly for one year; M bar between female and male cycles marks courting and mating period; EL bar denotes egg-laying period. Daily patterns (from top): nocturnal, *Peltophryne peltocephala*, Cuban giant toad (Sampedro-M. et al., 1982); nocturnal with morning component, *Triturus alpestris*, alpine newt (Martin et al., 1989); continuous but emphasis on diurnal, *Sphenops sepsoides*, a fossorial skink (Andrews and Kenney, 1990); diurnal and crepuscular, *Carphophis amoenus*, wormsnake (Barbour et al., 1969); single diurnal in spring and double diurnal in summer, *Gopherus berlandieri*, gopher tortoise (Rose and Judd, 1975). Seasonal patterns (from top): *Lacerta schreiberi*, Iberian green lizard (Salvador, 1987); *Sceloporus virgatus*, striped plateau lizard (Rose, 1981); *Clemmys guttata*, spotted turtle (Ernst, 1976).

irregularity of activity is hidden by such statements as: species "A" is active from May through September. One or more individuals may be active during this period, but their relative abundance fluctuates widely. Perhaps 70–80% of the adults are active each day during the breeding season, only 10–30% during the nonbreeding season, and fewer or none during bad weather.

Weather and climate affect the activity intensity and cycles of species and individuals. Arid-land reptiles (snakes, lizards, and turtles) commonly show unimodal activity patterns in the milder or cooler periods, and as the daily heat regime becomes increasingly warmer, midday activity slowly declines until no individuals are active at midday, yielding bimodal patterns of activity of morning and late afternoon (see Fig. 9.4, *Gopherus*). This bimodality may be displayed by some individuals, while others may appear in either the morning or the afternoon. Moisture has a similar effect on amphibians. Whether temperate terrestrial salamanders (*Plethodon*) or tropical terrestrial frogs (*Bufo*), the numbers of individuals and the duration of their nocturnal foraging activity change with availability of moisture. As their habitats dry, fewer and fewer individuals appear each night and are active for a shorter duration until surface conditions are totally unsuitable and no individuals appear. If it rains, there is a burst of activity (many individuals and more movement), and if rain occurs regularly (but not excessively), nightly abundance and activity decline to a steady state.

The preceding patterns show the direct effect of temperature, moisture, and light on activities of individuals and populations. These environmental factors also influence the internal (endogenous) physiological rhythms that become visible through behavior and timing of activity. These internal rhythms (circumannual) are apparent in reproductive behaviors (see Chapter 8) and appear to be influential, if unconfirmed, in the daily and other seasonal activity patterns of amphibians and reptiles. A few examples reveal the potential diversity of circadian rhythms. Many crocodilians alternate between shore during the day and water at night. Although this daily rhythm could be temperature driven, experimentally shifting the light cycle on captive American alligators also shifted the activity pattern independently of the temperature cycle. Constant dark experiments with frogs, salamanders, and lizards show many species with free-running circadian rhythms, but these rhythms stay in phase with the natural light cycle only if synchronized (entrained) by light. Since a short pulse of light will often reset the endogenous cycle, sunrise is likely the natural cue.

Our labeling focuses attention on the active phase of the daily and seasonal cycles; however, implicit with the use of "cycle" is the existence of an inactive phase. The daily inactive period is devoted to sleep and seasonally to torpor or hibernation. Both exist within amphibians and reptiles, although both lack the experimental attention given to them in mammals. Both share a reduced metabolic activity and the general suppression of physiological mechanisms although at different levels. The alternation between physiological states likely follows endogenous rhythms in all animals. Further, the universality of a daily inactive phase strongly suggests an evolutionary advantage. Literally, "burning the candle at both ends" allows little energy or time to be directed at preparation for and to reproduction.

CHAPTER 10

Homeostasis: Air, Heat, and Water

Metabolic processes operate in a narrow environmental band. Too cold or dilute, the processes stop; too hot or dry, processes abort and chemical compounds change. Survival requires a stable internal milieu within this narrow band, yet many amphibians and reptiles live in seemingly inhospitable environments. Survival is possible because these ectotherms have evolved a variety of homeostatic devices capable of maintaining their internal environments within tolerance (operational) limits in spite of the excesses of the external environments. The effects of temperature, gas availability and exchange, and water and ion balance on normal and stressed organisms are keys to understanding the survival and success of individuals and populations.

WATER BALANCE

A major portion (70–80%) of an amphibian's or reptile's mass is water and must remain so. In addition to remaining hydrated for proper physiological function, the ionic concentration of intra- and extracellular fluids must remained balanced, and the nitrogenous by-products of metabolism must be removed from the body. Osmoregulation services all three aspects and maintains a balance among them.

Water balance is challenged differently in fresh water, saltwater, and aerial (terrestrial) environments (Fig. 10.1). In fresh water, an amphibian or reptile is hyperosmotic (ionic concentration of the body is greater than that of the environment, or hypertonic). The physics of fluids demands that adjacent fluids have identical ionic concentrations, and osmotic pressure shifts water and ions between the two fluids until they are identical. Thus, a freshwater animal loses ions (salts) and gains water. To avoid excessive hydration, animals can change integument permeability to reduce the influx of water and the efflux of salts and increase urinary output but conserve salts without reducing excretion of nitrogenous waste. Marine species face the opposite osmotic challenge. They are hypo-osmotic (ionic concentration less than the environment, or hypotonic) and lose water and gain salt. To avoid dehydration, animals reduce integument permeability and also must reduce urine output yet eliminate nitrogenous waste. Terrestrial forms share the dehydration challenge and address its reduction in similar fashion as do marine species (Fig. 10.1).

Because nitrogenous wastes are toxic when concentrated, their removal is vital; however, removal requires a liquid medium and is at odds with the osmotic needs for marine and terrestrial forms. Ammonia, the basic by-product, is highly toxic. It is readily soluble in water and easily disposed of by freshwater forms with an excess of water. At some energetic expense, ammonia converts to urea, a less toxic but still water-soluble compound, and at a greater expense ammonia converts to uric acid, a nontoxic and insoluble compound. The relative availability of body water determines the urea–uric acid choice for marine and terrestrial species.

Mechanisms for Control

Adaptations for osmotic balance are compromises, as are all adaptations, among competing selective forces. Each species has evolved a set of behavioral, physiological, and morphological traits that operate as an integrated whole to resist environmental challenges. The following discussion emphasizes the dehydration aspect of water balance control.

Amphibian skin is an ineffective barrier to water flow on land or in water; this water flow is largely extracellular through channels and intercellular spaces. In fresh water, this permeability imposes no threat. Excess water and a mixture

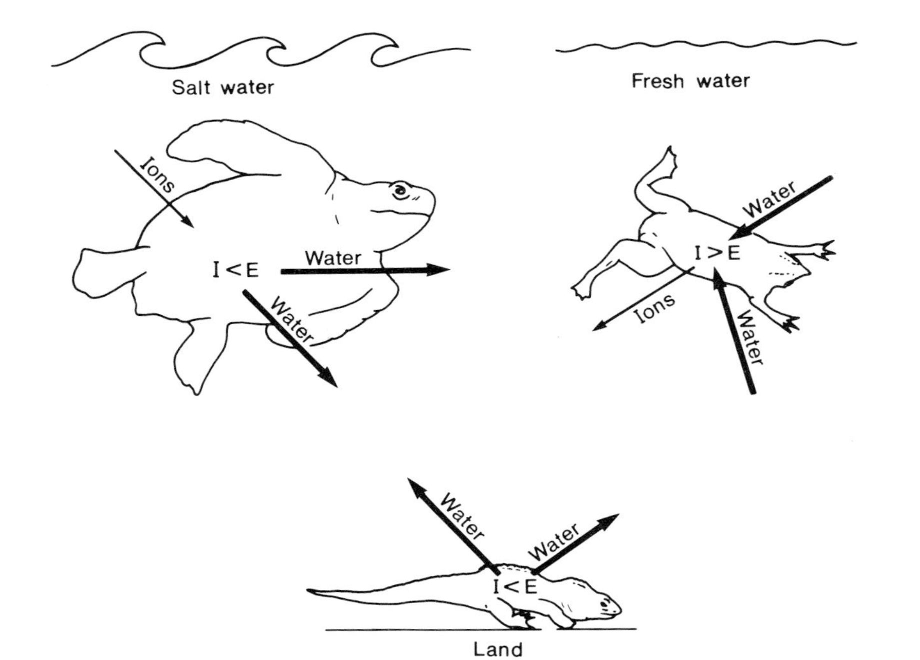

FIGURE 10.1 Osmotic challenges of amphibians and reptiles on land and in fresh and salt water. In salt water and on land, an animal is hyposmotic with an internal ionic concentration less than the surrounding medium (I<E), and water flows outward. In fresh water, it is hyperosmotic with an internal ionic concentration greater than the external medium (I>E), and water flows inward.

of ammonia and urea are expelled as urine. Sodium loss through the skin is low and matched by cutaneous resorption; sodium is also resorbed in the urinary bladder. Survival in salt water is less easily accomplished, and only the crab-eating frog (*Rana cancrivora*) lives regularly in marine salt water, although about 80 species of salamanders and frogs occur occasionally in brackish water (Table 10.1). *Rana cancrivora* avoids the dehydrating effects of salt water by increasing urea production and retention. Urea and the influx of sodium elevate the osmolarity of the intercellular fluids and the animal remains hyperosmotic, ensuring a continual influx of water. Intracellular osmolarity matches the plasma by elevated potassium, chloride, and amino acid concentrations.

Reptiles also have no difficulties in fresh water. Their skin is relatively impermeable, and water influx and salt efflux are relatively low and countered by increasing urine output and ionic resorption in kidney, urinary bladder, and colon. In salt water, the impermeable skin remains effective; however, salt overload (mainly from food and accidental ingestion) occurs. Because reptiles

TABLE 10.1
Amphibians Known to Live or Tolerate Brackish Water

Ambystomatidae	Hylidae
Ambystoma subsalsum	*Acris gryllus*
Dicamptodon ensatus	*Hyla regilla*
Plethodontidae	Leptodactylidae
Batrachoseps major	*Eleutherodactylus martinicensis*
Plethodon dunni	*Pleuroderma tucumana*
Salamandridae	Microhylidae
Taricha granulosa	*Gastrophryne carolinensis*
Triturus vulgaris	
	Pelodytidae
Sirenidae	*Pelodytes punctatus*
Siren lacertina	
	Pelobatidae
Bufonidae	*Pelobates cultripes*
Bufo boreas	*Scaphiopus hammondii*
Bufo viridis	
	Pipidae
Discoglossidae	*Xenopus laevis*
Bombina variegata	
Discoglossus sardus	Ranidae
	Rana cancrivora
	Rana clamitans
	Rana cyanophlyctis

[Modified from Balinsky (1981).]
Note: This list includes all families but only selected species.

are incapable of producing a hyperosmotic urine, marine species reduce urine output and have specialized glands for excreting excessive salt. These salt glands have cells (ionocytes) producing a concentrated brine; the cells typically secrete a sodium chloride solution, but in a few herbivorous lizards they excrete potassium and bicarbonate, the ions prevalent in plants. The salt glands have arisen independently in at least ten lineages (turtles: cheloniids, emydids, tortoises; lizards: iguanines, agamids, scincids, varanids; snakes: acrochorids, homalopsines, hydrophiines; all crocodylids). In all, the glands are paired and cephalic but derive from different glandular precursors in each major group.

The high permeability of amphibian skin is shockingly apparent in air. The skin's evaporative resistance coefficient (R) is 1 sec cm^{-1} (roughly equivalent to an uncovered bowl of water) for most amphibians in contrast to 300–500 s cm^{-1} for reptiles. The difference relates to amphibians' reliance on cutaneous respiration and the necessity of a moist surface for gaseous exchange. (Moist not because oxygen and carbon dioxide must transfer from air to epidermis through water but to maintain functional cell/tissue surfaces.) Amphibians have existed on the "horns of a dilemma" for eons; by reducing the skin's permeability for water, its permeability for gaseous exchange is also reduced. While reduced permeability occurs in some amphibians, other adaptations compensate for water loss. Foremost, amphibians limit their activity to moist or wet habitats or, if aerial, to daily and/or seasonal periods of highest humidity, and use their skins' permeability to gain water. Most amphibians do not drink, instead they absorb water through the skin. In terrestrial anurans, the ventral surface of the thighs and abdomen are heavily vascularized and capable of extracting water from soils that are only moderately moist.

Anurans also store large quantities (20–30% of body mass) of water in their bladders and resorb it as they desiccate. This trait allows semiaquatic and terrestrial species to forage away from water, then return to pond or burrow and rehydrate. Anurans also tolerate high dehydration, withstanding water loss of about 25% body mass in semiaquatic species and >40% in arid-land, terrestrial ones. In these circumstances, amphibians also tolerate a buildup of urea, which is expelled when the animal is rehydrated. In addition to differential tolerance of water loss, anurans also display differential rates for water loss, being slowest in arid-land species. Although these species have somewhat less permeable skins, rate reduction is mainly behavioral; they stop activity sooner and assume water-conserving postures. Two arid-land frogs (*Chiromantis*, *Phyllomedusa*) have wax glands, and by coating themselves they reduce their water loss ($R = 200$ s cm^{-1}) to near reptilian levels. They also excrete uric acid.

Reptiles are better equipped for terrestrial life. Their skin is thicker and heavily keratinized, but its ability to resist water loss derives mainly from heavy concentration of lipids in the mesolayer of the epidermis (see Fig. 3.2). Although highly impermeable, the skin is still the major water loss surface, representing about two-thirds of the total loss. Respiratory water loss can also be significant, particularly at higher temperatures when a reptile is using its respiratory surfaces for evaporative cooling. Some water is lost through the feces, but the colon in most species actively resorbs water. Similarly urinary loss is low, with the excretion of uric acid.

In general, water loss is not a problem for terrestrial reptiles with their highly impermeable skin and hyperosmotic urine. Water loss is sufficiently low that they can match loss with intake (drinking, free water in food, metabolic water from breakdown of food). Reptiles display some of the same trends as amphibians. Arid-land species are commonly more resistant to desiccation (behaviorally and physiologically reducing water loss) and more tolerant to dehydration (survive greater water loss and higher ionic concentrations). Turtles and lizards also store water in the bladders and elsewhere and can resorb it to reduce desiccation stress. Reptiles cannot, however, absorb water through their skin. Some arid-land species have integumentary adaptations and behaviors that capture free water (rain, dew) on the body surface and direct it to the mouth.

Stress on Eggs and Embryos

Eggs and embryos are less resistant and tolerant of water loss than juveniles and adults, particularly in amphibians. Amphibians must lay their eggs in water or a moist site, or provide them with water. The amphibian species with direct-developing, terrestrial eggs invariably have parental care. By lying on or curling around the eggs (*Amphiuma*, *Desmognathus*), the parent reduces water loss, and in *Eleutherodactylus*, the parent rests on the eggs and transfers body water to the developing embryos.

Reptile eggs and embryos are somewhat more resistant and tolerant of water loss. The extra-embryonic membranes of the embryos and the maternal membranes and shell provide an enclosure of variable permeability. The flexible, parchment-shelled eggs (lacking a well-developed calcareous layer) of most turtles, lizards, and snakes have the most permeable eggshells, which easily desiccate. These eggs must be laid in a moist environment, where they readily absorb the water required for normal development. Though moist surroundings are necessary, too much moisture prevents gas exchange and embryos suffocate. The hard-shelled eggs of geckos, crocodiles, and some turtles are moderately resistant to water loss. Their eggs have a high water content when laid and do not require additional water for normal development.

Drying Pools

Many frogs use temporary pools for spawning. These pools range in size from slight depressions no larger than a hoof print to large ponds of a hectare or more. Pools appear after heavy rains and soon disappear without additional rain. Being ephemeral, these pools lack permanent aquatic predators, but the pools' short existence impose a time constraint on development. Frogs must lay eggs when the pools appear, and larvae must hatch quickly and grow rapidly to exit (metamorphose) before the pools dry. The time constraint is most critical in arid habitats and frogs from many families depend on explosive breeding. The spadefoot toads (*Scaphiopus*/Pelobatidae) of southwestern North America appear and complete their breeding in a single night. Although their tadpoles have been reported to go from hatching to metamorphosis in 12–14 days, the tadpole stage is commonly two to three times longer and pools often dry before metamorphosis occurs. Two tadpole morphs develop from a single *Scaphiopus* egg clutch; most are small microphagous morphs with a few larger cannibalistic morphs. The higher protein diet of the latter permits faster growth and a higher probability of metamorphosis from a drying pool.

Foam nests produced by a variety of tropical frogs are another means of protecting eggs from dehydration and predation. These nests serve as shelter for the eggs and early larval stages, disintegrating after a few days. Some *Leptodactylus* tadpoles produce similar foam masses when stressed by drying pools. When only a thin film of water remains, the tadpoles become a seething mass; their constant rubbing against one another stimulates the release of mucus, which is beaten into a foam by their agitated behavior.

Acidic Waters

Acid rain and the resulting lowering of pH in aquatic habitats focus attention on the development and survival of amphibian larvae in acidic environments. Some habitats are naturally acidic (sphagnum and cedar bogs) from the release of tannins by natural decomposition of vegetation and the inability of some soils to neutralize (buffer) these naturally occurring acids. In the Pine Barrens of eastern North America, *Hyla andersoni* and *Rana virgatipes* breed in the tannic waters ($pH < 4$) of the cedar bogs. Their larvae develop normally in these acidic waters. *Bufo americanus* and *B. fowleri* occur sympatrically, but their larvae die in $pH < 4.5$ water, and even at pH 5, half of the larvae show abnormal development. Similar tolerance differences occur in salamanders; *Ambystoma jeffersonianum* larvae can hatch and metamorphose in pH 4–8 water, *A. maculatum* larvae in pH 6–10 water. In many instances, acid rains drop the pH of breeding waters below the tolerance of the eggs and larvae; this threat is greatest for temporary pond-breeders because they depend on rain and snow, which is little buffered by soils prior to filling the breeding pools.

Studies of larval response to acidic environments have been concentrated in Europe and North America, because acid rain is having a pronounced and highly visible effect in some forests. The tropics are, of course, not free from acid rain or from naturally occurring acidic waters; black water streams and swamps are widespread throughout the world's tropics and they have pH's similar to those of temperate bogs. The physiological adaptations of tropical amphibians in acid habitats will likely match those of temperate-zone, acid-adapted species.

GAS EXCHANGE

Gas exchange includes two functionally related but operationally different processes: transfer and utilization. Transfer encompasses the structures and processes for moving gases to and from the cells and the external medium. Utilization concerns metabolism, oxygen (O_2) consumption, and carbon dioxide (CO_2) production.

Respiratory Surfaces and Mechanics

Gas transfer (inflow of O_2 and outflow of CO_2) occurs by passive diffusion across cell membranes (intracellular), and, like water, gases flow from areas of high to low concentration. However, cellular diffusion is inadequate for large organisms because only the externalmost cells obtain adequate O_2 and can eliminate CO_2. In vertebrates, the blood vascular system carries O_2 to and CO_2 from each cell to the respiratory surfaces for exchange with the outside medium.

Amphibians have four respiratory or diffusion surfaces: gills, skin, buccopharynx, and lungs. Reptiles use the latter three and possibly the cloaca. Gills are strictly for aqueous respiration, lungs for aerial respiration, and skin and buccopharynx for aquatic and aerial respiration. Most amphibians and many reptiles rely on more than one respiratory surface, simultaneously in some situations and alternately in others. Although the respiratory surfaces derive from different anatomical systems, they share several traits because efficient (rapid and maximum) gas exchange requires a steep concentration gradient and a minimum membrane thickness between the two exchange media. Thus, respiratory surfaces are heavily vascularized with one or few cell layers between the capillaries and the water or air and have mechanisms for movement of water or air across the exchange surfaces to prevent gradient stagnation at the interface.

The medium imposes different conditions on gas exchange. Water holds significantly less O_2 than air, and O_2 content of water decreases with rising temperatures. The CO_2 capacity of water and air are little different, and CO_2 concentrations are generally low outside of the body, enhancing the outward diffusion of CO_2. However, the high viscosity of water relative to air promotes concentration-gradient stagnation at the exchange interfaces and requires regular ventilation, which is energetically more expensive owing to water's viscosity. Further, the viscosity slows diffusion in the vicinity of the animal, resulting in local depletion of O_2 and concentration of CO_2. Air is much lighter; gradient stagnation and local depletion and concentration are less likely, and ventilation is energetically less costly. Water has a higher heat capacity and conductance than air; the disadvantage is rapid heat loss and the lowering of metabolic rates, but conversely temperatures fluctuate less and more slowly in water. The major disadvantage of air is its drying effect on the respiratory surfaces, and these surfaces must remain moist for gas exchange; drying alters the surface structure and reduces permeability.

Gills are present in larval amphibians and some aquatic adult salamanders (proteids, sirens, neotenic ambystomatids and plethodontids), but gills are not the exclusive respiratory surfaces in any amphibian. In most taxa, CO_2 exchange via the gills exceeds O_2 absorption. Functional gills are heavily vascularized and have long, thin filaments. Salamanders have external gills and swing them back and forth for ventilation. In anuran tadpoles, the gills lie within an opercular chamber and a buccopharyngeal pump drives water over them. Tadpoles rely mainly on cutaneous respiration, and species living in hypoxic waters develop lungs early and pulmonary (aerial) respiration dominates.

The buccopharynx is a minor respiratory surface in a variety of taxa and becomes vital in long-term submergence only in a few species. It has a limited role in most tadpoles, although in hypoxic conditions, filter feeding and its associated mucus production are suppressed, thus improving gas transfer across the buccopharyngeal surfaces. Similarly, some turtles (*Apalone*, *Sternotherus*) can extract sufficient O_2 by buccopharyngeal exchange for survival during long-term submergences, such as hibernation. In the lungless plethodontids, buccopharyngeal respiration is minor. Nearly 100% of O_2 and CO_2 transfer is cutaneous in plethodontids, and though less in other amphibian families, cutaneous respiration retains a major role in many. For aquatic and semiaquatic taxa, the skin is the main CO_2 exchange surface. Cutaneous gas exchange occurs in reptiles and may be as much as 20–30% of total gas exchange. It occurs most prominently in aquatic species (*Acrochordus*, *Sternotherus*) with CO_2 emphasized, but it also occurs in measurable amounts in terrestrial taxa (*Lacerta*, *Boa*). In these scaled species, the exchange occurs at the scale hinge-interscalar spaces.

Lungs are the principal respiratory surfaces in the aerial-terrestrial amphibians and all reptiles. Being internal, lung ventilation becomes a highly visible respiratory phenomenon. Amphibians and reptiles use different mechanisms—the buccopharyngeal force-pump and thoracic aspiration, respectively—for moving an air mass in and out of the lungs (see Digestive and Respiratory Structures in Chapters 1 and 3 for functional descriptions). The force-pump mechanism requires more energy to fill the lungs than thoracic aspiration, and yet air pressure and volume are proportionately less. The force-pump is, however, especially efficient and effective for anuran vocalization. The more efficient thoracic aspiration is associated with larger lung volume and increased lung partitioning and vascularization in many reptiles.

Respiration and Metabolism

Gas exchange is a direct function of metabolism. Metabolic activities, whether anabolic or catabolic, require the energy derived from oxidation, so O_2 is required even in a resting or hibernating state. Metabolism can occur in oxygen's temporary absence but an oxygen debt develops that must be repaid. As noted earlier (Chapter 5, Energetics), metabolic rate is measured by O_2 consumption (O_2 ml g^{-1} hr^{-1} = V_{O_2}) or CO_2 production (same units of measurement); metabolism and gas exchange are inseparable.

Body size and temperature influence gas exchange. In general as mass increases, O_2 consumption (and CO_2 production) increases although the consumption rate (V_{O_2}) declines with increasing mass. This mass-specific relationship ties to the general physical principal that mass increases as a cube of length, whereas surface area increases as a square of length. Thus, respiratory surfaces may be unable to meet the metabolic needs without modifications. Modifications range from increasing surface by additional folds (skin) or partitions (lungs), increasing vascularization and/or placing blood vessels closer to the surface, increasing gas transport capacity of blood and increasing flow rate, and/or similar respiratory enhancing devices. Such changes occur ontogenetically and are apparent when comparing taxa of different sizes.

Aerobic metabolism is strongly temperature dependent, and O_2 consumption increases two- to threefold for every 10°C elevation. Even though metabolic activity reacts similarly in amphibians and reptiles, different groups possess different basal metabolic rates (Fig. 10.2), for example, anurans typically have higher and more temperature sensitive rates than salamanders. The temperature–metabolism relationship is linear in most ectotherms, but a few snakes and lizards have decoupled metabolism from temperature over a 3–5°C range (usually within their preferred activity temperatures) in which metabolism remains constant. Gas exchange and metabolism are influenced in varying amounts by a host of other factors. Some species show daily and/or seasonal fluctuations of the basal rate, indicating endogenous rhythms. Temperate species of amphibians have metabolisms that can be acclimatized and adjusted to seasonal temperature changes. Health and physiological state affect basal metabolic rates; for example, the rate is two times greater in an alligator fasting for 1 day than in one fasting for 3–4 days.

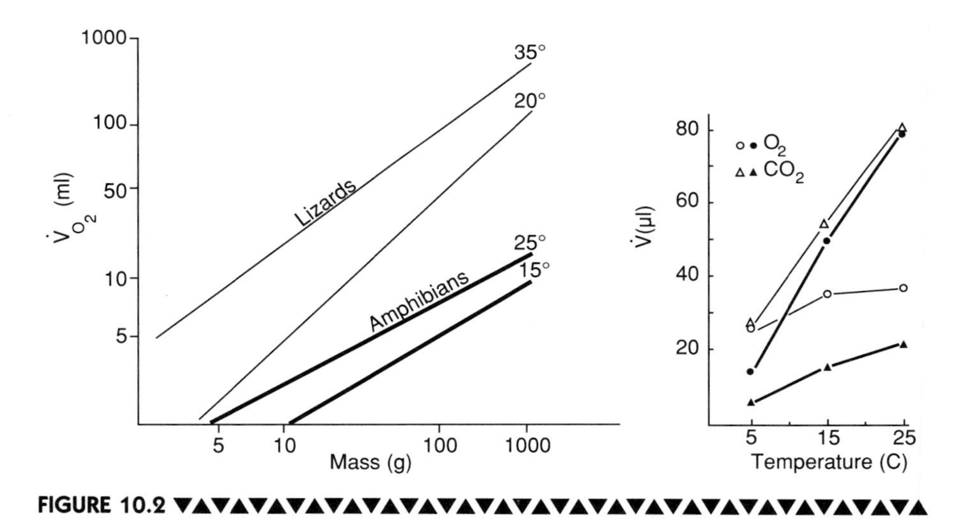

FIGURE 10.2 ▼▲▼▲▼▲▼▲▼▲▼▲▼▲▼▲▼▲▼▲▼▲▼▲▼▲▼▲▼▲▼▲▼▲▼▲

Effects of body mass and temperature on gas exchange in amphibians and reptiles. Comparison (left graph) of metabolic rates in temperate-zone amphibians and lizards highlights the lower rate of amphibians [Data from Bennett (1982) and Whitford (1973).] Effects of temperature (right graph) on cutaneous (open symbols) and pulmonary (solid symbols) gas exchange in anurans and lunged salamanders shows the efficiency of skin for CO_2 removal and lungs for O_2 pickup. [Adapted from Whitford, (1973).]

Aerobic and Anaerobic Metabolism

An animal's normal activities are fueled by energy from aerobic metabolism, that is, cellular metabolism supported by oxygen. Energy can also be obtained by anaerobic mechanisms in the absence of oxygen. Anaerobiosis is vital for providing quick energy during a burst of activity or during prolonged activity, such as escaping from a predator or surviving an anoxic event, such as prolonged submergence by a lung-breather. Although vital for survival, anaerobiosis is energetically costly and prolonged use is debilitating. During burst activity, anaerobiosis provides energy at 5–10 times the aerobic level but rapidly depletes energy stores and within a few minutes the animal becomes visibly fatigued (Fig. 10.3). Recovery may require hours or even days; the O_2 debt and lactic acid removal can proceed very rapidly if anaerobiosis was not excessive, but since anaerobic metabolism is highly inefficient (requiring as much as 10 times the food input for an equivalent amount of aerobic work), total nutrient and energy replacement requires much longer. An important feature of anaerobiosis for escape behavior is its temperature independence within a major portion of a species' temperature activity range, thus permitting an equally rapid escape response at a low temperature as at a high one (Fig. 10.3).

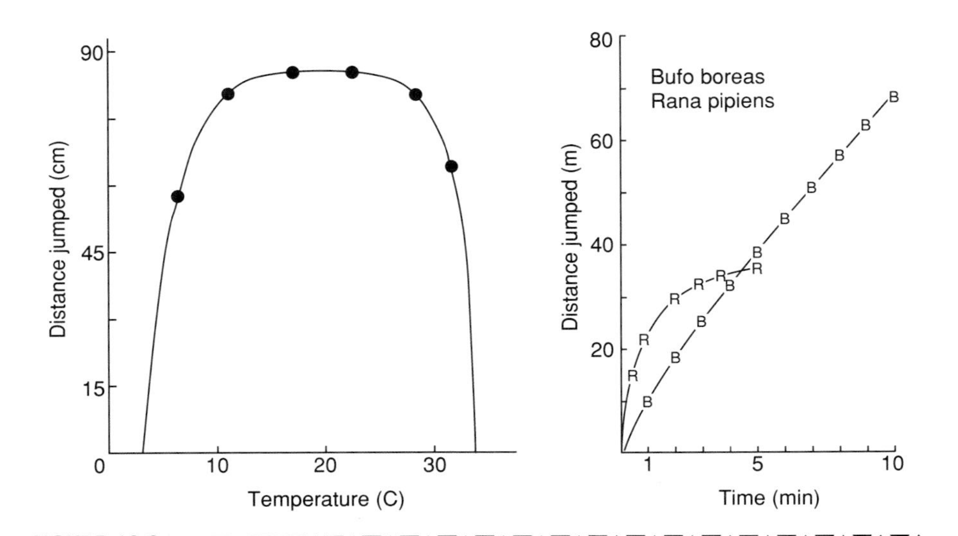

FIGURE 10.3 Anaerobic versus aerobic performance in anurans. Jumping performance (anaerobic-powered) of green frog, *Rana clamitans* (left graph), is temperature independent and remains nearly constant over 20°C range [Adapted from Huey and Stevenson (1979).]. Comparing performance (right graph), *Rana* (anaerobic-powered jumping) initially outdistances *Bufo* (aerobic-powered jumping), but fatigue quickly incapacitates the frog and the toad hops on without fatigue. [Adapted from Putnam and Bennett, (1981).]

THERMAL INTERACTIONS

Amphibians and reptiles are ectothermic. For most species, their body temperatures depend entirely on external heat sources. They generate metabolic heat, but heat production is low and they have no effective insulation to retain this heat. Insolation is the ultimate heat source but not the direct one for the majority of amphibians and reptiles, even though basking is the most visible heat-gain behavior. Heat sources (and sinks because losing heat is as important as gaining it) and a taxon's dependency on them are variable. The sources and sinks depend on a species' habits and habitats, and an individual's needs at a particular moment.

Heat exchange with the environment (Fig. 10.4) occurs via radiation, convection, and conduction. An ectotherm receives radiant energy from the sun directly or indirectly from reflected solar radiation and the heat of substrate and air. Sunlight striking a surface is variously absorbed and reflected; the absorbed solar radiation converts to heat and raises the temperature of the object. No natural object is totally absorptive or reflective of solar radiation. Most organisms have a mixture of absorptive and reflective surfaces, and many can change the absorptive–reflective nature of their surfaces. Dark surfaces are strongly absorptive, light ones reflective; an animal's colors and pattern express a balance between thermal needs and crypsis. In the context of an animal's heat gain and loss, reflected radiation is not lost heat, rather it is heat not captured. Solar radiation heats the fluids (air, water) and solids (substrate, vegetation) of an animal's surroundings. Heat exchange occurs between fluids and solids and the individual ectotherm in the same manner as water and gas exchange — from high "concentration" to low until the temperatures are in equilibrium. Exchange from solid to solid is conduction, from fluid to solid or vice versa is convection. Conduction and convection strongly affect the thermal balance of ectotherms with their poor insulation. Heat can be rapidly gained or lost via conduction and convection. Ectotherms use both in thermal regulation; however, for fossorial taxa broadly in contact with the soil and aquatic ones totally immersed in water, regulation is largely impossible and their temperatures rapidly equilibrate with their surroundings. Because soil and water are voracious heat sinks, they are also superb heat reservoirs giving up or gaining heat slowly and provide refuges from excessive heat gain or loss.

Two additional phenomena affect an ectotherm's thermal balance: evaporation and body size. Evaporation is synonymous with heat loss, offering no opportunity for heat gain. Body size affects the rate of heat gain or loss through the mass–surface area relationships (see foregoing explanation under Gas Exchange). Heat gain or loss slows with increasing mass because there is proportionately less surface area for heat exchange. In terrestrial lizards, the critical mass is about 20–25 g, at which size physiological mechanisms can have some effect on temperature control. Behavioral control of thermal interactions is possible even in the smallest ectotherms.

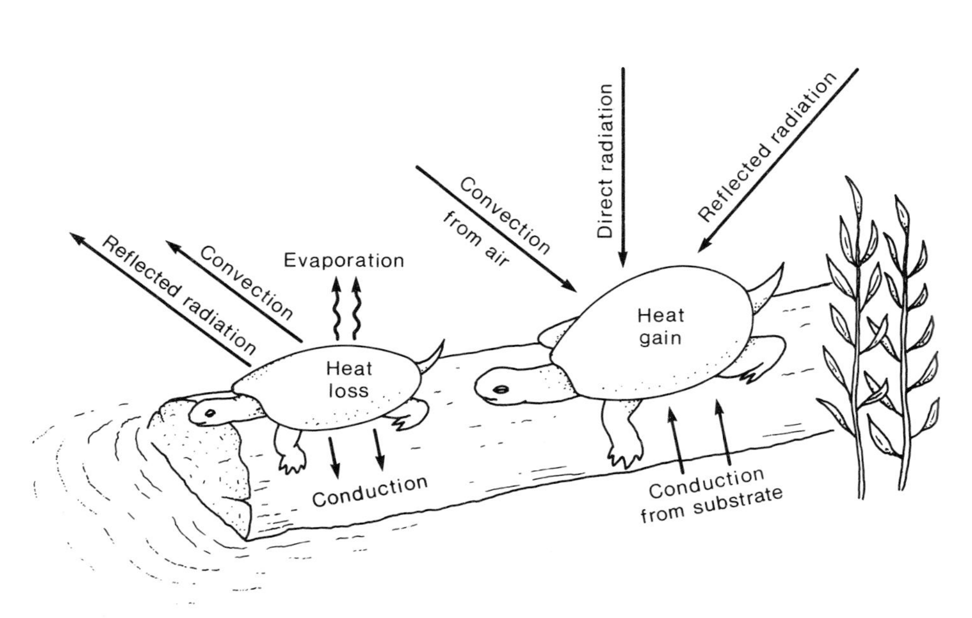

FIGURE 10.4 Heat exchange in ectotherms. Heat gain and heat loss are segregated for convenience of demonstration. [Adapted from Tracy (1976).]

Temperature Ranges and Tolerances

Each species has a range of body temperatures (Fig. 10.5) in which its individuals survive (tolerance range, TR) and another in which they pursue normal life activities (activity temperature range, ATR). The TR is bounded by the lethal minimum and maximum temperature. As an individual's body temperature approaches these temperatures, its tissues die. Body temperatures just above and below the lethal temperatures are the critical thermal minimum (CTMin) and maximum (CTMax) temperatures, respectively; an individual at either of these temperatures is physically immobilized. In many cases, several minutes or more at CTMin or CTMax can also be fatal through physiological failure. The CTMin-Max temperature range defines an individual's tolerance limits, but individuals limit their daily activity to a narrower temperature range, that is, the ATR with its voluntary minimum and maximum temperatures that bracket the limits of normal performance. Many taxa select and attempt to

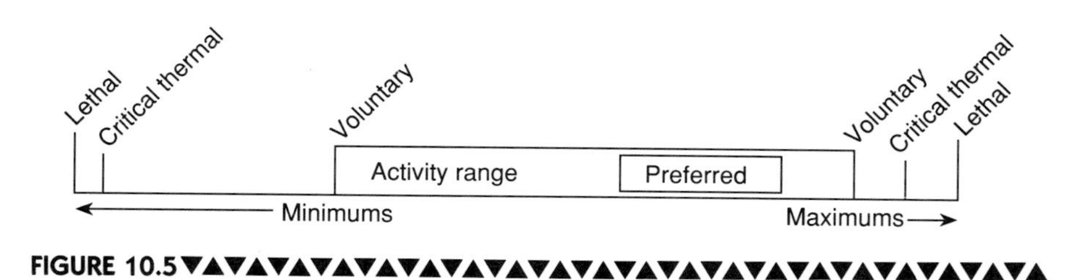

FIGURE 10.5 Profile of body temperature characteristics of an ectotherm.

maintain their body temperatures in an even narrower range (preferred or selected temperature range, PTR).

ATRs and tolerances reflect (and permit) the habits and habitat selections of a taxon. The broad diversity of amphibians yields an equally broad array of ATRs. Typically amphibians have lower ATRs than sympatric reptiles (Table 10.2) whether the comparison is made of tropical or temperate species. And of the amphibians, salamanders are active, on the average, at lower temperature than anurans. In the temperate zones, many salamanders and anurans breed at temperatures lower than 15°C; other species, particularly anurans, require warmer (20–32°C) summer nights for breeding. Most tropical amphibians have similar ATRs to the summer-breeding, temperate taxa. A major difference between temperate and tropical taxa is the breadth of TRs and ATRs. Because temperate-zone taxa experience a wide seasonal and daily fluctuation of temperatures, they have broader temperature ranges than lowland tropical species experiencing slight seasonal and daily fluctuations.

Few reptiles have ATRs ranging below 20°C; for most, 26–28° is the lower limit of the ATRs and 34–36° is not uncommon for the upper limits. These high ATRs are characteristic of species using solar basking to elevate body temperature and using a shuttling behavior in and out of the sun to maintain a narrow preferred temperature range. ATRs ranging below 20°C are not uncommon in fossorial, aquatic, nocturnal, and limbless or reduced-limbed species. These taxa live in high-conductance environments and/or environments lacking a direct radiant energy source. As in amphibians, reptiles from environments of widely fluctuating temperatures usually have broader ATR and TR ranges.

ATRs and the other parameters of temperature tolerance are not invariant for a species or an individual. Variation is of two general types: seasonal/environmental adjustment and genetic/evolutionary adjustment. While these parameters are not genetically fixed at specific set-point temperatures, an individual's biochemical and cellular processes operate optimally within a particular temperate range. Because there is individual variation, selection occurs to adapt populations to the local temperature regime. For example, green anoles (*Anolis carolinensis*) from northern populations have lower CTMins values than do anoles from southern populations, and similar shifts of temperature parameters are

TABLE 10.2 ▼▲▼▲▼▲▼▲▼▲▼▲▼▲▼▲▼▲▼▲▼▲▼▲▼▲▼▲▼
Active Body Temperatures (ATR) for Representative Amphibians and Reptiles

Taxon	ATR (°C)	Habits
Bolitoglossa subpalmata	7–13	Nocturnal/terrestrial
Dendrotriton bromeliacia	9–17	Nocturnal/arboreal
Necturus maculosus	5–30	Nocturnal/aquatic
Plethodon cinereus	3–22	Nocturnal/terrestrial
Pseudosiren striatus	8–26	Nocturnal/aquatic
Acris crepitans	8–35	Diurnal/terrestrial
Phyllomedusa sauvagei	22–41	Nocturnal/arboreal
Litoria caerulea	20–28	Nocturnal/semiarboreal
Taudactylus diurnus	14–24	Diurnal/terrestrial
Telmatobius culeus	10–16	Diurnal/aquatic
Clemmys guttata	3–32	Diurnal/semiaquatic
Dermochelys coriacea	22–32	Diurnal/aquatic
Geochelone gigantea	23–32	Diurnal/terrestrial
Terrapene ornata	15–35	Diurnal/terrestrial
Alligator mississippiensis	23–37	Nocturnal/aquatic
Sphenodon punctatus	6–25	Diurnal/terrestrial
Amphibolurus barbatus	25–40	Diurnal/terrestrial
Anolis tropidolepis	15–24	Diurnal/arboreal
Chamaeleo pumilus	4–37	Diurnal/arboreal
Cnemidophorus labialis	30–42	Diurnal/terrestrial
Gehyra variegata	24–35	Nocturnal/arboreal
Carphophis amoenus	14–31	Diurnal/semifossorial
Crotalus cerastes	14–41	Nocturnal/terrestrial
Morelia spilota	23–30	Nocturnal/semiarboreal

Sources: Salamanders—Bs, Cp, Pc, Ps, Feder et al., 1982; Nm, Hutchinson and Hill, 1976. Frogs—Ac, Brattstrom, 1963; Lc, Johnson, 1971; Ps, Shoemaker et al., 1987; Td, Johnson, 1971; Tc, Hutchinson, 1982. Turtles—Cg, Ernst, 1982; Dc, Friar et al., 1972; Gg, Swingland and Frazier, 1979; To, Legler, 1960. Crocodilians—Am, Avery, 1982. Tuatara—Sp, Avery, 1982. Lizards—Ab, Gv, Heatwole and Taylor, 1987; At, Cp, Cl, Avery, 1982. Snakes—Ca, Cc, Avery, 1982; Ms, Heatwole and Taylor, 1987.

Note: Temperatures given in Celsius and rounded to nearest integer.

observed in altitudinal transects. These parameters also shift seasonally (i.e., acclimatization, natural adjustment), so that the ATR, CTMax, etc., of a population will shift upward with increasing spring temperatures and then reverse as the temperatures decline in late summer and autumn. Similar shifts can be laboratory induced (acclimation) by holding specimens in constant-temperature regimes for several days; in most cases, acclimation mimics acclimatization adjustments.

Regulation and Nonregulation

While sharing ectothermy, amphibians and reptiles manage their thermal dependency differently. Amphibians' maintenance of a moist skin requires a different approach than that of the reptiles. Rising temperatures increase evaporation and speed water and heat loss.

The integument of most amphibians lacks an effective resistance to evaporative water loss and so experiences high water loss under all but the most humid conditions. Evaporative water cools the surface initially and gradually the core body temperature as well. Thus, the necessity of a moist skin and its evaporative water/heat loss limit the control of body temperatures (T_b) in amphibians; generally their active T_b are at the ambient temperature or below. Amphibians rely principally on behavior (postures, location) to obtain some degree of temperature stability. The exception to thermal conformity occurs in a few arboreal treefrogs that control evaporative cooling to stabilize T_b during high ambient (environmental) temperatures (T_e). The best-known examples are the waterproof frogs *Phyllomedusa* and *Chiromantis*. These frogs allow T_b to track T_e until T_b reaches 38–40°C, then skin glands begin secretion and evaporation allows the frog to maintain a stable T_b even if T_e attains 44–45°C. In other amphibians without cutaneous evaporative control, T_b follows T_e although T_b is usually 1–2°C less because of evaporative cooling.

The highly impermeable integument of reptiles permits direct exposure to sunlight without excessive water loss, and temperature control or regulation is common in reptiles. Of course, not all reptiles attempt temperature control, and some allow their T_b to track T_e. The spectrum of thermal interactions spans the entire range from temperature conformity to temperature regulation and near endothermy. As noted in the preceding tolerances discussion habitats and habits may preclude thermoregulation (i.e., control of T_b within a narrow range during the activity period and potentially at other times), but such assumptions must be carefully examined. Geckos may be thermal conformists during their nocturnal feeding forays, but at least some species carefully select diurnal retreats and shift locations therein to maintain their T_b within a narrow and elevated range.

Mechanisms for thermoregulation involve behavior, physiology, and morphology. Basking (see below) is the most visible behavioral mechanism, commonly used in shuttle thermoregulation. A lizard or snake alternates between sun and shade and often precisely ($\pm$2°C) maintains its T_b within the preferred range. Timing of activity and selection of different microenvironments for different activities are common thermoregulatory behaviors for all animals as are also posture and orientation. In seasonal activity (Chapter 9), we noted the shift from unimodal midday activity to bimodal early–late day activity as a pattern to select the least stressful temperatures for feeding in those areas of intense summer sunlight and heat. Similarly, using burrows or switching from terrestrial to arboreal foraging to avoid rising substrate temperatures is a common behavior for many arid-land lizards. All adjustments of behavior are a balance between the individual's physiological tolerances and the need to gather adequate food and avoid predators. It is this balance that invokes thermal conformity in many tropical forest reptiles. Sunlight patches are available within their preferred habitats but following shifting patches of light increases an individual's visibility to predators and likely reduces its feeding rate; natural selection has favored the thermal conformists.

Use of direct solar radiation is not strictly behavioral. Many lizards can lighten or darken their integument, and characteristically they begin basking with a dark skin to speed absorption and lighten the skin in midday to reflect radiation. Changes in heart rate and peripheral circulation also aid rapid absorption or radiation of heat. Tolerance for heat overload (hyperthermia) is less than for cold shock (hypothermia), and reptiles rapidly mobilize physiological mechanisms and behavior to suppress rising T_b approaching the CTMax. The overheating animal seeks shade and lower T_e sites, elevates body to increase convection heat loss, and increases evaporative cooling by increasing peripheral-surface circulation, panting (to expose moist respiratory surfaces), salivation, and urination and defecation on limbs and body. Physiological mechanisms for avoiding hypothermia are less evident; cold acts like a narcotic (both amphibians and reptiles) and if cooling occurs too quickly the animal becomes immobilized before it can reach shelter. To retard cooling, reptiles reduce heart rate and peripheral circulation. A few are capable of thermogenesis via microcontraction of muscles (e.g., incubating pythons). Physiological tolerance is the major adaptation for surviving extreme heat loss (see Freeze Tolerance section later).

Basking

Many reptiles and some amphibians raise their T_b by basking. Basking is behaviorally simple in the sense of locating a site in the sun and actively absorbing heat by postural adjustment to maximize surface exposure to sunlight. Physiologically, peripheral circulation increases to transfer heat from the surface to the body core. Basking seems to be used in two different fashions. In taxa living in high-conductance environments (water, soil), elevated T_b is possible when basking since T_b drops rapidly to T_e when basking stops. These species tend to bask after eating, presumably to raise core T_b to boost digestion. In aerial/terrestrial taxa, heat loss is slower, and after basking individuals can feed, court, and pursue other activities while remaining within their ATR. These baskers begin their activity period with basking to elevate T_b then proceed with normal activities. The latter pattern is the common one for terrestrial and arboreal snakes and lizards, the former one for crocodilians, aquatic turtles, and frogs. Aquatic species often use both atmospheric and aquatic basking. In the latter, an individual maintains an elevated T_b by floating in the warmer surface layer.

Inertial Endothermy

The benefit of large size is a proportionately slower heat loss. A few reptiles use this physical principle to maintain an elevated T_b. The most amazing example is the leatherback seaturtle (*Dermochelys coriacea*), which approaches mammalian endothermy on a diet of high-test gelatin (i.e., jellyfish). They maintain a T_b of 25–26°C in 8° seawater. Their dark skin may permit some heat gain through solar radiation, but their primary heat source is muscular activity. This heat is retained by a thick, oil-filled skin (an equivalent insulator to the blubbery skin of whales) and a counterflow circulatory system in the limbs. The latter adaptation reduces heat loss in cold waters and promotes the shedding of excess heat in tropical waters.

DORMANCY

When environmental conditions exceed an individual's capacity for homeostasis, retreat and inactivity offer an avenue for survival. Regular cycles of dormancy are major features in the lives of many amphibians and reptiles. Climatic fluctuations are the principal force for cyclic dormancy—hot and dry in desert regions and near or below freezing for temperate-zone areas. Seasonal fluctuations in food resource may drive dormancy in some tropical areas, although this remains unproven. Dormancy behaviors are commonly segregated into hibernation for avoidance of winter cold and estivation for all other dormancy, including acyclic drought-caused dormancy. Depending on a species' distribution, individuals may be dormant longer than they are active (e.g., about 1 mo yr^{-1} active for Arizona *Scaphiopus*, less than 4 mo yr^{-1} for Manitoba *Thamnophis*).

Physiological studies of amphibian and reptilian hibernators indicate that many species alter cardiovascular function and suppress metabolic activities to conserve energy and ensure adequate O_2 to vital organs. The metabolic rates are lower than if rates were slowed by temperature effects. The physiology of estivation is less clear; metabolic rates generally do not drop below temperature-normal rates, although water loss rates are variously reduced.

Hibernation

As winter approaches, most temperate-zone amphibians and reptiles seek shelter where the minimum T_e will not fall below freezing. Some amphibians and turtles avoid subfreezing temperatures by hibernating on the bottoms of lakes and streams. Since water reaches its greatest density at 4°C and sinks, animals resting on or in the bottom usually will not experience $<4° T_e$. Amphibians and reptiles hibernating on land are less well insulated by soil and must select sites below the frostline or be capable of moving deeper as the frostline approaches them. Terrestrial hibernators, the few that have been followed, do move during hibernation. Box turtles (*Terrapene carolina*) begin hibernation near the soil surface, reach nearly 0.5 m deep during the coldest periods, and then inch toward the surface as T_e moderates. Hibernating snakes (*Elaphe*, *Crotalus*) move along a thermal gradient in their denning caves/crevices, always staying at the warmest point. As an aside, the balled masses of snakes in hibernation dens are not conserving heat by huddling; lack of insulation and low metabolic and muscular activity are inadequate for maintaining T_b above T_e; balling may be a water conservation behavior.

Many aquatic hibernators rest on the bottom of ponds or streams rather than lie buried in the bottom. Though such sites might expose them to predation, hibernation in open water permits aquatic respiration (extrapulmonary) and apparently is sufficient to meet some or all of the oxygen expenditures of dormant amphibians or reptiles. In normoxic water, the O_2 demands of lunged anurans and salamanders are easily met by cutaneous respiration during hibernation. Cutaneous respiration also provides sufficient O_2 for some hibernating reptiles (*Chrysemys picta*, *Sternotherus odoratus*, *Thamnophis sirtalis*). Experiments on garter snakes hibernating submerged in a water-filled hibernaculum demonstrated that the submerged snakes used aerobic metabolism but at a more energy conservative rate than terrestrial hibernating conspecifics. In normoxic waters, turtles also remain aerobic through cutaneous and perhaps buccopharyngeal respiration; however, if buried in the mud, the hibernating animal must switch to anaerobiosis in this anoxic or hypoxic environment. Survival is possible because of high tolerance for lactic acid buildup, and in some instances, submerged turtles may shuttle between normoxic and anoxic sites. When in the normoxic ones, they can shift to aerobiosis and to some extent flush excess lactic acid. Turtles swimming below the ice of a frozen pond have been a common anecdotal observation, and now there is a physiological explanation.

Freeze Tolerance

Most temperate-zone amphibians and reptiles are able to survive brief periods of supercooling (-1° to -2°C); freezing (formation of ice crystals within the body) is lethal to all but a few species. Ice crystals physically damage cells and tissues. Intracellular freezing destroys cytoplasmic structures and cell metabolism. Extracellular freezing also causes physical damage, but the critical factor is osmotic imbalance. As body fluids freeze, pure water freezes first, increasing the extracellular osmotic concentration and dehydrating the cells. Intracellular dehydration disrupts cell structure and, if extreme, causes cell death. Extracellular freezing also blocks fluid circulation and the delivery of oxygen and nutrients to the cells. The damage from freezing causes the animal's death upon thawing.

A few species of frogs (*Hyla crucifer*, *H. versicolor*, *Pseudacris triseriata*, *Rana sylvatica*) and turtles (*Terrapene carolina*, hatchling *Chrysemys picta*) are "freeze tolerant" and can survive extracellular freezing. These frogs hibernate in shallow shelters, and although snow may insulate them, T_b still drops to -5 to -7°C. They freeze. Ice crystals appear beneath the skin and interspersed among the skeletal muscles; a large mass of ice develops in the body cavity. As much as 35–45% of the total body water may become ice and yet the frogs survive. When frozen, a frog's life processes are suspended; breathing, blood flow, and heartbeat stop. These frogs tolerate the large volume of body ice by producing and accumulating cryoprotectants (= antifreeze) within the cells. The cryoprotectants are either glycerol (*Hyla versicolor*) or glucose (the three other species), which protect and stabilize cellular function and structure by preventing intracellular freezing and dehydration. These freeze-tolerant species also possess specialized proteins that control extracellular freezing and adjust cellular metabolism to function at low temperatures and under anaerobic conditions.

The frogs do not physiologically anticipate winter and begin to produce the cryoprotectants. Rather, ice forms peripherally and triggers synthesis. The rate of freezing is slow, permitting the production and distribution of cryoprotectants throughout the body before any freeze damage can occur. As soon as the body begins to thaw, the cryoprotectants are removed from general circulation. Freeze tolerance extends into early spring at the time when the frogs begin reproductive activities. For the early spring breeders such as the spring peeper (*H. crucifer*) and the wood frog (*R. sylvatica*), this extended tolerance permits survival under the highly variable and occasionally subzero temperatures occurring during their breeding season. Freeze tolerance appears to be lost gradually and in association with the beginning of feeding.

Estivation

Amphibians in desert and semidesert habitats face long periods of low humidity and no rain. To remain active is impossible for all but a few species; death by dehydration occurs quickly. Arid-land species retreat to deep burrows with high humidity and moist soils, become inactive, and reduce their metabolism. Inactivity may dominate an anuran's life. *Scaphiopus hammondii* in the deserts of southwestern North America spend >90% of their life at rest; they appear explosively and breed with the first heavy summer rains, then feed for 2–3 weeks before becoming inactive for another year. When retreats become dehydrating, some anuran species (e.g., *Cyclorana*, *Neobatrachus*, *Lepidobatrachus*, *Pternohyla*, *Pyxicephalus*) produce epidermal cocoons. The cocoon forms by a sequential shedding of the stratum corneum; the successive layers form an increasingly impermeable cocoon that completely encases the frog except around the nostrils. *Siren* burrows in the mud of drying ponds and produces a similar epidermal cocoon.

PART V

POPULATIONS AND THE ENVIRONMENT

But you should not be deceived into believing that ecology is founded on exact quantitative laws that serve to predict events with the same authority as the equations of, say, physics or physical chemistry. An ecosystem is vastly more complex than a gas-filled balloon or a flask of reagents. Proceed cautiously . . .

E. O. Williams and W. H. Bossert, 1971

The complexity of ecosystems remains unchanged, but ecology has become increasingly rigorous in its mathematical analyses and the predictive power of its mathematical models have improved. Improved although not exact, the models are called on to address local to worldwide changes brought on by burgeoning human populations and the resulting environmental havoc as a single species exceeds the capacities of its local ecosystems.

Amphibian and reptilian populations have played a critical role in this development and testing of ecological models. Frog and lizard populations are especially amendable to experimental manipulations in natural and semi-natural habitats and have revealed the complex and changeable nature of inter-specific interactions. In spite of the research attention directed at these two groups and at the other amphibians and reptiles as well, our knowledge of their ecologies remains rudimentary. Their diversity seems to defy generalities.

CHAPTER 11

Population Structure and Dynamics

A biological population is a group of individuals of the same species living in the same area. Potentially, all adult members of a population are capable of interbreeding, and each population represents a single gene pool and a unit of evolution. Although the potential for random interbreeding is seldom, if ever, realized within a single generation, panmixia may occur over generations in small, localized populations.

Populations can be variously delimited, being as inclusive as all box turtles in eastern North America, or more limited to all hellbenders in the Ozark River drainage, or locally to all *Bufo exsul* in Deep Springs Valley. Each of the preceding is a biological population, but the local population (= deme) is the unit of special interest to biologists. The local population responds to local conditions: growing; shrinking or even disappearing (extinction); and evolving. Each local population is semi-isolated from other similar populations by minor or major habitat discontinuities, but few are totally isolated (closed) and most receive occasional immigrants from nearby or distant populations and commonly lose members via emigration.

CHARACTERISTICS OF POPULATIONS

Even though each population can be viewed as a single entity with birth, juvenile and mature phases, and eventually death, such a view oversimplifies and misrepresents the complexities of a population's history and persistence. The characteristics and stability of a population derive from the lives and deaths of its members. Each population has a size, an age distribution, a sex ratio, birth and death rates, and gene frequencies. These and other populational traits vary temporally and spatially within a population as well as among populations of the same species. The variability of populations' responses arises from differences in the nature and intensity of local environmental conditions (abiotic and biotic), the state of a population when experiencing these changing conditions, and the life-history characteristics of the population.

A snapshot of a population's structure is provided by a life table (= actuary table or survivorship/mortality schedule). Each table gives an age-specific summary of mortality from a cohort's origin (birth or deposition as eggs) to the death of all cohort members (Table 11.1A). These data are presented as the actual number of individuals observed or converted to a standard cohort of 100 or 1000, or to decimal proportions. Often the life table is combined with the population's fecundity schedule to provide an age-specific summary of mortality and reproduction (Table 11.1B,C,D).

Combined survivorship and fecundity schedules yield a summary image of a population's current state and likely near future and assist intra- or interspecific comparison of populations. The significant aspects are the average age at sexual maturity (i.e., duration of juvenile, nonreproductive life) and the age-specific mortality and fecundity. A number of other measures of a population's state can be derived from this pair of schedules, such as mean generation time (T), net reproductive rate (R_o, also called replacement rate), reproductive value (v_x), intrinsic rate of natural increase (r), and others. R_o is especially informative; it ranges between 0 and 10± for vertebrates. A value of 1.0 indicates that the population is stable (births = deaths), <1.0 decreasing and >1.0 increasing.

Survivorship (l_x) and mortality (d_x and q_x) are different aspects of the same population phenomenon, the rate of a cohort's disappearance. Survivorship maps the cohort decline from its first appearance to the death of its last member. Age-specific mortality (q_x) records the probability of death for the surviving cohort members during each time interval. The pattern of a cohort decline is often shown by plotting survivorship against time. Four hypothetical survivorship curves (Fig. 11.1, insets) represent the general types of possible survivorship patterns. In Type I (rectangular convex curve), survivorship is high (i.e., mortality low, $q_x < 0.01$) through juvenile and adult life, then all cohort members die nearly simultaneously ($q_x = 1.0$). Type III is the opposite pattern (rectangular concave curve), where mortality is extremely high ($q_x > 0.9$) in the early life-stages and then abruptly reverses to low mortality ($q_x < 0.01$) for the remainder of the cohort's existence. Type II patterns occupy the middle ground, with either a constant number of deaths (d_x) or a constant death rate (q_x). Although these idealized patterns are never matched precisely by natural populations, the patterns offer a convenient descriptive shorthand for comparing actual populational data.

Most amphibians with indirect development and turtles have Type III survivorship (Fig. 11.1). Amphibian eggs and larvae commonly experience high predation. Growth may temporarily make the older larvae safe from most aquatic predators, but predation is again high during metamorphosis and early terrestrial life. For those species breeding in temporary ponds, death of entire cohorts is a regular threat because ponds may dry prior to metamorphosis. Many turtle populations suffer high nest predation; freshwater species and seaturtles often have 80–90% of their nests destroyed within a day or two of egg-laying. The majority of the remaining amphibians and reptiles have Type II-like patterns with moderate and fluctuating mortality during early life and then a moderate to low and constant death rate during late juvenile and adult life. Weather (e.g., too wet or too dry) and its effect on food availability appear to be the major causes of juvenile mortality in these Type II species. No amphibian or reptile attains a close match to a Type I survivorship (*Xantusia* shows the closest approach known among reptiles; Fig. 11.1). Species with parental care may have an initial low mortality, but even crocodilians cease parental care well before their offspring are fully predator- and weatherproof.

Populations do not have fixed survivorship patterns. Annual patterns will be most similar in those populations with a nearly constant age structure, but even these populations can shift from one pattern to another as a result of a catastrophic event or an exceptional year of light or heavy predation. Males and females in the same cohort may have different survivorship curves, and if the difference is great, the resulting population will have an unequal sex ratio.

Natality and fecundity are used interchangeably and denote the reproductive effort or output (number of eggs or neonates) per individual (female and male) in the population; however, most fecundity schedules are for females only and the cohort's average output is halved. Average clutch and litter sizes range from 1 to 10,000+ in amphibians and reptiles (see Fig. 7.4), but these values are not an accurate measure of a cohort's fecundity. First, these reproductive outputs are without reference to time. Fecundity addresses output in a very specific time

TABLE 11.1 ▼▲

Life Tables and Fecundity Schedules

A. Life table for a French population of wall lizards, *Podarcis muralis*[a]						B. Survivorship and fecundity schedule for a South Carolina population of female sliders, *Trachemys scripta*[b]					
Age	Survivors		Mortality		Life expectancy						
x	n_x	l_x	d_x	q_x	e_x	x	l_x	m_x	l_xm_x	q_x	
0–1	570	1000	376	0.66	1.01	0	1.000	0.00	0.000	0.89	
1–2	194	340	146	0.75	0.99	1	0.105	0.00	0.000	0.46	
2–3	48	84	23	0.48	1.48	2	0.057	0.00	0.000	0.17	
3–4	25	44	13	0.52	1.36	·					
4–5	12	21	*6*	*0.50*	*1.31*	6	0.026	0.00	0.000	0.186	
5–6	*6*	*11*	*3*	*0.50*	*1.05*	7	0.021	1.28	0.027	0.186	
6–7	*3*	*5*	*2*	*0.50*	*0.70*	·					
7–8	*1*	*2*	*1*	*1.00*		10	0.011	1.28	0.014	0.186	
						·					
						15	0.004	1.28	0.005	0.186	
						·					
						20	0.001	1.28	0.002	0.186	
						·					
						22	<0.001	—	—	—	R_0 = 0.137

C. Survivorship and fecundity schedule for a North Carolina population of Applachian dusky salamanders, *Desmognathus ochrophaeus*[c]

x	l_x	m_x	$l_x m_x$	
0	1.000	0.0	0.000	
.				
4	0.087	4.5	0.392	
5	0.055	4.5	0.248	
6	0.034	4.5	0.153	
.				
8	0.013	4.5	0.058	
.				
10	0.005	4.5	0.022	
.				
12	0.002	4.5	0.009	$R_0 = 1.026$

D. Survivorship and fecundity schedule for a Utah population of female western yellow-bellied racers, *Coluber constrictor*[d]

x	l_x	m_x	$l_x m_x$	
0	1.000	0.0	0.000	
1	0.170	0.0	0.000	
2	0.125	0.2	0.023	
3	0.102	1.8	0.188	
4	0.078	1.8	0.178	
5	0.068	2.4	0.160	
.				
10	0.010	2.7	0.055	
.				
15	0.006	2.9	0.017	$R_0 = 1.187$

[a] Data from Barbault and Mou (1988); italicized values are hypothetical, assuming constant q_x for adults.

[b] Data from Frazer et al. (1990).

[c] Data from Tilley (1980).

[d] Data from Brown and Parker (1984).

Abbreviations and explanations: d_x, number of cohort members dying during age interval; e_x, average life (yr) expectancy for members alive at beginning of interval; l_x, proportion of cohort alive at beginning of interval; m_x, age-specific fecundity rate (average number of offspring produced by surviving cohort during each interval); $l_x m_x$, total fecundity of surviving cohort members in each interval; n_x, actual number of members alive at beginning of age interval; q_x, age-specific death rate (proportion of individuals dying during interval that were alive at beginning of interval); R_0, net reproductive rate (average lifetime fecundity for each cohort member); x, age interval (1 yr).

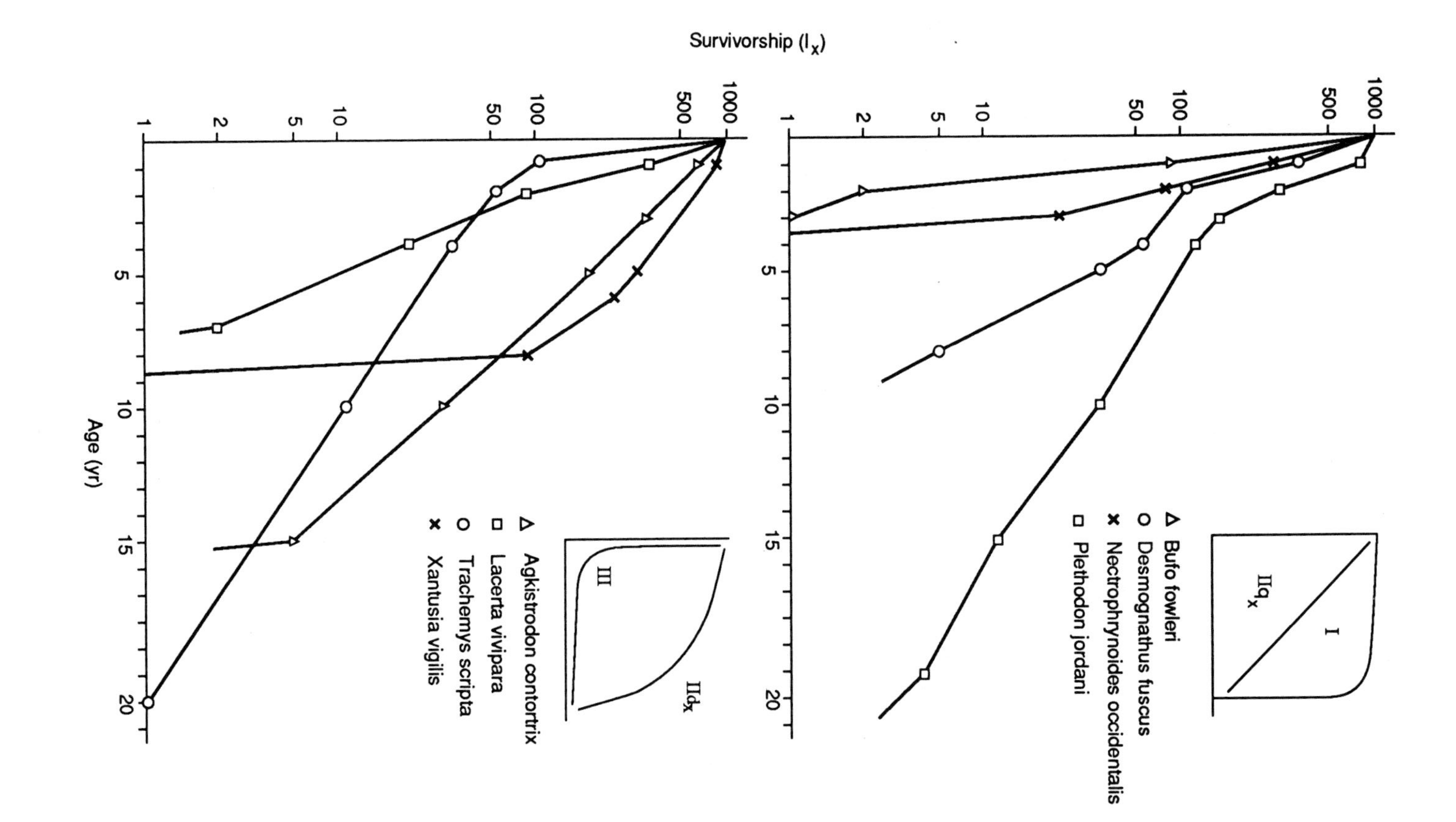

Survivorship (l_x)
Age (yr)
1000
500
100
50
10
5
2
1
5
10
15
20
I
II_{q_x}
Bufo fowleri
Desmognathus fuscus
Nectophrynoides occidentalis
Plethodon jordani
II_{d_x}
III
Agkistrodon contortrix
Lacerta vivipara
Trachemys scripta
Xantusia vigilis

domain: average offspring production per cohort member per unit time (typically years, but weeks or months may be the appropriate interval for short-lived species). The reproductive output of a cohort, thus, must be adjusted to account for clutch size, the frequency of production, and the number of members reproducing in each age interval. For example, many lizard populations produce multiple clutches each year, although the number and size of clutches differ among different age classes and usually increase with body size (= age). Additionally, clutch number and size can differ seasonally owing to differential availability of energy resources. The resulting fecundity value must account for these variables and estimate the offspring production for each time interval. Such data are not easily or quickly gathered, and even in the best field studies, fecundity estimates may rest on untested assumptions. The difference between a species' average clutch size and its fecundity is typically great, for example, 6.25 versus 1.28 eggs per year (Table 11.1), respectively, in *Trachemys scripta*. Also, since fecundity may increase/decrease with time (Table 11.1, *Coluber constrictor*), a population's time-specific net fecundity ($l_x m_x$) potentially can increase, although it seldom does since the survivorship (l_x) usually declines at a faster rate than fecundity (m_x) increases.

The estimates of time-specific reproductive effort are useful demographic parameters that permit ecologists to predict a population's future (R_0) and assess its mean generation time (T). As noted earlier, R_0 or net reproductive rate shows whether population size is increasing or decreasing; R_0 is the sum of the age-specific net fecundity ($l_x m_x$). Mean generation time is also calculated from the $l_x m_x$ values and denotes the average age of parenthood for the population. T is never less than the age at sexual maturity, but it nearly equals maturity in short-lived species and shifts upward in long-lived species, especially those species with high adult survivorship and increasing fecundity (m_x).

Age at sexual maturity (first reproduction) is explicit in fecundity schedules. Its critical nature is also evident by the amount of mortality occurring prior to the beginning of reproduction. Only in Type I populations does a major portion of the cohort reproduce. In the other patterns, half or less, commonly less than a tenth in amphibians and reptiles (Table 11.1 and Fig. 11.1), of the cohort reproduces. The trade-offs between maturing quickly or slowly and clutch/litter size are addressed below in Life-History Patterns.

FIGURE 11.1

Representative survivorship curves for amphibians and reptiles. Insets show the hypothetical survivorship patterns: Types I and IIq_x in amphibian graph; Types IId_x and III in reptile graph. [Data from: Amphibians — Bf (Fowler's toad), Breden (1988); Df (dusky salamander), Danstedt (1975); No, Lamotte (1959); Pj (red-cheeked salamander), Hairston (1983). Reptiles — Ac (copperhead), Vial et al. (1977); Lv (viviparous lizard), Barbault and Mou (1988); Ts (yellow-bellied slider), Frazer et al. (1990); Xv (desert recluse lizard), Zweifel and Lowe (1966).]

STRUCTURE AND GROWTH

Demography is the study of age structures and growth rates among populations. Age of maturity, survivorship, and fecundity are the principal determinants of population size, structure, and growth.

Size: Abundance and Density

Of all population parameters, size is the simplest to measure, although counting all individuals in a population is not an easy task. The abundance or number of individuals in a population imparts little information without reference to the delineation of a population. A nesting population of 500 Kemp's ridley seaturtles compared to 250 green seaturtles indicates that the ridley population is larger than the green population, until the reader realizes that the 500 turtles are the total adult female ridley population compared to the number of females present one night on one of many green seaturtle nesting beaches. In addition to defining the population censused, most comparisons convert abundance to density, the number of individuals per square meter or hectare. Density values need not be restricted to areal units and may be more meaningful if given as number of *Rana* per linear meter of shoreline or *Anolis* per tree. The goal is to have a density value that gives an immediate impression of the rarity or commonness of animals and is biologically meaningful in comparison with other populations. The simplicity and convenience of areal values promote the regular use of these density values and the disregard of potential miscomparisons of one-dimensional (shoreline), two-dimensional (forest floor), and three-dimensional (canopy) populations. This fault resides in many multiple-species density comparisons.

The contrasts in densities are often striking (Table 11.2). Explanations and associations of density variation are not always apparent or rigorously tested. Densities typically reflect the relative availability of resources but are also influenced by climatic events or levels of predation. The presence of abundant food and shelter and low predation permit high densities in many populations. The 27,200 red-backed salamanders (*Plethodon cinereus*) per hectare in a New England forest are particularly impressive; even more so when you discover that the combined biomass of this and three other salamander species exceeds the combined total biomass for birds and mammals in this forest. High densities of resident populations are not uncommon for small amphibians and reptiles (Table 11.2), and where comparative data are available, reduced predation often appears as a major factor. For example, anole populations are commonly two to four times larger on West Indian islands than on the Central American mainland. Competition for food, especially with insect-eating birds, may be less intense as well; however, site-specific diversity of anoles is also greater on the bigger islands. The notable feature of these >1000 individuals ha^{-1} is the advantage conferred

TABLE 11.2
Population Densities of Some Amphibians and Reptiles

Taxon	Density	Body size	Habit/habitat
Bolitoglossa subpalmata	4790	42	Terrestrial/trop. forest
Plethodon glutinosus	8135	63	Terrestrial/temp. forest
Arthroleptis poecilonotus	325	20	Semiaquatic/trop. savanna
Bufo marinus	160	90	Terrestrial/trop. scrub
Eleutherodactylus coqui	100	36	Terrestrial/trop. forest
Eleutherodactylus coqui	23,000	36	
Geochelone gigantea	27	400	Terrestrial/trop. scrub
Sternotherus odoratus	194	66	Aquatic/temp. lake & river
Apalone mutica	1257	210	Aquatic/temp. lake & river
Alligator mississippiensis	0.2	1830[a]	Semiaquatic/temp. marsh
Lacerta vivipara	784	56	Terrestrial/temp. forest
Mabuya buettneri	17	78	Arboreal/trop. savannah
Uromastyx acanthinurus	0.15	110	Terrestrial/subtrop. desert
Varanus komodensis	0.09	1470	Terrestrial/trop. scrub
Xantusia riversiana	3200	70	Terrestrial/temp. scrub
Agkistrodon contortrix	9	540	Terrestrial/temp. savannah
Coluber constrictor	0.3	630	Terrestrial/temp. scrub
Enhydrina schistosa	0.9	730	Aquatic/trop. tidal-river
Opheodrys aestivus	429	360	Arboreal/temp. forest
Regina alleni	1289	400	Aquatic/subtrop. marsh

Sources: Salamanders—Bs, Vial, 1968; Pg, Semlitsch, 1980. Frogs—Ap, Barbault and Rodrigues, 1979; Bm, Zug and Zug, 1979; Ec, Stewart and Pough, 1983; Turtles—Gg, Coe, and Bourn, 1978; So, Mitchell, 1988; Tm, Plummer, 1977. Crocodilians—Aa, Turner, 1977. Lizards—Lv, Pilorge, 1987; Vk, Auffenberg, 1978; Xr, Fellers and Drost, 1990. Snakes—Ac, Fitch, 1960; Cc, Brown and Parker, 1984; Es, Voris, 1985; Oa, Plummer, 1985; Ra, Godley, 1980.

Note: Density is the mean number of individuals per hectare; body size is length (SVL; CL for turtles in mm) of adult females.

[a] Total length.

by the relatively low energy demands of ectotherms. High densities (large populations) reduce the likelihood of population extinction by environmental stochasticity (random, unpredictable events). High local densities may also be a response to the patchy distribution of resources. Limited breeding area is a common concentrating factor. Anuran breeding densities commonly exceed 500 frogs ha^{-1} in temperate and tropical areas. Such high densities and competition for reproductive space are not limited to small-bodied species; before heavy human predation, many seaturtle and *Podocnemis* populations experienced high levels of nest destruction by subsequent nesting females digging into the nests of earlier females. High densities also arise from environmental fluctuations, such as the seasonal concentration of caimans and pleurodiran turtles in the drying pools of the Venezuelan llanos or single- and mixed-species aggregations of snakes in hibernacula.

High densities of resident populations are not restricted to just small-bodied ectotherms. Crocodilian and giant tortoise populations may also be dense (Table 11.2). In these taxa, the relatively low energy needs combined with longevity (physiologically long-lived and predator-proof) of several decades can produce large populations of adults. Not all crocodilian populations are dense and similarly the density of other large reptilian predators may be 1–2 individuals ha^{-1} or less, depending on the availability of resources — both food and space. Adult *Alligator mississippiensis* or *Crocodylus porosus* do not tolerate other adults or large juveniles within their territories and prey on the smaller individuals. Such density regulation by limited availability of space occurs in a variety of amphibians (*Atelopus varius*, *Rana catesbeiana*) and reptiles (*Sceloporus undulatus*) of various sizes.

Since densities reflect the availability of resources and the intensity of predation, population densities can show considerable fluctuation from year to year and site to site. The density of a Puerto Rican frog, *Eleutherodactylus coqui*, was nearly doubled by experimentally adding nest boxes. Density of an Australian gecko, *Oedura monilis,* varied ninefold among sites in the same forest; local density was totally dependent on the availability of loose bark for shelter. Populations of newts, *Notophthalmus viridescens*, differed nearly a 100-fold in nearby forest pools, apparently as a result of lower food availability and higher parasite densities in some pools.

Age Distribution

Life tables chart the size of each age class of a cohort through the life of the cohort. Age distribution analysis examines the size of each age class within a population at a single moment in time (Fig. 11.2). Age-class size can be either the actual number of individuals in the class or the proportion of the total population. The age distribution pattern for a population may be stable through time if its survivorship (l_x) and fecundity (m_x) schedules remain constant. In a stable age distribution, the proportion of individuals in each class remains constant. Some salamander populations (plethodontids in climax Appalachian forest) appear to have nearly stable age distributions where annual population loss (through mortality, emigration, and aging) in each class is matched by recruitment (aging and immigration). The equilibrium for these salamanders derives from a moderate longevity (> 10 yr), stable environment, the adults' partition of and occupation of all suitable habitat, and low predation.

A stable (or predictable) environment, adequate resources, and low predation are the likely requirements for the appearance of stable age distributions in all amphibian and reptilian populations. For most populations, climate, resource availability, and predation differ widely each year, thus survivorship and fecundity vary from season to season and from year to year. Changing age structures are most visible in annual populations. In annual populations, the population consists of a new generation each year, not uncommon for small species of lizards and frogs in seasonally cyclic climates. Juveniles hatch and grow during an equable season, have slow or no growth during a harsh season (dry or cold), and are sexually mature, mate, and lay eggs at the beginning of the next equable season. These adults usually lay several clutches and die, although they may survive until the next breeding period. Thus, at any instant, the age structure of an annual population is uniform, that is, all members have identical ages (within 4–8 weeks of one another), and the age of the population matches the age of the cohort (Fig. 11.2, *Uta*).

Another aspect of age distribution is the sex ratio (i.e., proportion of males in the population, with equality expressed as 1:1 or 0.5). The sex ratio is a critical demographic feature, as it directly affects the growth potential ($\geq l_x m_x / R_0$) of a population; simply put, more females produce more offspring. For bisexual diploid species with chromosomal sex determination, the primary sex ratio (number of males to females at fertilization) is 1:1, because the heterogametic parents produce equal numbers of male and female gametes. Other reproductive modes (e.g., unisexual, polyploidy) deviate from 1:1 because of altered gametogenic and fertilization mechanisms producing unequal numbers of female and male gametes. Differential gamete mortality or fertilization may produce a skewed sex ratio in bisexual species. Primary sex ratios have not been determined directly in amphibians and reptiles (this requires karyotypic examination of early zygotes) but rely on secondary sex ratios (at birth or hatching), and even these data are difficult to obtain because of the absence of external secondary sex characteristics in many juvenile amphibians and reptiles. In the majority of bisexual species with the exception of reptiles with temperature-dependent sex determination, the secondary sex ratios approximate 1:1.

Although many populations display sexual equality, unequal sex ratios are regularly reported for populations of adult amphibians and reptiles. Assuming equality for primary and secondary sex ratios, unequal adult sex ratios require differential mortality during the juvenile and/or adult stage. The inequality variously favors males or females. Only in lizards (based on a very small sample) is a trend evident, and it shows a higher mortality among males. For other reptiles and amphibians, there are roughly equal numbers of species or populations with female-dominated ratios as those with male-dominated ratios. There are also populations in which the inequality has shifted from one sex to the other in different years. The apparent lack of trends and annual shifts emphasize that many of the reported inequalities are more apparent than real, reflecting the

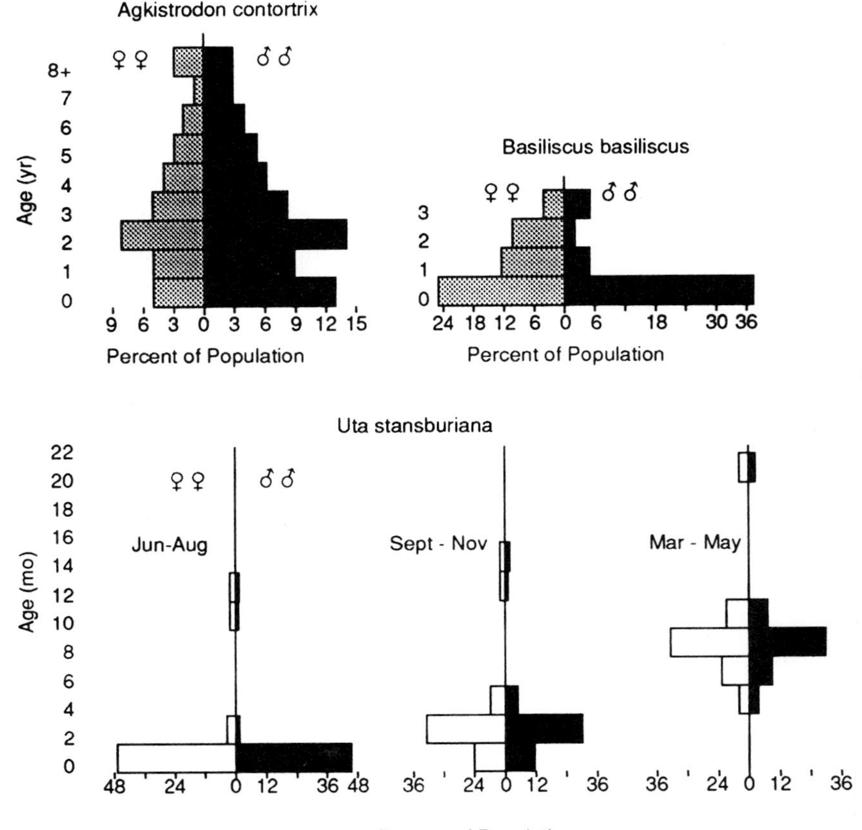
Agkistrodon contortrix
♀♀
♂♂
Age (yr)
Percent of Population
Basiliscus basiliscus
♀♀
♂♂
Percent of Population
Uta stansburiana
♀♀
♂♂
Age (mo)
Jun-Aug
Sept - Nov
Mar - May
Percent of Population

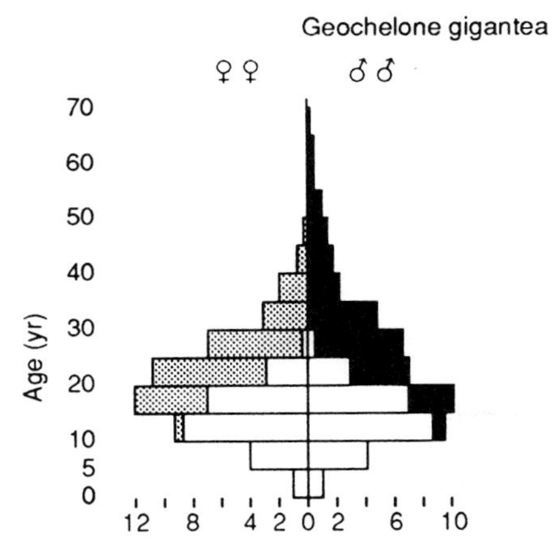
Geochelone gigantea
♀♀
♂♂
Age (yr)

difficulties of accurately censusing amphibian and reptilian populations and not the actual numbers of females and males in these populations.

Growth

Population growth may be positive or negative, with a population's size increasing or decreasing. Few amphibian and reptilian populations are stable (at equilibrium) with size remaining constant from year to year. Short-lived species, such as many small-bodied anuran and lizard species, have mean generation times (T) of a year or less, and their population sizes vary monthly (Fig. 11.3). Population size in these semiannual and annual species is highest at the end of the reproductive season when eggs hatch and juveniles join the population; the number of individuals slowly declines until the next reproductive season. As average longevity or generation time increases, population stability becomes increasingly possible, particularly in the adult segment of the population that has reached the carrying capacity of their habitat. This is possible because amphibians and reptiles commonly have Type III survivorship curves, and in long-lived species, the annual survivorship of the adults is high and recruitment (i.e., new individuals joining the population), although low, balances adult mortality.

Population growth is the change in number (N) of individuals per unit time (dN/dt) and is the sum of a population's recruitment and loss of members. Recruitment occurs through hatching or birth and immigration, and loss via death and emigration. Two models are relevant to the examination of population growth: the exponential equation, $dN/dt = rN$; and the logistic equation, $dN/dt = rN((K-N)/K)$. Both are oversimplifications of the factors affecting populational growth, but both contain parameters (r and K) that offer insights into and permit comparison of the dynamics of different populations. The intrinsic rate of increase or r measures the balance between recruitment and loss and is approximated by the relationship of the net reproductive rate to mean generation time ($\ln\frac{R_0}{T}$). When $r > 0$, the population is increasing, and when $r < 0$, the population is decreasing. Since a population can increase or decrease exponentially for only a very brief period of time without becoming unrealistically large or going extinct, the logistic equation better matches the growth of most

FIGURE 11.2 ▼▲

Age distribution patterns of some reptilian populations. Seasonal patterns for an annual species, *Uta stansburiana*, the side-blotched lizard (Tinkle, 1969). Point-in-time patterns for a moderate-lived species, *Agkistrodon contortrix*, copperhead (Vial et al., 1977), a short-lived species, *Basiliscus basiliscus*, common basilisk (Van Devender, 1982), and a long-lived species, *Geochelone gigantea*, Aldabran tortoise (Bourn and Coe, 1978). The bars denote the percentage (of total population) of males or females present in each age class; open bars indicate unsexed individuals; stippled for females and solid for males.

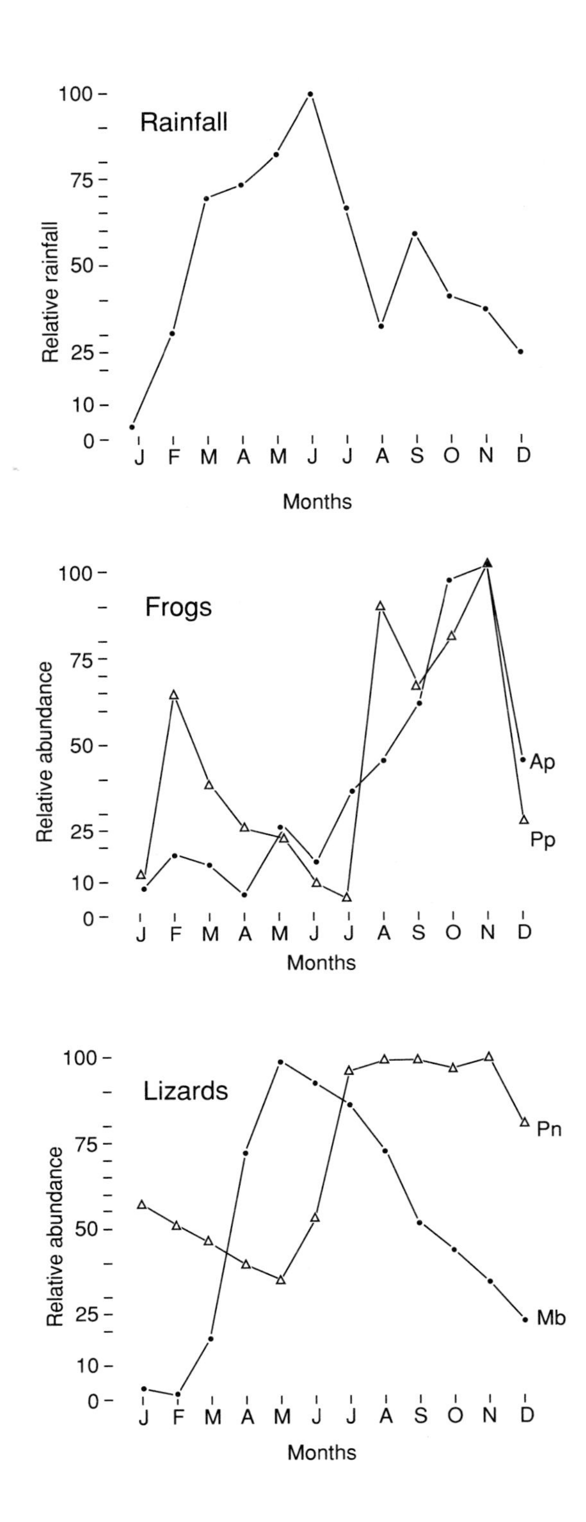
Rainfall
Relative rainfall
0
10
25
50
75
100
J F M A M J J A S O N D
Months
Frogs
Relative abundance
0
10
25
50
75
100
J F M A M J J A S O N D
Months
Ap
Pp
Lizards
Relative abundance
0
10
25
50
75
100
J F M A M J J A S O N D
Months
Pn
Mb

biological populations. In the population growth model, K represents the carrying capacity (i.e., resource limits) of the environment. When population size (N) exceeds the carrying capacity, the population will decrease, and when $N < K$, the population increases; thus, K represents an equilibrium population size.

K and r do not remain constant for a population of organisms. They are simply estimates of resource availability of the environment and the population's growth rate during the period of study. K and r vary seasonally or yearly as the abiotic and biotic forces impinging on the individuals and population change. However, their relative values and ranges of variation reflect a species' reproductive potential, adult body size, longevity, the type and pattern of resource use, and other aspects of a species' life history. Because of this association of r and K with different life-history patterns, species are often categorized as either r- or K-selected "strategists" (Table 11.3). No "strategy" (in a planning sense) by individuals, populations, or species is involved, only general trends matching small, short-lived species with high population turnover and high r, and larger, long-lived species with large, stable adult populations and low r. Indeed, recent studies on lizards continually emphasize that viewing species as r- or K-selected entities oversimplifies the evolution of life-history characteristics. K and r are more explanatory than predictive elements of growth models.

The direction and speed of growth show the interaction between recruitment and loss. Recruitment into a population may be viewed in two ways: joining the absolute population (i.e., all members irrespective of age) or the reproductive population (i.e., consisting of only sexually mature individuals). The manner and timing of recruitment depends on the reproductive pattern of a population. In populations with a single and brief reproductive period each year, all members of the new generation hatch or are born simultaneously (Fig. 11.3). In contrast, as the length of the breeding period increases, recruitment will occur over an increasingly longer interval.

These generalizations hold for both amphibians and reptiles, although different reproductive patterns and life-history characteristics affect how and when recruitment occurs. In amphibians with complex life cycles, recruitment is simultaneous in explosive breeders and longer to continuous in extended breeders. It is important to emphasize that recruitment begins when eggs are laid or young are born, that is, when the fate of the offspring is independent of the mother. For the explosive breeding spadefoot toads (*Pelobates*, *Scaphiopus*), a population will surge from a few dozen adult toads to several hundreds or thousands of individu-

FIGURE 11.3 ▼▲▼▲▼▲▼▲▼▲▼▲▼▲▼▲▼▲▼▲▼▲▼▲▼▲▼▲▼▲▼▲▼▲▼▲

Seasonal fluctuations in abundance of frogs (*Arthroleptis poecilonotus*, *Phrynobatrachus plicatus*) and lizards (*Mabuya buettneri*, *Panaspis nimbaensis*) living in a highly seasonal tropical environment (savanna/Lamto, Ivory Coast). Relative rainfall and abundances are scaled to the highest monthly average rainfall and densities. [Data from Barbault (1972, 1975) and Barbault and Rodriques (1978, 1979).]

als after a single night of breeding. The population immediately begins to decline through deaths by predation and developmental failures. If the pond dries prior to metamorphosis, the population returns to its prebreeding size and structure; if metamorphosis occurs, the new cohort adds a juvenile component to the age structure and the potential of recruitment into the breeding population in a year or two. For annual species with extended breeding seasons (e.g., whether for a frog, *Acris crepitans*, or a lizard, *Uta stansburiana*), recruitment occurs over a 1- to 3-month breeding season and is maximal near the end of the season, just prior to a surge of adult deaths; thereafter population size declines until the following year's breeding season. Few if any adults survive for a second year, hence the annual recruitment of juveniles into the adult population is essential for survival of the population. Taxa with 10+ years of longevity (e.g., salamanders, turtles) have relatively stable adult populations but variable-sized and -aged cohorts of maturing individuals; the breeding season may be brief (1–2 nights for some mole salamanders, *Ambystoma*) or extended (2–3 months for seaturtles). Survival of each new cohort to the following year is variable, typically less than 20% to none. Because of the extended and variable period of maturity in long-lived taxa, members of a single cohort join the adult population over several years (e.g., 2–8 yr for the spotted newt, *Notophthalmus viridescens*, and 6–20 yr in Blanding's turtle, *Emydoidea blandingi*). The size stability of adult populations is maintained by the high survivorship of the adults and continual, although low, annual recruitment.

Dispersal also affects population growth through emigration (loss) and immigration (gain). Emigration is difficult to distinguish from mortality, because disappearance by emigration can be confirmed only if individuals reappear or are found elsewhere. Immigration is easier to recognize but only if the entire population is marked. Because of these technical difficulties, immigration and emigration are assumed to balance one another and in survivorship and fecundity schedules are subsumed in the natality and mortality estimates. Dispersal is important in two other aspects: colonization or birth of new populations, and gene flow by the potential introduction of new alleles into populations and changing the gene frequencies (see Dispersal and Transient Movements, Chapter 9).

Factors affecting recruitment and loss segregate into density-dependent and density-independent ones. In the former, there is a direct correlation between a population's density and its rate of recruitment or loss. Viewed from the perspective of loss, density-dependent factors increase the rate of loss (mortality and/or emigration) as density increases. Predation and competition are the most apparent density-dependent ones; as density increases, individuals are more often encountered and captured by predators so that mortality increases in absolute numbers as well as proportionately. Increasing density eventually reaches a situation where one or more resources are in short supply and intrapopulational competition for resources intensifies; inadequate food leads to malnutrition or starvation for some or all members of the population, and lack of shelter exposes more individuals to predators or displaces some individuals into suboptimal habitats. Field experiments on larval frog and salamander populations (single- and multiple-species protocols) regularly demonstrate that growth rates of individuals are drastically reduced by crowding. Crowding leads to increased predation because the larvae require longer to grow to predator-proof sizes and, because of slower growth, increases the likelihood that the pond will dry prior to the cohort's metamorphosis (density-independent mortality resulting from a density-dependent effect).

Density-independent factors, such as flood, fire, volcanic eruption, and other catastrophic events, affect all members of a population equally and without regard to population size. Such factors regularly affect populations of amphibians and reptiles. Pond drying annually threatens temporary-pond breeding amphibians, and drowning of nests by storm-driven high tides threatens most seaturtle nesting beaches. These are only a few examples of density-independent mortality.

Population size and growth are regulated by a complex interaction of density-independent and -dependent factors. For example, a small isolated population of fence lizards, *Sceloporus undulatus*, seldom reached a density where resource availability was exceeded. Population size appeared to be regulated by summer and fall predation on egg-laying females that was proportionately greater at higher densities. Stability did not occur because irregular flooding drowned nests and adults, and low spring temperatures would periodically increase individuals' susceptibility to predation. Each year, density-independent mortality appeared to have a greater effect on annual survivorship than density-dependent mortality.

LIFE-HISTORY PATTERNS

Life history encompasses all aspects of an animal's life cycle from conception to death. The preceding sections and chapters touched on many life-history traits and their great diversity within amphibians and reptiles. Species range from tiny frogs and lizards of less than 1 g to giant turtles of 1 metric ton, from species with external fertilization depositing thousands of eggs to those with internal fertilization and development of one or two eggs, from nocturnal arboreal forms to diurnal aquatic ones, from species with life spans of less than a year to others living for decades, and a multitude of other life-history traits.

The hows and whys of such diversity are addressed by a wealth of models and hypotheses that fall under the umbrella of life-history theory. The theory offers explanations of an individual's survival and fitness (success in placing progeny in the reproductive pool of the next generation) relative to the individual's abiotic and biotic environment and links life-history traits (for examples, see Table 11.3). The key traits — age-specific mortality (q_x), net age-specific fecundity ($l_x m_x$), and age at maturity — emphasize demography, because populations,

TABLE 11.3 ▼▲▼
Demographic and Life-History Attributes Associated with *r*- and *K*-Type Populations of Amphibians and Reptiles

Attributes	*r*-type	*K*-type
Population size (density)	Seasonally variable; highest after breeding season, lowest at beginning of breeding season	High to low, but relatively stable from year to year
Age structure	Seasonally and annually variable; most numerous in younger classes, least in adults; usually ≤ 3 year classes	Adult age classes relatively stable; most numerous in adult classes, multiple year classes
Sex ratio	Variable, often balanced	Variable, often balanced
Survivorship	Almost always Type III	Types II and III
Mean generation time	Equivalent to age of sexual maturity	Often exceeds age of sexual maturity
Population turnover	Usually annual, rarely beyond 2 yr	Variable, often >1.5 times age of sexual maturity; to decades
Age at sexual maturity	Usually ≤2 yr	Usually ≥4 yr
Longevity	Rarely ≥4 yr	Commonly >8 yr
Body size	Small, relative to taxonomic group	Small to large
Clutch size	Moderate to large	Small to large
Clutch frequency	Usually single breeding season, often multiple times within season	Multiple breeding season, usually once each season
Annual reproductive effort	High	Low to moderate

[In part from Pianka (1970).]

not individuals, evolve and how the population as a whole adjusts its reproductive effort to yield a new generation of reproducing adults is critical for the continuation of the population. Nonetheless, selection operates on individuals. (In spite of the importance of the demographic traits, most life-history studies examine reproductive parameters owing to the difficulty of obtaining demographic data.)

An individual's life revolves around a triad of activities and processes: the gathering of resources (Chapter 5); the allocation of these resources to maintenance, growth, storage, and reproduction (Chapters 7–10); and survival to reproduction by avoiding predators, parasites, and disease (Chapter 6). The effect of each component within the triad depends on its instantaneous benefits and costs relative to an individual's survival and potential reproductive success. The decisions (trade-offs) are not conscious ones, rather they are physiological and behavioral responses selected across generations by evolution as appropriate and modified by the animal's current physiological and physical condition. For example, a gecko cannot decide whether it will lay two or four eggs since gecko physiology allows the production of two eggs at a time; thus healthy geckos produce two eggs, but if one recently lost its tail to a predator or it is underfed because prey are scare, it will not reproduce owing to inadequate energy reserves. Life-history theory examines not these "decisions of the moment"; instead it addresses the cost/benefits and trade-offs of different sets of life-history traits. The following examples are a tiny fraction of recent investigations and the life-history patterns revealed.

Four life-history traits (age of maturity, clutch/litter size, reproductive effort, and longevity) are used regularly to segregate populations into one of two states, generally recognizing but ignoring that a continuum exists between the two. The contrasting states are: short-lived populations characterized by individuals with short lives, early sexual maturity, large broods, and high annual reproductive effort (effort estimated by the proportion of egg-clutch mass to female's mass); and long-lived populations characterized by individuals with long lives, delayed sexual maturity, small broods, and low annual reproductive effort. (Populations persist over many generations; short- and long-lived refer to cohort generation time.) Short-lived populations are a common feature of anurans and lizards, particularly the smaller-bodied species with individual longevity seldom exceeding 4 yr. Turtles and crocodilians are the paragons of long-lived populations, but many salamanders, snakes, and large-bodied lizards also fall into this category. Life-history theory provides two opposing models for the evolution of these two population types. The deterministic model (r and K selection; see preceding section on population growth) postulates that high levels of density-independent mortality result in fluctuating population densities, which favor a high intrinsic growth rate (r) and produce short-lived populations. In the stochastic model (optimization), high density-independent mortality produces variable, and usually low, juvenile survivorship, which favors a stable and long-lived adult population. The conflict arises from the generality of the models and their attempt to explain life-history evolution of diverse organisms.

Life-history studies within discrete taxonomic groups and among conspecific populations have been somewhat more successful in uncovering the relationships among life-history traits and different environmental regimes. Even in these cases, difficulties are encountered that make conclusions less robust than desirable. A major difficulty is the discrimination between a trait's variation arising from selective pressures to adapt individuals to their present habitat/location and variation resulting from a proximate response to current perturbation in local conditions. The adaptive component of variation may be a response to predation, to competition for similar food or space, to a regular climatic event, to soil acidity or hardiness, or to other abiotic and biotic conditions. Proximate variation arises in response to unusually wet or dry weather, high or low food abundance, and other conditions that improve or degrade the health of individuals. Phyletic constraints set limits on trait variation; for example, egg size in small turtles is limited to the width of the pelvic canal and in amphibians by holoblastic cleavage. Awareness of the difficulties in the data and interpretation promotes caution but does not deny the evolution of life-history patterns.

Population densities of Caribbean island and Central American mainland anoles (equal-sized, different species) are often two to four times or greater in the island species. Such differences in densities suggest a different manner of population regulation and likely differences in life-history traits. The island populations are food-limited; mainland sites have three times the food resources of island sites. Population turnover (mortality) is much higher in mainland anoles, thus predator pressure is likely much greater on the mainland than in the islands. Anoles in both areas mature at the same size, but mainland ones at half the age of the island anoles. This difference is not predator driven, but rather related to food availability and the resultant slower growth of island anoles. However, energy limitation has selected for a lower reproductive effort in island females with island hatchling weight to female weight being significantly less than that for mainland anoles. Thus, low prey abundance produced proximate response in age at maturity and an evolutionary response in reproductive investment.

In another island–mainland comparison, chuckwallas (*Sauromalus*) also have higher densities on two Gulf of California islands compared to their Mexican mainland congeners; in addition the island *Sauromalus* are nearly two times the size (SVL) of the mainland species. Island *Sauromalus* are food-limited like the island anoles and in part compensate for this low food abundance (and periodic total absence of food) by feeding for longer periods of time and in a larger variety of places. Island and mainland *Sauromalus* invest equal energy in each offspring but island species reproduce much less frequently, although it is uncertain whether this feature is a proximate or evolutionary response. Larger size, however, appears to be an evolutionary response to both food limitation and predation. Larger size removes adult *Sauromalus* from the diet of the smaller predators occurring on islands and provides a larger energy reserve during periods of starvation.

Life-history studies (often labeled reproductive tactics) within the major groups of amphibians and reptiles reveal a variety of evolutionary trends, although the trends must be accepted cautiously since they are based commonly on assumptions of demographic responses to presumed selective (environmental) regimes. Presumably the ancestral condition for aquatic-breeding amphibians was the production of numerous small eggs, once (semelparity) or two to three times (multiple clutches = iteroparity), and short life spans in response to high mortality in all age classes. This pattern persists for many small anurans in seasonal environments. Two divergent trends away from this pattern occur in anurans. In those retaining aquatic reproduction, increasing longevity and body size produce even larger clutches and higher lifetime reproductive effort (e.g., large-bodied bufonids and ranids). The other direction is the reduction or elimination of aquatic reproduction and encompasses the multiple and independent origins of parental care, direct development, and extrauterine ovoviviparity. In all of these reproductive trends, egg number reduces with increasing egg size (producing larger offspring and often skipping the vulnerable free-living larval stage) and adult longevity increases. Viviparity in caecilians represents the same trend to reduce larval and early juvenile mortality. Salamanders largely lack the live-bearing adaptation but have emphasized parental care (large, aquatic clutches in cryptobranchids; small terrestrial clutches in plethodontids), direct development (many plethodontids), and adult longevity (widespread among salamanders).

Among the reptiles, crocodilians epitomize large body size, adult longevity, and parental care. They produce reasonably large hatchlings, but not proportionately large compared to adult size. Their reproductive emphasis is on multiple clutches over a long adult life, and apparently most species attain sexual maturity quickly (4–8 yr) by a combination of relatively rapid growth and maturity at half or less of average adult size. Turtles also use longevity to increase individual reproductive output, and in many species egg-laying may continue for more than two decades. However, unlike crocodilians, they grow much slower and mature later. Even in small species (kinosternids, emydids; available data derive largely from temperate species) maturity is seldom attained in less than 5 yr and may require more than 10 yr. The extreme condition occurs in seaturtles, where most species require a minimum of 20 yr and as much as 40–50 yr to attain sexual maturity. Clutch size regularly increases with body size in small-bodied turtle species, occasionally in larger species.

In squamates, the association between clutch and body size is variable. Numerous lizard species (e.g., geckos, anoles, some skinks) have fixed clutch sizes; whereas in other lizards and many snakes, clutch size is a function of body size. Among the lizards, many small-bodied species have short life spans (1–2 yr) and produce multiple clutches of eggs. With increasing body size (approx. >75 mm adult SVL), lizard longevity tends toward a positive association with body size. Among the longer-lived species, many have a single clutch per season. Live-bearing lizards typically produce a single litter of a few neonates each year. Many live-bearing snakes are also single clutch per season breeders, but commonly have significantly larger clutches (>40 neonates in large natricine species). Viperines commonly become pregnant on biennial or longer cycles. Among the egg-laying snakes, a single clutch per season is the most common pattern. No snake species is known to have an annual population turnover; adult life spans appear to be commonly >4 yr but likely seldom exceed 20 yr except in the giant species.

The preceding generalizations emphasize reproductive characteristics. Other life-history traits also show selection to local environmental conditions. For example, in amphibians with their complex life cycles, selective pressures can operate on growth rates, developmental time, and other larval traits to reduce the probability of mortality.

KIN RECOGNITION AND DISCRIMINATION

Populations consist of multiple families, each representing a single matriarchal lineage. Large populations have many such lineages and typically high genetic diversity; conversely, small populations have fewer families and less genetic diversity. For example, unisexual populations are genetically homogeneous and small bisexual demes may be nearly so. The influence of genetic relatedness on population structure and dynamics has often been overlooked by biologists, although recent studies on social evolution are emphasizing the effect of an individual's presence and actions on their genetic relatives (kin).

For amphibians and reptiles, few data are available on how kinship affects population dynamics and the fitness and evolutionary potential of different genetic lineages. The available data feature tadpole social behavior and growth rates, but kinship surely will prove important in the ecology and behavior of other amphibians and reptiles, although it will be difficult to demonstrate in most. Evolutionarily, kinship concerns those traits (primarily although not exclusively behavioral ones) that do not benefit or may even decrease an individual's survival and reproductive potential but do benefit the survival and reproduction of kin. Benefits can accrue to kinship if and only if an individual can recognize its kin and assess the degree of relationship. The possible benefits include the avoidance of reproduction with kin to promote outbreeding, reduction of intraspecific competition for resources, enhancement of food gathering, and development of social predator warning systems.

Tadpoles are excellent experimental animals for identifying the presence and effects of kinship recognition owing to their large clutch sizes and adaptability to laboratory rearing and field manipulations. Kinship recognition and discrimination varies among the species tested, pointing to species-specific adaptations associated with different ecologies. *Rana cascadae* tadpoles preferentially associate with groups of unfamiliar siblings over familiar groups of mixed siblings and nonsiblings. Further they show the ability to recognize and discriminate between full siblings, half-maternal siblings, half-paternal siblings, and nonsiblings; given a choice, they select full siblings over the other three groups, half-maternal siblings over half-paternal or nonsibling groups, and half-paternal siblings over nonsiblings. In contrast, *R. aurora* tadpoles can recognize siblings when raised with their kin but not if raised in isolation or with nonsiblings; *R. pretiosa* apparently lack the ability to recognize kin for they do not associate with full kin even when raised with siblings. *Rana cascadae* occur naturally in small cohesive aggregations (schools), and the other two *Rana* only intermittently. *Bufo* tadpoles (*B. americanus*, *boreas*) associate preferentially with kin when raised with siblings but not if raised in a mixed group of siblings and nonsiblings, although American toads can recognize kin when raised in isolation. Such observations indicate the possibility of three different recognition mechanisms: kin recognition cue learned by early association within a kin group; recognition by matching of unfamiliar individuals with phenotypes of earlier associates; and recognition of unknown kin with shared alleles. The nature of cues is unknown although they are probably chemical and possibly by-products of excretory or fecal materials.

The value of discrimination (preferential association) of kin is little tested as yet; however, kin association enhances growth and early metamorphosis in *Bombina variegata*. Individual tadpoles grew faster and siblings showed more uniform growth rates when reared with kin than with nonkin. *Pseudacris triseriata* tadpoles also had enhanced growth rates when raised with kin versus nonkin, but association with kin may be neutral (*Hyla gratiosa*) or detrimental (*Rana arvalis*). The benefits of kin association, not unexpectedly, will vary depending on the ecology of the species and the past history of the population.

CHAPTER 12

Population and Species Interactions

Energy flow in and through an ecosystem and its communities is the major process maintaining community structure (i.e., composition and relative abundance of component species). Although critical, energy flow through and within amphibian and reptilian populations is not emphasized in this chapter. Rather, the factors associated with the use and partitioning of energy and other resources by individuals and populations comprising the local communities are examined. Aspects of energetics are presented in Chapters 5, 10, and 11.

The term "community" implies an organized structuring of groups of species with regard to composition and interaction. Further implicit in the concept of biological communities is the evolution of the structure. As yet, no evidence exists to support community evolution, although species–species interactions have led to adaptations (coevolution) in one or more of the interacting species. Also implicit in the community concept is that the community is the entire complex (trophic network) of interacting organisms in one location from the energy-capturing plants to the tertiary predators and decomposers. Community is used rarely in this all-encompassing sense, rather in a limited sense of a potentially interacting assemblage (association) of species (often of the same taxonomic group; Fig. 12.1) at a single site. This narrower concept is the one used most frequently in current ecology.

LOCAL SPECIES ASSEMBLAGES

The environment, even within an area as small as a hectare or 0.1 ha, offers each organism a mosaic of choices for shelter, feeding, and reproduction. An organism's selection and survival depends on its inherent tolerances and requirements, and as observed earlier (Chapter 9), this selection results in a nonrandom distribution of individuals and species within an area. Further, the local distribution of resources greatly affects how we view and define the local communities. Plant communities and/or geological features are the most common delimiters of herpetological communities, so we commonly recognize and name these

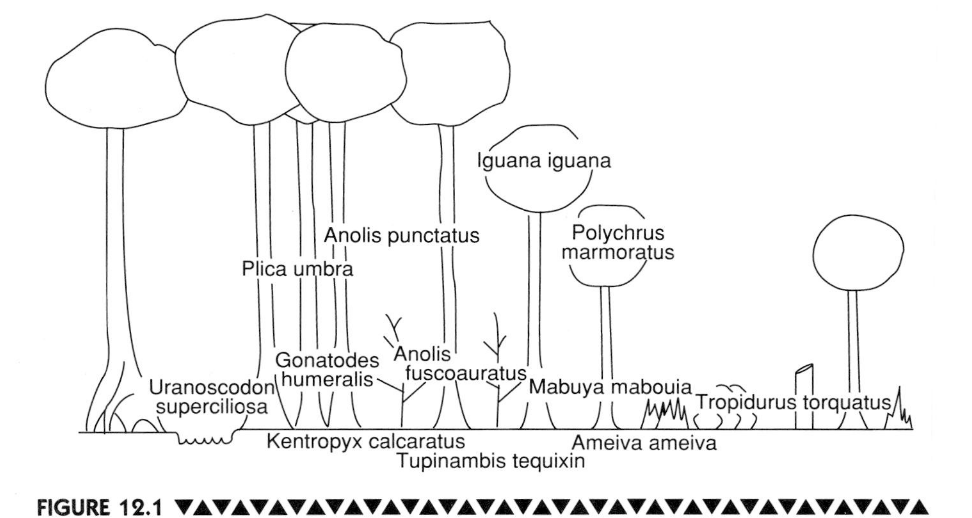

FIGURE 12.1 ▼▲▼▲▼▲▼▲▼▲▼▲▼▲▼▲▼▲▼▲▼▲▼▲▼▲▼▲▼▲▼▲▼▲▼▲▼▲
Lizard assemblage of Belém, Brazil. The schematic cross section traverses habitats from rain forest through secondary growth forest to grassland-agriculture areas. Each species is located in its area of maximum abundance. [Modified from Rand and Humphrey (1968).]

communities by habitat (e.g., pine flatwood amphibians, eucalyptus savanna lizards) or topography (Mississippi River turtles, Sonoran Desert herpetofauna).

Herpetofauna is a neutral term for all the amphibian and reptilian species occurring within an area, as large as a continent or as small as a pond, irrespective of the distribution and utilization patterns of resources. Assemblages or communities reflect both these patterns and the biases of the observers. Thus species composition of a fauna varies broadly. For example, the herpetofauna of the Great Smoky Mountains National Park may be alternately viewed as the herpetological community of the entire park or of a single mountain or ravine within the park.

Characteristics of Communities

Patterns of co-occurrence of two or more species are common and foster the community concept. No matter how broadly or narrowly defined, a community's structure is its species composition and the abundance of each species; all other aspects (e.g., trophic organization, energy flow) of community ecology derive from these two features. Even though patterns of co-occurrence are evident, the actual causes for these associations are not. The factors governing the presence or absence of a species and the abundance of its members are numerous and sort into abiotic and biotic ones. Abiotic factors are a function of the physical environment and each species' physiological tolerances (Chapter 11; also later sections of this chapter). Biotic ones are resource-related and concern interactions with other species. These interactions may be direct (catching prey or being prey) or indirect (shade from a tree or high humidity of a forest); they may have a positive, negative, or neutral effect on an individual's survival and reproduction, hence controlling a population's persistence or extinction. Direct interactions include predation, mutualism, and competition and are major factors shaping community structure; historical factors including colonization and extinction events also determine species composition.

The basic organization of all communities follows energy flow through the various life-forms from plants through consumers and decomposers. Life's energy derives entirely from the sun. Plants capture this radiant energy and convert it into plant tissue; herbivores eat the plants and convert the energy to animal tissue and so on through one or more levels of carnivores. At each step (trophic level; see inset, Fig. 12.2) in this food or energy chain, energy is lost through organisms' inability to assimilate all food obtained and through metabolic activities (i.e., respiration). Trophic pyramids reflect sequential energy loss with declining biomasses from producers (plants) through herbivores to the various carnivore levels. Amphibians and reptiles are mainly primary and secondary carnivores, that is, eating other consumers and in turn being eaten, thus herpetological communities generally occupy the middle region of the trophic chain or food web (Fig. 12.2).

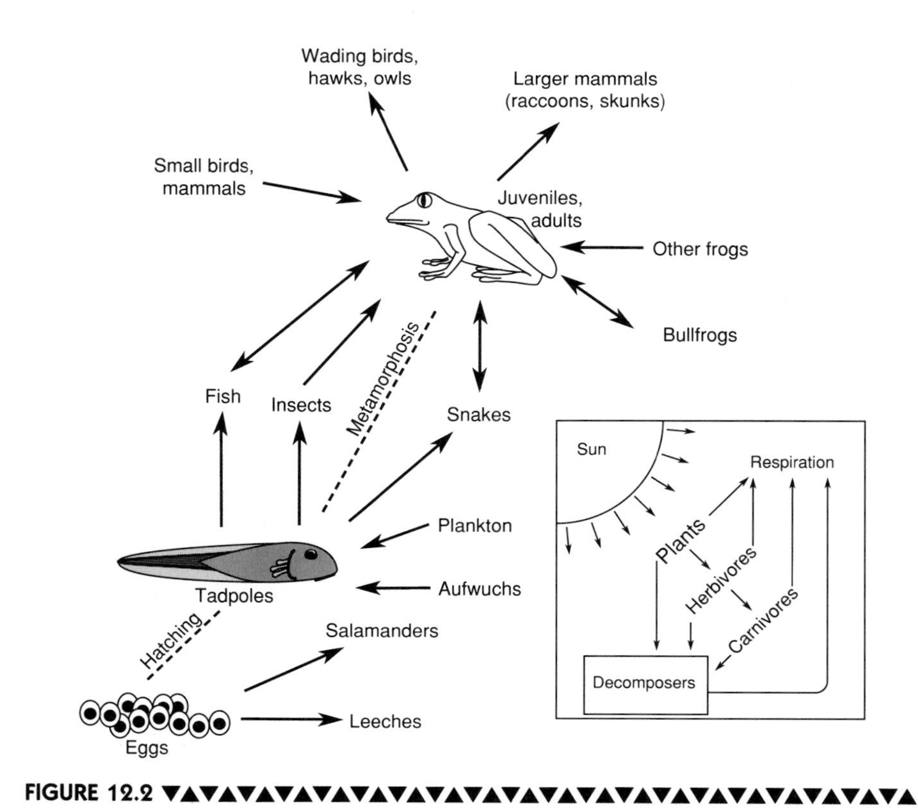

FIGURE 12.2 ▼▲

A generalized food web for an eastern North American pond with *Rana catesbeiana* (bullfrog) as the focal point. Arrows denote direction of energy flow, that is, consumption. The inset (lower right) shows the general pattern of energy flow through communities. [Data from Bury and Whelan (1984).]

Amphibian and reptilian assemblages (i.e., limited-membership communities) are diverse (Table 12.1), reflecting the availability of energy and shelter, species source pool from which to draw colonists, and rigors of the abiotic-biotic environment. The trend is for the number of species (species density) to increase from cold-temperate to tropical areas (possibly total biomass as well, although comparable data are unavailable); however, countertrends occur in the various groups. For example, turtle communities are more diverse in the midlatitudes (Northern Hemisphere) than in the tropics; plethodontid salamander communities, whether tropical or temperate, share similar species numbers when forest microclimates are matched. Lizard communities at the same latitude (within a single continent) commonly display increasing diversity with increasing aridity; the reverse trend occurs in amphibians and turtles. Such generalities have many exceptions and often the differences reflect sampling success and protocol (e.g., duration, completeness, and area of sampling). The Savannah River site (Table 12.1) has a richer herpetofauna compared to some herpetofaunas of more tropical sites. This richness is real. The Savannah River site has numerous aquatic and terrestrial habitats, is within an area of high species diversity, is not isolated by natural and man-made barriers from other natural areas, and has been exceptionally well sampled (continuously for 20+ yr). No tropical mainland sites have had comparable sampling, although some Thailand and Borneo sites were thoroughly sampled for 1–2 yr, and Barro Colorado, Panama (an island with a water barrier preventing free interchange of fauna with adjacent mainland forest) has had continual, but sporadic, sampling for 30+ yr. A striking feature of all subtropical and tropical assemblages is the high diversity of frogs and snakes; both groups are specialists, with frogs finely partitioning space/time and snakes partitioning food.

Herpetological communities are dynamic with changing species composition (also interactions) and abundances of each species. This dynamism is most evident in anuran breeding associations and the resulting tadpole assemblages. In eastern North America, anurans segregate into late winter-spring and summer breeders, and although there is some overlapping of calling males, the major choruses and egg-laying of the spring and summer breeders is regularly 1–2 months apart. For example, the spring association at a North Carolina pond included two salamanders (*Ambystoma maculatum, A. opacum*) and five frogs (*Bufo americanus, Hyla crucifer, Pseudacris triseriata, Rana palustris, R. sphenocephala*), whose larvae commonly metamorphose prior to the major breeding activity of the summer association of five anuran species (*Acris crepitans, H. chrysoscelis, R. catesbeiana, R. clamitans, Gastrophryne carolinensis*). A study of the larval assemblage of a Maryland pond with a similar but reduced species composition showed that the species composition, the relative abundance, and spatial distribution of each species varied each year in both spring and summer associations. Amphibian breeding and larval communities in the tropics are no less dynamic with monthly and yearly changes in species composition and abundance.

More sedentary amphibian (i.e., with direct development and parental care) and reptilian communities may experience less seasonal alterations in species composition, but few populations are stable in size or age structure so these associations have seasonal and annual changes in relative abundances of the component species. Change occurs through seasonal shifts in activity patterns of each species and the appearance of offspring. Lizard and snake assemblages commonly possess species with distinct seasonal activity patterns, so that the relative abundances change with the seasons (e.g., in southwestern North America iguanid species are most abundant in spring and fall and teiids in the summer; see also Activity Cycles, Chapter 9).

Environmental factors also alter the species composition and relative abundances within herpetological assemblages and can even eliminate them. Atypical rainfall or temperature patterns alter breeding patterns in amphibians or modify

TABLE 12.1

Composition of Continental Herpetological Assemblages[a] from Different Climates

Site	Caecilians	Frogs	Salamanders	Turtles	Lizards	Snakes	Crocodilians	Total	Latitude
Andrew Experimental Forest (forest)	0	3	7	0	3	3	0	16	44°N
U.K. Nature Reserve (grassland and forest)	0	9	1	4	7	16	0	37	39°N
Prince William (forest)	0	10	10	4	4	13	0	41	38°N
Savannah River site (swamp and forest)	0	23	16	12	9	35	1	95	33°N
Barro Colorado (forest)	1	29	0	5	22	39	2	98	9°N
Santa Cecilia (forest)	3	81	2	6	28	51	2	173	0°N
Tucumán (forest)	0	16	0	1	26	24	0	67	28°S
Vienna (fields and forest)	0	12	5	1	5	5	0	28	48°N
Rota (grassland and forest)	0	6	2	1	10	5	0	24	37°N
Lamto (savanna)	0	17	0	0	10	12	0	39	8°N
Kivu (forest)	0	29	0	2	10	38	1	80	2°S
Lochinvar (grassland and forest)	0	22	0	2	6	14	1	45	16°S

V. Crookes Reserve (grassland and forest)	0	17	0	0	8	14	0	39	30°S
Lazo Nature Reserve (forest)	0	6	0	0	1	6	0	13	43°N
Chitwan (grassland and forest)	0	11	0	7	10	24	2	54	28°N
Sakaerat (fields and forest)	1	24	0	2	30	47	0	104	14°N
Ponmudi (forest)	2	24	0	0	16	14	0	56	9°N
Nanga Tekalit (forest)	1	47	0	0	40	47	0	135	3°N
Kakadu (grassland and forest)	0	14	0	1	14	10	1	40	13°S
Big Desert (grassland and scrub)	0	4	0	0	18	2	0	24	35°S

Sources: Andrew Experimental Forest, Oregon, USA, Bury and Corn, 1988; Prince William National Forest, Virginia, USA, Mitchell et al., unpubl.; University of Kansas Natural History Reservation, Kansas, USA, Fitch, 1965; Savannah River Plant, Georgia, USA, Gibbons and Semlitsch, 1991; Barro Colorado Biological Station, Canal Zone, Panama, Myers and Rand, 1968; Santa Cecilia, Ecuador, Duellman, 1978; Tucumán (bosques chaqueros), Argentina, Laurent and Teran, 1981; Vienna, Austria, Tiedemann, 1990; Rota, Spain, Busack, 1977; Lamto, Ivory Coast, Barbault, 1976; Kivu, Zaire, Laurent, 1954; Lochinvar National Park, Zambia, Simbotwe and Friend, 1985; Vernon Crookes Nature Reserve, Natal, Bourquin and Sowler, 1980; Lazo State Nature Reserve, Maritime Territory, Russia, Shaldybin, 1981; Royal Chitwan National Park, Nepal, Mitchell and Zug, unpubl.; Ponmudi, India, Inger et al., 1984; Sakaerat Experiment Station, Thailand, Inger and Colwell, 1977; Nanga Tekalit, Sarawak, Lloyd et al., 1968 (island/no continent at this latitude in Asia); Kakadu National Park, Northern Territory, Australia, Simbotwe and Friend, 1985; Big Desert, Victoria, Australia, Woinarski, 1989.

Note: The data are the number of species, excluding introduced or exotic species. The assemblages are arranged in three north–south transects: the Americas, Europe–Africa, and Asia–Australia.

[a] Each assemblage represents the taxa likely to be present in a 25-km^2 area and represents multiple habitats in most cases.

food resources, thereby increasing or decreasing survivorship. Prolonged droughts eliminate short-lived amphibian species, particularly those dependent on temporary ponds for breeding. A brief period of freezing temperatures removed several species from a subtropical anuran community in Brazil. A snake community in Utah disappeared following a grass fire that resulted in soil erosion and drifting sand closing the snakes' hibernacula. Successional changes of vegetation also alter herpetological communities, for example, Florida sand-scrub communities declined from 16 to 9 species as the scrub matured.

Stable herpetological communities may exist, but as yet none have been identified. However, it is noteworthy that the species composition of North American fossil herpetofaunas (Pleistocene) closely matches the composition of the faunas occurring in the same areas today, indicating a persistence of species assemblages through major climatic fluctuations.

Intra- and Interspecific Interactions

Community structure derives from a mixture of abiotic and biotic factors (Table 12.2). As noted earlier, the nature of the physical environment, composition of the species source pool, and interactions of members are major factors. A tropical or boreal climate, a rocky or sandy soil, and a myriad of other physical features set the conditions for a species' survival and reproduction. Community membership also depends strongly on what species are available to colonize an area. Numerous amphibian and reptilian species are adapted for survival in arid

TABLE 12.2 ▼▲▼▲▼▲▼▲▼▲▼▲▼▲▼▲▼▲▼▲▼▲▼▲
Properties Determining a Species' or Population's Membership, Position, and Persistence in a Community

Organismic	Environmental
Body size	Severity of physical environment
Diet (trophic position)	Spatial fragmentation
Mobility	Long-term climatic variation
Homeostatic ability	Resource availability
Generation time	Partitionability of resources
Number of life stages	
Recruitment	

[After Schoener (1986).]

Note: Schoener proposes these properties to examine the structure and dynamics of assemblages (e.g., intertidal algae, island anoles). These properties also highlight factors affecting an organism's survival and reproductive success, hence a population's or species' niche and community affiliations.

lands, but dispersal limitations prevent Australian *Moloch* from colonizing in the Chihuahuan Desert or even shovel-nosed snakes (*Chionactis*) from the nearby Sonoran Desert reaching the Chihuahuan Desert. Once a species tolerant of the physical environment reaches a site, its persistence depends on its interactions with the current community members and future colonists.

Community organization and its species membership have become increasingly recognized as multifactored (Table 12.2). Interactions are not singled out here because they are more important than other factors, but because recent research on herpetological assemblages has emphasized interactions. The major interactions are predation, symbiosis, and competition, and of these, competition has occupied center stage for the past 50 years.

Indeed, competition was accepted as the major biotic factor establishing animal-community structure and driving the morphological and behavioral divergence between similar or closely related species living in the same habitat. However, competition has lost its universal acceptance as "the explanation" for community organization. In part, this change results from an increasing recognition that structure is greatly and regularly influenced by stochastic events (e.g., arrival order of colonists, adverse weather) more than by constant or intermittent competition among component species. Additionally, predation has been recognized as a prominent factor in influencing community structure, particularly population densities. Perhaps more influential for the declining explanatory importance of competition is the tendency of competition theory (e.g., niche, compression hypothesis, species packing, resource partitioning) to accept most cases of nonoverlapping resource use as evidence of competition and the difficulty of clearly demonstrating competition among vertebrates.

No single factor is responsible for community structure. The factors mentioned here and others operate (often intermittently) singly, in combination, in opposition, or synergistically depending on the circumstance to produce the observed structure. This structure is only a temporary aspect of the population dynamics of the component species.

Prey–Predator Interactions

Most amphibian and reptilian species are both predator and prey, usually simultaneously, although within a food chain, eggs and juveniles may be the major prey and adults the major predators. Indeed, predation is the major mortality factor in most amphibian and reptilian populations and a major factor in their population structures and dynamics (Chapter 11). The importance of predation in their lives is outlined in Chapter 6. In spite of this importance, the role of predation has only recently been addressed relative to its effect on community structure and interactions, largely in amphibian larval associations, which are admirably suited to field observation and experimental manipulation.

The winter-spring temporary pools of eastern North America harbor diverse communities of breeding frogs and salamanders and their larvae, containing 1–4 species of salamanders (3 *Ambystoma* spp., *Notophthalmus viridescens*) and one or more species of frogs (*Bufo americanus,* 1–3 *Hyla* spp., 1–2 *Pseudacris* spp., 1–2 *Rana* spp.). The salamander larvae prey on each other, tadpoles, and invertebrates. *Notophthalmus* adults eat eggs and larvae of all amphibians including their own. The tadpoles are planktivores, but if sufficiently dense, they compete intra- and interspecifically, affecting their growth rates and times to metamorphosis. Under natural conditions, various predator interactions can and do occur among the salamander larvae. *Ambystoma opacum* is a fall migrant and lays its eggs in the dry basin of the pond; the larvae hatch as the pond fills in midwinter and feed on zooplankton and insect larvae. *Ambystoma maculatum* lays its eggs in late winter, and its larvae hatch in early spring. *Ambystoma opacum* larvae immediately begin to prey on the *A. maculatum* larvae, often decimating the *A. maculatum* larval population and occasionally totally eliminating it. Infrequently because of winter droughts or low plankton densities, *A. opacum* larvae are no bigger than hatching *A. maculatum* larvae and the two species then compete for prey. In ponds with *A. tigrinum,* which lays eggs as soon as the ponds fill, *A. opacum* larvae are initially larger than *A. tigrinum* larvae and eat them; however, *A. tigrinum* grow faster and soon begin to prey on the *A. opacum*.

Competition among tadpoles for food occurs in experimental enclosures within natural ponds and likely occurs naturally although perhaps not regularly. At moderate to high population densities, growth is slowed and metamorphosis delayed in single-species and multispecies enclosures. Since many of the insect predators can prey only on the smaller tadpoles, slow growth increases predation on all species. Experiments with predatory *Notophthalmus* on six species of tadpoles showed that predation pressure greatly altered the relative abundances of the tadpoles. At high predation, the competitively inferior *H. crucifer* survived best and became the most abundant species; at moderate predation, *H. gratiosa* survived best; and at all levels of predation, the four competitively superior species experienced the greatest mortality. Since amphibians and reptiles are common predators of one another, similar interactions can be expected in most herpetological assemblages.

Amphibians and reptiles prey widely on other animals, and these activities serve to regulate the dynamics of prey populations and communities. Predation stabilizes communities. Tentative observations on caiman predation in nutrient-poor lakes illustrate the possible benefits (Fig. 12.3) to this assemblage. In tropical forests, nutrients are held largely in the plants and rapidly recycled into plants following decomposition, thus streams and lakes are often nutrient poor. Annual floods inundate low-lying forests and enlarge forest lakes; concurrently fish of the main channel migrate into the forest lakes to spawn. Fish diversity and populations have declined with the increasing harvest and decimation of caiman populations. Studies showed that caiman feeding on the adult fish nearly doubled the amounts of calcium, magnesium, phosphorus, potassium, and so-

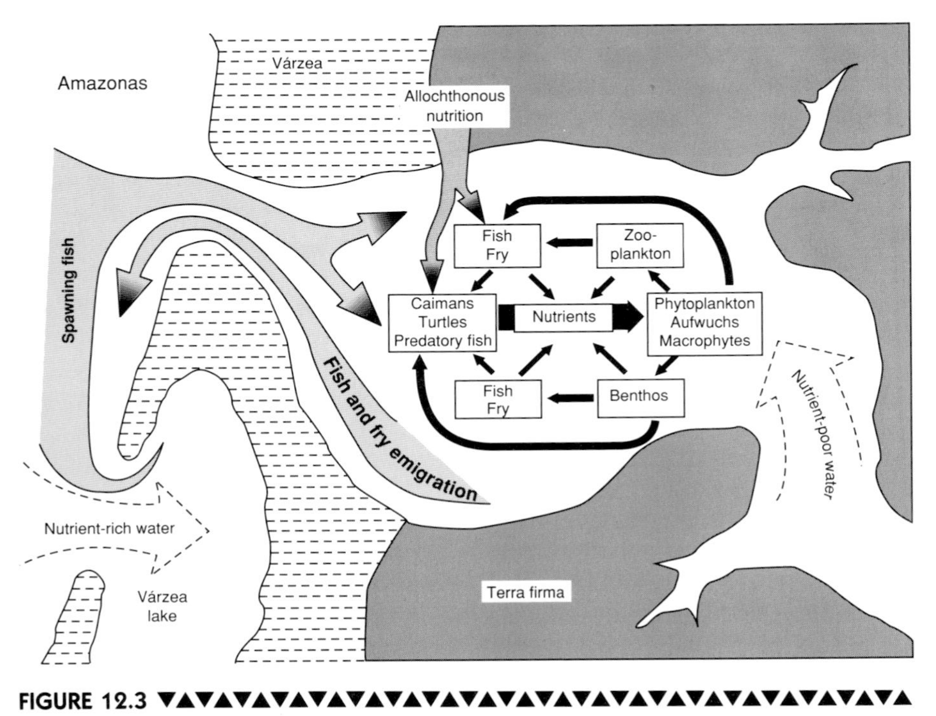

FIGURE 12.3 ▼▲
Caiman predation and the enhancement of the nutrient cycle in mouth-lakes of small Amazon tributaries. [From Fittkau (1970).]

dium in the water, making it a much more productive system for hatchling fish and other aquatic organisms.

Symbiosis

Symbiosis encompasses intimate associations where one species occurs on or in the body or habitation of the host species. The association may be detrimental (disease, parasitic) to the host or beneficial (mutualism and commensalism) to one or both species. The effect of symbiotic interactions on the composition of herpetological communities is unknown and largely so for population structure and dynamics as well. Disease and parasites variously affect the performance and survival of individual amphibians and reptiles (Chapter 6), but whether disease or parasites can or do exclude an amphibian or reptile species from a community is an open question.

In beneficial symbiosis, interactive species may be dependent on each other with both benefiting from the association (mutualism) or with both capable of living independently and the association benefiting one or both species (commensalism). Mutualistic interactions have not been identified between reptilian and/

or amphibian species or, for that matter, between an amphibian/reptile and a microorganism (i.e., gut microflora). Commensal associations are better known, although the actual interactions (neutral, beneficial, or detrimental) are unknown. Multiple-species snake hibernacula and gopher tortoise (*Gopherus polyphemus*) burrow associations likely represent commensalisms. In the former, all residents can benefit from reduced water loss if the hibernating individuals form a "snake-ball" thereby reducing the exposed surface areas of each snake. In the latter, associates (snakes, frogs) gain shelter from predators and a more equitable and stable environment for homeostasis.

Competition

Organisms compete for resources when resources are in short supply or the quality of the available resources is variable. Competition occurs mainly for nourishment (energy and nutrients), space (spatial and temporal), and mates (Chapters 7 and 8). Competition occurs among members of the same population or species (intraspecific) or among species (interspecific). Both forms of competition alter population and community structure and greatly influence the spacing of individuals (Chapter 9). Competition occurs in two ways; interference and exploitation. Interference competition limits the access of one competitor to existing resources, for example, territories and dominance hierarchies. In exploitation, the resource is used by one competitor so it is unavailable to the other competitor, for example, capture and consumption of prey.

The complexities of competition can be seen in the interaction of the slimy salamander (*Plethodon teyahalee,* member of the *glutinosus* complex) and the red-cheeked salamander (*Plethodon jordani*) in the southern Appalachian Mountains of North America. Both species are confined to the southern Appalachians; slimys occur mainly at low elevations and red-cheeks at high elevations. They have varying degrees of elevational overlap and co-occurrence. In the Great Smoky Mountains, they occur together (syntopic) in a narrow zone of 70–120 m, and in the Balsam Mountains, the zone of syntopy is over 1200 m wide. The Smokies populations were assumed to be more competitive than the Balsam ones. The differing levels of competitive interaction in these two areas provided an opportunity to manipulate population densities and test for the presence of competition. In one set of experiments, single-species plots were created in each overlap zone by the selective removal of one species. If competition were more intense in the Smokies, the population in each plot should show a greater increase in density in the Smokies experimental plots than in the Balsam plots. That prediction proved correct. On the Smokies plots, removal of *P. teyahalee* allowed a significant increase in *P. jordani*—in younger age classes as would be expected—and the same results occurred for *P. teyahalee* with the removal of *P. jordani.* The higher densities became evident after three years. The densities increased in the Balsam experimental populations but significantly less than in the Smokies populations.

Another set of experiments tested the "competitiveness" of the two populations of *P. jordani*. Since the Smokies population (recognized by red-cheeks) lives in a narrow zone of overlap, its individuals are likely stronger competitors than Balsam population members (gray-cheeks). Red-cheeks were replaced with gray-cheeks in two-species plots in the Smokies and the opposite replacement in the Balsam plots. *Plethodon jordani* suppressed the density of *P. teyahalee* populations in both sets of experimental plots; however, red-cheeked *P. jordani* caused a striking reduction in the Balsam *P. teyahalee,* demonstrating the superior competitive ability of the red-cheeks.

Similar population manipulations have been performed on anoles in the West Indies and arid-land iguanids of southwestern North America. Results have been variable; some demonstrate competition with density increasing in single-species plots, but not in others. One set of field experiments with *Sceloporus merriami* and *Urosaurus ornatus* showed competition to be intermittent and one-sided. In years of high prey abundance, no competition occurred, and during low prey conditions, *U. ornatus* showed significant changes in individual growth rates and population densities in single-species plots, whereas *S. merriami* did not.

Resource Partitioning and Niche

When Elton proposed the niche concept, he viewed the niche simply as where animals live and what they eat. This simple two-dimensional resource concept has since become multidimensional and with multiple definitions. Yet, the niche concept remains a measure of resource use and is equally applicable to patterns of an individual, population, or species resource use. With very few exceptions, niche research focuses on competitive interactions and ignores other interactions by measuring niche breadth/niche overlap and examining how the resources of space, time, and food are partitioned among the age classes and sexes of a population and/or among different sympatric populations (species).

A review of resource partitioning in herpetological communities (Table 12.3) reveals that space or habitat specificity is of prime importance for all amphibians and reptiles except amphibian larvae. In larval associations, species segregate by seasons, whether in temperate-zone or tropical habitats. Prey types are of equal or greater importance for partitioning species in snake assemblages. Snakes have the greatest number of food specialists, and many are further specialized for capturing and consuming large prey. In contrast, frogs, salamanders, and lizards typically partition their prey resources by size—a not unexpected axis for insectivores feeding on relatively small prey. In all assemblages studied (Table 12.3), partitioning occurs along two or more axes, with habitat and food partitioning common to most. Time is a less important means for partitioning resource use, although approximately 50% of the lizard and snake assemblages show some diel segregation of species and snakes show an even higher seasonal partitioning. Competition is only partially responsible for these resource-

TABLE 12.3
Relative Importance of Food, Space, and Time in Amphibian and Reptilian Communities

	Most important			Partitioned			
Taxa	Food	Time	Habitat	Food	Time diel	Time season	Habitat
Salamanders	25	0	75	100	15	8	87
Frogs	0	0	100	94	23	23	88
Amphibian larvae	6	88	6	75	0	88	83
Turtles	100	0	0	100	100	50	0
Lizards	53	36	11	94	45	24	91
Snakes	44	56	0	93	100	50	71

[After Toft (1985).]

Note: The values are the percent of studies in which one resource was the most strongly partitioned and in which a resource was partitioned.

partitioning patterns; predator avoidance and phyletic constraints (e.g., physiological tolerances) undoubtedly interact with competition to adjust a species position along a resource axis.

The guild is another resource utilization concept with a competition bias. The bias is intentional, because as originally defined, a guild includes all members (regardless of their taxonomic position) of a community exploiting the same class of resources in a similar way, thus overlapping significantly in their niche requirements. Like the community concept, the guild concept is seldom used in its original all-encompassing sense but confined to a single taxonomic unit (e.g., forest-floor salamanders, frog-eating snakes). The taxon-guild is a useful tool for testing the actual interactions within groups chosen on the basis of a presumed utilization of one or more shared resources (presumed overlap of niche requirements). For example, six salamanders (*Desmognathus ochrophaeus, D. wrighti, Eurycea bislineata, Plethodon jordani, P. serratus, P. teyahalee*) occur sympatrically in the southern Appalachians. Adults of all live on the forest floor, forage for small insects and invertebrates in the leaf litter, and rest in similar burrows. Their shared use of resources qualifies them as members of the same guild (although an incomplete one) and as potential competitors. Removal of both *P. jordani* and *P. teyahalee* was not accompanied by an increase in the abundance of the other four species, in contrast to the response of *P. jordani* or *P. teyahalee* (see earlier discussion) when the other one was removed. The potential competitors proved not to be competitors, at least not with the two larger species. Further removal experiments are required to demonstrate whether the four smaller species compete with one another. These results highlight the fallacy of the commonality of interspecific competition within taxon-guilds, an assumption often made by researchers when they are considering the evolution of life-history characteristics.

GEOGRAPHY OF POPULATIONS

Determinants of Species Distributions

Each organism has an inherent set of tolerances and requirements. Populations have similar sets that are a composite of those of their individual members. If an individual's tolerances are not exceeded, its requirements fulfilled, and it avoids predators and disease, it survives. However, survival of the individual does not ensure survival of the population; the individual must reproduce and so must its offspring for the population to persist. It is this latter aspect that makes age-specific mortality and age-specific fecundity the key life-history traits (Chapter 11) and, ultimately, the key determinants of a population's occurrence and persistence in any geographical area.

Since a species consists of multiple local populations, a species' distribution represents the total occurrences of its populations. The borders of each species' distribution mark the areas where populations waver between extinction and self-perpetuation. At one season or year, conditions allow reproduction and survival of the young and the population grows; in the next, reproduction is unsuccessful and the population drifts toward extinction. The factors affecting these population cycles are climate (micro- and macro-) relative to physiological tolerances, availability and access to resources, and interspecific interactions (including human activities). The final factors, historical accident and dispersal ability, determine what species are likely to occur in an area and the probability of their reaching an area.

Among the thousands of amphibian and reptilian species, many can live in locations outside of their natural distribution. The successes of numerous exotic lizards in the Miami area, of the marine toad (*Bufo marinus*) in the West Indies and the Southwest Pacific, and of the brown treesnake (*Boiga irregularis*) in Guam show this ability. Normally, however, the species occupying an area came from nearby areas, with vacant habitats "filled" by a few migrants crossing barriers (geographical or unsuitable habitats) or by the slow expansion of a population into less hospitable areas. Dispersal abilities and opportunities are variable. Typically, small and fossorial amphibians and reptiles have poor dispersal abilities, and large and aquatic ones have good dispersal. Coastal and riverside species are more likely to be transported elsewhere than are inland species. Amphibians rarely cross saltwater barriers; reptiles commonly do. These generalities have numerous exceptions and indicate only a few of the factors associated with a species' dispersal and its tolerances and preferred habitats. Dispersal ability and the nature of barriers are also critical in determining the level of gene flow among populations and local population differentiation.

It is hardly surprising that climate affects species occurrence. An animal will not survive in an area where one or more of its physiological tolerances are regularly or constantly exceeded. Temperature, rainfall, and their periodicity establish the climatic regimes under which individuals and populations must operate. Tolerance limits (Chapter 10) are species specific (although often species are locally adapted to temperature or moisture extremes not experienced by distant populations) and define the limits of each species' distribution. The edges of ranges often match closely the isograms of rainfall and temperature (Fig. 12.4). Frequently, the effects of temperature and/or rainfall are greater on one lifestage than on another, but the survival of each stage is critical for the survival of the population. Spring droughts may prevent temporary-pond amphibians from breeding or metamorphosis of the larvae; while the adults may survive

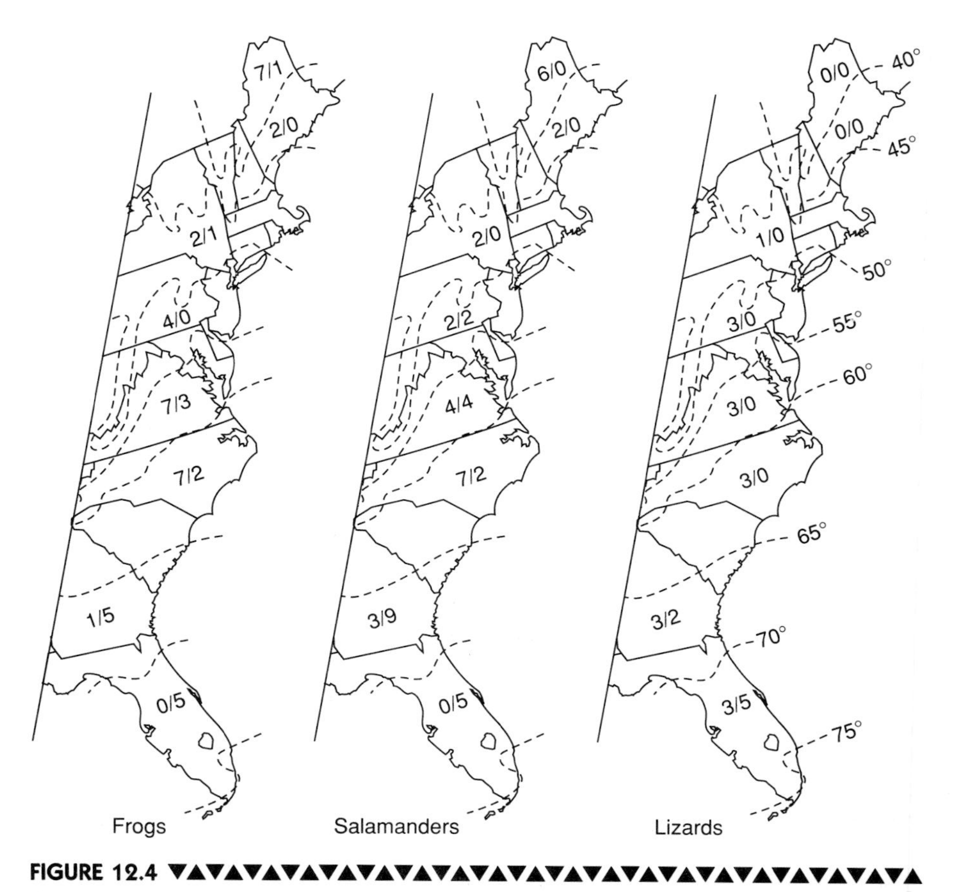

FIGURE 12.4 ▼▲

Temperature isotherms and the northern and southern termini of frog (n = 30 spp.), salamander (26 spp.), and lizard (16 spp.) distributions in eastern North America (piedmont and coastal plains). The isotherms are mean annual temperature (°F); the integers in each zone are number of species with ranges terminating in the interval (northern terminus/southern terminus, respectively). [Isotherms from USDA (1941), distributional data from Conant and Collins (1991).]

(tolerate) the drought, sequential years of drought prevent recruitment and, in short-lived species, the populations go extinct. In turtles with their temperature-dependent sex determination (Chapter 7), cooler summers may not prevent hatching or increase juvenile mortality but will produce all-male cohorts and, if this hatchling sex ratio continues, an all-male population results and eventual extinction.

Availability and access to resources are the symbiotic and competitive facets of interspecific interactions. Earthworm specialists, whether a snake or salamander species, cannot survive in an area lacking earthworms or where earthworm abundance is so low that the cost of finding and eating them is greater than the energy consumed. Low prey abundance even for generalist predators will affect their ability both to survive and to reproduce. Space in the form of shelter and refuge is also a critical resource. Symbiotic relationships, such as the gopher frogs' (*Rana capito*) dependency on gopher (mammal and tortoise) burrows, may be necessary to provide adequate and proper resources for survival. Geological features, such as rock outcrops in northern prairie, are necessary for snake hibernacula, and numerous other habitat features (e.g., vegetation structure or type, soil friability) can influence the availability of a resource necessary for survival or reproduction. Interference competition for shelter has been proposed for the restriction of *Plethodon shenadoah* to a few isolated talus areas surrounded by the forest-floor filled with *P. cinereus*. Most space–resource associations are so obvious that they often go unmentioned, for example, friable soil for burrowers, springtime ponds for temporary-pond breeders, etc. The other facet of interspecific interactions is where the population (species) is a resource for another species. Disease, parasitism, and predation affect and are affected by population dynamics and likely the presence or absence of a species in an area, although the exclusion or elimination of an amphibian or reptilian species by pathogen or parasite is unconfirmed. Some observations indicate the possibility of such events. In the late 1960s and early 1970s, northern leopard frog populations (*Rana pipiens*) from Vermont to Minnesota suffered major population declines that appeared to be associated with viral or bacterial disease. The eastern newt (*Notophthalmus viridescens*) shares mountain ponds with a small leech, and if leech density becomes too high, the newts leave the pond. Fish predation on larval amphibians prevents many species from breeding successfully in permanent-water ponds. Predation on amphibian larvae by adult or other larval amphibians may eliminate the larval population of one species one year, and if this predation occurred repeatedly, it could cause the local extinction of the prey species. Might not the spotty occurrence of the tiger salamander in eastern North America be the result of larval predation by the marbled salamander?

Climate, resources, and interspecific interactions vary from area to area, and each population adjusts (adapts) to its local conditions. Since conditions are nowhere the same, each population adapts differently and diverges genetically (evolving) from other populations. If this divergence continues, speciation can occur; however, speciation is a rare outcome of evolution because adjacent populations exchange individuals. The migrants bring new genes into the population and this gene flow tends to homogenize the gene pools of adjacent populations. The rate of gene flow is a function of the closeness of the populations and the dispersal tendency of the species. The rate can be quite low, yet maintain the genetic continuity of distant populations. But while maintaining continuity, local populations can adapt to local conditions. Typically, these adaptations involve traits associated with reproduction (Fig. 12.5; also see Life-History Patterns, Chapter 11).

Biogeography—Patterns of Distribution

Different geographical areas contain different sets of plants and animals. These sets (floras and faunas) can be relatively uniform in species number and composition in adjacent areas. This uniformity of flora and fauna can persist over

FIGURE 12.5 Geographical variation in clutch size for populations of *Chelydra serpentina* (snapping turtle) in North America. Each integer represents the mean clutch size for the area occupied by the integer. [Data from Fitch (1985), Ernst (unpubl.), and Mitchell (unpubl.).]

large or small areas then gradually or abruptly change to a new flora and fauna. Biogeography is the study of these patterns of changes and uniformity in life-groups and attempts to identify the causes for the change and uniformity. Biogeography has two general approaches. Geographical ecology or ecological biogeography examines the structure of the different communities from a resource utilization perspective, either from a numerical basis of how many species exist in an area or how different species partition the available resources. Historical biogeography examines the structure of the communities based on the relationship and origins of the taxa, emphasizing the phylogenetic affinities of the species and their evolutionary histories.

Life Zones and Biogeographical Realms

Current worldwide distributions of communities form two patterns: life zones or biomes and biogeographical realms. The former emphasizes the similarity of plant-community appearance relative to climate, and the latter the phylogenetic relatedness of the component taxa. The biome concept ignores animals and recognizes sets of communities based on plant structure (e.g., height and shape of plants, leaf structure, deciduous or evergreen vegetation). The major terrestrial biomes are: tundra; boreal coniferous forest (taiga); temperate deciduous and rain forest; temperate grassland; chaparral; desert; tropical grassland; tropical scrub forest; and tropical deciduous and rain forest. These biomes can be further subdivided and have been in multiple ways. In all cases, the life zones reflect the annual cycle of temperatures and rainfall; animal distributions match the zones in general but rarely specifically for amphibians and reptiles. Few amphibians or reptiles occur in the boreal forest and only marginally for the few that do. Species-poor assemblages are widespread in the boreal forest zone and are dominated by amphibians. Species and community diversity (herpetofaunas) expands within the temperate biomes and increasingly so into the tropics, but unlike how plant form matches climate, animal form follows taxonomic affinities more closely.

Biogeographical realms (Fig. 12.6) map patterns of taxonomic affinities with close phylogenetic ties within each realm and more distant relatives between realms; realms tend to more accurately reflect animal than plant distributions. The realms derive from higher-order relationships, typically subfamily and above, and reflect past geological events (barriers and connections for species dispersal). Indeed, the present distribution of many amphibian and reptilian families matches the past continental connections and fragmentations proposed by the plate-tectonic (drifting continents) theory of geologists. (Compare Fig. 12.6 with the distribution maps in Chapters 14–18.) For example, salamanders occur mainly in the holarctic (nearctic + palearctic) and frogs mainly in the Southern Hemisphere, and these distributions match the Mesozoic split of the supercontinent of Pangaea into the northern Laurasia and southern Gondwana. Ancient groups still retain a Laurasian or Gondwanan distribution. Pipid frogs occur in both Africa and South America. Myobatrachid-heleophrynid-leptodactylid frogs

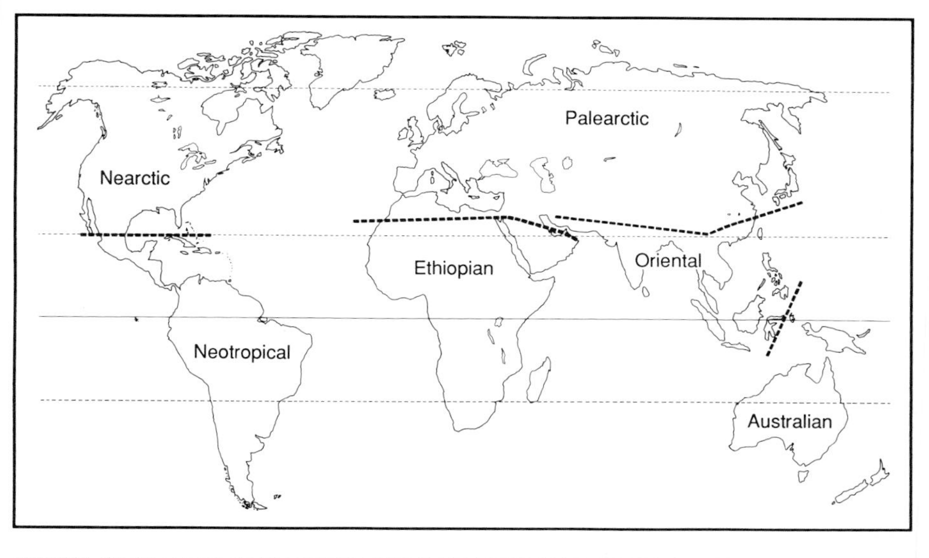

FIGURE 12.6 Biogeographical realms of the world.

are each others' closest relatives and link Australia, southern Africa, and South America. Chelid and pelomedusid turtles display a Gondwanan distribution. Cryptobranchid, plethodontid, and salamandrid salamanders occur in both North America and Eurasia, suggesting an ancient distribution throughout Laurasia. Just as these interfamilial relationships match ancient topographies, intergeneric and interspecific distributions match more recent (but still ancient) geological events and climates. Each continent has been divided (by biologists) into biogeographical provinces that are delimited by more or less abrupt terminations in species distributions and shifts from one community to another. These discontinuities in community structure likely reveal a prior isolation of the two communities and speciation of member populations.

Geographical Ecology

Species richness (diversity) is a geographical comparison of community structure. The striking differences in number of species at different localities and latitudes (see Table 12.1) has long intrigued biologists. Most attention has been directed at explaining the tendency for diversity to increase from high-latitude habitats to tropical ones and, as for most other studies of communities, to examine the changing diversity in limited-taxonomic assemblages. Although emphasis centers on the number of species, the abundance of each species is not unimportant. However, accurate and comparable data on abundance are largely unavailable (see later).

Numerous explanations have been proposed to account for the differences in richness. The primary explanations are evolutionary time, ecological time, climatic stability, climatic predictability, spatial heterogeneity, amount of primary productivity, and species interactions. These explanations are not mutually exclusive, and likely multiple causes operate in different combinations at different locations and at different times. Older communities presumably have had more time for species to adapt to the local environment (evolutionary time), or there has been insufficient time for species to colonize new or modified habitats (ecological time). Amphibian and reptilian species of northern sites (Andrew Experimental Forest and Vienna; see Table 12.1) are all wide-ranging species with distributions >1000 km^2 in contrast to many species at the tropical sites with small geographical distributions. Although temperate-zone amphibians and reptiles may have individuals with limited dispersal abilities, their populations move apace with their preferred habitats. Such movement is evident from the reoccupation of glaciated portions of North America in the last 10–15 thousand years, with the current ranges of some species (*Ambystoma laterale, Rana septentrionalis, Elaphe vulpina, Emydoidea blandingi*) now totally within former glaciated areas.

Climatically stable areas have little seasonal or long-term change in temperature or rainfall. Such locations are limited to a few rain forest areas of the world, and one such area is a belt of Amazonian forest on the eastern face (midaltitude) of the Andes. Santa Cecilia has 173 species and a site in Peru has >200 species. Climatically predictable habitats are far more numerous with regular cycles of wet–dry or hot–cold seasons, but show variable species richness depending on latitude. Relative length and harshness of the cold or dry seasons are rarely considered and can be quite influential in limiting species membership. Climate predictability may be more imagined than real; climate records of this century emphasize the great irregularity in the beginning, end, and length of seasons, and this predictability is no less or more regular in the tropics than in the temperate zone.

Habitats with a greater spatial heterogeneity have more species. This explanation does not account for the diversity of the vegetation that creates the spatial heterogeneity but states the direct correlation between the number of animal species and the number of microhabitats provided by vegetation. Structurally complex forest habitats have more species than the structurally simpler grassland and desert habitats. Within the same area, complexity affects species diversity. Three habitats exist in the Sakaerat area, Thailand (see Table 12.1): gardens and fields, deciduous forest, and evergreen forest. Herpetofaunal diversity increases proportionately with spatial heterogeneity at Sakaerat (54, 67, and 77 species, respectively) and elsewhere. Productivity is not unrelated to spatial heterogeneity and shows the same positive correlation. Simply put, high food availability and variety translate into more and different consumers. For example, snake diversity shows a direct correlation of snake species number with prey diversity, irrespective of latitude.

Species interactions are assumed to operate in several ways. Competition for resources supposedly drives the evolution of narrower and more specialized niches (= niche compression and species packing) in areas of high and relatively stable productivity. Body size differences among sympatric congeners (regularly seen in frogs and lizards) may have resulted from competition for food resources and resulted in a partitioning of prey resources by creating different-sized congeners eating different-sized prey. Similarly, the habitat partitioning of sympatric congeners (e.g., anoles) divides access to the resources spatially. Predation presumably influences the coexistence of multiple species consuming overlapping resources by maintaining all species below the carrying capacity of the resources.

A common assumption is that the abundance of each species is less in a species-rich community than in a species-poor one. This conjecture may be true for the number of tree species in a rain forest compared to a temperate-zone deciduous forest but is probably not true for most herpetological communities (see density estimates, Table 11.2), although actual comparisons do not exist. Such comparisons would examine the actual abundance (density) of each member species. Another abundance comparison examines the equability or evenness of the relative abundance of each species within the assemblage. Both comparisons are confounded by several factors: body size, trophic position, seasonal and annual fluctuation in population densities, and the widespread lack of accurate population censuses (particularly for tropical populations). Ignoring these difficulties, it is unlikely that any community or assemblage possesses high equitability. More likely, one or a few species have high densities with all other species at low densities. The Sakaerat skink assemblages (Fig. 12.7) show abundance patterns that are likely repeated or even more extreme between the common and rare species in other herpetological assemblages whether they are from tropical or temperate areas.

Exceptions to equal abundance between species-rich and species poor assemblages occur between mainland and island anole assemblages. Island populations (species- poor) have densities 2–10 times those of the mainland ones. A few other lizards (congeneric comparisons) also show higher densities on islands, but such comparisons for other amphibians and reptiles are lacking. These differential densities appear to be associated with differences in predation (see Life-History Patterns, Chapter 11).

Species richness also differs markedly between island and mainland assemblages. Islands have fewer species when compared to comparable-sized areas on the mainland. Further, a positive relationship exists between island size and species richness, and a negative relationship between distance of an island from a colonizing source area and species richness. These species–area relationships led R. H. MacArthur and E. O. Wilson to develop the equilibrium theory of island species diversity. The equilibrium theory proposes that a balance exists between the number of species colonizing an island and the number of species going extinct. The colonization or immigration rate is a function of the island

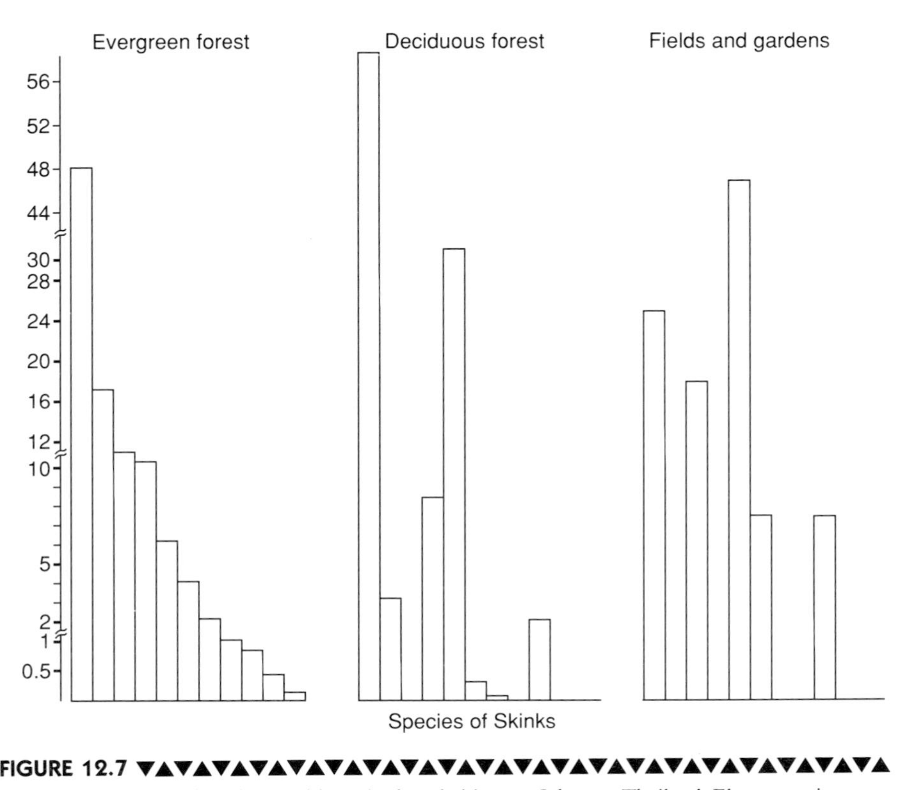

FIGURE 12.7

Relative abundance of skink assemblages in three habitats at Sakaerat, Thailand. Eleven species occur in the evergreen forest, and subsets of them occur in the deciduous forest and agriculture areas. The species are ranked in order of decreasing abundance for the evergreen-forest assemblage, and that order is retained for the two other assemblages. Skink abundance is unequal between habitats; respectively, 53, 43, and 4% of the total for the three habitats. [Data from Inger and Colwell (1977).]

distance from a source area and the extinction rate is a function of island size. Immigration and extinction are assumed to be continuous, thus species number may reach an equilibrium value and remain constant even though the composition of the species assemblages changes continually.

Although the linearity between island size and species number exists for lizards (Fig. 12.8; few other herpetological groups have been examined), island assemblages deviate from several theoretical predictions. Lizard assemblages commonly have a higher species diversity than predicted (suggesting supersaturation) and possess a constant number of species over a wide range of small island sizes. These deviations likely result from lower dispersal and extinction rates than those of the birds and insects on which the theory was developed.

A species–area effect has been proposed for peninsulas as well as islands.

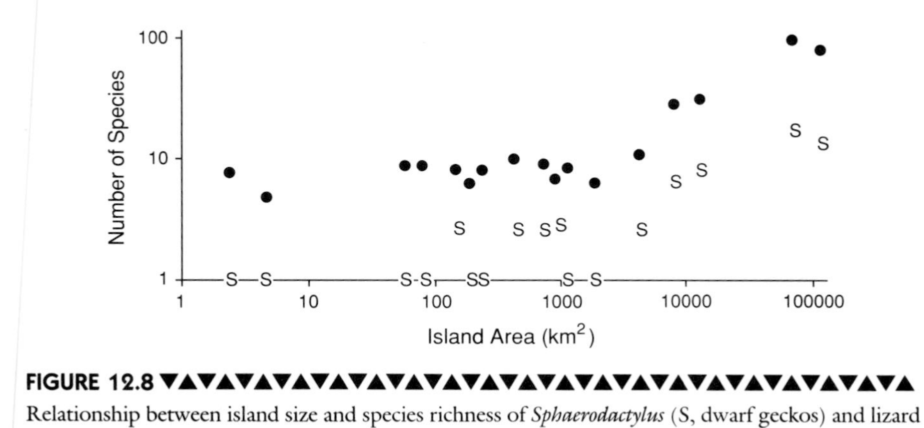

FIGURE 12.8 ▼▲▼▲▼▲▼▲▼▲▼▲▼▲▼▲▼▲▼▲▼▲▼▲▼▲▼▲▼▲▼▲▼▲

Relationship between island size and species richness of *Sphaerodactylus* (S, dwarf geckos) and lizard (dots) faunas in the West Indies. [Data from Schwartz and Henderson (1991).]

The peninsula effect predicts a decline in species richness from a peninsula's base to its tip. Its applicability to amphibians and reptiles remains uncertain. Species diversity does decline for some herpetofaunas (peninsular Florida) but not for others. The reptiles of Baja California are as diverse at the tip as at the base.

Historical Biogeography

In historical biogeography, the perspective shifts from the near past of ecological biogeography to the deep time of geological history, and from intra-community interactions to phylogenetic relations for reconstruction of species and higher taxa distribution patterns. Current theoretical debate focuses on dispersal versus vicariance explanation, although many studies fall somewhere between the two extremes. The dispersal viewpoint dominated biogeographical studies until the early 1970s when the geological worldview shifted from static continents to drifting continents (plate tectonics). The vicariance viewpoint also grew with the cladistic emphasis on rigorous and explicit phylogenetic analysis.

Dispersal theory rests on two basic premises: taxonomic groups have a center of origin and each group disperses from its origin-center across barriers, with the resulting communities or biota deriving from one to several centers and dispersal events. The vicariance theory premises are: taxonomic groups or biotas are geographically static and geological events produce barriers and the biotas diverge (allopatric speciation of community members) subsequent to their isolation. Both theoretical approaches require knowledge of phylogenetic relationships to discern the ancient dispersal routes or the areas occupied by the ancient biotas. Because allopatric speciation appears to be the dominant mode of speciation and the fragmentation of a biotic unit is more parsimonious than a biota dispersing as a single unit, vicariance interpretations are generally preferred over dispersal explanations. Vicariance explanations are also more applicable to testing. Dispersal explanations are required, however, to account for the evolution of oceanic island biotas, for example, for the Galápagos and Hawai'ian islands.

The geological histories of most areas and their herpetofaunas are so complex that a single theory is inadequate. The Seychelles' herpetofauna is a good example of this complexity. This fauna comprises several levels of endemicity (Table 12.4) that strongly indicates multiple origins with components arriving at different times. The oldest elements are the sooglossid frogs and caecilians. Both amphibian groups have only distant (and somewhat uncertain) affinities with African taxa. Both are confined to the high granitic islands of the Seychelles that have been emergent since the Mesozoic and further appear to be fragments of the Indian tectonic plate that broke free from the current African plate and moved northward to collide with the Asian plate. Since amphibians are noted for their inability to cross large saltwater gaps, these amphibians and perhaps the gecko *Ailuronyx* are likely derived from ancestors living on the original African-Indian plate. The rhacophorid frog *Tachycnemis* and some reptiles also appear to be derived from an early Seychellan herpetofauna but likely from taxa that arrived via island hopping across short water barriers. The day-geckos (*Phelsuma*) and others are more recent arrivals and show closer affinities with Malagasian and African taxa, but presumably arrived prior to human colonization. More recent components have arrived via human transport (*Gehyra*).

A vicariance explanation generally fits the distribution patterns of chelid turtles (Fig. 12.9). Excluding *Chelodina, Pseudemydura* and the ancestor of all other extant chelids arose from an unknown vicariance event. The ancestral

TABLE 12.4 ▼▲

Relative Ages of Select Components of the Seychelles' Herpetofauna

Ancient (>60 MA)	Near ancient (<60–10 MA)	Near recent (<10 MA)	Recent (<1000 yr)
Grandisonia			
Sooglossus	*Tachycnemis*	*Ptychadena*	
Ailuronyx	*Urocotyledon*	*Phelsuma*	*Gehyra*
	Janetscincus	*Mabuya*	
	Lycognathophis	*Boaedon*	*Ramphotyphlops*
	Pelusios seychellensis	*P. subniger*	

[Data in part from Nussbaum (1985).]

Note: Taxa are arranged vertically: caecilians, frogs, geckos, skinks, and turtles. The age of the taxa are based on their degree of taxonomic differentiation and endemicity. The ages are arbitrary estimates beginning immediately prior to the Seychelles separation from Gor[illegible]vana (Ancient) and mark the islands' progressive isolation from faunal source areas.

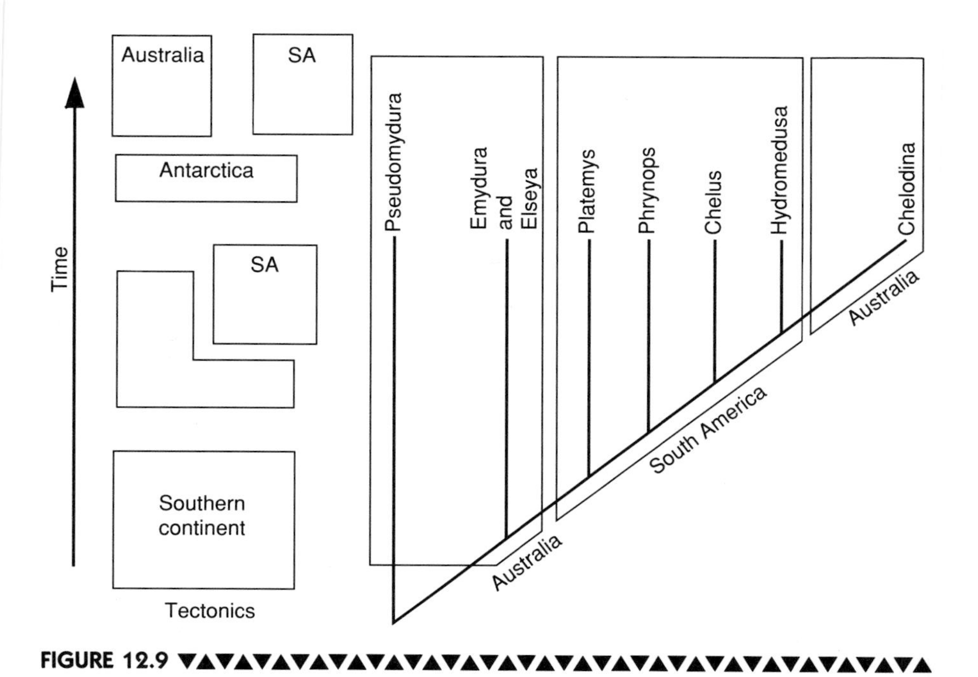

FIGURE 12.9 ▼▲▼▲▼▲▼▲▼▲▼▲▼▲▼▲▼▲▼▲▼▲▼▲▼▲▼▲▼▲▼▲▼▲▼▲

Comparison of phylogenetic relationships of chelid (sideneck) turtles and their distributions to the tectonics of southern continents. [Data from Gaffney (1977).]

species occurred throughout a large southern continent (Australia-Antarctica-South America) that fragmented into two and then three continents. On Australia, the ancestor became *Emydura;* on Antarctica, it became extinct; and on South America, the ancestor radiated (via vicariance events) into the four extant South American chelid genera. If *Chelodina* is the sister-group of *Hydromedusa,* the preceding hypothesis is false. The *Chelodina–Hydromedusa* sister-group relationship requires even earlier vicariance events to produce the extant genera with their joint occurrences on the large southern continent and subsequent differential extinction to restrict four genera to South America and Australia. Such difficulties with the vicariance explanation illustrate the difficulties of working with organisms for which the actual phylogenetic relationships are unknown, and past geological events are uncertain. The strength of the vicariance theory is in providing biogeographical hypotheses that are testable.

The Seychelles and chelid examples highlight the necessity of a pluralist approach to biogeographical analysis and the need to provide explanations (hypotheses) that can be tested. Multiple levels of interpretations are likely required for the patterns of most herpetofaunas and their component species.

CONSERVATION

Herpetofaunas are threatened everywhere by the continuing growth of human populations. Only those species of plants and animals that can tolerate (in some instances, thrive) in human-modified habitats seem likely to survive if our population growth continues unabated. Human-commensal species exist among amphibians and reptiles, but they are few compared to the diversity of the world's herpetofauna. Preservation of a significant fraction of this diversity is possible only through the preservation of entire biotas. Preservation is no less or more important whether the biotas are temperate or tropical, marine or terrestrial; only the urgencies vary from site to site, from species to species.

Effective conservation requires knowledge of both the genetics and ecology of the populations and communities to be preserved and the use of these data with sound ecological principles. The task is immense. A few aspects are addressed briefly here.

Endangered Species

From an ecological perspective, the world's biotas are endangered owing to competition with and predation by a single species, *Homo sapiens*. The competition is direct through our growing need for space and the conversion of natural habitats for agriculture and other human uses, and indirect via acid rain and other forms of pollution (including accidental or intentional introductions of nonnative species) that interfere with health, food procurement, and reproduction. Predation occurs through intolerance and indiscriminate killing of amphibians and reptiles, and through harvesting them for food, leathers, medicines, and pets.

How many amphibians and reptiles are endangered? We do not really know. We can identify some, because we have seen their populations continue to decline for the past several decades. However, the recent and seemingly sudden decline in many amphibian populations throughout the world suggests that many more species and populations are in a precarious state than predicted five or ten years ago. This viewpoint may seem to be an alarmist and pessimistic one, but the evidence continues to mount that even populations of some common and widespread species are declining, changing, and disappearing. For example, *Bufo boreas* (western toad) was until recently widespread throughout the northern Rockies of western North America; it is now absent or has greatly reduced populations throughout broad areas on the eastern face of the Rockies. The causes for its decimation remains unknown. The causes for other species are more obvious. *Rana aurora* and other Californian *Rana* populations have disappeared owing to the introduction and spread of the predaceous bullfrog *Rana catesbeiana*. Crocodilian populations throughout the world and large snake populations have been decimated by leather-trade harvesting. Where and when protected, some crocodilian populations have made spectacular recoveries (e.g., *Alligator mississippiensis,* Australian *Crocodylus porosus*), but only as long as harvesting is closely regulated.

Seaturtle protection has been less successful. Seaturtle demography and biology handicap effective protection. The populations are mobile and move across political boundaries with ease; their protection is less easily transferred from country to country. Seaturtles persistently return to the same nesting beaches on a regular annual schedule, and most major nesting beaches are in economically depressed countries where the turtles have historically been harvested for food. Critically, seaturtles require a minimum of 20 yr (perhaps as much as 50 yr) to reach sexual maturity (*Lepidochelys kempii* maybe somewhat less), and now they experience high predation throughout this long growth interval, whereas historically mortality decreased greatly after a few years of growth. Mortality has been elevated by "incidental" predation from human fishing activities and pollution (e.g., ingestion of oil-tars and plastics). All populations of all species of seaturtles are declining, some precipitously so.

Refuges and Captive Breeding

Commensal species can survive without their "natural habitats"; other species cannot. Two options are open for species preservation: refuges and captive breeding. Refuges provide the best, if not the ideal, solution. Captive breeding is a viable solution only for the short term and with the goal of reintroduction into reconstructed natural habitats. To achieve the greatest success, both strategies require a thorough understanding of a species' population genetics, demography, and community interactions. Successful conservation measures can still be effective even though we lack these data for most species, but we must reevaluate our conservation measures regularly. Such assessments require that endangered species (populations) are not treated as icons; instead investigations on their biology, genetics, and demography must be encouraged (even within the protective confines of refuges and national parks).

The necessity of continued research on endangered species is illustrated by the well-meaning hatchery operations in seaturtle conservation. Because nest mortality is high in some areas, egg clutches have been moved from female-dug nests to sand-filled styrofoam containers in a protected shelter. Unfortunately under these circumstances, the incubation temperatures were lower than in natural nests, thus the protectors were unwittingly producing all-male cohorts of hatchlings. As soon as it was learned that seaturtles had temperature-dependent sex determination, such artificial incubation was abandoned. Now egg protection practices attempt to match the temperature regime of natural nests while providing protection from predators.

Similar mistakes will occur in other conservation efforts but delay only exacerbates the threat to a population. Indeed, controversy surrounds all conservation choices. Refuges are a necessity, but how large, what shape, and how many continue to be debated. The debate concerns both demographic and genetic issues. Refuge size, shape, and number are critical because they directly impact population size, the number of subpopulations, and the interchange of population members. Large populations are necessary to maintain genetic diversity and to avoid demographic and environmental fluctuations (stochasticities) that can result in extinction.

Environmental fluctuations include catastrophic events such as floods and fires, which can wipe out a population in a single instance, or drought, exceptionally cold winters, or similar climatic fluctuations, which cause death directly or affect food supply, shelter availability, or changes in predator pressure. Demographic stochasticities are random (unpredictable) events that influence which individuals live or die, reproduce or not, and how many young are produced. Environmental fluctuations affect demography, but survivorship and fecundity fluctuate without outside influences. Large and numerous populations have more resilience and are less likely to drift to extinction. Population exchange adds new members to a population or provides colonists to reestablish populations in areas where populations have disappeared. These factors enforce the adage that bigger is better for refuges and also the importance of having numerous refuges with corridors for population exchange among them. Whether to establish numerous small reserves connected via corridors versus larger reserves, however, remains a vigorously debated topic.

Large and numerous refuges are equally important for maintaining genetic diversity and avoiding the genetic problems of inbreeding depression, outbreeding depression, drift, and bottlenecks. As a population decreases, it becomes increasingly likely that all members are kin, and in many species, kin tend to avoid mating with one another so the population continues to decline. Outbreeding depression is a converse situation where mating of unrelated individuals from distant populations adapted to different environments results in less fit offspring. This depression is a potential outcome of restocking efforts using individuals from different populations. Similar depression may also occur naturally if closely related species hybridize owing to the absence of mates, the deterioration of habitats, or other disturbances removing natural isolating features in small refuges. Drift (random loss of alleles) and bottlenecks (sudden population reduction with few survivors) result in loss of genetic diversity and leave populations less able to adapt to changing or new selective pressures.

These genetic problems are compounded in captive breeding programs, because breeding programs begin with few individuals and usually from a single area. With limited stock, inbreeding is the only option and immediately begins (unintentionally) the further reduction of genetic diversity and the increase of homozygosity. The loss of genetic diversity and the selection for individuals capable of flourishing in captivity increase the probability that the stock will lose its ability to survive in the wild and will lack the genetic flexibility to evolve to meet future environmental perturbations. As homozygosity increases, the lethal recessive alleles are expressed and the health of the stock declines. The difficulties and uncertainties of maintaining a viable stock for reintroduction argues against captive breeding except in the most dire circumstances, and then only if the program is well designed and includes regular genetic monitoring. Although often well intended, captive breeding by individual petkeepers will not yield a genetically healthy stock for reintroduction (especially when breeding is for specific color morphs). More critically, the pet trade is a serious and continual drain on wild populations. Even multiple institutional programs have difficulties maintaining genetically healthy stocks.

Pessimism is difficult to dismiss. It is not, however, an acceptable excuse for inaction. Every effort must be made to conserve species and natural areas, reduce pollution, and learn more about the biology and ecology of plants and animals.

PART VI

CLASSIFICATION AND SYSTEMATICS

Biological nomenclature forms a beautiful system. I doubt whether it is fully appreciated even by the average biologist, who is apt to be irritated by the trivial inconsistencies that turn up through daily use.

M. Bates, 1950

Classifications and skeletons are functional analogs and, not unsurprisingly, are incorrectly perceived by most people. Both provide the fundamental support for dynamic systems. Our common view of support is the stasis of architectural constructs, giving under pressure but remaining structurally unchanged, and therein lies the misperception. Classifications and skeletons change with new demands, yet remain totally functional during the remodeling.

CHAPTER 13

Systematics: Theory and Practice

Systematics is the practice and theory of biological classification. Classifying objects is part of human nature and has its origin in prehistory. The earliest human societies began to name and recognize plants and animals for practical reasons, such as what is good or bad to eat and what will or will not eat humans. This partitioning of objects places them into conceptual groups and is practiced daily by all of us. In modern systematics, we attempt to discover the full diversity of life, to understand the processes producing this diversity, and to classify organisms in a manner expressing their phylogenetic relationships (i.e., evolutionary history).

BASIC CONCEPTS

Evolution, the concept of descent with modification, is the glue that unites the diverse aspects of modern biology. Therefore, our classification should reflect the evolutionary history of organisms as closely as possible. Since we can never know the actual evolutionary history of a group, our classifications are inferences or estimates of phylogeny and will change as our knowledge of a group's phylogeny improves. Our classifications are schemes for naming and categorizing like organisms into the same groups. Each name identifies an organism or group of organisms and provides an index to information associated with that name. Biological classification is traditionally hierarchical (a system of nested sets), with each ascending level potentially containing more subgroups and characterized by the shared similarities of the included subgroups of organisms. Although the hierarchical system was introduced because of its efficiency for categorization and retrieval and prior to the recognition of organic evolution, hierarchies can be used to reflect evolutionary history.

Species are the basic units of our classifications and the only real units, existing not as a human-made category but as real entities. Each species is a set of unique, genetically cohesive populations of organisms, reproductively linked to past, present, and future populations as a single evolutionary lineage. Our hierarchical classifications place closely related species together in the same genus, combine related genera into the same family, and so forth up through the various levels of the classification. At each level, we proceed backward in time to points of evolutionary divergence, specifically to a speciation event giving rise to new lineages (Fig. 13.1).

For hierarchical classification to reflect evolutionary history, conceptual rules are necessary. A principal rule is that the grouping of organisms be monophyletic (i.e., unique history of descent) and represent a single lineage with the ancestor and all descendants. This goal would seem to be easy to achieve if the members of the group are adequately known and studied; however, aside from the difficulties of accurately recognizing relationships of divergent species, there is the difficulty of tradition. Many taxonomic groups were delimited before organic evolution was recognized. Because of tradition, there is a reluctance to discard the old group concepts. The present, widespread use of Reptilia is a case in point, as noted in the Preface and Chapter 3. This common usage makes Reptilia a paraphyletic group because it includes the ancestor and many but not all descendant groups, specifically excluding birds and mammals (Fig. 13.2). The Amphibia is considered by some biologists to be a polyphyletic group (Fig. 13.2), one in which several monophyletic lineages (but not the common ancestor of these lineages) are combined into a single taxon. In time, the paraphyly and polyphyly will be corrected and the contents of the Reptilia and Amphibia will be much different from their current ones. In fact, the entire arrangement of the vertebrates needs to change to obtain a truly monophyletic classification (e.g.,

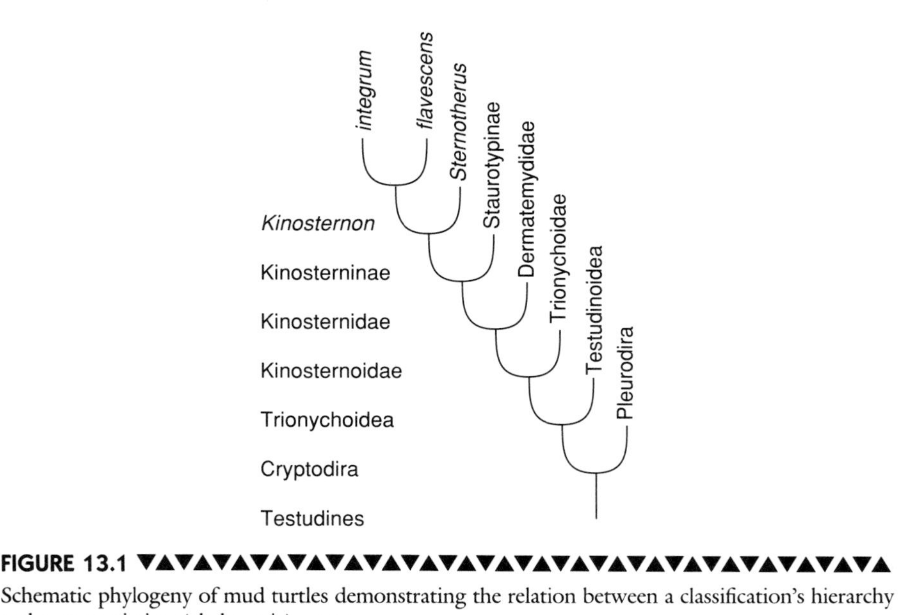

FIGURE 13.1 Schematic phylogeny of mud turtles demonstrating the relation between a classification's hierarchy and past speciation (cladogenic) events.

Table 13.1). The tradition and conservatism of biologists resist the adoption of such a classification.

The theory and practice of systematics have experienced a revolution since the early 1960s. Techniques and concepts have been proposed, discussed, modified, or discarded. Systematics remains an active arena of conceptual development, and there is an increasing urgency in its practice as human activities speed the destruction of biological diversity. The following topics provide a brief and incomplete introduction to systematics.

SYSTEMATIC ANALYSIS

Systematic research is a search for evolutionary patterns. Investigations are of two general types: populational and species patterns of relationships, and phylogenetic patterns. The first type concerns the recognition of species through the analysis and definition of variation within and among populations and/or closely related species. The second type is directed at the resolution of genealogical relationships among species, genera, and higher taxonomic groups. Although they represent the two poles of a continuum, most systematic studies fit easily into one of these two categories.

Species limits and relationships are discerned by examining individuals, and any attribute of an organism can serve as a character. Characters provide a means

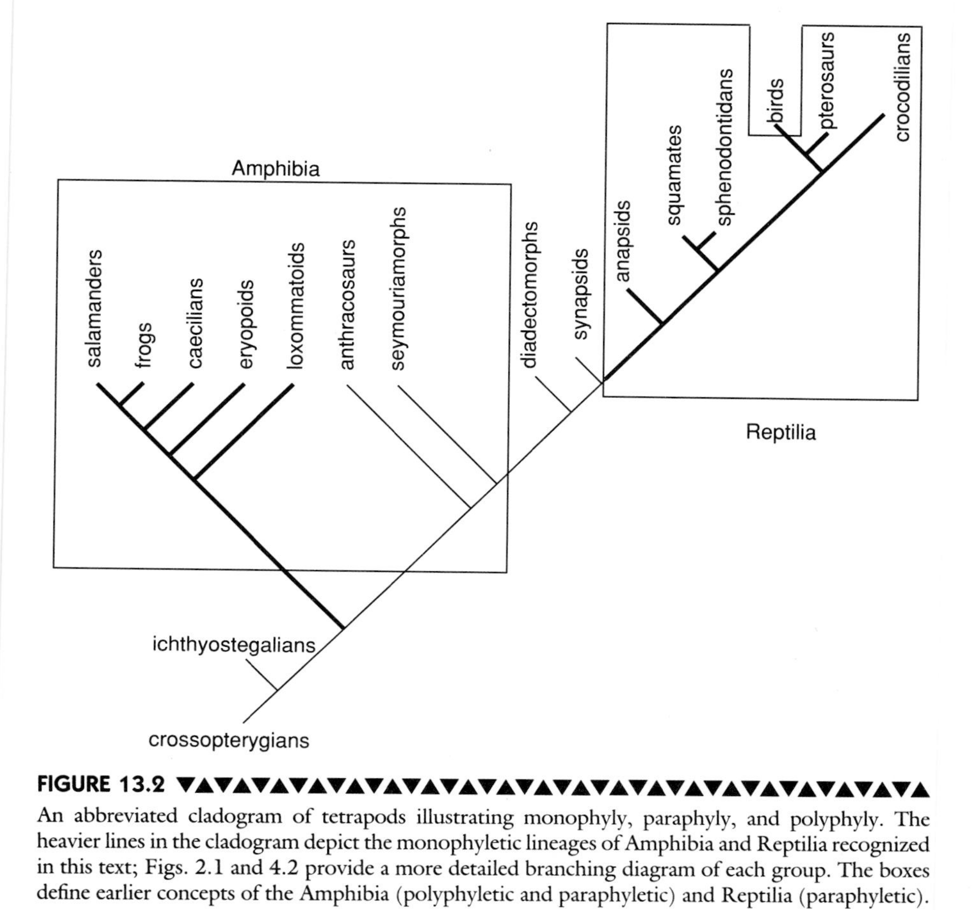

FIGURE 13.2 ▼▲
An abbreviated cladogram of tetrapods illustrating monophyly, paraphyly, and polyphyly. The heavier lines in the cladogram depict the monophyletic lineages of Amphibia and Reptilia recognized in this text; Figs. 2.1 and 4.2 provide a more detailed branching diagram of each group. The boxes define earlier concepts of the Amphibia (polyphyletic and paraphyletic) and Reptilia (paraphyletic).

to infer the affinities of one individual (or larger group) to another and to recognize species and other taxa (singular, taxon). These inferences permit our examination of evolutionary processes and the diversity arising therefrom. This diversity occurs at many levels, from the variety of genotypes (individuals) within a deme to the number of species within a genus (or higher group) or within a habitat or geographical area. Only through the recognition of which group of individuals is a species and which ones are not are we able to address other biological questions.

Types of Characters

Characters can be anatomical (e.g., a process on a bone, number of scales around midbody, snout–vent length), physiological (resting metabolic rate,

TABLE 13.1 ▼▲▼▲▼▲▼▲▼▲▼▲▼▲▼▲▼▲▼
Strict Cladistic Classification of Vertebrates

Subphylum Vertebrata
  Infraphylum Petromyzontia
  Infraphylum Gnathostomata
    Superclass Chrondrichthyes
    Superclass Teleostomi
      Class Actinistia
          Order Coelacanthiformes
      Class Euosteichthyes
        Subclass Actinopterygii
            Superdivision Halecomorphi
            Superdivision Teleostei
        Subclass Sarcopterygii
          Infraclass Dipnoi
          Infraclass Tetrapoda
            Superdivision Amphibia
              Division Lissamphibia
            Superdivision Anthracosauria
              Division Batrachosauria
                Subdivision Cotylosauria
                  Supercohort Amniota
                    Cohort Synapsida
                    Cohort Reptilia

[After Wiley (1979); abbreviated and some alternate taxonomic names used to conform with Table 3.1.]

Note: Each monophyletic group is named, hence lineages arising later in time are in a lower category.

thyroxine-sensitive metamorphosis), biochemical-molecular (composition of venom, DNA sequence), behavioral (courtship head-bobbing sequence), or ecological (aquatic). The type of characters used depends on the sort of questions being asked. Only a sampling of characters can be presented here.

Systematic studies involve the comparison of two or more samples of organisms through their characters. This comparison has led to the development of two procedural concepts: the operational taxonomic unit (OTU) and character states. OTUs are the units being compared and can be single individuals, populations, species, or higher taxa. The actual conditions of a character are its states, for example, eye iris—blue or green—or body length—25 or 50 mm. The assumption of homology is implicit in the comparison of character states, that is, all states of a character derive from the same ancestral state. The preceding characters and their states also illustrate that characters are either qualitative (descriptive) or quantitative (numeric). Qualitative characters have discrete states (e.g., either/or states: either feet webbed or not webbed; either 17, 18, 19, or 20 scales around midbody); quantitative ones have continuous states (e.g., skull length of the same specimen can be recorded as 2, 2.3, 2.34, or 2.339 cm).

To be useful for systematics, a character's states must have low variation within samples yet vary among samples. A character with a single state (invariant condition) in all OTUs yields no systematic information. A highly variable character with numerous states in each subsample adds confusion to an analysis and should be examined more closely to identify the cause of the high variability or be excluded from the study.

Knowledge of the sex and state of maturity of each specimen is critical for recognition of variation between females and males, and among ontogenetic stages. Both aspects must be considered whether the characters are anatomical, behavioral, or molecular in order to avoid confounding intraspecific variation with variation at an interspecific or higher level.

Morphology

Anatomical characters include three classes of characters: (1) mensural or morphometric characters are measurements or numeric derivatives thereof (e.g., ratios, regression residuals) that convey information on size and shape of a structure or anatomical complex; (2) meristic characters are those anatomical features that can be counted, such as number of dorsal scale rows or toes on the hindfoot; and (3) qualitative characters describe appearance, for example, presence and absence of a structure, its color, location, or shape.

1. The most common mensural character in herpetology is snout–vent length (SVL). This measurement gives the overall body size of all amphibians, lepidosaurs, and crocodilians, and differs only slightly from group to group depending on the orientation of the vent (transverse or longitudinal). Because of their shells, carapace length (CL) and plastral length (PL) are the standard body size measurements in turtles. Numerous other measurements are possible and dozens of different ones have been employed to characterize differences in size and shape. Mensural characters are not confined to aspects of external morphology but are equally as useful in the quantification of features of internal anatomy, for example, skeletal, visceral, or muscular.

As in all characters, the utility of measurements depends on the care and accuracy with which they are taken. Consistency is of utmost importance, so each measurement must be defined precisely and each act of measuring done identically from specimen to specimen. The quality of the specimen and the nature of the measurement also affect the accuracy of the measurement. Length (SVL) of the same specimen differs whether it is alive (struggling or relaxed) or preserved (shrunk by preservative; positioned properly or not), thus the researcher may wish to avoid mixing data from such specimens. Similarly, a skeletal measurement will usually be more accurate than a visceral one because soft tissue can be compressed when measured and the end points are often not as sharply defined. Differences can also occur when different researchers measure the same characters on the same set of animals. Thus within a sample, variation of each character includes the "natural" differences between individuals and the researcher's measurement "error." The latter is not serious and is likely encompassed within the natural variation if the researcher practiced a modicum of care during data taking. The use of adequate samples (usually >20 individuals) and central tendency statistics subsumes this "error" into the character's variation and further offers the opportunity to assess the differences among samples and to test the significance of the differences, as well as providing single, summary values for each character.

2. Meristic characters are discontinuous (= discrete). Each character has two or more states and the states do not grade into one another. There are 2, 3, or 4 teeth on the premaxillary bone, not 2.5 or 3.75 teeth. Meristic characters encompass any anatomical feature (external and internal) that can be counted. The possibilities seem endless. Researcher measurement error is also possible with meristic characters but likely less so. These characters can also be examined and summarized by basic statistical analysis.

3. Qualitative characters encompass a broad range of external and internal features, but unlike mensural and meristic characters, they are categorized in descriptive classes. Often a single word or phrase is adequate to distinguish among the various discontinuous states, for example, pupil vertical or horizontal, coronoid process broad and truncate or narrow and attenuate, carotid foramen in occipital or in quadrate. Qualitative characters may have multiple states (>2), not just binary states. Even though these characters are not mensural or meristic, they can be coded as numeric, simply by the arbitrary assignment of numbers to the different states.

The preceding characters emphasized the gross aspect of anatomy, but microscopic aspects also provide characters. One of the more notable and widely used microscopic (cytological) aspects is karyotype or chromosome structure. The most basic level is the description of chromosome number and size: diploid (2N) or haploid (N) number of chromosomes, and number of macro- and microchromosomes. A slightly more detailed level identifies the location of the centromere (metacentric with centromere in center of chromosome, acrocentric with it near the end, and at the end in telocentric chromosomes) and the number of chromosomes of each type or the total number (NF, nombre fundamental) of chromosome arms (segments on each side of the centromere). Special staining techniques allow the researcher to recognize specific regions (bands) on chromosomes and to more accurately match homologous pairs of chromosomes within an individual and homologous pairs between individuals.

Molecular Structure

The preceding characters were largely visible to the unaided eye or with the assistance of a microscope. Chemical and molecular structures also offer suites of characters for systematic analyses. The nature of these characters may involve the actual structure of the compounds (e.g., chemical composition of the toxic skin secretions of poison-dart frogs or nucleotide sequence of DNA fragments) or comparative estimates of relative similarity of compounds (e.g., immunological assays).

The techniques from molecular biology have been widely and enthusiastically adopted by many systematists. Their use in systematics rests on the premise that a researcher can assess and compare the structure of genes among individuals, species, or higher taxa through the examination of the molecular structure of proteins and other compounds that are only a few steps removed from the gene. Molecular data have proved valuable, but like other characters, they have their own special problems in interpretation. They offer a different perspective, sometimes yield new insights, and, in many instances, permit us to answer questions that cannot be addressed with other kinds of characters. Importantly, whatever the nature of a character, the fundamental assumption is that the character being compared between two or more OTUs is homologous, and this requirement applies to molecular characters as well as gross anatomical ones.

The techniques of molecular systematics are varied and complex. Different techniques are selected to investigate different levels of relationships. Electrophoresis is especially good (and relatively inexpensive) for examining genetic relationships of individuals within and among populations and of populations within and among species. Other techniques may be used at this level of comparison, but often they are more effective in the examination of higher-level relationships, which represent older divergence and speciation. The following sections offer a glimpse at molecular systematic techniques.

Electrophoresis

Proteins are the major structural components and chemical regulators of cells. Their structure is a direct reflection of the DNA sequence of genes coding their formation—a gene's DNA sequence is transcribed to make mRNA, and mRNA is translated into a chain of amino acids, the protein. One group of proteins, the enzymes, are catalysts of the cell's chemical reactions, and commonly each enzyme functions in a single type of reaction. This high specificity makes them critical and key components in cellular metabolism and potentially useful systematic characters. This potential is further enhanced by the occurrence of structurally different forms (allozymes) of the same enzyme. The different allozymes arise from gene mutations that have altered the structure of the DNA, and this alteration is translated directly into an altered structure of the enzyme.

Electrophoresis can identify different allozymes (also other chemical compounds) via their differential mobility in an electrical field. Each enzyme is a chain of amino acids and has a specific size, shape, and net charge (positive, negative, or neutral). Mobility is a function of the allozyme's charge, shape, and size; minor mutations may alter one or more of these facets of allozymic structure and affect the allozyme's speed of migration.

In biological studies, the electrophoretic apparatus consists of an electric-power pack, positive and negative electrodes to a buffer tray, and a sheet of gel matrix stretching between the positive- and negative-charged buffer trays. Tissue samples (fluid homogenates, often of muscle or liver tissue) are placed in a row at one end of the gel, and when the power is turned on, the allozymes begin to migrate toward the positive or negative electrodes. After several hours, the gel is removed and stained (specific biological dyes for each type of enzyme) to reveal the position of the allozymes for each sample (Fig. 13.3). Most current studies assay for 15–30 enzymes.

Each enzyme is a character, and its allozymes are its states. The stained gel (zymogram) provides the raw data and shows whether each sample shares the same allozymes. Electrophoretic data are variously coded and can be simple counts of allozyme matches and mismatches, frequencies, or presence/absence of heterozygotes. The manner of coding and analysis depend on the problem addressed, whether examining intrapopulational or intergeneric relationships.

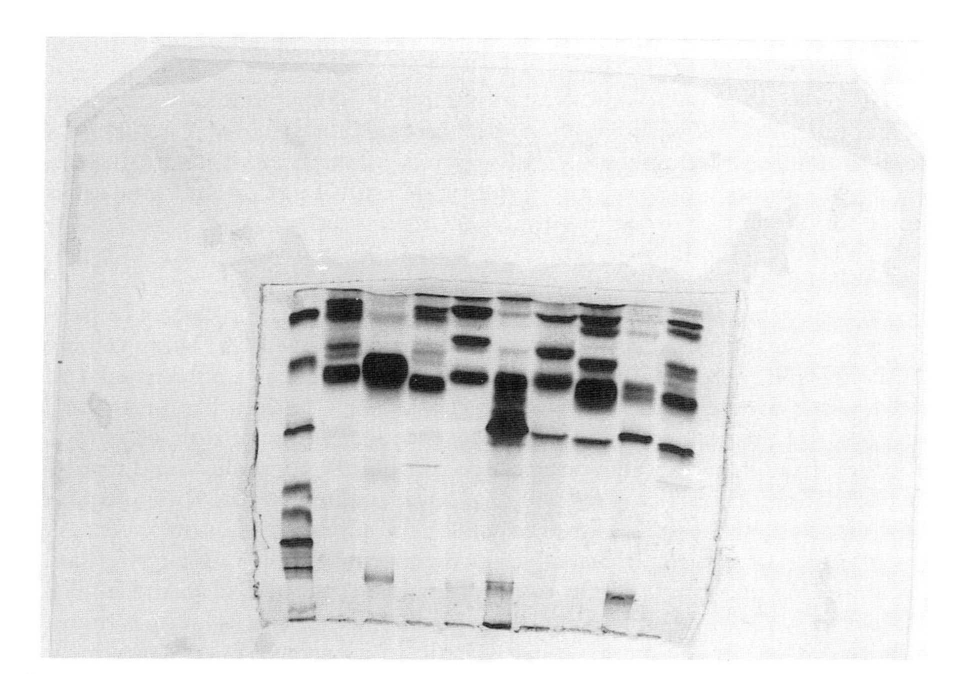

FIGURE 13.3 Variation in the proteins of cloacal (scent) gland secretions from pythons (left) and boas (right). Each column on the polyacrylamide gel represents a single individual and species; the column on the far left shows the molecular weight standards. Gel stained with Coomassie blue. (Courtesy of P. Weldon and T. Leto.)

Immunology

The antigen–antibody or immunological reaction provides a mechanism for estimating the genetic affinities of species. The concept is simple. Homologous proteins of closely related species are structurally similar, and as relationships become distant (increasing divergence from a common ancestor), protein structure becomes increasingly different. When a foreign protein (the antigen) is introduced into a host animal, the host's normal reaction is to produce antibodies, specifically constructed to intercept and deactivate the antigen. By using these antibodies to test the level of the immunological reaction with the antigens of many different species, the researcher obtains an estimate of the similarity of each test antigen to the antigen of the donor OTU. A strong reaction indicates a high similarity in protein structure, a weak reaction a low similarity. The antibody "recognizes" (attaches to) specific amino acid sequences of the donor's antigen, and fewer and fewer sequences are recognized as the structural differences of the test antigens increase. The basic protocol requires the introduction of the antigen into a host animal (rabbit, goat, etc.), time for the host's immunological system to produce antibodies, removal of blood serum from host and purification of antiserum, and performing in vitro comparisons of antisera reactions of the antigens from a series of test OTUs.

Of the several immunological tests, immunodiffusion, immunoelectrophoresis, and micro-complement fixation (MCF) are used in systematic studies, and blood albumins are the usual proteins compared. All three tests translate the level of antigen–antibody reaction into a numerical estimate of protein similarities and hence the relative similarity of the OTUs.

Nucleic Acids

Molecular biology is providing systematics with a growing arsenal of techniques for assessing relationships. None appears more powerful than the ability to examine and compare the structure of DNA and RNA. The attractiveness of nucleic acids for inferring phylogenetic relationships is that their nucleotide sequences are the basic informational units encoding and regulating all of life's processes. Examination of nucleic acid structure among amphibians and reptiles became a common practice in the 1980s as advances in methodology and equipment made the techniques more accessible and affordable to systematists. A major feature of nucleic acid analyses is their broad comparative power and spectrum, ranging from the ability to examine and identify individual and familial affinities (e.g., DNA fingerprinting) through tracing matriarchal lineages (mitochondrial DNA) to estimating phylogenetic relationships among species or orders (nuclear DNA). While extremely valuable for systematic studies, nucleic-acid characters are not a panacea and have their own set of difficulties in analysis and interpretation.

Numerous techniques are available for comparing nucleotide sequence among different taxa. DNA–DNA hybridization was the first technique used on a large scale in vertebrate systematics although not in herpetology. It takes advantage of disassociation of the two strands of DNA at high temperatures and the reassociation of complementary strands as the temperature drops. DNAs from two OTUs are combined, disassociated, and then allow to reassociate. Complementary strands from the same as well as different OTUs will reassociate. The number of mismatched base-pairs (nucleotides) in the hybrid DNA molecules increases with evolutionary time of divergence, and this is also reflected in a depression of the melting or disassociation temperature. The difference in melting temperature between pure and hybrid DNA provides a measure for assessing the level of relationship.

More recently, the technology for directly determining the sequence of nucleotides (base-pairs) has become increasingly accessible and is generally preferred, because sequence data provide discrete character information rather than estimates of relative similarity between nucleic acids (e.g., as in DNA hybridization) or their products (immunological tests). Several sequencing protocols are available, and in all, it is necessary to select or target a specific segment of a particular nucleic acid owing to the enormous number of available sequences within the cell and its organelles. First, the nucleic acid to be examined is selected (e.g., mitochondrial DNA, ribosomal RNA) and then specific sequences within this molecule are targeted. Then the target sequence is amplified (e.g., cloning, polymerase chain reaction) to produce numerous copies of the sequence for each OTU being compared. The sequence copies are isolated and purified for sequencing. Sequence determination relies on site-specific cleavage of the target sequence into fragments of known nucleotide sequences and the separation and identification of these fragments by electrophoresis. The homologous sequences are then aligned and provide the data for analyzing the degree of relationship among the compared OTUs.

Methods of Analysis

The opportunities for analysis are as varied as the characters. Choice of analytical methods depends on the nature of the question(s) asked and should be made at the beginning of a systematic study, not after the data are collected. With the breadth of systematic studies ranging from investigations of intrapopulational variation to the relationships of genera or families, the need for a carefully designed research plan seems obvious.

Systematic research often begins when a biologist discovers a potentially new species, notes an anomalous distribution pattern of a species or a character complex, or is simply curious about the evolution of a particular group of organisms. Having formulated a research objective, a preliminary study will explore the adequacy of the characters and data collection and analysis protocols for solving the research question. A wide array of analytical techniques are available to evaluate the adequacy of the characters and protocols and then to examine the relationships among the characters and among the samples and OTUs. A small set of the available analytical techniques and their uses follows. These techniques divide into two groups, numeric and phylogenetic. Numeric analyses offer a wide choice of methods to describe and compare the variation of OTUs and/or their similiarity to one another. Phylogenetic analyses address common ancestory relationships of OTUs, specifically attempting to uncover the evolutionary divergence of taxa.

Numeric

Any study of variation will examine multiple characters and numerous samples. The resulting data cannot be presented en masse but should be summarized and condensed. Numeric analyses provide this service. The initial analysis examines single characters within each sample—univariate statistics. The next phase compares individual characters within subsamples (e.g., females to males), the relationship of characters to one another within samples, and character states of one sample to those of another sample—bivariate statistics. The final phase usually is the comparison of multiple characters within or among samples—multivariate analysis. Each phase (Table 13.2) yields a different level of data reduction and asks different questions of the data, for example, what is (1) the variability of each character, (2) the difference in the means and variance between sexes or among samples, and (3) the covariance of characters within and among samples.

Even the briefest species description requires univariate statistics. A new species is seldom described from a single specimen, so univariate analysis shows the variation of each character within the sample and provides an estimate of the actual variation within the species. Means, minima and maxima, and standard deviations are the usual statistics presented. An in-depth study of a group of species typically uses univariate and bivariate statistics to examine the variation within each species and one or more multivariate techniques to examine the variation of characters among the species and the similarities of species to one another.

Multivariate analysis (Table 13.2) has become increasingly important in the analysis of systematic data, particularly mensural and meristic data sets. They permit the researcher to examine all characters and all OTUs simultaneously and to identify patterns of variation and association within the characters, and/or similarities of OTUs within and between samples. Principal component analysis is often used in an exploratory manner to recognize sets of characters with maximum discriminatory potential and preliminary OTU groups. These observations may then be used in a discriminant function analysis to test the reliability of the OTU groupings. Because these statistical techniques are readily available on microcomputers, there has been a tendency to use them without an awareness of their limitations and mathematical assumptions. Users should be aware that mixing of meristic and mensural characters, using differently scaled mensural

TABLE 13.2 ▼▲▼
Examples and Definitions of Numeric Analytical Tools

Univariate

Frequency distributions. Presentation techniques to show frequency of occurrence of different data classes or character states. They (e.g., frequency tables, histograms, pie charts) permit easy visual inspection of the data to determine normality of distribution, range of variation, single or multiple composition, etc.

Central tendency statistics. Data reduction to reveal midpoint of sample for each character and variation around the midpoint. Mean (average value), mode (most frequent value), and median (value in middle of ranked values); variance, standard deviation, standard errors (numeric estimates of sample's relative deviation from mean); kurtosis and skewness (numeric estimates of the shape of a sample's distribution).

Bivariate

Ratios and proportions. Simple comparisons (A : B, % = B/A × 100) of the state of one character to that of another character in the same specimen.

Regression and correlation. Numeric descriptions (equation and value, respectively) of the linear relationship and association of one character set to another.

Tests of similarities between samples. A variety of statistical models (χ^2, Students' t, ANOVA/analysis of variance) test the similarity of the data between samples.

Nonparametric statistics. Statistical models containing no implicit assumption of particular form of data distribution. All other statistics in this table are parametric ones and most assume a normal distribution.

Multivariate

Principal components analysis/PCA. Manipulation of original characters to produce new uncorrelated composite variables/characters ordered by decreasing variance.

Canonical correlation. Comparison of the correlation between the linear functions of two exclusive sets of characters from the same sample.

Discriminant function analysis/DFA. Data manipulation to identify a set of characters and assign weights (functions) to each character within the set in order to separate previously established groups within the sample.

Cluster analysis. A variety of algorithms for the groupings of OTUs on the basis of pairwise measures of distance or similarity.

[In part, modified from James and McCullough (1985, 1990).]

characters, or comparing data sets of unequal variance, may yield meaningless results.

Clustering analysis is another multivariate technique but is not strictly statistical in the sense of being inferential or predictive. The numerous clustering algorithms use distance or similarity matrices and create a branching diagram or dendrogram. These matrices derive from a pairwise comparison of each OTU for every character to every other OTU in the sample (Fig. 13.4). The raw data in an OTU × Character matrix are converted to an OTU × OTU matrix in which each matrix cell contains a distance or similarity value between each pair of OTUs. The clustering algorithm uses these values to link similar OTUs and OTU groups to one another, proceeding from the most similar to the least similar.

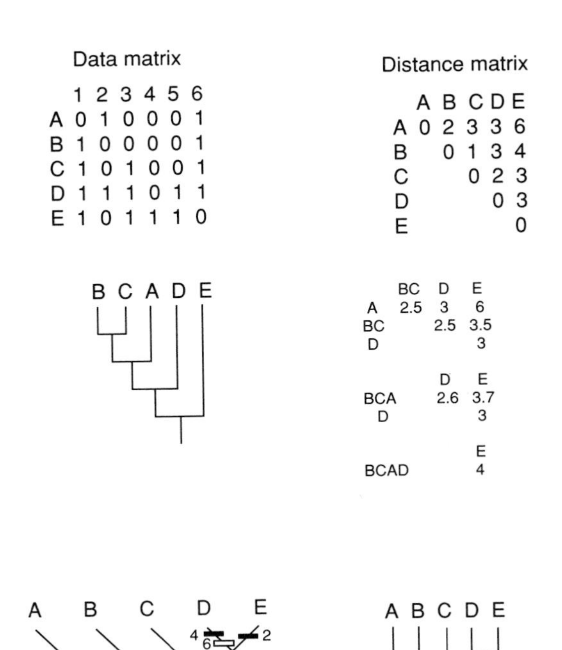

FIGURE 13.4 

Construction of branching diagrams by two methods: phenetics and cladistics. The OTU × Character matrix (upper left) contains five OTUs (A–E) and six characters (1–6). Each character has two states, 0 or 1 (e.g., absent or present, small or large, etc.). Pairwise comparison of OTUs creates an OTU × OTU matrix. The distance values are the sums of the absolute difference between states for all six characters. Zeros fill the diagonal because each OTU is compared to itself; only half of the matrix is filled with the results of a single analysis because the two halves are mirror images of one another. An unweighted pair-group method (UPGM) clustering protocol produces a phenetic dendrogram (phenogram, middle left); in UPGM, the most similar OTUs are linked sequentially with a recalculation (middle right) of the OTU × OTU matrix after each linkage. The cladogram (lower left) derives directly from the OTU × Character matrix. The solid bars denote a share-derived character state, the open bars an evolutionarily reversed state, and the character numbers. For comparison with the UPGM phenogram, the cladogram is converted to a stylistically identical dendrogram (lower right).

Phylogenetic

The preceding numeric techniques do not provide estimates of phylogenetic relationships, but rather summarize the level of similarity. Overall similarity has been argued as an estimate of phylogenetic relationship. This concept was the basic tenet of the phenetic school of systematics, which came into prominence in the late 1950s and then was replaced rapidly by phylogenetic systematics. Phenetics as a classification method has largely disappeared (although many of its analytical algorithms remain), because its basic premise of "similarity equals genealogical relationship" is demonstrably false in many instances, and the resulting classifications do not reflect accurately the evolutionary history of the organisms being studied. Another basic premise is that large character sets produce more robust and stable classifications; unfortunately, the addition of more characters usually changes the position of OTUs on the dendrogram and produces a dissimilar classification. This instability of OTU clustering arises from the use of unweighted characters and the swamping of useful characters by ancestral and nonhomologous ones.

Phylogenetic analysis has been variously practiced since the publication of Charles Darwins's *Origin of Species*. However, in the mid-1960s, coincidental with the publication of the English language edition of Willi Hennig's *Phylogenetic Systematics,* systematists began more rigorous and explicit character analyses and reconstructions of phylogenies (taxa genealogies). This approach gives repeatability to systematic practices and is broadly known as cladistics. The basic tenets of phylogenetic systematics are: (1) only shared-derived similarities are useful in deducing phylogenetic relationships; (2) speciation produces two sister species; and (3) speciation is recognizable only if the divergence of two populations is accompanied by the origin of a derived character state.

Character analysis plays a major role in phylogeny reconstruction, because it is necessary to determine the ancestral or derived status for each character state. A special terminology is associated with character state polarity: plesiomorphic, the same state as in the ancestral taxon; apomorphic, a derived or modified state from the ancestral condition; autoapomorphic, a derived state occurring in a single descendant taxon; and synapomorphic, a shared-derived state in two or more taxa. Sister-groups are taxa of equivalent rank uniquely sharing the same ancestor; synapomorphic characters identify sister-groups.

Determination of character state polarity can use one or more protocols. Outgroup analysis is generally considered the most reliable method. Operationally, the researcher identifies a candidate sister-group(s) (out-group) of the group being studied (in-group) and then examines the distribution of character states for each character in these two groups. If a state occurs only in the in-group (but not necessarily in all members of the group), it is hypothesized to be apomorphic; and if present in both in- and out-groups, it is considered plesiomorphic. Ontogenetic analysis examines the condition of a state relative to its appearance in a developmental sequence. Earlier-appearing states are hypothesized to be plesiomorphic and later ones to be apomorphic. These two analyses are complementary, nonoverlapping methods and can be used to test the conclusions of each other, although the ontogenetic method is argued to be less reliable because earlier developmental appearance does not always match ancient evolutionary appearance (primitive). Commonality and geological precedence are supplementary methodologies and are generally less reliable. In the former, widespread or common occurrence is considered the plesiomorphic state, and in the latter, occurrence in taxa from older or lower strata is assumed to be plesiomorphic.

Once the characters have been polarized, the researcher can construct a cladogram by examining the distribution of apomorphic states. Numerous computer algorithms are available for the evaluation of character state distributions, but the sequential linkage of sister-groups can be done by hand with small samples of OTUs and characters. Using the OTU × Character matrix in Fig. 13.4 and identifying all "1" states as apomorphic, the steps are: D and E are sister taxa, synapomorphic for character 5; C and D-E are sister-groups, synapomorphic for character 3; B and C-D-E are sister-groups, synapomorphic for character 1; and A and B-C-D are sister-groups, synapomorphic for character 6. Taxon E shows the plesiomorphic state for character 6, which might suggest that E is not a member of the ABCD lineage; however, it does share three other apomorphic characters, and the most parsimonious assumption is that character 6 underwent an evolutionary reversal in E. Similarly, the most parsimonious assumption for the synapomorphy of character 2 in taxa A and D is convergent evolution. Shared character states of independent origin are nonhomologous or homoplasic.

NOMENCLATURE

Another important aspect of systematics is the assignment and use of taxonomic names—nomenclature. All biologists must correctly identify the animal or plant studied and then must use the correct taxonomic names in reporting the results of their study. Failure to provide the correct scientific name and higher taxonomic assignments prevents other biologists from recognizing that the results may be important to them or cause others to compare the incorrectly labeled results with their own results derived from different organisms.

Brief History

Humans have classified and named plants and animals since language first appeared. However, our formal system of animal classification dates from Linnaeus's tenth edition of *Systema Naturae* in 1758. This catalogue gave a concise diagnosis of the known species of plants and animals and arranged them in a hierarchical classification of genus, order, and class. Importantly, this edition of Linnaeus's catalogue was the first publication to use consistently a two-part name (a binomial of genus and species), hence the reason for selecting it as the beginning of zoological classification. Scientific names of plants and animals remain binomials and in Latin (the language of scholars in the eighteenth century).

While Linnaeus proposed a set of rules for naming plants, he offered none for animals, except by example. Rules or not, later biologists began to describe and classify as they wished. With many groups and countries following their own rules of nomenclature, multiple names for the same species and higher taxa were adopted by different groups of naturalists. Classifications were becoming idiosyncratic and unintelligible to outsiders, even to those studying the same group of animals. To avoid the impending chaos, the botanical and zoological communities separately developed codes for the practice of nomenclature. The most recent code for zoologists is the *International Code of Zoological Nomenclature, Third Edition*. It is a bilingual edition (French and English) and now is being revised.

Rules and Practice

The Code is a legal document for the practice of classification, specifically for the selection and assignment of names to animals from species through family groups. Unlike our civil law, there are no enforcement officers. Enforcement of the Code occurs through the biological community's acceptance of a scholar's nomenclatural decisions. If the rules and recommendations are followed, the scholar's decisions are accepted; if the rules are not followed, the decisions are invalid and not accepted by the community. Where an interpretation of the Code is unclear or a scholar's decision uncertain relative to the Code, the matter is presented to the International Commission for Zoological Nomenclature (a panel of systematic zoologists), who, like the U.S. Supreme Court, provide an interpretation of the Code and accept or reject the decision, thereby establishing a precedent for similar cases in the future. The Code has six major tenets:

1. All animals extant or extinct are classified identically, using the same rules, classificatory hierarchies, and names where applicable. This practice avoids dual and conflicting terminologies for living species that may have a fossil record. Further, extant and fossil taxa share evolutionary histories and are properly classified together.

2. Although the Code applies only to the naming of taxa at the family-group rank and below, all classificatory ranks have latinized formal names. All except the species and subspecies epithets are capitalized when used formally; these latter two are never capitalized. For example, the major rank or category names (phylum, class, order, family, genus, species) for the green iguana of

Central America are Chordata, Euosteichthyes, Squamata, Iguanidae, *Iguana iguana*. The names may derive from any language, although the word must be transliterated into the Roman alphabet and then converted to a Latin form.

3. To ensure that a name will be associated correctly with a taxon, a type is designated—type genus for a family, type species for a genus, and a type specimen for a species. Such designations permit other systematists to confirm that what they are calling taxon X matches what the original author recognized as taxon X. Comparison of specimens to the type is critical in determining the specific identity of a population. Although the designation of a single specimen to represent a species is typological, a single specimen as the name-bearer unequivocally links a particular name to a single population.

Of these three levels of types, only the type of the species is an actual specimen, nonetheless this specimen serves conceptually and physically to delimit the genus and family. This results because a family is linked to a single genus by the designation of a type genus, which in turn is linked to a single species by a type species, hence to the type specimen of a particular species. The characterization at each level, thus, includes traits possessed or potentially possessed by the type specimen. An example of such a nomenclatural chain is: *Xantusia* Baird, 1859 is the type genus of the family Xantusiidae Baird, 1859; *Xantusia vigilis* Baird, 1859 is the type species of the genus *Xantusia;* and three specimens USNM 3063 (in the United States National Museum of Natural History) are syntypes of the species *Xantusia vigilis*.

Several kinds of types are recognized by the Code. The holotype is the single specimen designated as the name-bearer in the original description of the new species or subspecies, or the single specimen on which a taxon was based when no type was designated. In many nineteenth-century descriptions, several specimens were designated as a type series; these specimens are syntypes. (The more recent Codes do not approve of the designation of syntypes.) Often syntypic series contain individuals of more than one species, and sometimes to avoid confusion, a single specimen, a lectotype, is selected from the syntypic series. If the holotype or syntypes are lost or destroyed, a new specimen (neotype) can be designated as the name-bearer for the species. Other types (paratypes, topotypes, etc.) are used in taxonomic publications; however, they have no official status under the Code.

4. Only one name may be used for each species. Yet commonly, a species has been recognized and described independently by different authors at different times. These multiple names for the same animal are known as synonyms and arise because different life-history stages, geographically distant populations, or males and females appear differently, or because an author is unaware of another author's publication. Whatever the reason, the use of multiple names for the same animal would cause confusion, hence only one name is correct.

Systematists have selected the simplest way to determine which of many names is correct, namely, by using the oldest name that was published in accordance with rules of the Code. The concept of the first published name being the correct name is know as the "Principle of Priority." The oldest name is the primary (senior) synonym and all names published subsequently are secondary (junior) synonyms (Table 13.3). Although simple in concept, implementation of the Principle may not promote stability, especially so when the oldest name of

TABLE 13.3
Abbreviated Synonymies of the European Viperine Snake and the Cosmopolitan Green Seaturtle

***Natrix maura* (Linnaeus)**

1758	*Coluber maurus* Linnaeus, Syst. Nat., ed. 10, 1:219. Type locality, Algeria. [original description; primary synonym]
1802	*Coluber viperinus* Sonnini & Latreille, Hist. nat. Rept. 4:47, fig. 4. Type locality, France. [description of French population, considered to be distinct from Algerian population]
1824	*Natrix cherseoides* Wagler in Spix, Serp. brasil. Spec. nov. :29, fig. 1. Type locality, Brazil. [geographically mislabeled specimen mistaken for a new species]
1840	*Coluber terstriatus* Duméril in Bonaparte, Mem. Accad. Sci. Torino, Sci. fis. mat. (2) 1:437. Type locality, Yugoslavia. Nomen nudum. [= naked name; name proposed without a description so *terstriatus* is not available]
1840	*Natrix viperina* var. *bilineata* Bonaparte, Op. cit. (2) 1:437. Type locality, Yugoslavia. Non *Coluber bilineata* Bibron & Bory 1833, non *Tropidonotus viperinus* var. *bilineata* Jan 1863, non *Tropidonus natrix* var. *bilineata* Jan 1864. [recognition of a distinct population of *viperina;* potential homonyms listed to avoid confusion of Bonaparte's description with other description using *bilineata* as a species epithet]
1929	*Natrix maura,* Lindholm, Zool. Anz. 81:81. [first use of current combination]

***Chelonia mydas* (Linnaeus)**

1758	*Testudo mydas* Linnaeus, Syst. Nat., ed. 10, 1:197. Type locality, Ascension Island. [original description; primary synonym]
1782	*Tesudo macropus* Wallbaum, Chelonogr. :112. Type locality, not stated. Nomen nudum.
1788	*Testudo marina vulgaris* Lacépedè, Hist. nat. Quadrup. ovip. 1: Synops. method., 54. Substitute name for *Testudo mydas* Linnaeus.
1798	*T. Mydas minor* Suckow, Anfangsg. theor. Naturg. Thiere. 3, Amphibien :30. Type locality, not stated. Nomen oblitum, nomen dubium. [forgotten name, not used for many years then rediscovered; name of uncertain attribution, tentatively assign to *mydas*]
1812	*Chelonia mydas,* Schweigger, Königsber. Arch. Naturgesch. Math. 1:291. [present usage, but many variant combinations appeared after this]
1868	*Chelonia agassizii* Bocourt, Ann. Sci. nat., Paris 10:122. Type locality, Guatemala. [description of Pacific Guatemalan population as distinct species]
1962	*Chelonia mydas carrinegra* Caldwell, Los Angeles Co. Mus. Contrib. Sci. (61): 4. Type locality, Baja California. [description of Baja population as a subspecies]

[Modified from Mertens and Wermuth, (1960) and Catalogue of American Amphibians and Reptiles, respectively.]

Note: The general format of each synonym is: original date of publication, name as originally proposed, author, abbreviation of publication, volume number and first page of description, type locality. Explanations of the synonyms are presented in brackets.

a common species has been unknown for many decades and then is rediscovered. Should *viridisquamosa* Lacépéde, 1788 replace the widely used *kempii* Garman, 1880 for the widely known Kemp's ridley seaturtle *Lepidochelys kempii*? No! The goal of the Code is to promote stability of taxonomic names, so the Code has a 50-year rule that allows commonly used and widely known secondary synonyms to be conserved and the primary synonym suppressed. The difficulty with deviating from priority is deciding when a name is commonly used and widely known—the extremes are easy to recognize but the middle ground is broad. Under these circumstances, the case must be decided by the Commission.

In deciding whether one name should replace another name, a researcher determines whether a name is "available" prior to deciding which of the names is "valid." The concept of availability rests upon a taxonomic description of a new name obeying all the tenets of the Code in force at the time of the description. Some of the basic tenets are: published subsequent to 1758 (tenth edition of **Systema Naturae**), a binomial name for a species-group taxon, name in Roman alphabet, appearing in a permissible publication, and the description purports to differentiate the new taxon from existing ones. If the presentation of a new name meets these criteria and others, the name is available. Failure to meet even one of the criteria, such as publication in a mimeographed (not printed) newsletter, prevents the name from becoming available. Even if available, a name may not be valid. Only a single name is valid for each species, no matter how many other names are available. Usually, the valid name is the primary synonym. The valid name is the only one that should be used in scientific publications.

5. Just as for a species, only one name is valid for each genus or family. Further, a taxonomic name may be used only once for an animal taxon. A homonym (the same name for different animals) creates confusion and is also eliminated by the Principle of Priority. The oldest name is the senior homonym and the valid one. The same names (identical spelling) published subsequently are junior homonyms and invalid names. Two types of homonyms are possible. Primary homonyms are the same name published for the same taxon, for example, *Natrix viperina bilineata* Bonaparte, 1840 and *Tropidonotus viperina bilineata* Jan 1863. Secondary homonyms are the same names for different taxa, for example, the insect family Caeciliidae Kolbe, 1880 and the amphibian family Caeciliidae Gray, 1825.

6. When a revised Code is approved and published, its rules immediately replace those of the previous edition. This action could be revolutionary if the new code differed greatly from the preceding one, but most rules remain largely unchanged. Such stasis is not surprising, for the major goal of the code is to establish and maintain a stable nomenclature. Rules tested by long use and found functional are not discarded. Those with ambiguities are modified to clarify the meaning. Where a rule requires major alteration and the replacement rule results in an entirely different action, a qualifying statement is added so actions correctly executed under previous rules remain valid. For example, the first edition of the code required that a family-group name be replaced if the generic name on which it was based was a secondary synonym; the second and third edition do not require such a replacement, thus the latter two editions permit the retention of the replacement name proposed prior to 1960 if the replacement has won general acceptance by the systematic community. Such exceptions promote nomenclatural stability.

Naming a New Species

One product of many systematic studies is the recognition and description of species. Exploration of new areas and faunas, use of new techniques, and larger samples often reveal new patterns of continuity and discontinuity among populations and species. This new information may require the recognition of new species with the proposal of a new name or the resurrection of a junior synonym.

When a researcher recognizes that a population or set of populations represents a distinct species, the first task is to discover whether this species already has a name, possibly residing in the synonymy of another species. Good original descriptions are obviously important in this search, but the type specimens are even more important because they can be compared directly to the other specimens now recognized as representatives of a distinct species. These representatives need not match the type in every detail but must, in the researcher's opinion, all belong to the same species. Once the match is made the junior synonym of another species becomes the senior synonym of the newly rediscovered species, assuming that the synonym is available (in a nomenclatural sense).

If no name is available, the researcher will prepare a new species description. The format for this description may vary but should contain the following elements: the new name; type designation; diagnosis of the new species; description of the type; summary of the variation within the new species; discussion of relationships; etymology of the new name; distribution of new species; and list of specimens examined. The new name is, of course, a binomial consisting of a genus name and a species epithet. The epithet is the new part, and it must be totally new—never having been used previously within this genus. Name selection requires a thorough search of the literature to ensure that the new name is neither a synonym or homonym.

CHAPTER 14

Caecilians and Salamanders

GYMNOPHIONA

Caecilians are amphibian earthworm look-alikes. They have blunt, bullet-shaped heads attached to long cylindrical, limbless bodies. Their bodies end bluntly immediately behind the vent or with a very short tail. All caecilians are distinctly segmented by encircling primary grooves, and in some taxa, the primary annuli (segments formed by the primary grooves, usually one annulus for each vertebra) are further partitioned into secondary and even tertiary annuli by additional encircling grooves. Dermal (bony) scales may be present, lying deep within the tissue of the annular grooves.

The earthworm appearance derives from their

burrowing habits. The head is the digging tool, hence the skull is compact and tightly knit to withstand the pressures of compression-tunneling. The eyes have degenerated into small, dark-pigmented spots, lying beneath the skin and, in some cases, beneath skull bones. There are no ear openings, and the jaw is underhung to permit prey capture in narrow tunnels and yet out-of-the-way for tunneling. Prey are located and identified by the retractable sensory tentacles, one on each side of the head between the nostril and the eye. Caecilians move through their tunnels by a combination of serpentine and concertina locomotion. The limbs have been completely lost; not even a vestige of the pectoral or pelvic girdle remains in the body wall.

All caecilians have internal fertilization, which is accomplished by the male's eversible phallodeum (copulatory organ), a modified portion of the cloacal wall. Development may be internal or external, indirect or direct.

Extant caecilians are entirely tropical, occurring worldwide except for Madagascar and Papua–Australia. There are over 150 species, currently divided among six families (Table 1.1). Their biology is largely unknown, and what is known derives from observations made during capture and from captive or museum specimens.

Ringed caecilian *Siphonops annulatus* (Caeciliaidae). (K. Miyata)

Family Caeciliaidae

The Caeciliaidae is the largest family of caecilians, with nearly 90 species in 23 genera. They are widespread in tropical America, eastern and western Africa, the Seychelle islands, and India (Fig. 14.1). This speciose family is difficult to characterize precisely and probably contains two or more distinct evolutionary lineages. Two subfamilies (Caeciliainae, Siphonopinae) are recognized by some researchers, but each is likely paraphyletic. Our knowledge of intergeneric relationships is still too rudimentary to recognize and classify monophyletic groups within the Caeciliaidae.

All caeciliaids have distinct primary annuli. Some species have secondary grooves dividing the primary annuli, but none has tertiary grooves. Scales are present in the annular grooves of some genera and absent in others. The rear end of the body is capped with a terminal shield but lacks a true tail. True tails have caudal vertebrae and myomeres. Eyes may or may not be visible externally; in some genera, eyes lie in bony sockets beneath the skin; in others, they lie beneath bone. Tentacles are variously positioned; in some taxa, the tentacles are adjacent to the nostrils, whereas in others, they are near the eyes. The middle ear contains a columella.

Caeciliaid genera are either oviparous (e.g., *Grandisonia*, *Hypogeophis*) or viviparous (*Caecilia*, *Dermophis*). Some oviparous taxa lay eggs in or near water and have free-living larvae. Others (*Hypogeophis*, *Idiocranium*) display direct development and parental guarding of the eggs. Reproduction appears to be seasonal

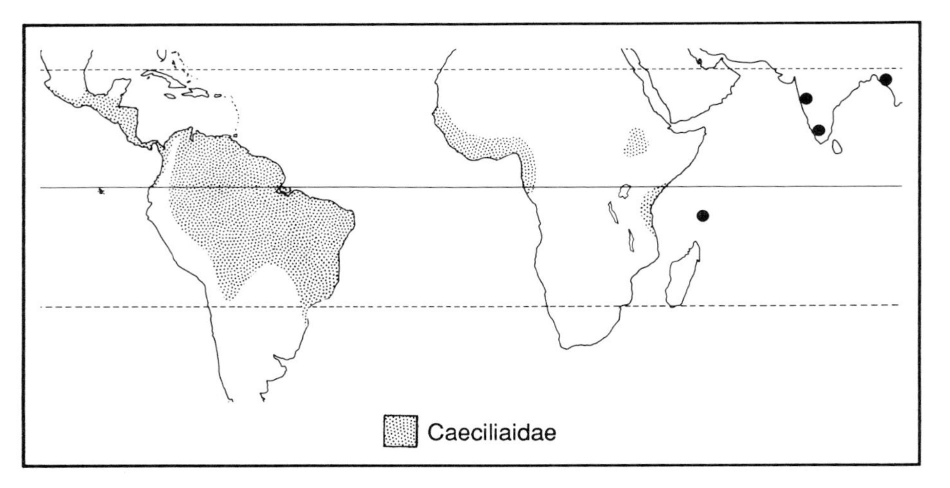

FIGURE 14.1 Distribution of the extant Caeciliaidae.

or near-continuous, depending largely on the climatic pattern of the area of occurrence.

Family Ichthyophiidae

The ichthyophiids are the principal caecilians of India and Southeast Asia (Fig. 14.2), consisting of two genera (*Caudacaecilia*, *Ichthyophis*) and about 36 species. They are strongly annulated with the primary annuli divided by secondary and tertiary grooves. Scales are present in the annular grooves, although occasionally absent from the anteriormost grooves. The body ends in a short but true tail (possessing caudal vertebrae and myomeres). Their eyes are visible externally and lie beneath the skin. Each tentacle lies between the eye and the nostril, commonly closer to the eye. The middle ear contains a columella.

Both genera are oviparous. In the few known examples, development is indirect. *Ichthyophis* lays eggs in its burrows near water. When the eggs hatch, the larvae exit the burrows and crawl overland to a nearby pond or stream.

Family Rhinatrematidae

The rhinatrematid caecilians are a group of two genera (*Epicrionops*, *Rhinatrema*; nine species) in northern South America (Fig. 14.2). All are strongly annulated with the primary annuli divided by secondary and tertiary grooves. Numerous scales are present in the annular grooves. The body ends in a short, true tail. The eyes are visible externally and lie beneath the skin in bony sockets. The tentacles are immediately anterior to the eyes. The middle ear contains a columella. Both genera are oviparous.

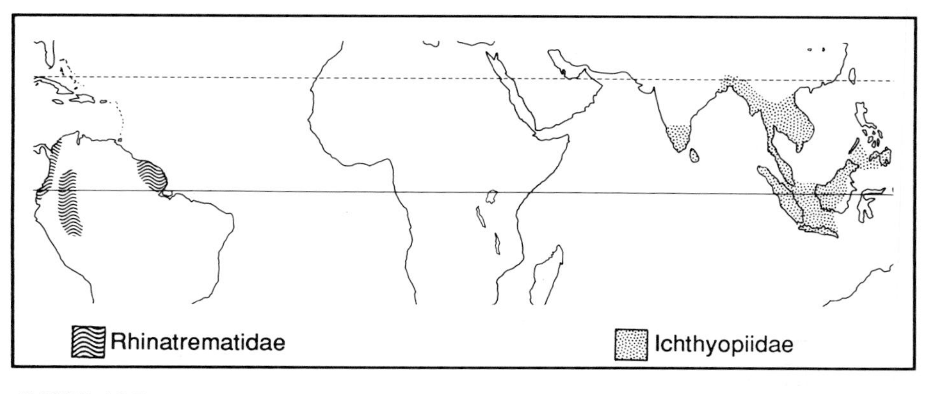

FIGURE 14.2 ▼▲
Distribution of the extant Ichthyopiidae and Rhinatrematidae.

Family Scolecomorphidae

Scolecomorphids consist of two genera (*Crotaphatrema*, *Scolecomorphus*; five species) from eastern and western equatorial Africa. (As yet, no caecilians have been found in inland Central Africa.) They possess only primary annuli, and only a few vestigial scales occur in the posteriormost annuli. They lack a true tail. Their eyes lie beneath the skull bones, and the tentacles are near the eyes. The middle ear lacks a columella. Of the two genera, one appears to have oviparous species and the other viviparous ones.

Family Typhlonectidae

The typhlonectids comprise four South American genera (*Chthonerpeton*, *Nectocaecilia*, *Potomotyphlus*, *Typhlonectes*; 12 species). They have only primary annuli; although in a few species, some primary annuli are partially dissected by false secondary grooves. They lack a true tail. Their eyes are always visible beneath the skin and never covered by bone. The sensory tentacles are small and usually closer to the nostrils than to the eyes. Columellae are present. All typhlonectids are viviparous. The embryos' gills fuse into saclike structures that serve as placentae for gas and perhaps waste exchange.

Potomotyphlus and *Typhlonectes* are strongly aquatic. Their bodies are laterally compressed and bear a middorsal fold or fin.

Family Uraeotyphlidae

This family contains a single genus, *Uraeotyphlus* (four species) in southern India (Fig. 14.3). *Uraeotyphlus* has primary annuli divided by secondary but not tertiary grooves, and the annular grooves do not completely encircle the body. Scales are present in the annular grooves. The body ends in a short, but true tail. The eyes are visible externally and lie beneath the skin. The tentacles are far forward beneath the nostrils. The middle ear contains a columella. All species are probably oviparous.

Phylogenetic Relationships of Caecilians

The caecilians are the least known of all living amphibians. Their predominantly subterranean existence has effectively hidden them from the attention of most biologists. They existed as seldom mentioned oddities, all lumped in the family Caeciliaidae until Edward H. Taylor began his survey of them in the 1960s. His "Monograph of Caecilians," published in 1968, called attention to their diversity and how little was known about them. Although they have become better known since then, their phylogenetic relationships still remain unclear.

The rhinatrematids appear to be the most primitive group and the caeciliaids and typhlonectids the most derived. Although the relationships (Fig. 14.4) are uncertain, the dendrogram is consistent in a number of features. Chromosome

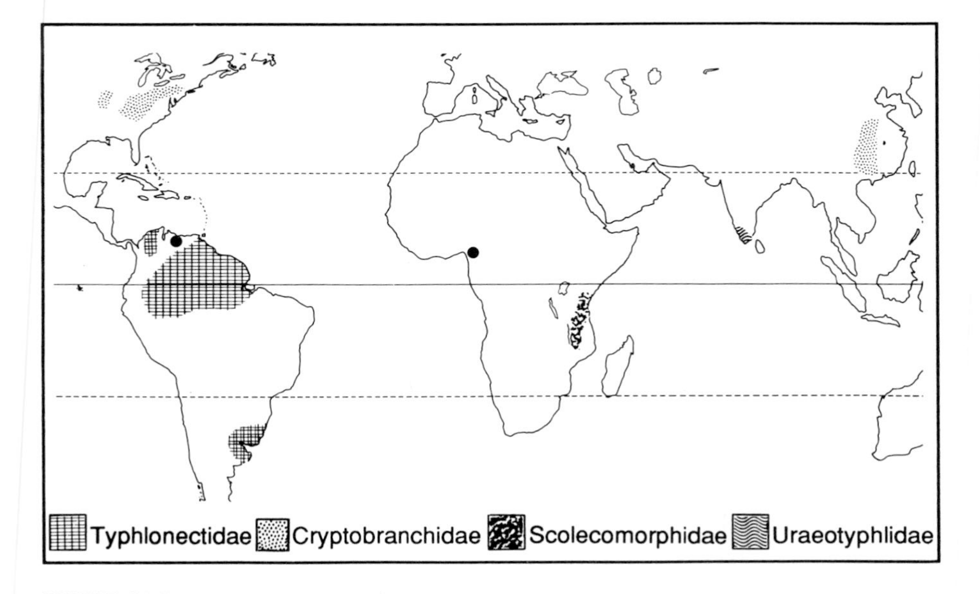

FIGURE 14.3 ▼▲
Distribution of the extant caecilian families Scolecomorphidae, Typhlonectidae, and Uraeotyphliidae, and the salamander family Cryptobranchidae.

number decreases from Ichthyophiidae (2N = 42) to Caeciliaidae (24–30) and Typhlonectidae (20). Scolecomorphidae contains oviparous and viviparous members; all taxa below the scolecomorphids are oviparous and all above are viviparous. Scolecomorphids, caeciliaids, and typhlonectids lack tails; the remaining genera have tails, albeit very short ones.

CAUDATA

Salamanders, as their group name states, are tailed amphibians. All have long tails, elongate cylindrical bodies, and distinct heads. Most have well-developed limbs, often short relative to body length, but two aquatic families have reduced their limbs. Their skulls are reduced by the loss of many elements; other elements remain cartilaginous or partly so.

Most have internal fertilization, although none has a copulatory organ. Internal fertilization occurs via a male-deposited spermatophore from which the female plucks a packet of spermatozoa with her cloacal lips. With the exception of a few species, development occurs externally, either indirectly via a larval stage or directly.

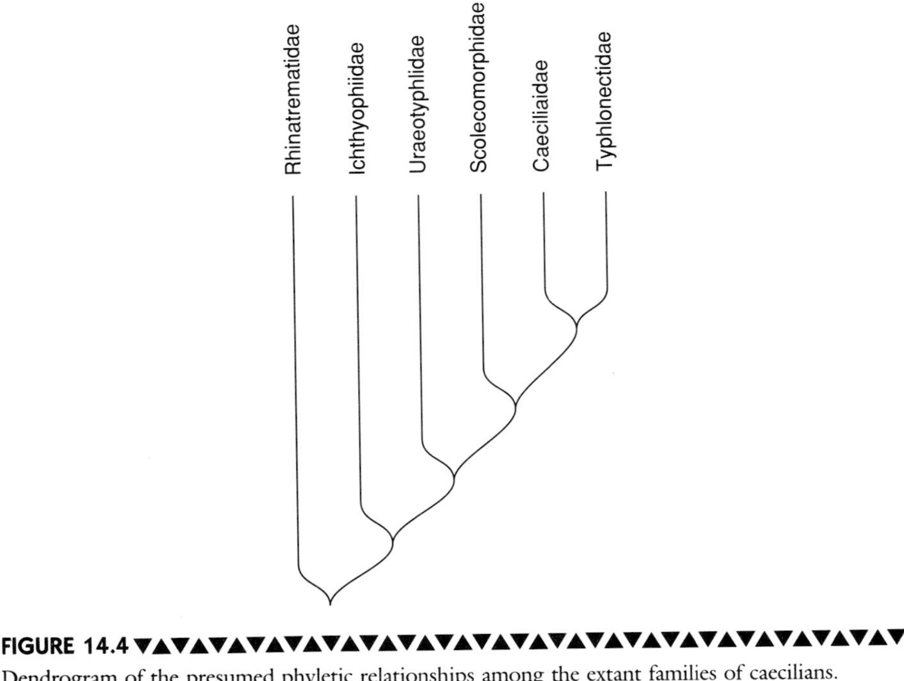

FIGURE 14.4 Dendrogram of the presumed phyletic relationships among the extant families of caecilians.

Salamanders are principally temperate-zone amphibians and occur throughout the forested areas of the cool and warm temperature parts of the Northern Hemisphere. One lineage has dispersed into tropical America, but most of these tropical species occur in cool mountain forests. Salamanders range from fully aquatic species to totally terrestrial or arboreal ones. There are about 390 extant species divided among eight or more families (see Table 1.1). These families group into the primitive salamanders (Cryptobranchoidea), the advanced salamanders (Salamandroidea), and the sirens (Meantes). All cryptobranchoids are limbed and have external fertilization. The salamandroids are fully limbed but have internal fertilization. The sirens are totally aquatic salamanders, lacking hindlimbs and a pelvic girdle. Their mode of fertilization is unknown.

Cryptobranchoidea

Family Cryptobranchidae

The cryptobranchid salamanders are large neotenic salamanders of eastern Asia and central and eastern North America (Fig. 14.3). Although aquatic, they lack gills as adults, retaining a single pair of gill slits (open in *Cryptobranchus*,

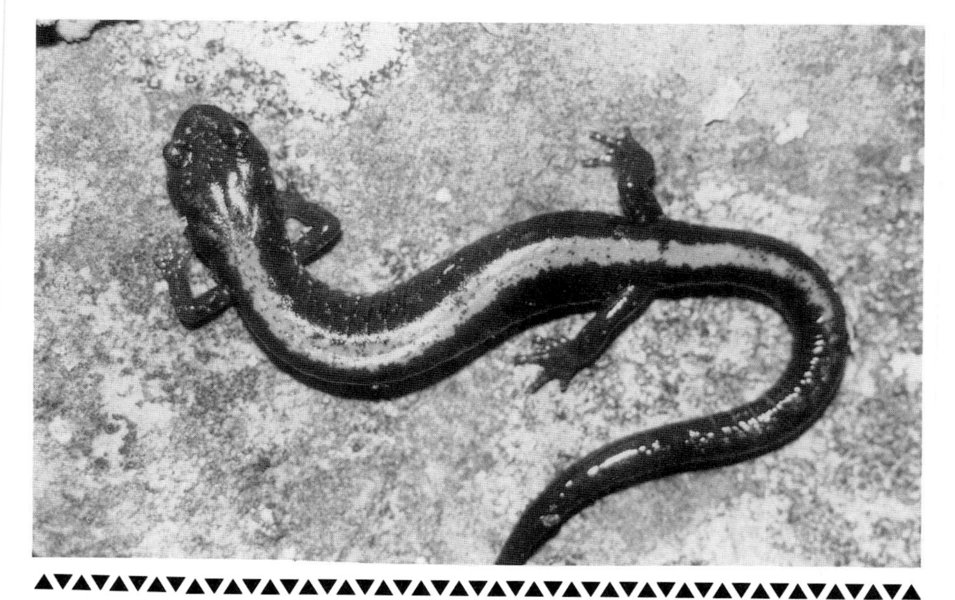

Shenandoah salamander *Plethodon shenadoah* (Plethodontidae). (G. Zug)

closed in *Andrias*). Their skin appears two sizes too large for them and lies in numerous folds and wrinkles over their somewhat flattened body. The skin is the nearly exclusive respiratory surface, because gills are absent and the lungs are small and largely nonfunctional. They retain the larval absence of eyelids as an adult trait.

Andrias is the larger of the two genera; the Japanese *A. japonicus* reaches 1.4 m TL and the Chinese *A. davidianus* grows to 1.5 m. *Cryptobranchus alleganiensis* is much smaller at 0.75 m but still one of the world's largest salamanders. All three species are confined to clear, cold mountain streams. They are largely nocturnal, hiding beneath rocks and sunken logs during the day, sometimes foraging on heavily overcast days or diurnally searching for mates during the breeding season. Even though aquatic, they are not good swimmers but bottom-walkers, using only undulatory locomotion for short-distance escapes to hiding places. They are carnivores and feed on a wide variety of invertebrate and vertebrate prey; crayfish are a preferred food of *Cryptobranchus*.

Cryptobranchoids lack the stereotypic courtship displays of the salamandroids. *Cryptobranchus* males search for females; when a male finds a female, he "drives" her into his nest chamber until she has oviposited. The eggs are laid in two gelatinous strings (one from each oviduct), and the male sheds semen over them in a fishlike manner. A male may sequentially attract two or more females to his nest chamber, after which he guards the multiple egg clutches. Whether in breeding season or not, adult males and females appear to be territorial (site specific) and drive other individuals from their hiding places.

Family Hynobiidae

The hynobiid salamanders are the only exclusively Asian salamanders (Fig. 14.5). There are 32+ species divided among 8 to 12 genera; the number of species and genera is uncertain owing to disagreement among current researchers. The hynobiids segregate into a temperate group (e.g., *Batrachuperus*, *Liua*, *Paradactylodon*) occurring in a zone paralleling the Himalayas and a cool temperate-subarctic group (*Hynobius*, *Ranodon*, *Salamandrella*) in northern Asia from the Urals to Japan (Fig. 14.5). Most hynobiids are small (<10 cm TL), although one species, *Ranodon sibiricus*, may reach 25 cm TL. Most species complete metamorphosis, hence they have eyelids and lack gills or gill slits in the adult stage. Most are stout-bodied, robust-limbed salamanders, usually with a smooth skin. Their skulls contain a pair of lacrimals and septomaxillae (absent in cryptobranchids), and the palatal teeth are in patches or transverse rows. All have large, functional lungs, except *Onychodactylus* (absent) and *Ranodon* (reduced).

Hynobiids display little evidence of courtship. Most species are terrestrial except during the breeding season when they migrate to breeding ponds or streams. Lacking any apparent attractant or courtship, the female begins to lay her eggs. Male *Hynobius* are attracted to the females by the appearance of eggs extruding from the females' vents, and the eggs are fertilized externally. Presum-

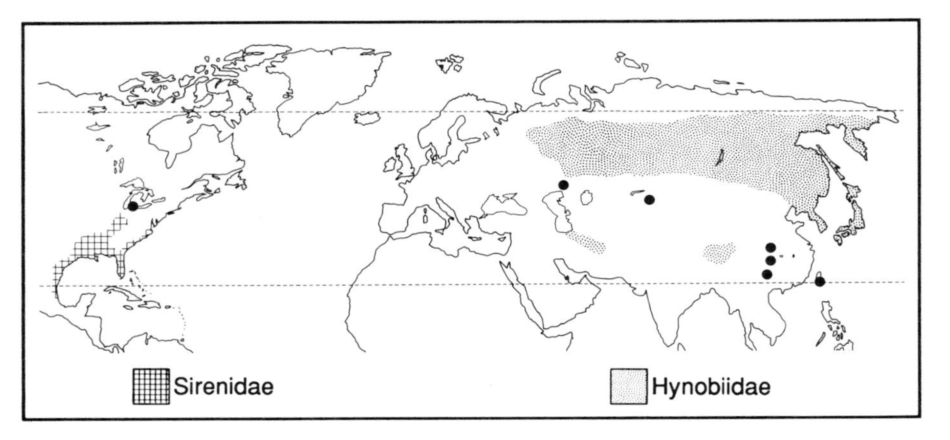

FIGURE 14.5 ▼▲▼▲▼▲▼▲▼▲▼▲▼▲▼▲▼▲▼▲▼▲▼▲
Distribution of the extant Hynobiidae and Sirenidae.

ably a similar mechanism operates for the other hynobiids; all of them use external fertilization. Even though *Ranodon sibiricus* produces a rudimentary spermatophore, the female lays her eggs on it instead of taking it into her cloaca. Females shed their eggs in a pair of gelatinous masses or tubes, one from each oviduct. Males then shed their sperm directly on the egg masses. Development is indirect with a free-living larval stage. Neoteny, likely facultative, occurs in *Batrachuperus* and *Hynobius lichenatus*. Overall, the biology of the hynobiids remains poorly studied.

Meantes

Family Sirenidae

The eellike sirens of southern North America (Fig. 14.5) are neotenic. They have lidless eyes, large external gills and gill slits, and a larval skin structure. Their eellike appearance is enhanced by small forelimbs, only a little larger than the gills, and the absence of hindlimbs and a pelvic girdle. They are unlike any other salamanders in many aspects. For example, they have an interventricular septum and a horny beak instead of premaxillary teeth .

The sirens consist of two genera, *Siren* (true sirens, two species) and *Pseudobranchus* (dwarf sirens, one species). All three sirens typically live in heavily vegetated, slow-moving water habitats, such as lakes, marshes, and swamps. They are active predators, preying on a variety of aquatic invertebrates, which

Lesser siren *Siren intermedia* (Sirenidae). (R. G. Tuck)

they capture by suction feeding. The larger *Siren* (adults 30–60 cm TL, *S. intermedia*; 50–90 cm, *S. lacertina*) readily captures crayfish. The dwarf sirens (15–20 cm TL) eat principally insect larvae, small crustaceans, and worms.

In spite of their locally high abundance and widespread distribution, their biology is poorly known. Females lack spermatheca for storage of sperm, and males have no obvious structure for producing spermatophores, thus external fertilization is the assumed reproductive mode. Courtship behavior has not been observed; eggs are laid singly or in small clusters attached to vegetation.

Salamandroidea

Family Ambystomatidae

The ambystomatids are stout-bodied, robust-limbed, and thick-tailed salamanders of North America (Fig. 14.6); adults are usually >18 cm TL. Most species of the two genera (*Ambystoma*, 30± species; *Rhyacosiredon*, 4 species) are terrestrial during adulthood and return to water only for reproduction. However, some species and/or populations are neotenic: *Ambystoma mexicanum* complex (6 species); *A. gracile*; *A. tigrinum*; *A. talpoideum*; and *Rhyacosiredon altamirani*. The *mexicanum* complex contains the axolotl, and all members are permanently neotenic although metamorphosis can be experimentally induced in some, but not all, members of a population. Neotenic populations of the other species are

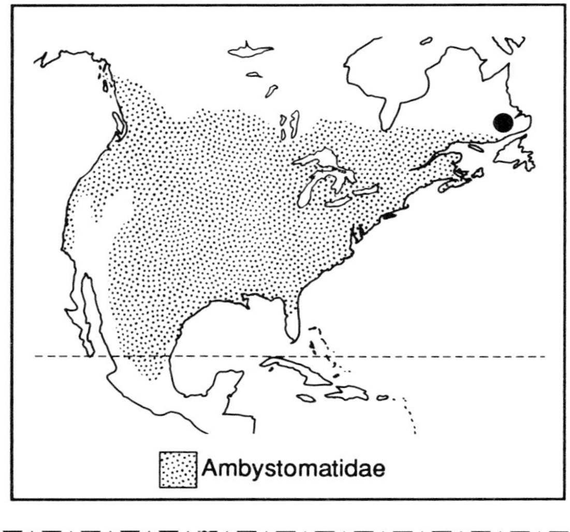

FIGURE 14.6 Distribution of the extant Ambystomatidae.

Blue-spotted mole salamander *Ambystoma laterale* (Ambystomatidae). (R. W. Barbour)

facultative neotenes and some or all adult members of a population metamorphose naturally when their pond dries.

Structurally, ambystomatids share many characteristics with plethodontids, although they possess large and functional lungs. Also their skull is less reduced and their vertebrae are amphicoelous, thus retaining some ancestral traits which are modified in plethodontids. They are predominantly winter breeders, migrating to ponds during brief midwinter warm (<10°C), rainy periods. The first wave of migrants are males, which await the females on subsequent nights. Courtship occurs in water with the males "dancing" and nudging the females (Fig. 8.4) and then depositing numerous spermatophores. Each female picks up one or more sperm packets and, during the next several days, lays eggs. The adults leave the ponds and return to their subterranean homes until the next year. Marbled salamanders (*Ambystoma opacum*) deviate from this reproductive pattern by reproducing in late autumn and having a terrestrial courtship. They lay their eggs in nests in forest-floor depressions; the larvae hatch when the nests are flooded by midwinter-early spring rains or snowmelt water.

Family Amphiumidae

The amphiumids are eellike, permanently neotenic salamanders of southern North America (Fig. 14.7). During a partial metamorphosis, they lose their external gills but retain one pair of gill slits and lidless eyes. Their fore- and hindlimbs and girdles are greatly reduced and inconspicuous on their thick cylindrical bodies. The three species of *Amphiuma* are recognizable by the numbers of toes on each limb, *tridactylum* with three toes, *means* with two, and *pholeter* with one. The former two species are large salamanders with adult lengths

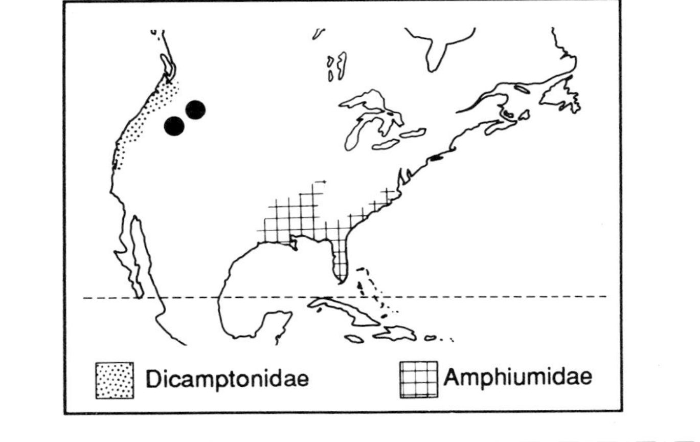

FIGURE 14.7 ▼▲
Distribution of the extant Amphiumidae and Dicamptodontidae.

exceeding 1 m TL, whereas the latter species is considerably smaller, <30 cm. All species are aquatic, although *A. means* has been found foraging on land during rainy nights.

Field observations indicate that males court several females simultaneously or that multiple females contend for the attention of a single male. Since females in other salamandroid genera are passive or even rebuff the male's efforts, these observations require confirmation. Sex is not easily determined and the observations may have been of several males vying for a single female. Courtship ends with the male depositing his spermatophore directly into the female's cloaca.

Family Dicamptodontidae

Two genera (*Dicamptodon*, *Rhyacotriton*) of ambystomatidlike salamanders live in the moist coastal forests of western North America (Fig. 14.7). They differ from the ambystomatids by several characteristics, for example, free columella, unfused exoccipital-prootic-opisthotic complex, lacrimal present, and spinal nerves that exit intravertebrally. These dissimilarities are largely ancestral features, but since these two genera share no unique features with ambystomatids, they are unlikely to be related to ambystomatids. The two genera are also dissimilar and segregated into two subfamilies (Dicamptodontinae, Rhyacotritoninae).

Dicamptodon contains four species: *aterrimus*, *copei*, *ensatus*, and *tenebrosus*. Although all species show some neoteny, *D. copei*, the smallest species (13–17 cm TL), is permanently neotenic. Populations or individuals are facultatively neotenic in the other three species; members of these species are also larger (20–30 cm TL) as adults. *Rhyacotriton* consists of a single, streamside species (*olympicus*) of moderate size (9–12 cm TL). Even though *Rhyacotriton* is fairly uniform morphologically throughout its range, at least four distinct populations are recognizable biochemically and likely represent distinct species. All dicamptodontids deposit eggs singly in water beneath stony rubble. Larval development is slow (>1 yr) likely owing to the coolness of the aquatic nesting sites.

Family Plethodontidae

The plethodontids comprise a diverse group of more than 300 species in tropical America, North America, and northern Italy (Fig. 14.8). Both the smallest and largest plethodontids are Mexican bolitoglossins, ranging from the tiny *Thorius* (25–30 mm TL) to the large *Pseudoeurycea bellii* (32 cm TL). Body shape is diverse, but all have four limbs. Some plethodontids are stocky and short-limbed, and others are elongate and slender-limbed; some have tails equal to body length, and in others, the tails are twice body length. Although the morphologies are diverse, plethodontids are a monophyletic group and uniquely possess nasolabial grooves. Their skulls are reduced; the exoccipital-prootic-opisthotic complex is fused, and the lacrimal, pterygoid, and opercular bones are absent. All vertebrae are opisthocoelous, and except for three anteriormost ones, spinal nerves exit intravertebrally. Plethodontids lack the ypsiloid cartilage and lungs.

Plethodontids are not the only lungless or reduced-lunged salamanders,

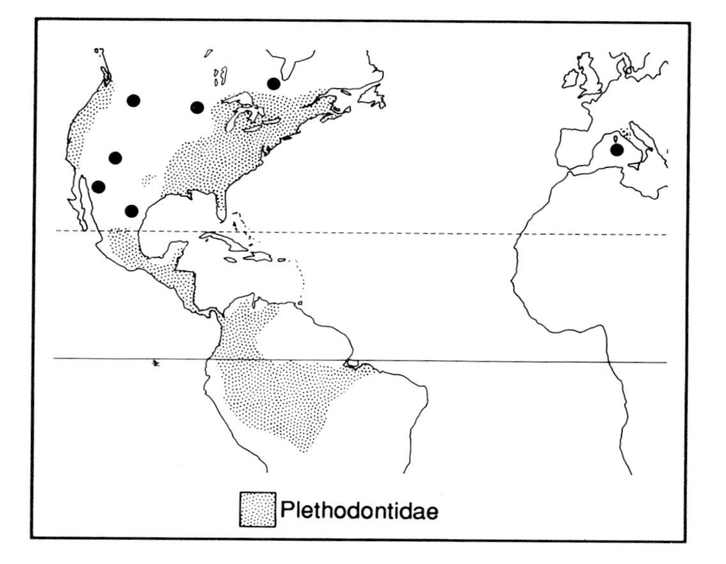

FIGURE 14.8 ▼▲▼▲▼▲▼▲▼▲▼▲▼▲▼▲▼▲▼▲▼▲▼▲▼▲▼▲▼▲▼▲▼▲
Distribution of the extant Plethodontidae.

although lung loss is independently derived in plethodontids and each of the other taxa (e.g., the hynobiid *Onychodactylus* lacks lungs, and vestigial lungs occur in cryptobranchids and the dicamptodontid *Rhyacotriton*). Absence or reduction of lungs correlates with occurrence in rapid, turbulent mountain streams. The reduction of lungs possibly occurred because, as hydrostatic organs, lungs tend to maintain salamanders at neutral buoyancy. Neutral buoyancy allows an animal to be swept off the bottom in rapid-moving water; reduction and loss of lungs make an animal negatively buoyant, hence weighting the animal and keeping it on the bottom. Other hypotheses suggest that lunglessness (in plethodontids) arose in association with reduction in head width and an increased dependency on cutaneous respiration in semiaquatic ancestors or in association with the evolution of terrestrial courtship and mating, which supposedly is energetically less demanding.

Two major, presumably ancient, lineages are evident within the plethodontid salamanders. The Desmognathinae (*Desmognathus*, *Leurognathus*, *Phaeognathus*) is confined to eastern North America and has a unique jaw-opening mechanism in which the lower jaw is held stationary and the skull swings upward. The Plethodontinae includes all the other genera, and they have the typical vertebrate jaw mechanism with the lower jaw swinging downward. Desmognathines display both indirect and direct development, and aquatic to terrestrial species. One group of plethodontines (tribe Hemidactyliini — *Eurycea*, *Gyrinophilus*, *Haideotriton*, *Hemidactylium*, *Pseudotriton*, *Stereochilus*, *Typhlomolge*, *Typhlotriton*) has indirect development (aquatic, free-living larvae) and aquatic to terrestrial adults. Two other tribes (Plethodontini — *Aneides*, *Ensatina*, *Plethodon*; Bolitoglossini — all other plethodontines) have direct development. The bolitoglossins include all the Latin American genera, *Batrachoseps* of western North America, and *Hydromantoides* of western North American and southern Europe.

In plethodontids, neoteny occurs only in the hemidactyliin salamanders: *Eurycea* species of the Edwards Plateau in Texas, *E. tynerensis*, *Gyrinophilus palleucus*, *Haideotriton*, *Typhlomolge*, and *Typhlotriton*. All these neotenes are spring or cave species. In addition to incomplete metamorphosis and the retention of gills, most are slender-bodied and -limbed (Fig. 14.9), and many have degenerate eyes and reduced skin pigmentation.

Family Proteidae

Living proteids consist of two, externally dissimilar genera of neotenic salamanders. Both groups are totally aquatic, but the North American *Necturus* dwells in surface waters whereas the European *Proteus* is a cave species (Fig. 14.10). Superficially *Proteus* appears more similar to the cave-dwelling hemidactyliins than to *Necturus*, since it has a slender body and limbs, reduced eyes beneath the skin, and pigmentless skin. In other aspects, *Proteus* and *Necturus* share a similar habitus; their bodies are depressed and heavily muscled, the limbs are short with well-developed feet, the heads are broad and flattened, and their

FIGURE 14.9
Texas cave salamander *Eurycea neotenes* (Plethodontidae). (R. W. Barbour)

tails are stout and laterally compressed with dorsal and ventral fins. Uniquely among salamanders, they lack maxillae and have two pairs of larval gill slits.

Proteids are moderate-sized salamanders. The monotypic *Proteus* ranges from 20 to 25 cm TL. Three *Necturus* share this size range, and another one (*N. punctatus*) is distinctly smaller (<19 cm maximum TL). *Necturus maculosus* is the

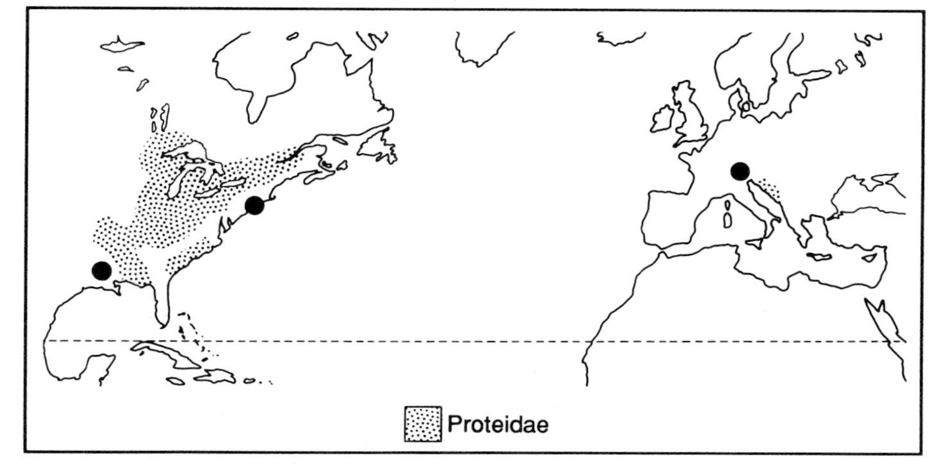

FIGURE 14.10
Distribution of the extant Proteidae.

largest species; adults usually range from 25 to 35 cm, but occasionally reach 45–48 cm TL (Fig. 14.11). It also has the largest distribution, including the entire Mississippian drainage and some northeast Atlantic drainages. All prefer clear water. They are nocturnal foragers and eat a variety of prey with a preference for crayfish. *Necturus maculosus* courts in the autumn, but egg-laying does not occur until the subsequent spring. Up to 50 eggs are attached to the roof of the female's shelter, and whether or not they receive active care, they are protected by her presence.

Family Salamandridae

Salamandrids have a quadripartite distribution in central and eastern Asia, Europe, eastern North America, and western North America (Fig. 14.12). The genera of each area are distinct. The American fauna is the least speciose with three species in the east (*Notophthalmus*) and three in the west (*Taricha*). In contrast, nine genera (20+ species) occur in Europe and six genera (15+ species) in Asia. Most salamandrids are small with total lengths seldom exceeding 20 cm (<10 cm SVL), and the larger taxa (e.g., European *Pleurodeles* and *Salamandra*) are less than 35 cm TL. Salamandrids share a number of features with the plethodontids, such as fusion of the exoccipital-prootic-opisthotic com-

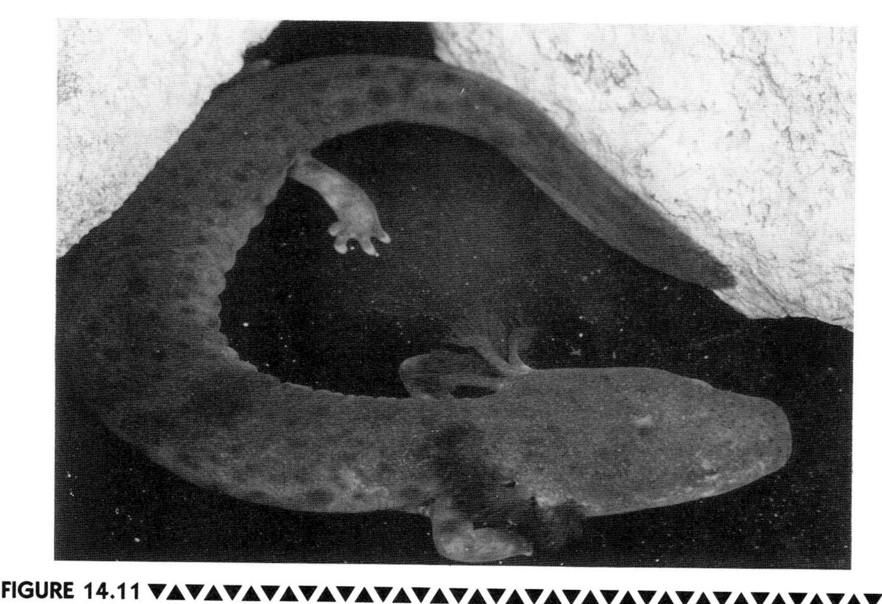

FIGURE 14.11 ▼▲▼▲▼▲▼▲▼▲▼▲▼▲▼▲▼▲▼▲▼▲▼▲▼▲▼▲▼▲▼▲
Mudpuppy *Necturus maculosus* (Proteidae). (R. W. Barbour)

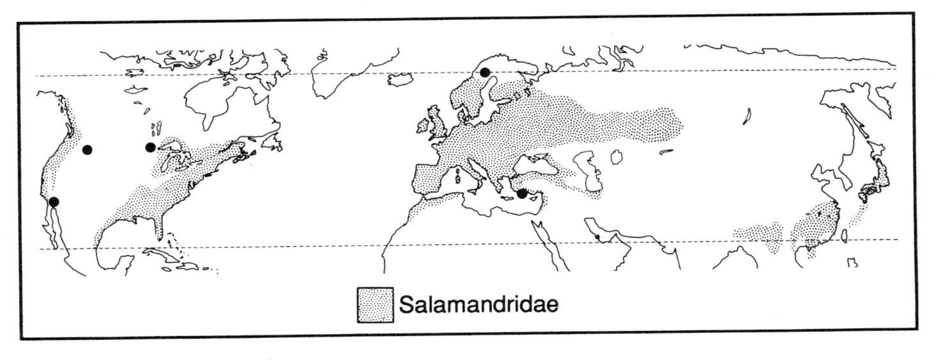

FIGURE 14.12 ▼▲▼▲▼▲▼▲▼▲▼▲▼▲▼▲▼▲▼▲▼▲▼▲▼▲▼▲▼▲▼▲▼▲
Distribution of the extant Salamandridae.

plex and opisthocoelous vertebrae; however, salamandrids possess lacrimals, lungs, and ypsiloid cartilages. Also only their two anteriormost spinal nerves exit intervertebrally. Their skin is typically granular or rugose because of numerous poison glands (Fig. 14.13), and the secretions of these glands are the most toxic of all salamanders. In association with their high toxicity, many salamandrids are brightly colored, at least ventrally or seasonally. Their bright coloration advertises their toxicity to potential predators.

FIGURE 14.13 ▼▲▼▲▼▲▼▲▼▲▼▲▼▲▼▲▼▲▼▲▼▲▼▲▼▲▼▲▼▲▼▲▼▲
Burmese newt *Tylototriton verrucosus* (Salamandridae). (K. Nemuras)

Two salamandrids, *Salamandra atra* and *Mertensiella luschani*, are livebearers (see Chapter 7); a few other species lay terrestrial eggs, but the majority deposit eggs in the water and have a free-living larval stage. All have courtship displays in which the male circles the female and nudges or rubs her, and in a few species, the male grasps the female and deposits his spermatophore in or near her cloaca. Three life cycles are evident among the taxa with aquatic larvae. In some (e.g., *Cynops*, *Pleurodeles*), the larvae metamorphose into aquatic juveniles, and this aquatic existence persists throughout adult life. Others (*Taricha*, *Triturus*) have aquatic larvae; upon metamorphosis, the salamanders become terrestrial and return to water only to breed. In northern populations, *Notophthalmus* has a triphasic life cycle: aquatic larvae, terrestrial juveniles (efts), and aquatic adults. Facultative neoteny occurs in some populations of a few species, *Notophthalmus viridescens*, *Triturus alpestris*, *T. cristatus*, and *T. helevaticus*.

Phylogenetic Relationships of Salamanders

Many, likely all, salamander families arose in the Mesozoic. All the families are morphologically distinct, and the known species and genera are assigned easily and unquestionably to a single familial lineage. The only exceptions are *Necturus* and *Proteus*, whose inclusion in a single family remains a minor controversy. They share a few unique traits, indicating a sister relationship, but since both are neotenes, the significance of the shared traits is questioned. Indeed, it is the high level of paedomorphosis (see Chapter 7) in salamanders that creates difficulties in recognizing phylogenetic relationships among the salamander families. The suprafamilial grouping used here is one of both convenience (i.e., reflecting the general consensus of the herpetological community) and a conservative representation of sister-group relationships. The three groups (sirens, cryptobranchoids, salamandroids) do appear to represent separate evolutionary groups (Fig. 14.14) with long individual histories for each family.

Nearly a dozen different phylogenies have been proposed (based on morphological traits) during the last two decades. Although they are strikingly different in some branching patterns, concordance (matching) of branches is also evident. This concordance strengthens the likelihood that some branches reflect actual phylogenetic relationships. Cryptobranchids and hynobiids, and ambystomatids and plethodontids consistently form pairs. Dicamptodontids are often linked with the ambystomatids, and the salamandrids are commonly a sister-group of the ambystomatids-plethodontids. Numerous primitive characteristics place the cryptobranchoids at the base of the phylogenetic dendrograms, and the other groups form the terminal groups. This branching pattern yields a dendrogram in which the basal members show external fertilization, high number of chromosomes, and the presumed primitive spinal nerve morphology, and the terminal families share internal fertilization, reduction in chromosome number, and a derived spinal nerve pattern (nerves exiting intravertebrally on all but the anterior two or three vertebrae).

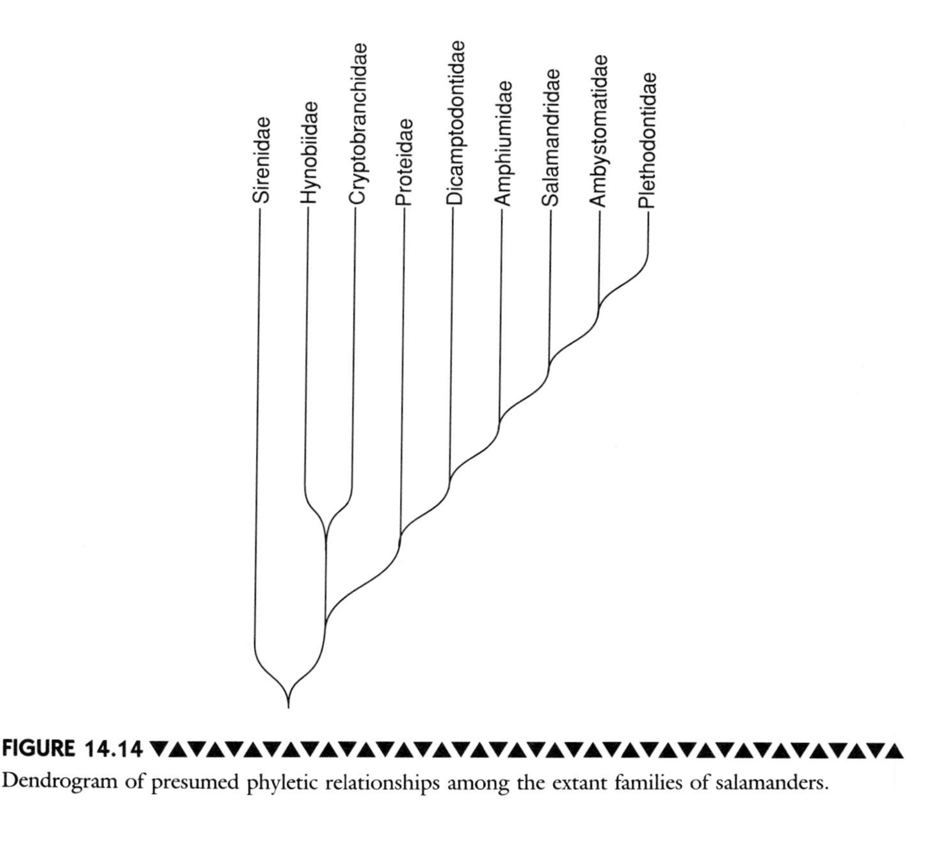

FIGURE 14.14
Dendrogram of presumed phyletic relationships among the extant families of salamanders.

The placement of sirens, amphiumas, and proteids on this dendrogram varies widely among the various proposed phylogenies. Sirens are the most enigmatic of this triad of neotenic families, because they share a mix of highly derived and primitive traits (e.g., the derived spinal nerve pattern but high chromosome number and presumably external fertilization). For this reason, sirens are usually proposed as arising early, either immediately before or after the divergence of the cryptobranchoids; the alternative is a sister-group relationship with salamandrids owing to the shared spinal nerve pattern. The amphiumids are variously placed as a sister-group of all or most of the advanced families with the exception of proteids or as a sister-group of the plethodontids. The divergence of the proteids is proposed most often to have occurred subsequent to the divergence of the cryptobranchoids.

A recent study based on ribosomal RNA sequences suggests some profoundly different divergence times, although retaining many of the previously proposed sister-groups. From earliest to latest divergences, the sequence is amphiumids-plethodontids as the first divergence; then the rhyacotritonines;

followed by sirens-cryptobranchoids, proteids, and salamandrids; the dicamptodontines; and the ambystomatids as the terminal groups. The most striking aspect of this proposal is that it requires a major reinterpretation of the primitive-derived polarity of the states of many morphological traits and independent (parallel) evolution of others. For example, the RNA dendrogram requires internal fertilization with spermatophores and the intravertebral spinal nerve-exit pattern as ancestral states. These interpretations are quite contrary to the current consensus, but that does not make them incorrect. Only further testing of the various phylogenetic hypotheses will resolve these issues.

CHAPTER 15

Frogs

ANURA

All extant frogs share a similar body form, no matter how big or small the species or whether they are aquatic, fossorial, or arboreal. All have a greatly shortened vertebral column of nine or fewer presacral vertebrae, eight in most species. All presacral vertebrae, except the atlas (first vertebra), have transverse processes, and ribs are absent (in most families) or reduced and usually confined to the second through fourth vertebrae in some primitive families. The presacral vertebrae are firmly articulated, allowing only slight lateral and dorsoventral flexure, and the postsacral vertebrae are fused into a rod-shaped urostyle lying within the ilial canal.

These skeletal traits give frogs their short, stout-bodied appearance that is further enhanced by the large, muscular limbs and the absence of a tail. The hindlimbs are usually longer and heavier than the forelimbs. Both limb girdles are robust and well ossified; the pelvic girdle is uniquely modified by an elongation of the ilia and a compaction of the pubes and ischia.

All these features are associated with jumping locomotion. The long hindlimbs provide the propulsive force to lift and propel the frog forward. The short body provides a compact mass to be hurled forward, and the shortened vertebral column, robust pectoral girdle, and sturdy limbs readily absorb the shock of landing. Hindlimbs extend synchronously and jump the frog forward. Frogs regularly leap 2–10 times their body length, and leaps can be as much as 30–40 times body length. The length of a jump depends on the morphology and physiology of a species and the urgency of movement. Some species (usually ones with toad-type bodies) seldom jump but use the typical vertebrate walking gait, and totally aquatic species use the synchronous hindlimb extension for swimming.

Their heads are flat and broad with large mouths. The anuran skull has lost many elements (e.g., lacrimals, jugals, supra- and basioccipitals, ectopterygoids) but remains well ossified. The premaxillae and maxillae bear teeth in most frogs (absent only in a few, e.g., bufonids, *Rhinophrynus*); conversely, dentaries lack teeth in most species. The tongue is large and posteriorly free in most frogs and is the prey-capture mechanism. Vocal sacs occur in the males of most species.

With few exceptions, frogs have external fertilization, and males amplex females in order to juxtapose the cloacae of the two sexes to ensure fertilization of the eggs as they emerge from the females' cloacae. Indirect development with free-living larvae is common but so are direct development and extrauterine ovoviviparity. Indirect development of anurans differs from that of caecilians and salamanders by differences in adult and larval morphology and life-styles and the necessity of a major structural and behavioral reorganization at metamorphosis. Also unlike the two other amphibian groups, frogs have no strictly neotenic species, and they have been much more successful in adapting to the full spectrum of terrestrial and freshwater environments and occur on all continents and many islands.

Considerable differences of opinion still exist on the relationships of various groups of frogs and has resulted in many different classification schemes. The classification adopted here (see Table 1.2) is an amalgam of several recently proposed classifications (principally Frost's and Dubois') and reflects both a consensus of recent usage and presumed relationships of extant subfamilies and families.

Primitive Frogs

Family Ascaphidae

Ascaphus truei (37–50 mm SVL) is the sole member of this family (Fig. 15.1). The tailed frog lives in and along fast-moving mountains streams in

FIGURE 15.1 ▼▲▼

True's frog *Ascaphus truei* (Ascaphidae). (R. W. Van Devender)

northwestern North America (Fig. 15.2) from sea level to near the tree line. It shares many characteristics with *Leiopelma* and is often placed in the same family; however, *Ascaphus* has several unique features (e.g., copulatory organ) and a long independent history. Much of ascaphid morphology reflects a life in rapid-flowing streams. They lack aerial voices and tympana, and have copulatory organs for internal fertilization, streamlined tadpoles with oral discs modified as ventral suckers, and strongly webbed hindfeet.

Tailed frogs are mainly nocturnal, hiding beneath streamside rocks and debris during the day and emerging at night to feed. They are active at low temperatures (10–17°C) and require high humidity, venturing from streamside only during rainy weather. Courtship and copulation occur in late summer (September), but egg-laying does not occur until the subsequent summer (July); the sperm remains viable in the lower oviducts for over 10 months.

Family Discoglossidae

Discoglossids are a small group of European and Southeast Asian frogs (Fig. 15.2), commonly divided into two subfamilies (Bombinatorinae — *Alytes*, *Barbourula*, *Bombina*; Discoglossinae — *Discoglossus*). Adults are small, ranging from 40–55 mm SVL in *Alytes* and *Bombina* and 60–70 mm in *Discoglossus* to a maximum of 80–85 mm in *Barbourula*. With the exception of the terrestrial

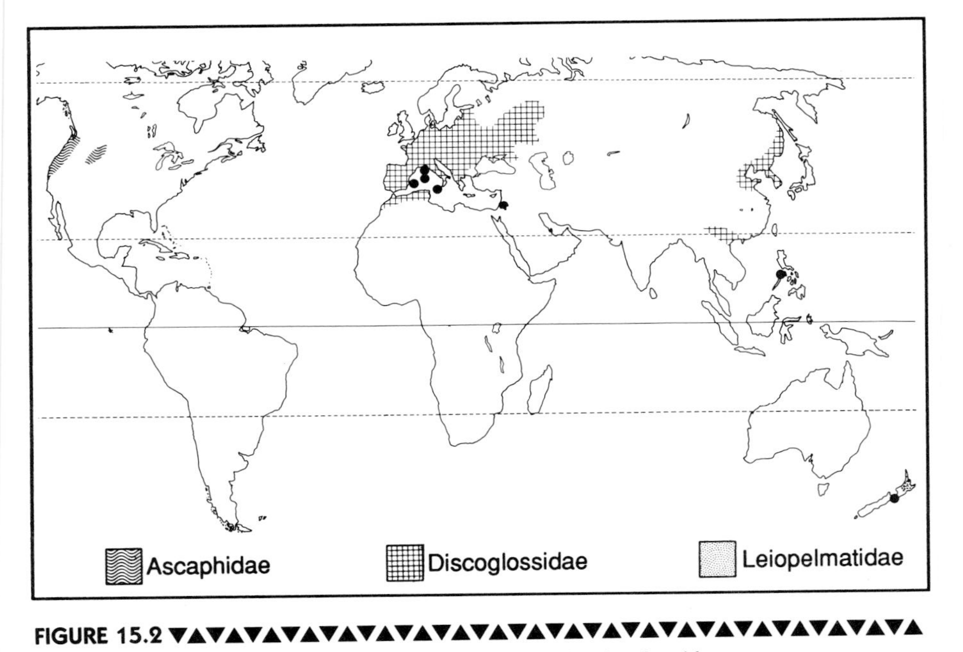

FIGURE 15.2 ▼▲
Distribution of the extant Ascaphidae, Discoglossidae, and Leiopelmatidae.

Alytes, the discoglossids are aquatic species of shallow streams and ponds. Males of all have voices and inguinal amplexus. Eggs are laid in water, except for *Alytes*, and hatch into free-living larvae (indirect development). In *Alytes*, males wrap the eggs around their hindlimbs and body, carry the eggs until they are about to hatch, and then return to water for the larvae to swim free. All discoglossids share free ribs in adults, eight stegochordal-opisthocoelous presacral vertebrae, expanded sacral diapophyses, arciferal pectoral girdles, astragalus and calcaneum fused at ends, and premaxillary and maxillary teeth. All discoglossids have inguinal amplexus and indirect development.

The fire-bellied toads, *Bombina*, are diurnal and, although dark and camouflaged above, are readily seen because of their high activity and tendencies to aggregate. They are protected from many predators by a warty, glandular skin. When attacked, they advertise their unsuitability as food by the unken-reflex. This arching reflex (see Fig. 6.5) displays their bright venters of black mottling on yellow, orange, or red backgrounds. The other genera rely on crypsis.

Family Leiopelmatidae

This family contains three species of frogs, *Leiopelma archeyi*, *L. hamiltoni*, and *L. hochstetteri*, in northern New Zealand (Fig. 15.2). All are small (31–49 mm SVL, with *L. hamiltoni* the largest), cryptically colored frogs. All three are secretive frogs living only in a few areas along the borders of cool forest creeks and seepage areas. During the day, they hide beneath rocks, logs, and other forest litter and emerge at night to feed. Small clusters of large, yolky eggs (5–20) are laid in small depressions beneath rocks or logs, and the male broods the egg clutch. Development is best described as incomplete direct. Their larvae hatch at a relatively late stage of embryogenesis and are nearly immobile in *L. archeyi* and *L. hamiltoni*. Males continue to guard the larvae at least in the former species. The larvae of *L. hochstetteri* have typical tadpole bodies and are mobile.

Leiopelma represents one of the most primitive groups of living frogs, appearing little changed from two fossil taxa from the Jurassic of Argentina. *Ascaphus* and *Leiopelma* share a number of characters such as a pair of "tail-wagging" muscles and free ribs in adults, nine ectochordal-amphicoelous presacral vertebrae, slightly expanded sacral diapophyses, arciferal pectoral girdle, astragalus and calcaneum fused at ends, premaxillary and maxillary teeth, and inguinal amplexus. Most are primitive characters and indicate only that both genera have retained some morphological features of the Jurassic frogs. Phylogenetically, the divergence of these two lineages is as distant in time (Cretaceous or earlier) as their current geographical separation.

Transitional Frogs

Family Pelobatidae

Pelobatids consist of three subfamilies of small- to moderate-sized, terrestrial frogs in North America, Europe, and Himalayan and Southeast Asia (Fig. 15.3). The Leptobatrachiinae (most 50–80 mm SVL) contains four Asian genera: *Leptobrachella*, *Leptobatrachium*, *Leptolalax*, and *Scutiger*. *Scutiger* are predominantly high-elevation frogs (30+ species), occurring commonly above 1000 m with a record of 5030 m for one specimen of *Scutiger boulengeri*; the other genera are predominantly lowland-low montane frogs of subtropical and tropical Asia. The Megophryinae (*Megophrys*, *Ophryophryne*; 90–120 mm SVL) is also a tropical Asian group; these forest-floor frogs are large-headed, often with fleshy horns over the eyes. The Pelobatinae (50–100 mm SVL) contains the spadefoot toads of Europe and North America (*Pelobates* and *Scaphiopus*, respectively)(Fig. 15.4). They have short, squat, toadlike bodies and warty, though soft, skins. Their colloquial name derives from a large, keratinous tubercle on the outside edge of each hindfoot; all are fossorial (subterranean) and burrow rearfirst with an alternate shuffling movement of the hindlimbs. In spite of their remarkable similarities in anatomy, ecology (e.g., sandy soils and dry habitats), and behavior (explosive breeders), some researchers argue that they represent two convergent groups; the similarities more likely reflect sister-group relationships although with an ancient separation.

All pelobatids share the following characteristics: no free ribs; eight stegochordal-procoelous presacral vertebrae in adults; broadly expanded sacral di-

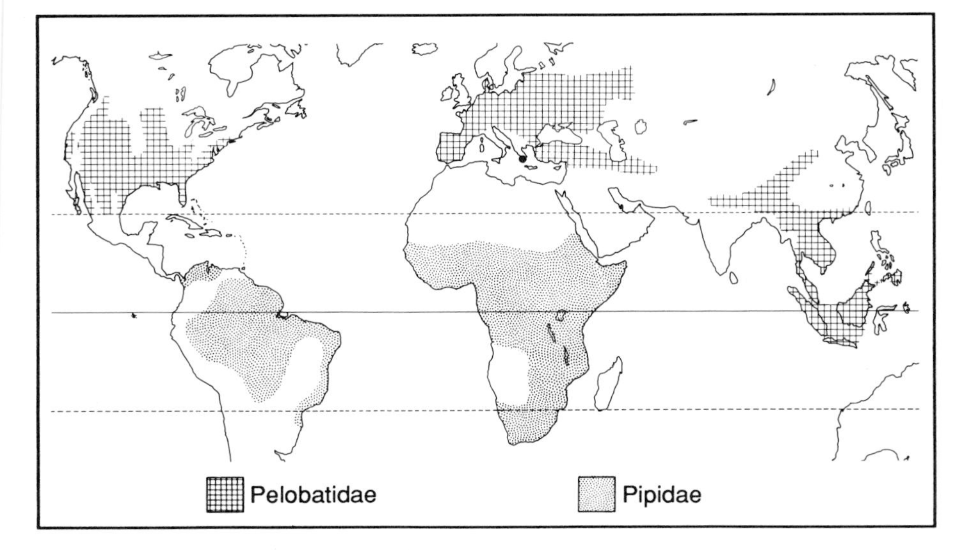

FIGURE 15.3 ▼▲▼▲▼▲▼▲▼▲▼▲▼▲▼▲▼▲▼▲▼▲▼▲▼▲▼▲▼▲▼▲
Distribution of the extant Pelobatidae and Pipidae.

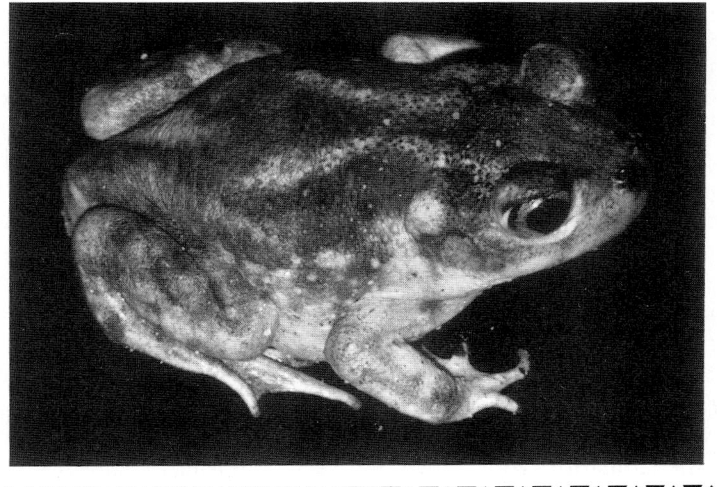

FIGURE 15.4 ▼▲▼▲▼▲▼▲▼▲▼▲▼▲▼▲▼▲▼▲▼▲▼▲▼▲▼▲▼▲▼▲
Holbrook's spadefoot toad *Scaphiopus holbrooki* (Pelobatidae). (G. R. Zug)

apophyses; arciferal pectoral girdle; astragalus and calcaneum fused at ends; and premaxillary and maxillary teeth. All also display inguinal amplexus and indirect development.

Family Pelodytidae

Pelodytes, the sole member of this Eurasian family, consists of two allopatric species (*P. punctatus*, *P. caucasicus*)(Fig. 15.5). Both parsley frogs are moderately small (45–55 mm SVL), terrestrial species living in moist habitats from sea level to midmountain. They share many characteristics with the pelobatids and have been included with them in former classifications. Several osteological features are distinct (e.g., fully fused astragalus and calcaneum, fused first and second presacral vertebrae) and suggest an ancient divergence from a common ancestor.

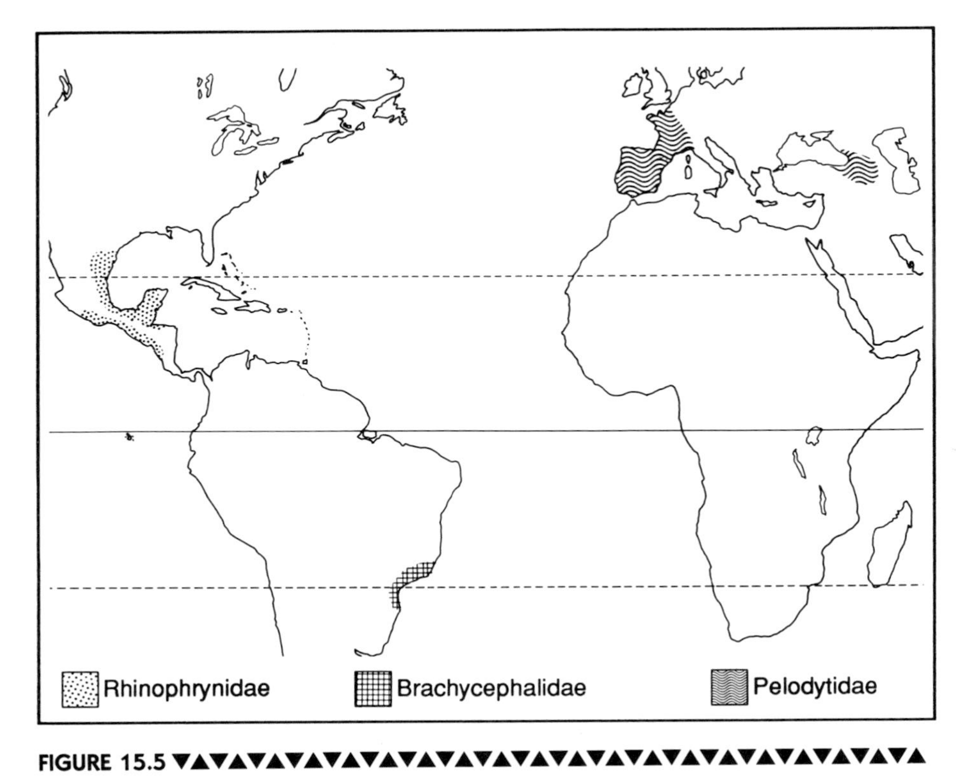

FIGURE 15.5 ▼▲
Distribution of the extant Brachycephalidae, Pelodytidae, and Rhinophrynidae.

Advanced Frogs I

Family Pipidae

Pipids consist of 25 + extant species of aquatic frogs, presently confined to Africa and South America (see Fig. 15.3). They are highly specialized frogs, even though their fossil history extends back to the Early Cretaceous. Many of their specializations are associated with an aquatic life. All retain the lateral line system as adults. They have streamlined bodies terminating in large, powerful hindlimbs for synchronized kick-swimming. The pelvic girdle shows several unique modifications presumably associated with their swimming behavior: enlarged crests on the ilia; ossified pubes; and fusion of the sacral vertebrae with the urostyle. The forelimbs and girdles also show unique modifications, but these appear to be associated with feeding. The fingers of the forefeet are elongate and have tactile-sensory tips. The forefeet find and sweep food into the tongueless mouth. Pipids also possess free ribs in juveniles, epichordal-opisthocoelous vertebrae, 6–8 presacral vertebrae, expanded sacral diapophyses, a pseudofirmisternal pectoral girdle, astragalus and calcaneum fused at ends, and premaxillary and maxillary teeth in some members.

There are five genera of pipid frogs: the neotropical *Pipa* and the African *Hymenochirus*, *Pseudhymenochirus*, *Silurana*, and *Xenopus*. Although *Xenopus* represents the most primitive pipid, it is an extremely specialized frog and it or close relatives occurred in the Cretaceous (60 MA). The other genera are no less specialized and likely diverged from the "xenopus" stock in the Cretaceous or early Tertiary. Their relationships are well expressed in a recent classification recognizing three subfamilies: Dactylethrinae—*Xenopus*; Siluraninae—*Silurana*; and Pipinae—*Hymenochirus*, *Pipa*, *Pseudhymenochirus*. All retain the primitive inguinal amplexus and indirect development (except *Pipa pipa* with extrauterine ovoviviparity), but their tadpoles are specialized midwater filter feeders and lack keratinous mouthparts. Further, Pipinae have elaborate courtship behaviors. Males attract females by sharp clicking sounds produced by snapping the hyoid apparatus. Upon amplexus, the pair performs a series of aquatic somersaults that allow the male to fertilize the eggs prior to their rolling onto the female's back in *Pipa* or being deposited on the water surface in *Hymenochirus*. *Xenopus* is the most speciose of the pipids with a dozen species, many of which are polyploids.

Family Rhinophyrnidae

Rhinophrynus dorsalis (Mexican burrowing frog) is a coastal lowland species of Central America (Fig. 15.5) and the sole living member of this family. This peculiar frog has a tiny cone-shaped head and four short, but robust limbs projecting from a large, somewhat flattened, globular body (75–85 mm SVL). Its skin is smooth except for the calloused snout. This globular microcephalic habitus is associated with a fossorial existence and termite diet; similar body forms have arisen independently in termite/ant-eating, burrowing frogs in several other families. Uniquely, *Rhinophrynus* has its tongue attached at the rear, and the tongue extends straight outward to ensnare prey on its tip.

Rhinophrynus dorsalis possesses an admixture of primitive and derived characters: no free ribs; ectochordal-opisthocoelous vertebrae; eight presacral vertebrae; expanded sacral diapophyses; an arciferal pectoral girdle; astragalus and calcaneum fused at ends; and no premaxillary and maxillary teeth. It has inguinal amplexus and indirect development; the tadpole lacks keratinous mouthparts.

Advanced Frogs II

Family Brachycephalidae

Brachycephalids contain a few small (<18 mm SVL) toadlike frogs in the humid Atlantic coastal forests of southeastern Brazil (Fig. 15.5). These tiny frogs are usually cryptically colored and are immediately recognizable by the reduced number of digits, two functional fingers on the forefoot, and three toes on the hindfoot. They live on and in the forest-floor litter.

Brachycephalids lack ribs and premaxillary and maxillary teeth, have seven holochordal-procoelous presacral vertebrae, expanded sacral diapophyses, a modified arciferal pectoral girdle, and the astragalus and calcaneum are fused only at their ends. One (*Brachycephalus*) of the two genera has a bony shield fused to

Green toad *Bufo viridis* (Bufonidae). (G. R. Zug)

the top of the vertebrae. All brachycephalids have a tiny buzzlike call, axillary amplexus, and presumably direct development.

Family Bufonidae

Bufonids, the true toads, occur on all continents (Fig. 15.6); however, their presence in Australia and New Guinea results from recent (mid-1930s) introductions of *Bufo marinus* for the control (unsuccessful!) of cane-beetles and other insect pests. The genus *Bufo* shares this wide distribution, although it is likely that *Bufo* is a polyphyletic genus (200+ species) representing different lineages in Asia, Africa, and South America. The complexity of this genus is hidden by the superficial similiarity of its species. Almost all share a short, robust habitus, and a thick, glandular skin with numerous warts. They range in size from dwarfs (18–20 mm SVL) to giants (>20 cm SVL). The divergent body-form species have attracted more attention and, as such, have been divided among several subfamilies, each likely derived from a different lineage of "*Bufo*."

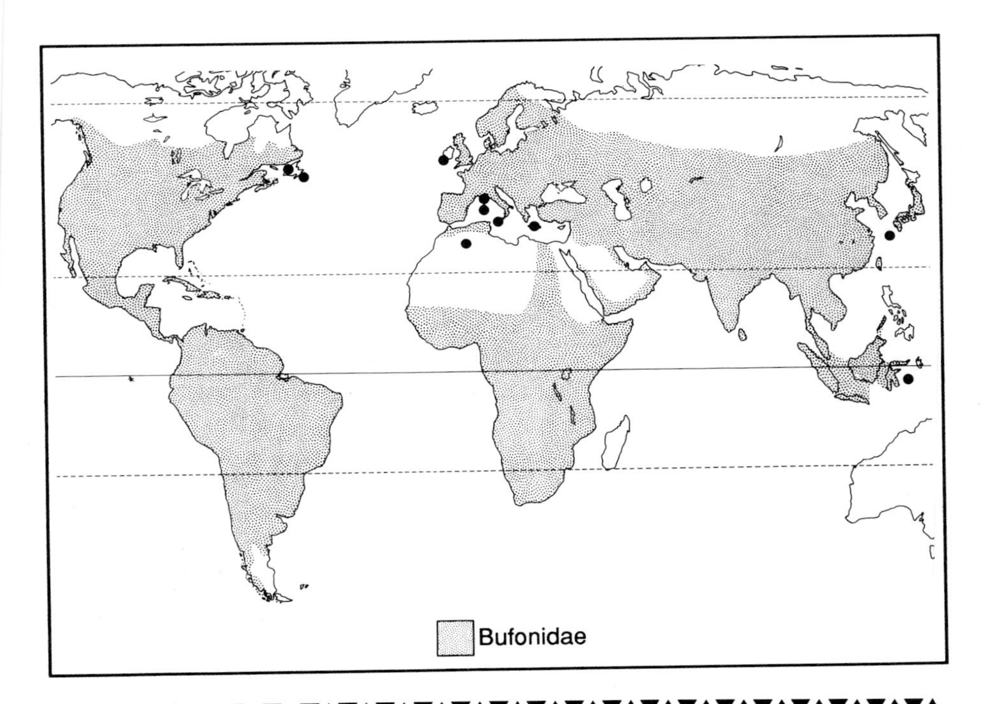

FIGURE 15.6 ▼▲▼▲▼▲▼▲▼▲▼▲▼▲▼▲▼▲▼▲▼▲▼▲▼▲
Distribution of the extant Bufonidae, excluding the introduced populations of *Bufo marinus*.

Commonly five subfamilies are recognized: Adenonominae, Allophryninae, Atelopodinae, Bufoninae, and Tornieriobatinae. These subfamilies share an absence of free ribs and an edentulous upper jaw, holochordal-proceolous vertebrae, 5–8 presacral vertebrae, expanded sacral diapophyses, an arciferal pectoral girdle, astragalus and calcaneum fused only at ends, and the presence of a Bidder's organ in males. Many toads have bony, firmly knitted skulls with the skin fused to the roofing bones. Bufonids use axillary amplexus, and the majority deposit strings of small eggs from which hatch small, generalized tadpoles with keratinous mouthparts.

The adenonomine toads are a small group of Asian genera (e.g., *Ansonia*, *Pedostibes*, *Pelophryne*). They are mostly small- to moderate-sized, streamside frogs of tropical forests (*Pedostibes* is semiarboreal) with indirect development. *Allophryne ruthveni* is the sole member of its subfamily. It is a small treefroglike anuran of northern South America. Its placement in the bufonids or hylids remains questionable. The atelopodines contain most of the peculiar neotropical bufonid genera (e.g., *Andinophryne*, *Atelopus*, *Dendrophryniscus*). The atelopodines are small- to moderate-sized (<80 mm SVL) anurans, often with an elongate body and long slender limbs; their skin ranges from smooth and shiny (although highly glandular) to a typical warty surface. Most atelopodines have free-living tadpoles, although *Osornophryne* may have direct development. The tornieriobatines contain the unusual African bufonids (e.g., *Capensibufo*, *Nectophrynoides*, *Stephopaedes*). Although most are more bufo like in appearance than the atelopodines, each genus has pecularities in larval or adult morphology. They show the full range of anuran reproductive adaptations from external fertilization and a free-swimming larvae through direct development of terrestrial eggs to internal fertilization and birth of toadlets. *Bufo* comprises the bufonines, the typical toads.

Family Centrolenidae

The glass frogs are a moderately diverse group (*Centrolene*, 20+ species; *Cochranella*, 30+ species; *Hyalinobatrachium*; 20+ species) of tropical American treefrogs (Figs. 15.7 and 15.8). Their colloquial name derives from their lightly pigmented or pigmentless venters that allow the viscera to be seen through the body wall. Dorsally they are shades of green, rarely brown, with yellow, white, blue, or red markings. They possess the typical treefrog foot structure with expanded digit tips and an intercalary cartilage between the terminal and penultimate phalanges of each digit. In other characteristics, they lack ribs, have premaxillary and maxillary teeth, eight holochordal-procoelous presacral vertebrae, expanded sacral diapophyses, an arciferal pectoral girdle, and complete fusion of astragalus and calcaneum.

Parental care is common, and a male may guard one to several clutches. Females deposit their eggs adjacent to calling males on leaves overhanging streams. Development is indirect, and when the eggs hatch, the tadpoles drop

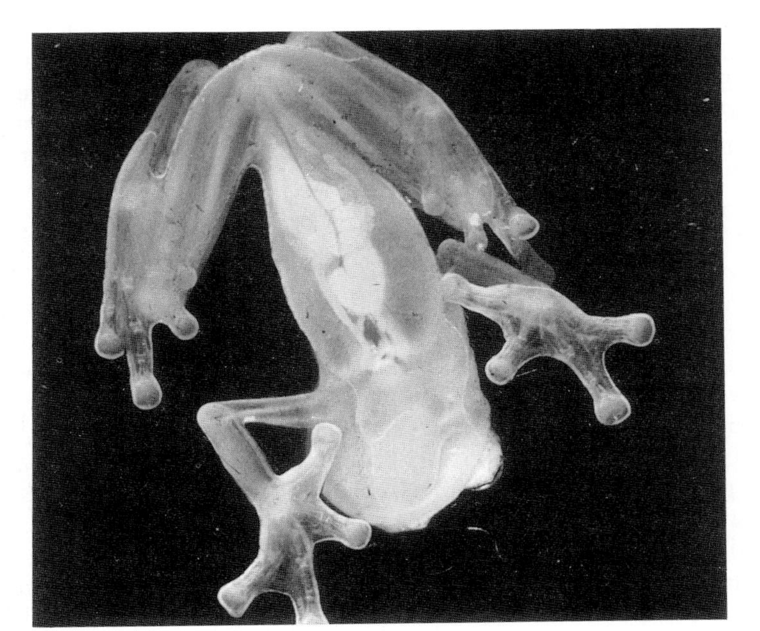

FIGURE 15.7
A glass frog *Hyalinobatrachium valerioi* (Centrolenidae). (R. W. McDiarmid)

into the water below. There the elongate tadpoles complete their development, commonly living within the leaf litter on the stream bottom.

Family Dendrobatidae

The poison-dart frogs (e.g., *Colostethus*, *Phyllobates*) are a moderately diverse group (6 genera, 120± species) of semiaquatic and terrestrial frogs of neotropical forests (Fig. 15.9). All are small (most <50 mm SVL), active frogs. Many are boldly colored. The bold colors advertise (aposomatism) the presence of toxic skin alkaloids; these alkaloids are especially toxic in the brightly colored *Dendrobates* and *Phyllobates* and least toxic in the more somber-colored *Colostethus*. As a group, they share absence of ribs, edentate or dentate upper jaws, eight holochordal-procoelous presacral vertebrae (occasionally anterior ones fused), cylindrical sacral diapophyses, a firmisternal pectoral girdle, and the astragalus and calcaneum fused only at their ends.

Parental care (female or male) apparently occurs in all dendrobatids. Males attract females by calling, amplexus is cephalic or absent, and the eggs are laid on the forest floor, streamside, or in arboreal retreats. One parent guards the

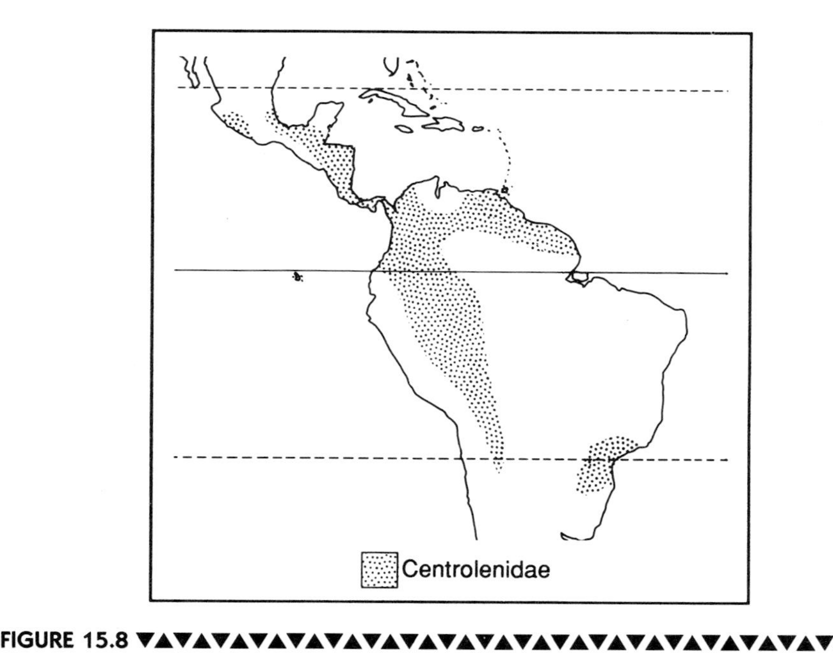

FIGURE 15.8
Distribution of the extant Centrolenidae.

eggs until they hatch, then the tadpoles slither onto the parent's back and are carried to a nearby pool or bromeliad for a standard tadpole development.

Family Heleophrynidae

The Heleophrynidal consists of a single genus (*Heleophryne*, five species) living in the mountain streams of South Africa (see Fig. 15.13). They are small (30–60 mm SVL), treefroglike anurans with expanded digit tips, which may assist their movements on wet boulders. They have indirect development with torrent-adapted tadpoles. The tadpole's oral disc is large, ventrally placed, and suctorial, allowing them to cling to rocks while grazing; remarkably, they lack a keratinous beak. They are characterized by the absence of ribs, teeth on premaxillae and maxillae, eight ectochordal-amphicoelous presacral vertebrae with a persistent notochord, cylindrical sacral diapophyses, an arciferal pectoral girdle, and the astragalus and calcaneum fused only at their ends.

Family Hylidae

Hylids are the treefrogs of temperate Eurasia and the Americas (Fig. 15.10). They exist in a variety of sizes (18–140 mm SVL) and shapes. Most hylids have

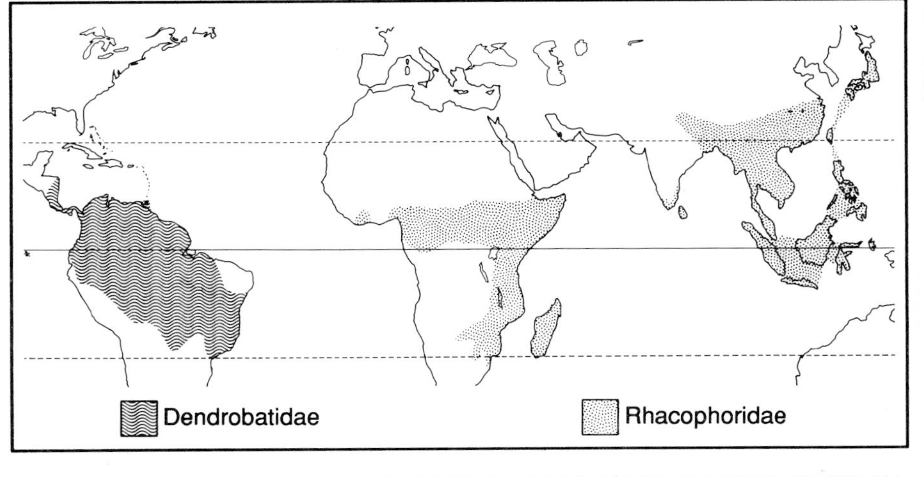

FIGURE 15.9 ▼▲▼▲▼▲▼▲▼▲▼▲▼▲▼▲▼▲▼▲▼▲▼▲▼▲▼▲▼▲▼▲▼▲
Distribution of the extant Dendrobatidae and Rhacophoridae.

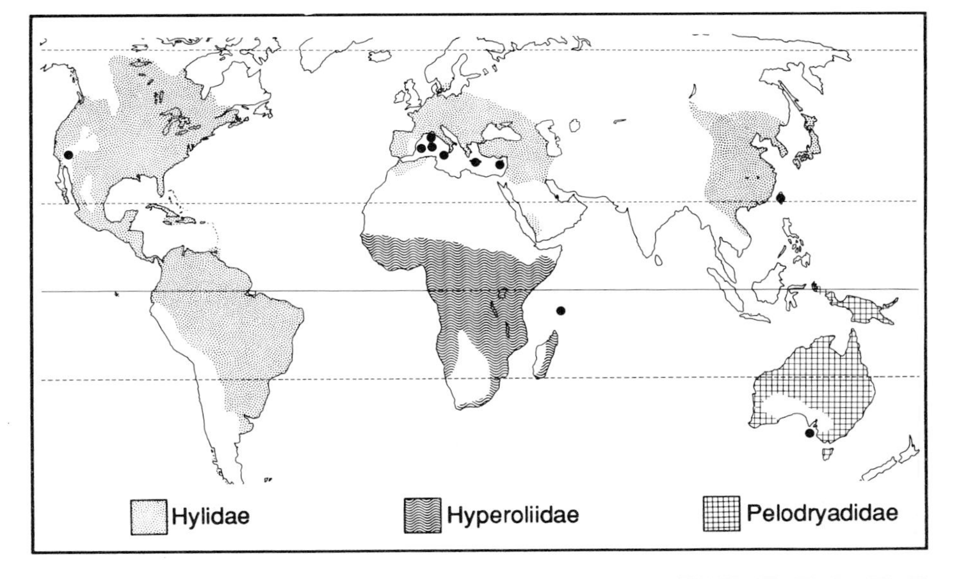

FIGURE 15.10 ▼▲▼▲▼▲▼▲▼▲▼▲▼▲▼▲▼▲▼▲▼▲▼▲▼▲▼▲▼▲▼▲▼
Distribution of the extant Hylidae, Hyperoliidae, and Pelodryadidae.

a slender-bodied and long-limbed treefrog habitus, but there are ranalike taxa (*Acris*) without expanded digit tips, and several groups of casque-headed species (*Osteopilus*, *Hemiphractus*). Hylids lack ribs, have premaxillary and maxillary teeth, eight holochordal-procoelous presacral vertebrae, expanded sacral diapophyses, arciferal pectoral girdles, and the astragalus and calcaneum fused only at their ends. Amplexus is axillary, and indirect development dominates.

Hylids are a diverse group (500+ species) with their greatest diversity in the New World tropics. Of the three subfamilies currently recognized (Hemiphractinae, Hylinae, Phyllomedusinae), hylines are the most speciose and widespread. *Hyla* (250+ species) occurs throughout much of the family's distribution and is the only hylid of Eurasia (3–4 species). Most hylines (e.g., *Acris*, *Hyla*, *Ololygon*, *Smilisca*) have typical anuran aquatic eggs and tadpoles. The phyllomedusines (e.g., *Agalychnis*, *Phyllomedusa*) of tropical America are highly arboreal frogs (one terrestrial exception) and lay eggs on leaves over water; upon hatching, the tadpoles drop into the water and develop in the usual tadpole manner. Hemiphractines (tropical South America) display various forms of extrauterine ovoviviparity. Some (e.g., *Hemiphractus*, *Stefania*) carry their eggs openly on their backs; others carry the eggs in an epidermal brood pouch on their backs. In the latter group, development is indirect (*Flectonotus*, *Fritziana*, some *Gastrotheca*), and the tadpoles are held briefly before being released into arboreal water containers (e.g, bromeliads, bamboo stems). Other *Gastrotheca* carry the eggs for full term and tiny froglets emerge from the dorsal pouches.

Family Hyperoliidae

The reed frogs are small- to medium-sized treefrogs (15–82 mm SVL), predominantly African with a few species on Madagascar and one species in the Seychelles (Fig. 15.10). Though less diverse (220± species) than the hylids, they have an equal variety of body forms and habits. Most are arboreal, but there are also terrestrial savanna and forest species, fossorial, rock-dwelling (saxicolous), grass-clinging, and even floating vegetation species. Many are brightly colored, often enameled in appearance, and others have more somber, cryptic colors. Hyperoliids share the absence of ribs, presence of premaxillary and maxillary teeth, eight holochordal-procoelous presacral vertebrae, cylindrical sacral diapophyses, a firmisternal pectoral girdle, the astragalus and calcaneum fused only at their ends, axillary amplexus, and direct development.

Three groups (Hyperoliinae, Kassininae, Leptopelinae) can be discerned in the hyperoliids. Leptopelines include only the speciose *Leptopelis* (40+ species) of equatorial Africa; most are arboreal forest species. Males call solitarily; eggs are laid in holes near water, and tadpoles must move to water upon hatching. Kassinines (*Kassina*, *Kassinula*, *Phlyctimantis*, *Tornierella*; 18–20 species) occur throughout sub-Saharan Africa and include terrestrial and arboreal species. Males commonly call in choruses in or near water; eggs are deposited in water and development is indirect. Hyperoliines (9 genera, e.g., *Afrixalus* with 20+ species; *Heterixalus* and *Hyperolius* with 120 species; *Tachycnemis*) occur throughout the range of the family and occupy a full range of habitats from fossorial to arboreal. Eggs are deposited in or over permanent water, on the ground, and in arboreal water containers. In all known cases, development is indirect, and if the eggs are outside of water, the tadpoles must reach water upon hatching.

Family Leptodactylidae

The leptodactylids are an exceptionally diverse group of anurans (50+ genera, 800+ species). This diversity is remarkable in that it occurs totally within the Neotropics (see Fig. 15.13); the few leptodactylid species in the Nearctic represent neotropical genera. Leptodactylids occur from sea level to above the tree line in the Andes, are fossorial, terrestrial, arboreal, and full aquatic, and range from some tiny *Eleutherodactylus* (12–15 mm SVL) of the West Indies to giants such as *Batrachophrynus* (14 cm SVL) and *Leptodactylus* (20 cm SVL). Even with this high diversity, leptodactylids appear to be a monophyletic group. All share the absence of ribs, have dentate premaxillae and maxillae, eight holochordal-procoelous presacral vertebrae, cylindrical sacral diapophyses, an arciferal pectoral girdle, and the astragalus and calcaneum fused only at their ends. The preceding "all" has a few exceptions for some traits, the exceptions usually occurring in a specialized member of a species group.

Leptodactylid diversity divides into four major groups: Ceratophryinae, Hylodinae, Leptodactylinae, and Telmatobiinae. Ceratophryines are the horned frogs (*Ceratophrys*, *Lepidobatrachus*) of equatorial and southern South America. They are moderate to large and have large triangular heads with wide gapes, large, robust bodies, and often fleshy horns over the eyes. They are terrestrial frogs and aggressive predators. The hylodines contain three genera (<20 species) from southeastern Brazil. *Crossodactylus* and *Hylodes* are small (<50 mm SVL), forest-streamside, diurnally active species; in contrast, the monotypic *Megaelosia* is a larger (110–120 mm SVL) streamside frog. The remaining genera are divided between the Leptodactylinae and Telmatobiinae, about two-thirds to the latter. Both subfamilies contain small to large species. Most leptodactylines are terrestrial, whereas telmatobiines range from aquatic to arboreal.

Eleutherodactylus is the most speciose (400+ species) telmatobiine. They range from tiny species (12–20 mm SVL) to moderate-sized ones (60–80 mm), and from semiaquatic to arboreal species. All have direct development, lay large-yolked eggs (10–100 per clutch depending on the species), and in many a parent guards the eggs in a terrestrial/arboreal retreat. A single species (*E. jasperi*, Puerto Rico) is ovoviviparous; it must have internal fertilization, but the mechanism is unknown. Females carry three to five embryos; their development is rapid with birth 32 days following amplexus. Their nourishment derives entirely from the egg yolk, although they have large fan-shaped tails for intrauterine gas exchange.

Eastern narrow-mouth toad *Gastrophryne carolinensis* (Microhylidae). (R. W. Barbour)

Family Microhylidae

Microhylids have a worldwide although discontinuous distribution (Fig. 15.11). This distribution reflects their propensity for tropical, moist habitats. Most are small (<40 mm, 12–100 mm SVL), usually stout-bodied with sturdy, short limbs and a small head, although some have a treefrog habitus. Skin texture varies from smooth to fine-warty. They have a unique tadpole with a single median spiracle and no keratinous mouthparts (except for one genus). With limited exceptions, microhylids are also characterized by two to three palatal folds, toothless premaxillae and maxillae, eight holochordal-procoelous presacral vertebrae, no ribs, broadly expanded sacral diapophyses, a firmisternal pectoral girdle (clavicles commonly reduced or absent), astragalus and calcaneum fused only at their ends, and axillary amplexus.

They are a moderately diverse group (60+ genera, 280+ species). This diversity is reflected in the recognition of nine subfamilies: Asterophryninae, Brevicipitinae, Cophylinae, Dyscophinae, Genyophryninae, Melanobatrachinae, Microhylinae, Phrynomerinae, and Scaphiophryninae. Each subfamily, with the exception of the dyscophines, melanobatrachines, and microhylines, is confined to a single tropical area. Asterophrynines and genyophrynines are the New Guinean radiations with a few genyophryine outliers in the rain forests of northeastern Australia. Both subfamilies have direct development, and parental

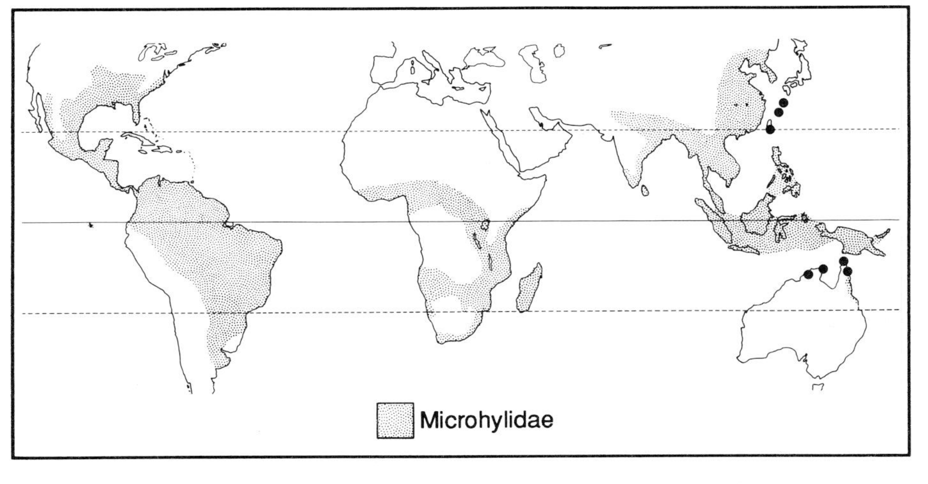

FIGURE 15.11 ▼▲
Distribution of the extant Microhylidae.

care appears to be common. Asterophrynines (e.g., *Barygenys*, *Xenobatrachus*) are terrestrial species, living in and on the forest floor or in grass tussocks above the tree line; all have the stout-bodied habitus. Genyophrynines (e.g., *Cophixalus*, *Oreophryne*, *Sphenophryne*) are treefroglike and have variously expanded digit tips; many are arboreal (Fig. 15.12), and others live among the boulders and cliffs of mountain streams.

Cophylines, scaphiophrynines, and dyscophines comprise the Madagascaran microhylid fauna; the first two are Madagascaran endemics. Cophylines are predominantly treefroglike in habitus and ecology. Scaphiophrynines and *Dyscophus* are terrestrial-semifossorial frogs, most with a stout-bodied habitus. Brevicipitines, melanobatrachines, and phrynomerines are the African microhylids. *Breviceps* and its relatives are nearly spherical in shape with head barely distinguishable and short, robust limbs protruding from the body. This globular appearance is further enhanced by their tendency to inflate the body when disturbed. They are burrowers with species adapted to forest through near-desert habitats. Males are commonly less than half the size of females; amplexus is impossible so the male becomes glued to the female's back during mating. Small clutches of eggs are laid in subterranean nests; development is direct. *Phrynomerus* (four species) is the sole member of the phrynomerines. They look like elongate, heavy-bodied *Dendrobates* with a similar skin texture and aposematic coloration. At least *Phrynomerus bifasciatus* has a toxic skin secretion, and all share diurnal-terrestriality and an ant diet with the dendrobatids. They have indirect development. Melanobatrachines show a peculiar disjunct distribution with two genera in the moun-

FIGURE 15.12 ▼▲
A treefrog *Cophixalus variegatus* (Microhylidae). (J. W. Lang)

tains of Tanzania, Africa, and one in southwestern India. These are small, toadlike anurans with indirect development.

Microhylines are predominantly terrestrial, stout-bodied, microcephalic frogs with 25 + genera divided nearly equally between the Americas and southern Asia. In both areas, species range from semiarid to rain forest habitats and usually are fossorial. Most have indirect development (e.g., American *Gastrophryne*, *Hypopachus*, Asian *Microhyla*), although a few have direct development (American *Myersiella*, Asian *Oreophryne*). Several American genera are tiny and have structural reductions, apparently because of dwarfing (e.g., *Syncope*, 13 mm SVL). The widespread Asian *Kaloula* (12+ species) has an elongate body and long limbs and is an active surface forager, even semiarboreal and scansorial in a few species.

Family Myobatrachidae

Myobatrachids are the dominant terrestrial frogs of Australia (Fig. 15.13), possessing the habits and occupying the habitats that elsewhere are occupied by bufonids, leptodactylids, microhylids, pelobatids, and ranids. These roles yield a diversity of genera (20), although they are not particularly speciose (100+). The genera are divided equally between two subfamilies: Limnodynastinae and Myobatrachinae. Both possess about the same diversity of life-styles (predominantly semiaquatic and terrestrial) and body sizes (20–80 mm SVL). Both also

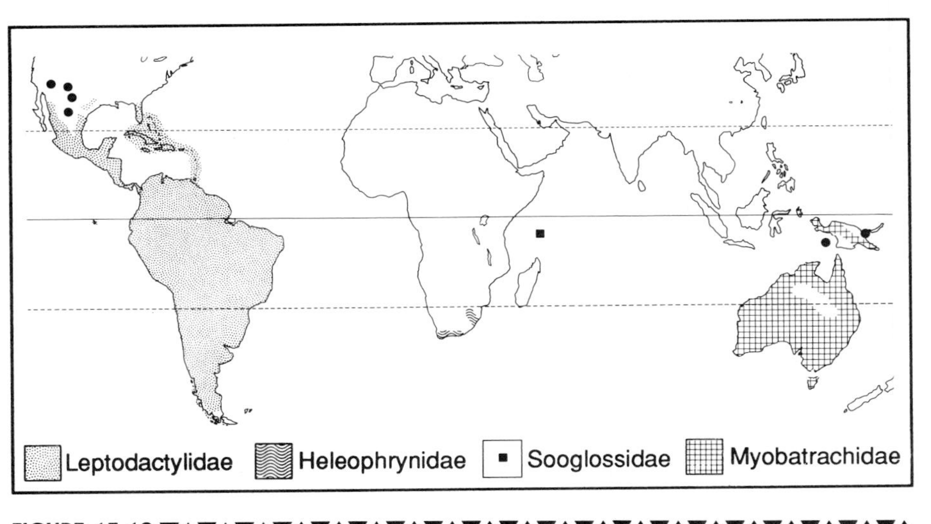

FIGURE 15.13 ▼▲▼▲▼▲▼▲▼▲▼▲▼▲▼▲▼▲▼▲▼▲▼▲▼▲▼▲▼▲▼▲▼▲▼▲

Distribution of the extant Leptodactylidae, Heleophrynidae, Myobatrachidae, and Sooglossidae.

share an absence of ribs, presence of premaxillary and maxillary teeth, eight amphicoelous presacral vertebrae with persistent notochords in juveniles (procoelous in many adults), expanded sacral diapophyses, an arciferal pectoral girdle, completely fused astragalus and calcaneum, and inguinal amplexus. Development is mainly indirect.

In the limnodynastines, some genera (e.g., *Mixophyes*, *Neobatrachus*) display the standard anuran breeding biology with aquatic eggs and free-living tadpoles; others (*Lechriodus*, *Limnodynastes*) produce a frothy egg mass that floats on the water until the tadpoles hatch and drop into the water; and a few (*Kyarranus*, *Philoria*) lay large yolked eggs in moist areas and have nonfeeding tadpoles. The myobatrachines *Crinia*, *Pseudophryne*, and others have a standard anuran breeding biology. *Assa* males carry eggs and larvae in inguinal pockets; female *Rheobatrachus* carry eggs and larvae in their stomachs. Both of the latter genera give "birth" to froglets.

Family Pelodryadidae

The New Guinean and Australian treefrogs (see Fig. 15.10) comprise the Pelodryadidae and are currently placed in three genera: *Cyclorana*, *Litoria*, and *Nyctimystes*. *Cyclorana* is terrestrial and rana-like, and the latter two genera are predominantly but not exclusively arboreal. They are moderately speciose (<140 species) and, with the exception of *Cyclorana*, have the typical treefrog habitus or a slight deviation thereof.

As close relatives of the hylids they share many familial characteristics, such as the absence of ribs, presence of premaxillary and maxillary teeth, eight holochordal-procoelous presacral vertebrae, expanded sacral diapophyses, an arciferal pectoral girdle, and astragalus and calcaneum fused only at their ends. All pelodryadids share a throat muscle that is absent in the hylids. All have the standard anuran biphasic life cycle: laying eggs in water, free-living tadpoles, distinct metamorphosis into terrestrial/arboreal froglets. As a terrestrial group, *Cyclorana* differs substantially and its affinities to the treefrogs has been confirmed only recently. A striking aspect of their life is highlighted by their colloquial name, water-holding frogs. They are capable of storing large volumes of free water in their bodies, and prior to the dry season, they burrow into the ground and, if available, absorb water from the soil. Under extreme dehydrating conditions, they reduce desiccation by lowering their metabolic activity (estivation) and by encasing themselves in epidermal cocoons.

Family Pseudidae

The four pseudid species are notorious for their giant tadpoles that transform into smaller frogs. Pseudids comprise two genera (*Pseudis* and *Lysapsus*) of semi-aquatic, small- to moderate-sized frogs of South America (see Fig. 15.14). They are stout-bodied with large hindlimbs and fully webbed feet. They possess eight holochordal-procoelous presacral vertebrae, arciferal pectoral girdles, and cylindrical sacral diapophyses; in addition, they lack ribs, have a dentate upper jaw, astragalus and calcaneum fused only at their ends, and axillary amplexus. The best-known member of this family is *Pseudis paradoxa*, because it has the largest tadpoles (to 220 mm TL, 98 g), yet the largest adult is no larger than 120 mm (SVL)(Fig. 15.15).

Family Ranidae

Ranids are an extremely diverse group (50± genera, 700+ species). They occur worldwide (Fig. 15.16) although their distribution in South America and Australia is limited and confined to a few species of *Rana* in each area. In fact, the distribution of the family is largely defined by the occurrence of *Rana* species (only *Rana* occurs in the Americas). *Rana* is a paraphyletic assemblage of species, which have proved resistant to phyletic analysis owing to high uniformity of morphology and parallelism. In contrast to this uniformity, ranids also possess divergent lineages; some that are likely very ancient and are occasionally recognized as distinct families. Arthroleptinae, Astylosterninae, and Hemisinae are three such groups; in addition, three other subfamilies (Mantellinae, Phrynobatrachinae, Raninae) fall more comfortably under the ranid mantle. Generally, all six subfamilies share the following traits: ribs absent, premaxillary and maxillary dentate (edentate in mantellines, hemisines, some arthroleptines), eight holochordal-procoelous presacral vertebrae, cylindrical sacral diapophyses, a

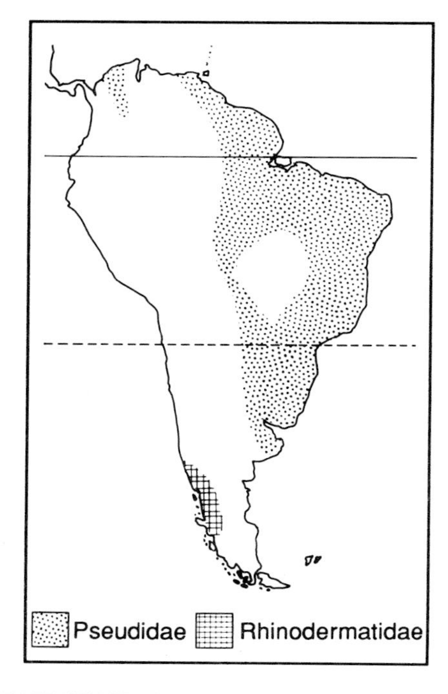

FIGURE 15.14 Distribution of the extant Pseudidae and Rhinodermatidae.

firmisternal pectoral girdle, astragalus and calcaneum fused only at their ends, axillary amplexus, and direct development.

Arthroleptines, astylosternines, hemisines, and phrynobatrachines are African endemics, occurring throughout sub-Saharan Africa. *Hemisus* with eight species (30–60 mm SVL) is the sole hemisine. These burrowers look like microhylids with their stout bodies and microcephalic heads. They are grassland or scrub forest inhabitants and lay small clutches of large eggs in burrows; supposedly when the tadpoles hatch, the guarding female digs an escape tunnel to a nearby pond. Arthroleptines (three genera) are miniature rana-like frogs (most <40 mm SVL) with direct development. Astylosternines (five genera) are small- to moderate-sized rana-like frogs, largely confined to western Central Africa. They have indirect development. Phrynobatrachines are the most diverse group (11 genera, 80+ species) of these African endemics. Most phrynobatrachines are small (<25 mm SVL), rana-like frogs with lightly warty skin. They are terrestrial but live in moist habitats, often with standing water. Some (e.g., *Cacosternum*, *Phrynobatrachus*, *Natalobatrachus*) have aquatic eggs and free-living tadpoles; others show various types of direct development, *Arthroleptella* with motile, nonfeeding tadpoles, *Anhydrophryne* with froglets hatching from terrestrial eggs.

FIGURE 15.15
Paradox frog *Pseudis paradoxa* (Pseudidae). (A. Cardoso)

Skipping frog *Rana cyanophylictus* (Ranidae). (G. R. Zug)

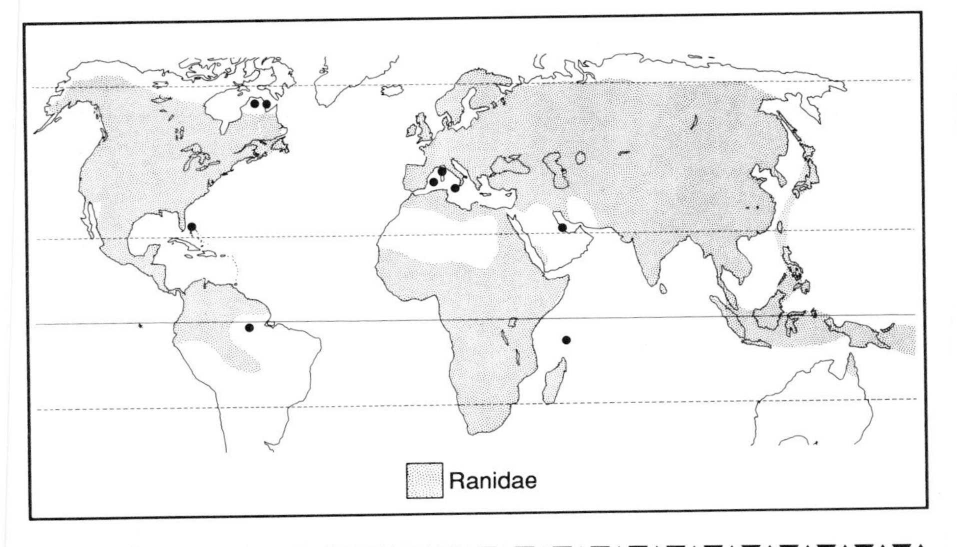

FIGURE 15.16 Distribution of the extant Ranidae.

Mantellines (three genera) are Madagascaran endemics. *Mantidactylus* is moderately speciose (50+ species). *Mantella* is less diverse (<12 species); most species are small and brightly colored. Ranines are diverse (25+ genera, including *Rana*) and occur worldwide. Many of the genera have the ranalike habitus (e.g., African *Conraua* and *Ptychadena*, Asian *Occidozyga*) and others are treefrog-like (Asian *Amolops*, some *Platymantis*), toadlike (*Tomopterna*), and aquatic (*Aubria*, *Staurois*). They range from the small Asian *Amolops* (20–30 mm SVL) to the giant African *Conraua* (30 cm SVL). Most ranines have indirect development but a few (e.g., *Discodeles*, *Platymantis*) have direct development with froglets hatching from terrestrial eggs (Fig. 15.17).

Family Rhacophoridae

Rhacophorids are the treefrogs of Asia, Madagascar, and Africa (see Fig. 15.9). They are moderately diverse (10 genera, <200 species) and range in size from 15 to 120 mm SVL with most >40 mm. Each geographical area has a distinct set of genera: *Chiromantis* in Africa; *Aglyptodactylus* and *Boophis* in Madagascar; and *Philautus*, *Polypedates*, *Rhacophorus*, and four other genera in Asia. With the exception of the rana-like *Aglyptodactylus*, rhacophorids are strongly arboreal and have the typical treefrog habitus and expanded digit tips. As a group, they lack ribs, have premaxillary and maxillary teeth, eight holochordal-procoelous presacral vertebrae, cylindrical sacral diapophyses, a firmisternal pectoral girdle, and astragalus and calcaneum fused only at their ends. Amplexus is axillary, and both indirect and direct development occurs.

FIGURE 15.17
Fijian treefrog *Platymantis vitiensis* (Ranidae). (G. R. Zug)

Two subfamilies have been proposed: Philautinae for *Philautus* and Rhacophorinae for all the other genera. Yet *Philautus* appears more closely related to the other Asian genera than to the Madagascaran or African taxa. *Philautus*, *Nyctixalus*, and *Theloderma* are reproductively similar, sharing tree-hole nest sites and nonfeeding larvae with a brief developmental period. *Chiromantis*, *Polypedates*, *Rhacophorus*, and others lay their eggs in foam nests in trees above water; hatching tadpoles drop into the water. The foam nests are often jointly created by two or more pairs of amplectant frogs, and at least in *Chiromantis*, unpaired males may assist. The Madagascaran species lay eggs in water and have a typical aquatic tadpole life cycle.

Family Rhinodermatidae

Rhinoderma has two species living in the temperate Pacific forests of Chile and Argentina (see Fig. 15.15). Both species are small (30 mm SVL) and bear fleshy appendages on the tips of their snouts. They are unique among anurans in laying eggs on land and then the male picks them up and carries them in his buccal cavity (*R. rufum*) or vocal sacs (*R. darwinii*). In the former, the tadpoles are released into water when they hatch and in the latter carried until the larvae metamorphose. Rhinodermatids lack ribs and premaxillary and maxillary teeth, have eight holochordal-procoelous presacral vertebrae, broadly expanded sacral diapophyses, a pseudofirmisternal arciferal pectoral girdle, and the astragalus and calcaneum fused only at their ends.

Family Sooglossidae

The sooglossids consist of three species of small frogs (11–55 mm SVL), occurring in mountain forests on two of the granitic Seychelle Islands (see Fig. 15.13). They are peculiar frogs. Males have voices and call individually not in choruses, but both females and males lack tympana (eardrums). They have eight procoelous presacral vertebrae with a persistent notochord, an arciferal pectoral girdle, expanded sacral diapophyses, and an extra bone in each hindlimb. The eggs are terrestrial. *Sooglossus gardineri* lays its eggs beneath leaves and stays with them for three to four weeks until they hatch into tiny froglets. *Sooglossus sechellensis* also lays its eggs beneath debris on the forest floor and guards them for two to three weeks; the eggs hatch as nonfeeding tadpoles that crawl up onto the female's back and remain until they metamorphose into froglets. The reproductive behavior of the third species, *Nesomantis thomasseti* (Fig. 15.18), is unknown; gravid females have large eggs, hinting at direct development.

Phylogenetic Relationships of Frogs

In the past, both Salientia and Anura have been used as the ordinal names for frogs. Both names still apply to frogs, but Salientia is used currently to

FIGURE 15.18 ▼▲▼▲▼▲▼▲▼▲▼▲▼▲▼▲▼▲▼▲▼▲▼▲▼▲▼▲▼▲▼▲▼▲▼
Seychelle rock frog *Nesomantis thomasseti* (Sooglossidae). (G. R. Zug)

encompass the Proanura, an ancient and extinct lineage, and the Anura, its sister lineage. Anura contain all extant frogs as well as their extinct relatives.

Our knowledge of the relationships of the living groups of frogs has improved considerably during the last decade; there are, nonetheless, still many areas of uncertainity. The most primitive (i.e., with the largest set of ancestral character states) are the ascaphids and leiopelmatids. These frogs represent very ancient divergences from the ancestral anuran stock of all extant frogs. *Ascaphus*, at least biochemically, appears more closely related to the discoglossids, with *Leiopelma* being the sole survivors of an earlier divergence.

Subsequently and presumably in the mid-Mesozoic, anurans split into the two major lineages of modern frogs, discoglossids-neobatrachians and pelobatoids-pipoids (Mesobatrachia). Although this divergence remains debatable, some of the proposed relationships within the two proposed lineages have strong supporting data and a long history of acceptance; other proposals remain less certain. The pipoid pair (Pipidae, Rhinophrynidae), in spite of their dissimilar appearances and behaviors, are monophyletic. This monophyly is supported by several suites of synapomorphic (uniquely shared-derived) traits, such as type of tadpole and chondrocranial morphology. Immunological data and feeding behavior do not support the pipoid pair relationships or the affinities of *Rhinophrynus* to any other anuran group or family; these data indicate only the ancientness of *Rhinophrynus* and the divergence of the major anuran lineages. Within the pipids, siluranine and pipines are sister-groups, and they are the sister-group to the more divergent dactylethrines (*Xenopus*).

Linking the pipoids and pelobatoids (Pelobatidae, Pelodytidae) is a recent proposal. Previously pelobatoids were considered an intermediate group between the primitive and advanced frogs. Their intermediacy derived from a mix of supposedly primitive and advanced character states. Reevaluation reveals that pelobatoids are not intermediate between ascaphids-discoglossids and neobatrachians; instead they are a specialized group with closer, albeit ancient, affinities to pipoids. Within the pelobatids, the megophryines are strongly morphologically divergent from the other pelobatids and are occasionally recognized as a distinct family.

The recent proposal of a discoglossids-neobatrachian lineage is novel, because earlier ones proposed the pelobatids as an intermediate grade in anuran evolution. Discoglossids do share a number of primitive traits with leiopelmatids and ascaphids; however, their apomorphic traits do not support a close affinity to these two families or indicate a close relationship to pelobatoids. Rather they are a sister group to all the neobatrachian families. Once again, this family and each subfamily represent ancient splits, nonetheless the bombatorines and discoglossines are each other's nearest living relative.

The neobatrachines are the dominant anurans today. Their diversity hides relationships and numerous different patterns have been proposed. In one scheme (Fig. 15.19), two main branches are evident: bufonoids (hyloids) includ-

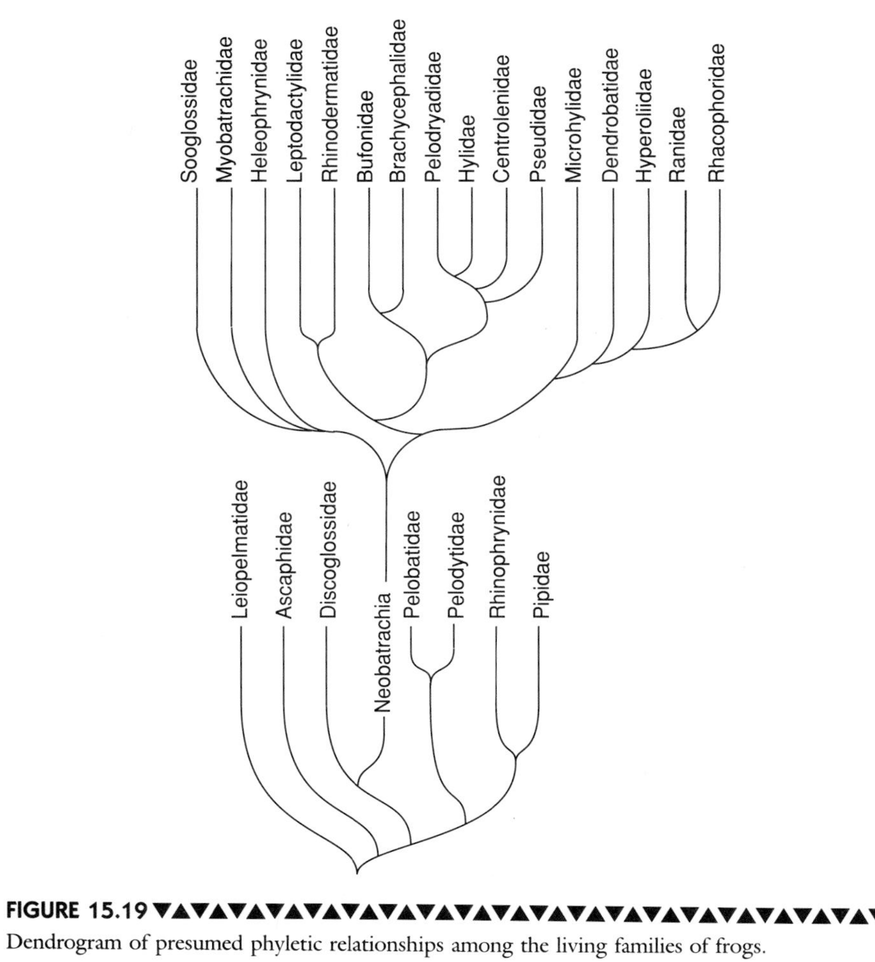

FIGURE 15.19 ▼▲▼
Dendrogram of presumed phyletic relationships among the living families of frogs.

ing leptodactylids, bufonids, hylids, and their relatives; ranoids with microhylids, ranids, and their relatives. These two branches are not novel, and their basic familial representation has long recognition. Past differences were mainly associated with the placement of the heleophrynids, myobatrachids, and sooglossids. The former two families were linked to the leptodactylids, either as sister-group families or contained within the leptodactylids as subfamilies. The sooglossids were considered to be ranids or derived from them. Several features, including vertebral morphology and amplexus type, support a relationship of sooglossids with leptodactylids. Other traits, however, indicate that these three families represent a lineage diverging early from the main neobatrachian stock and subsequently diverging from one another.

Among other features, the bufonoids and ranoids are distinguished by the nature of the pectoral girdle, arciferal versus firmisternal, respectively. The bufonoids are the most diverse group with radiations on each of the southern continents. The family composition of the three subgroups (leptodactylids-rhinodermids, toads, treefrogs) is seldom questioned, although enigmatic genera occur in each and have been shuffled among the family groups, for example, *Allophryne* between hylids and bufonids. All three subgroups show major adaptive radiations in Latin America. Only the toads show similar radiations elsewhere (Africa, tornierobattines; southern Asia, adenomines). The Australian treefrogs (pelodryadids), though speciose, are structurally and reproductively uniform.

Microhylids are clearly within the ranoid radiation but diverged early from other ranoids. With the exception of the microhylines, each subfamily represents a radiation within a limited geographical area. The microhyline exception is probably a taxonomic artifact, since the American genera appear to be a monophyletic group and the Asian microhylines are unstudied in this respect. The single American radiation contrasts sharply with the multiple radiations in Africa, Madagascar, and New Guinea.

Of all anuran groups, the ranid subgroup is the most controversial, not in what genera are included, but how they are divided and assigned taxonomically. Until the 1970s, Ranidae included all the aquatic-terrestrial genera, and Rhacophoridae had all the treefrogs. Since then, research has revealed that the intergeneric relationships are more complex than formerly imagined. Even the usage of Ranidae here makes it paraphyletic. The two treefrog guilds (hyperoliids, rhacophorids) are indeed distinct and have separate origins. The rhacophorid divergence is relatively more recent than that of the hyperoliids. Furthermore, the arthroleptines and astylosternines are as morphologically distinct from one another as from the hyperoliids, rhacophorids, and other ranid subfamilies, suggesting origins equally as old as that of the hyperoliids and recommending their recognition as separate families. *Hemisus* is also a distinct lineage and is sometimes recognized as a separate family. The situation with the other ranid subfamilies is less clear; the main difficulty is the absence of a phylogenetic analysis of all *Rana* species to determine how many lineages exist under this single name and how each relates to other subfamilies and ranine genera.

The relationships of dendrobatids remain unresolved. They have been linked by various researchers with bufonids, leptodactylids, or ranids. Although small suites of characters suggest each relationship, the evidence is equivocal. Recent immunological tests showed little or no cross-reactions with members of each of the preceding families, thus suggesting a very ancient origin (>100 MA).

CHAPTER 16

Turtles and Crocodilians

TESTUDINES

Turtles are nature's tanks. Their armor derives from elements of the axial and appendicular skeleton as well as neomorph skeletal plates. The carapace (dorsal shell) contains dermal bones (costal and neural elements) overlaying and fused to the trunk vertebrae (10) and ribs (8) and a continuous ring of neomorphic peripheral, nuchal, and pygal dermal elements. The plastron (ventral shell) contains a few elements of the pectoral girdle and sternum but is largely an articulated plate of dermal neomorphic plates. The carapace and plastron are joined laterally by bony bridges (mostly plastral bones) or ligament. The bony shell is a most effective predator defense because the limb girdles are within armor, and in most extant turtles, the limbs, head, and neck can be drawn within the shell. No other vertebrates have a comparable armor, and none has the girdles within the rib cage. The turtles' persistence for over 200 MA (Late Triassic and probably earlier) attests to the shell's effectiveness.

Although a most ancient group and the sole survivors of one of the earliest reptilian lineages (anapsids), turtles are highly specialized in many features in addition to the unique bony shell. Most noticeable is their abandonment of teeth and the substitution of an ever-growing keratinous jaw sheath, a feature independently evolved by the feathered archosaurians. Their highly mobile neck (eight cervical vertebrae) and its retraction mechanism are unique. Two different mechanisms have evolved: horizontal contraction or folding and vertical contraction with an S-shaped flexure. These retraction mechanisms proclaim the divergence of the two major groups of extant turtles: Pleurodira (sidenecks) and Cryptodira (hidden- or S-necks). Other unique features, such as the jaw-closing mechanism, indicate the monophyly of each group as well as their shared ancestory.

All turtles are oviparous. None have attained any form of live-bearing, and presumably all lay eggs in an early stage of embryogenesis. This latter aspect can result from delayed fertilization with a sperm-storage ability and/or development starting but then halting in the blastula or early gastrula stage until laid. Nesting is uniform in turtles. No matter what their hindfoot morphology, egg pits are dug with an alternate scooping motion of the left and right hindfeet or a minor deviation thereof.

Pleurodira

The sidenecks are the least diverse of modern turtles. Today, only two families (Chelidae, Pelomedusidae) survive. Both are strictly Southern Hemisphere groups and confined to freshwater habitats. Both share a jaw-closing mechanism levering over a pterygoid process, convex mandibular articular surfaces, pterygoid and basioccipital bones separated by the quadrate, a pelvic girdle fused to the plastron, and paired mesoplastral elements often present.

Family Chelidae

Chelids are moderately diverse (10 genera, 35+ species) and occur in tropical-temperate Australia and South America (Fig. 16.1). They range in size from 14 to 48 cm adult CL (carapace length) and in shape from smooth-shelled, ovoid-flattened, and ovoid-domed genera (most genera) to the rugose, knobbed-triridged, nearly rectangular shell of *Chelus*. All chelids appear to be carnivorous. They eat a variety of invertebrates and vertebrates, searching for their food on stream and lake bottoms. The matamata (*Chelus*) may stalk and herd prey, but more often remains hidden in the bottom litter and with a sudden neck thrust

Papuan snakeneck turtle *Chelodina parkeri* (Chelidae). (R. W. Barbour)

and a quick opening of the cavernous mouth suck in prey that come too close. Chelids are highly aquatic and good swimmers; a few bask, either at the water surface or aerially.

Chelids characteristically lack mesoplastral elements, usually have a cervical scute on the carapace, and show a reduction or loss of neural bones. Their head usually is covered with a smooth skin or small scales; skeletally, nasals and vomers are usually present and the premaxillaries and dentaries are unfused.

The Australian chelids contain three groups. *Pseudemydura* is the smallest (<15 cm CL) pleurodire and the rarest. It is confined to a small area of marshes and swamps in southwestern Australia. These are seasonal wetlands, and *Pseudemydura*'s annual life cycle is tied closely to the area's flooding and drying. During the latter, the turtles dig into the bottom and estivate. The snakenecks (*Chelodina*) and shortnecks (*Elseya*, *Emydura*, *Rheodytes*) are the other two groups; both are widespread in the watered portion of Australia. Snakenecks are unmatched in the length of their necks, which equal and often exceed the length of their shells; the shortnecks have the typical turtle-length necks. Both groups have tropical and temperate species groups. Reproduction in the former is keyed to the wet–dry cycle; nesting occurs in the dry season with hatching and emerging during the first heavy rains of the wet season; incubation, depending on species

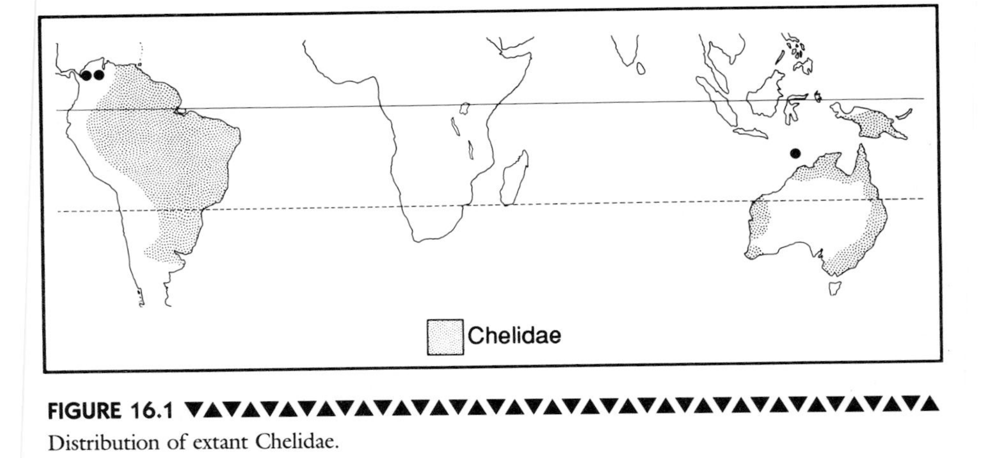

FIGURE 16.1 ▼▲

Distribution of extant Chelidae.

and weather conditions, is 75–325 days. In the temperate species group, nesting occurs during the warm spring and summer; hatching and emergence occur before the arrival of cool weather, with an incubation of 39–66 days.

The South American chelids also consist of three subgroups. The peculiar-appearing *Chelus* is the sole member of its subgroup, but unlike *Pseudemydura*, it is abundant and widespread throughout the major tropical drainages (Atlantic and Caribbean). *Hydromedusa* matches some of the Australian snakenecks in its neck length; *Phrynops*, *Acanthochelys*, and *Platemys* have typically turtle-length necks. The reproductive patterns of these turtles are sufficiently unknown to prevent generalities. Speciation for these four genera has occurred principally in temperate South America; only *Phrynops* and *Platemys* have tropical species.

Family Pelomedusidae

Pelomedusids (5 genera, 25+ species) occur in tropical-subtropical South America, Africa, and Madagascar (Fig. 16.2). They range in size from 12 to 107 cm CL and have smooth, domed, or flattened ovoid shells. There are two genera in South America (*Peltocephalus*, *Podocnemis*), two in Africa (*Pelomedusa*, *Pelusios*), and three in Madagascar (*Erymnochelys* and the two African genera). The American and Madagascaran pelomedusids are predominantly moderate to large (<30 cm CL) river turtles; they are active swimmers and herbivorous (uncertain for *Erymnochelys*). Dietary evidence is equivocal but hints at specialized diets for some of the American species, for example, *Podocnemis unifilis* and *P. erythrocephala* skim the water surface for tiny plants and vegetation fragments, whereas *P. expansa* feeds primarily, though not exclusively, in flooded forest and heavily on floating fruits, seeds, and flowers. The African species occur in a variety of aquatic habitats from seasonal ponds to rivers. They are poor swim-

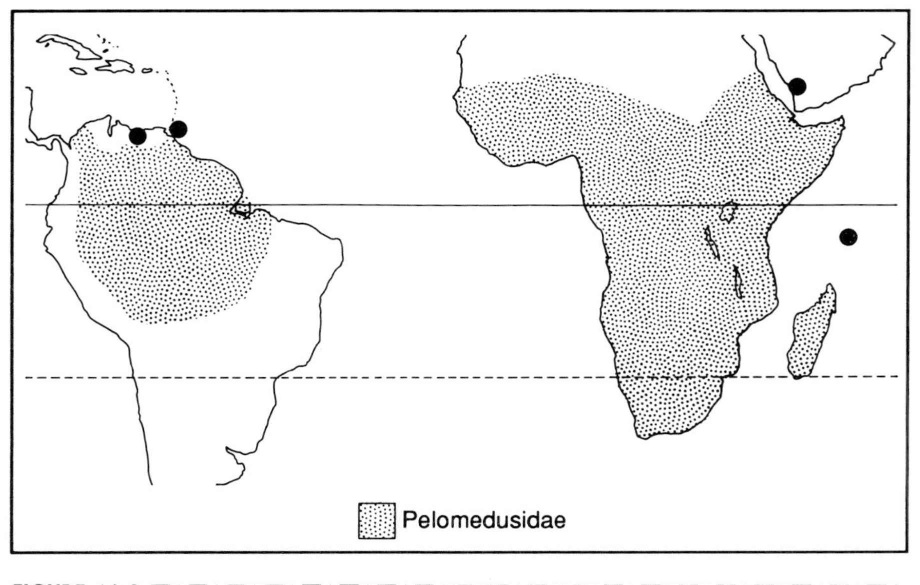

FIGURE 16.2 ▼▲▼▲▼▲▼▲▼▲▼▲▼▲▼▲▼▲▼▲▼▲▼▲▼▲▼▲▼▲
Distribution of extant Pelomedusidae.

mers, instead depend on bottom-walking. Most are small- to moderate-sized (<30 cm CL) and feed opportunistically on invertebrates and vertebrates.

Pelomedusids characteristically have a pair of mesoplastral elements, lack a cervical scute on the carapace, and possess a full series of neural bones. Their heads usually bear moderate to large scales; skeletally, nasals are usually present, vomers absent, and premaxillaries and dentaries fused.

The African *Pelusios* is the most speciose pelomedusid (15+ species). More than half of the species occur widely in several drainages and are sympatric with and very similar to one or more congeners, but it is unknown whether they occupy the same niches. *Pelomedusa* is monotypic (*P. subrufa*) and occurs throughout sub-Saharan Africa, frequently in the drier areas and even in temporary ponds.

Cryptodira

Cryptodires show a much greater diversity than the pleurodires. At least nine families and 40+ genera are currently recognized. The diversity includes marine, brackish, freshwater, and terrestrial (wet forest to desert) species. This diversity, however, occurs predominantly outside of the geographical range of pleurodires. Cryptodires are largely absent from freshwater and terrestrial habitats in the Southern Hemisphere, except for trionychids (Africa) and tortoises (Africa, South America). Both cryptodiran families have diversified in Africa; however, tortoises are terrestrial and have little, if any, interactions with pleurodires, but such a facile explanation is not available for the aquatic softshells (trionychids).

The nine families represent four separate evolutionary groups: seaturtles (Chelonioidea); snapping turtles (Chelydroidea); pond turtles and tortoises (Testudinoidea); and softshells and musk turtles (Trionychoidea). In addition to the vertical neck-retraction mechanism, all cryptodires share a jaw-closing mechanism levering over an otic process, a concave mandibular articular surface; pterygoid and basioccipital in contact, not separated by the quadrate; pelvic girdle with ligamentous attachments to plastron; and no mesoplastral element.

Family Cheloniidae

Cheloniid seaturtles are the hard-shelled marine turtles. They possess a bony carapace structurally identical to those of pond turtles and most other turtles, and their shells are covered by large, hard epidermal scutes. All bony elements of the plastron are present, although reduced. Shell shape is compressed-fusiform, an adaptation for the highly aquatic life of cheloniids. Similarly, the forelimbs are streamlined and compacted, forming aerofoil-style flippers for aquatic flight in which the flippers rotate and propel the seaturtle in both the up and down strokes. Their heads and necks are largely nonretractable, adding to their basic hydrodynamic body shape.

Green seaturtle *Chelonia mydas* (Cheloniidae). (G. R. Zug)

Cheloniids occur worldwide (Fig. 16.3) in subtropical-tropical seas, and many populations (either juveniles or juveniles and adults) migrate seasonally into temperate seas for feeding and/or nesting. The five extant genera form two subfamilies, Carettinae (*Caretta*, *Lepidochelys*) and Cheloniinae (*Chelonia*, *Eretmochelys*, *Natator*). Loggerheads, *Caretta*, are the most temperate cheloniids, and mostly nest on subtropical and temperate beaches. The other genera nest exclusively or predominantly in tropical areas. Seaturtles mature slowly and few populations contain females nesting before they are 20 years or older; some may require 40–50 years to attain sexual maturity and first reproduction. Typically, females nest every second to fourth year, but during a nesting season, they lay multiple clutches (usually two to five clutches per season). Green seaturtles (*Chelonia*) are herbivores, eating marine grass where available and otherwise subsisting on algae. The other seaturtles are carnivores and to some degree specialize on various invertebrates (e.g., *Caretta* on mollusks and crabs, *Eretmochelys* on sponges and other sedentary invertebrates).

Family Dermochelyidae

Leatherbacks, *Dermochelys coriacea*, are the largest living turtles (1.5–2.5 m CL, 500–1000 kg.) They occur in all major oceans (Fig. 16.4). Their nesting beaches are in the tropics, but at least adults migrate seasonally into cold seas, apparently reaching their northernmost (and presumably southernmost) points

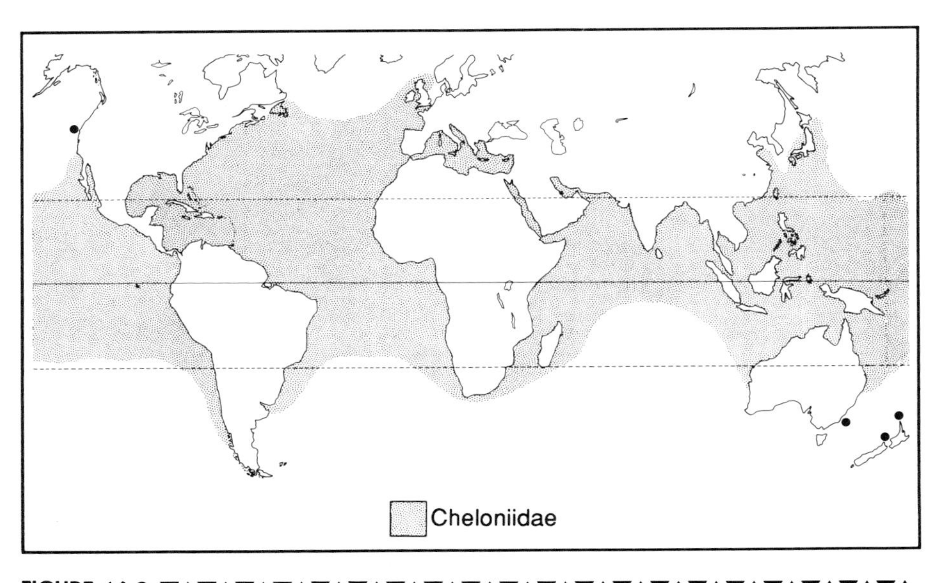

FIGURE 16.3 ▼▲
Distribution of extant Cheloniidae.

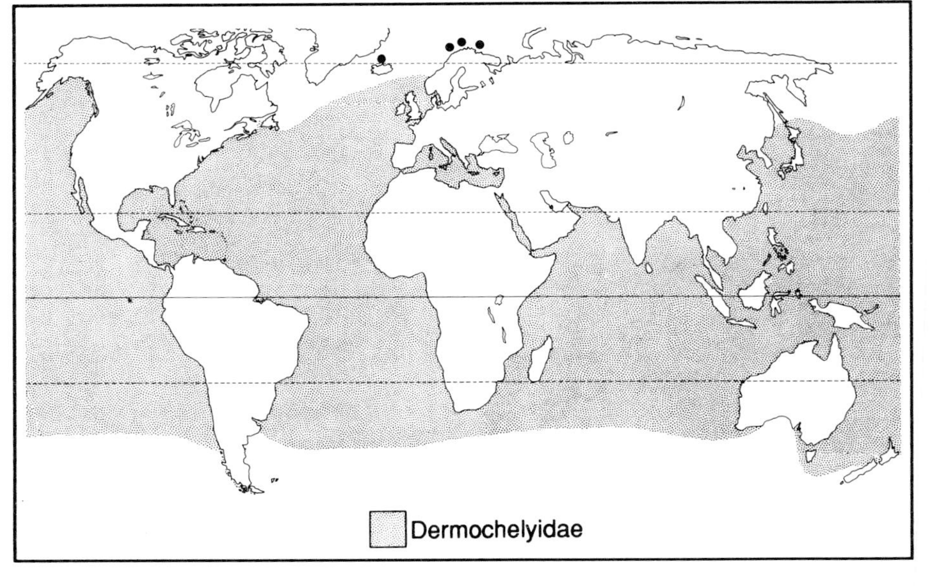

FIGURE 16.4 ▼▲▼▲▼▲▼▲▼▲▼▲▼▲▼▲▼▲▼▲▼▲▼▲▼▲▼▲▼▲▼▲▼▲▼▲
Distribution of the extant Dermochelyidae.

as the jellyfish blooms occur and then departing prior to the onset of sharply declining autumn temperatures. Their jellyfish diet is remarkable since it is 90% water, and of all the turtles, leatherbacks are the only ones that are functional endotherms. Part of their endothermic ability derives from their large size and the associated slow heat loss from such large bodies; however, they have an effective insulation in their oil-dense skin and a circulatory counterflow arrangement in their limbs that conserves body heat.

Dermochelys shares the depressed fusiform body and flipper forelimbs with the cheloniids. Their hydrodynamic shape and proportionately larger flippers enable them to swim at speeds in excess of 30 km hr^{-1}. They differ from the cheloniids in carapace structure. The large dermal bony plates have been lost and are replaced by a thin sheet of articulating osteoderms lying over but not fused to the ribs. Also, the hard keratinous scutes are absent (except briefly in hatchlings), and the shell surface is a smooth, tough skin. The bony plastral elements are present but further reduced from the cheloniid condition. Like the cheloniids, *Dermochelys* nests on a multiple-year cycle and several times within a nesting season.

Family Chelydridae

Chelydrids consist of three genera of American and Asian turtles (Fig. 16.5). All three are aquatic carnivores and range in size from the small Asian *Platysternon*

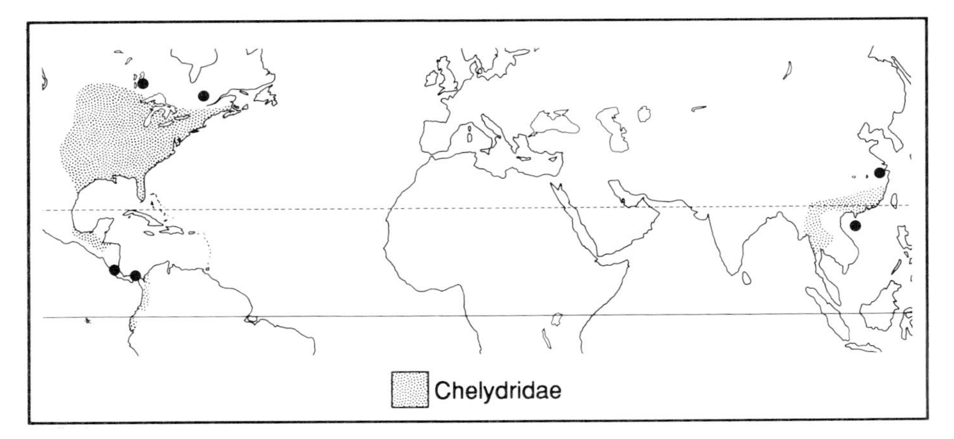

FIGURE 16.5 ▼▲▼▲▼▲▼▲▼▲▼▲▼▲▼▲▼▲▼▲▼▲▼▲▼▲▼▲▼▲▼▲▼▲
Distribution of the extant Chelydridae.

FIGURE 16.6 ▼▲▼▲▼▲▼▲▼▲▼▲▼▲▼▲▼▲▼▲▼▲▼▲▼▲▼▲▼▲▼▲▼▲
Big-headed turtle *Platysternon megacephalum* (Chelydridae). (R. W. Van Devender)

(10–18 cm CL) (Fig. 16.6) to the giant *Macroclemys* (40–80 cm). All are large-headed, large-bodied, and stout-limbed turtles; their carapaces are flattened, triridged, and nearly rectangular in outline, and the plastrons are reduced, greatly so in *Chelydra* and *Macroclemys*. Of all living turtles, these three have very long tails, nearly equaling the length of their carapaces.

Among freshwater turtles, chelydrids appear to be the most active nocturnal foragers. They eat a broad assortment of animal prey (living and dead). Fish are an important dietary item for all three, but invertebrates, including molluscs (especially for *Macroclemys*), worms, and crustaceans, are common items. *Macroclemys* is highly aquatic and seldom leaves the water except for egg-laying. In contrast, *Platysternon* regularly forages on shore, and *Chelydra* is intermediate, foraging mainly in the water but often moving overland between ponds or streams.

Family Carettochelyidae

Carettochelys insculpta, the pig-nosed turtle of southern New Guinea and northern Australia (Fig. 16.7), is the sole living member of this family. It is a moderately large (55 cm CL maximum), aquatic turtle. Its forelimbs are flipperlike and, like seaturtles, it swims via aquatic flight (see Cheloniidae). The shell is complete dorsally and ventrally, ovoid in outline, high-domed, and covered by a heavy, rugose skin. The head ends in a thick, snorkellike snout, hence its colloquial name. *Carettochelys* is a denizen of rivers and lakes, preferring slow-moving backwaters and lagoons. It is an omnivore, seemingly with a preference for vegetable matter, eating both the fruits and seeds of bankside trees as well as seeds, roots, and stems of aquatic vegetation.

Family Dermatemydidae

Dermatemys mawii is also the sole living member of its family and shares the complete, ovoid, thick-boned shell (65 cm maximum) of *Carettochelys*; however, thin epidermal scutes cover the shell. It inhabits the lakes and slow-running rivers of southern Mexico (Fig. 16.8) and feeds almost entirely on vegetable matter, both submergent aquatic plants and fruits and leaves of shoreside plants. It is

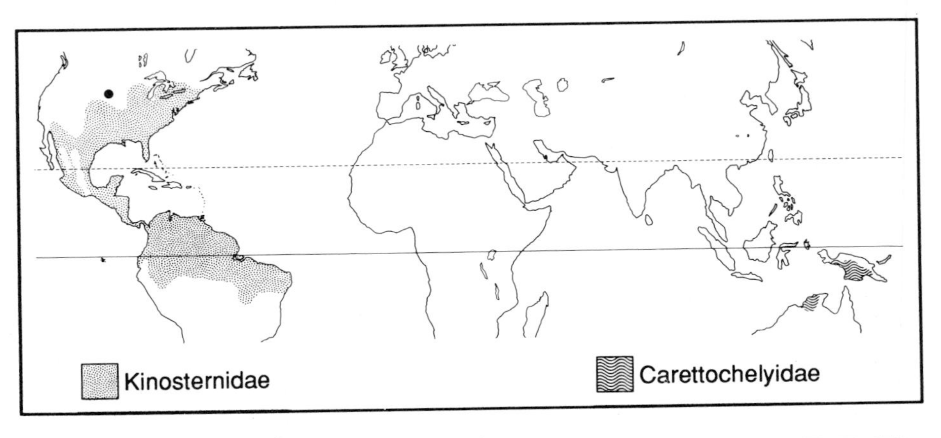

FIGURE 16.7 ▼▲
Distribution of the extant Carettochelyidae and Kinosternidae.

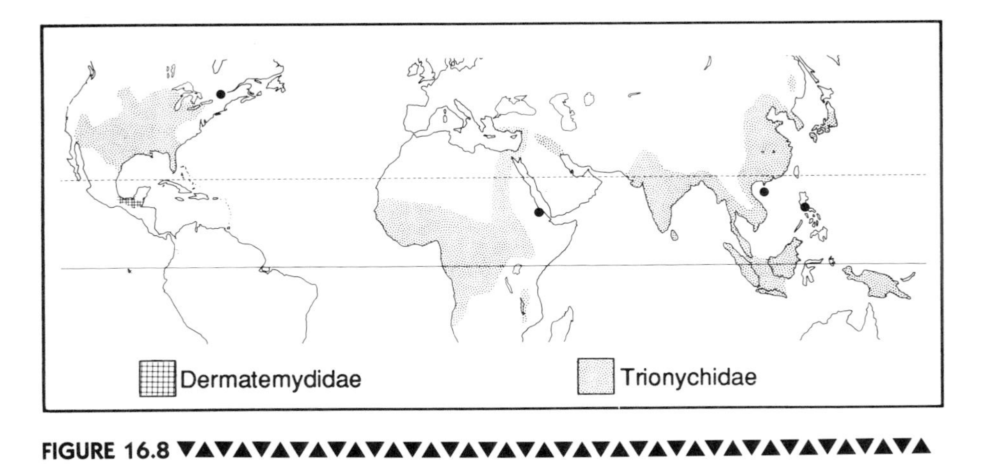

FIGURE 16.8 ▼▲

Distribution of the extant Dermatemydidae and Trionychidae.

highly aquatic and nearly incapable of movement on land; only the females come ashore to lay their eggs, apparently immediately adjacent to water.

Family Kinosternidae

Kinosternid turtles occur widely in the Americas (see Fig. 16.7). There are four genera representing two ancient groups: Kinosterninae—*Kinosternon*, *Sternotherus*; Staurotypinae—*Claudius*, *Staurotypus*. All four taxa are aquatic, but many species (*Kinosternon*) occasionally forage on land and use terrestrial retreats for summer estivation or burrow in the bottom of ponds during droughts (*Claudius*, *Kinosternon*). They are poor swimmers and forage actively by bottom-walking. Their carnivorous diets encompass a broad spectrum of invertebrate and vertebrate prey. A few species (e.g., *Sternotherus minor*) specialize on molluscs and have broad jaw surfaces and large jaw muscles to crush snail and bivalve shells.

Kinosternines are predominantly small species (<15 cm CL, one species to 25 cm) with high-domed, oblong shells. *Kinosternon* is the most speciose (15 species) and widespread of all kinosternids. They have large hinged (single or double) plastrons, which can fully close the shell in some species (*K. acutum*, *K. integrum*). They inhabit the full range of aquatic habitats from temporary desert ponds and rivers to large rivers and lakes. *Sternotherus* (4 species) includes smaller turtles (rarely >12 cm CL), usually living in permanent water and restricted to eastern North America. Staurotypines have greatly reduced, cruciform plastrons. They occur in slow-moving coastal streams of northern Middle America. *Staurotypus* with two species is a giant among the kinosternids, reaching nearly 40 cm CL in *S. triporcatus*. *Claudius* is much smaller (<15 cm).

Family Trionychidae

Softshells are a diverse (14 genera) and nearly cosmopolitan group (Fig. 16.8). These flattened, pancake-shaped turtles have reduced bony carapaces and plastrons (Fig. 16.9). The carapace lacks the ring of peripheral bones (except *Lissemys*), and costal plates are short with the rib ends projecting out of them. The plastral bones are reduced to a trellislike frame as in seaturtles. The surface of the shell is a thick, leathery skin with no hint of epidermal scutes. The head typically ends in a protruding snorkellike snout, and with their exceptionally long necks, trionychids can lie hidden on the bottom and extend their neck until only the tip of the snout projects above the water surface. A pancake shape would seem ill-suited for aquatic locomotion, but to the contrary, trionychids are excellent and fast swimmers, propelled by rowinglike strokes of all four strongly webbed feet. They are predominantly carnivorous, although some tropical species (e.g., *Trionyx triunguis*, *Chitra indica*) may seasonally include plant matter in their diet, and some populations may specialize on molluscs or other abundant prey. Prey are captured by both foraging and ambush.

There are two groups of trionychids: Cyclanorbinae of Africa and greater India and the cosmopolitan Trionychinae. They are easily recognized by the presence of femoral flaps on the plastron of cyclanorbines and their absence in trionychines. Cyclanorbines include only three genera (four species), the Indian *Lissemys* and the African *Cycloderma* and *Cyclanorbis*; they show no apparent

FIGURE 16.9 ▼▲▼▲▼▲▼▲▼▲▼▲▼▲▼▲▼▲▼▲▼▲▼▲▼▲▼▲▼▲
Smooth softshell turtle *Apalone mutica*. (R. W. Barbour)

differences in general behavior and ecology from the trionychines. The trionychines, although geographically more widespread, are not particularly diverse (17 species); their greatest diversity is in southern Asia and southeastern North America.

Family Emydidae

The emydids include a variety of turtles ranging from terrestrial (*Terrapene*) through semiaquatic (*Clemmys*) to freshwater (*Pseudemys*) (Fig. 16.10) and estuarine (*Malaclemys*) taxa. With the exception of the European-Southwest Asian *Emys*, emydids are New World turtles (Fig. 16.11), and there, only *Trachemys* occurs in tropical America. The diversity of their ecologies is matched by a diversity of body shapes and sizes, and they share body shapes and sizes with their batagurine sister group (Testudinidae). The differences between these two groups are subtle and internal. The differences include such traits as an angular bone of the lower jaw touching Meckel's cartilage and a narrow basioccipital. The converse conditions occur in testudinids.

Emydids are small- to moderate-sized turtles ranging from *Clemmys guttata*

FIGURE 16.10 Hieroglyphic cooter *Pseudemys concinna* (Emydidae). (R. W. Barbour)

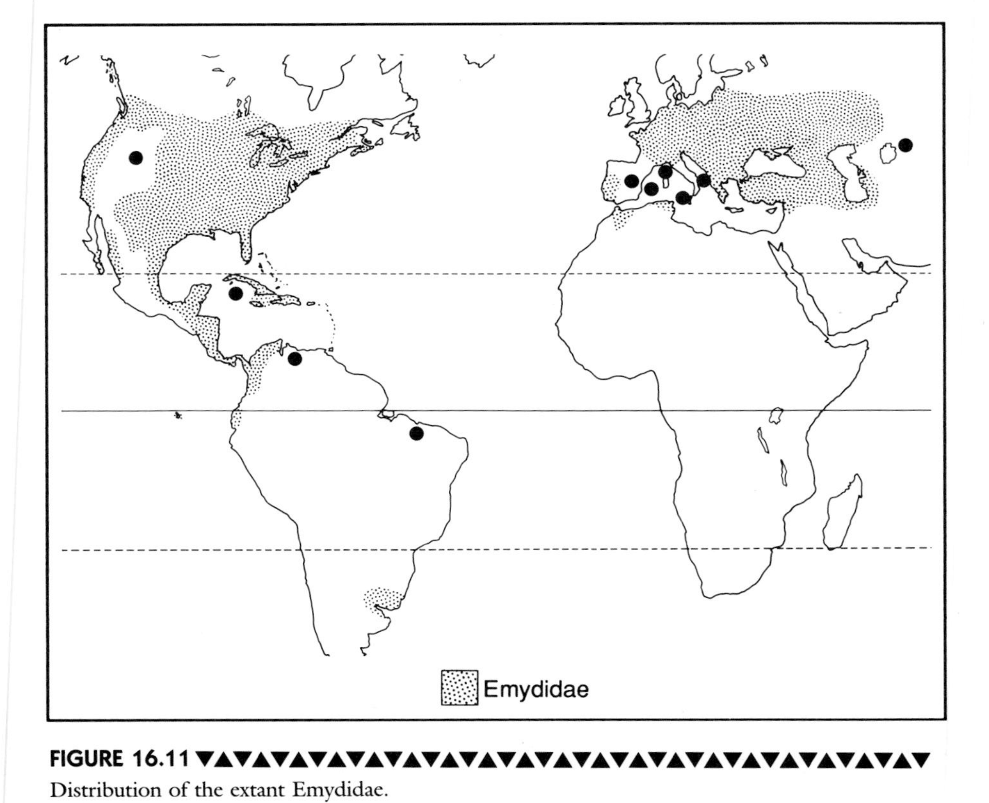

FIGURE 16.11 ▼▲▼▲▼▲▼▲▼▲▼▲▼▲▼▲▼▲▼▲▼▲▼▲▼▲▼▲▼▲▼▲▼▲▼
Distribution of the extant Emydidae.

and *C. muhlenbergii* (12 cm CL maximum) to *Trachemys scripta* (60 cm). Most emydids are omnivores by necessity but carnivores by proclivity when given the opportunity. *Graptemys* and *Malaclemys* prefer molluscs, and in some species, sexual size dimorphism results in the large adult females being molluscivores and the much smaller males predominantly insectivores. *Emydoidea* and *Deirochelys* specialize on crayfish and have evolved similar behaviors and morphologies to avoid injury during capture of these strongly clawed prey. *Pseudemys* and the larger *Trachemys* species are herbivores; their young are insectivorous and convert to herbivory as they grow larger.

Family Testudinidae

Testudinids are the most diverse (35± genera, 100+ species) and geographically widespread (Fig. 16.12) of the living turtles. Testudinids range from arid-terrestrial species to totally aquatic species, which emerge only to lay eggs. As noted earlier, they are not distinguishable from the emydids externally owing to

Aldabran giant tortoise *Geochelone gigantea* (Testudinidae). (G. R. Zug)

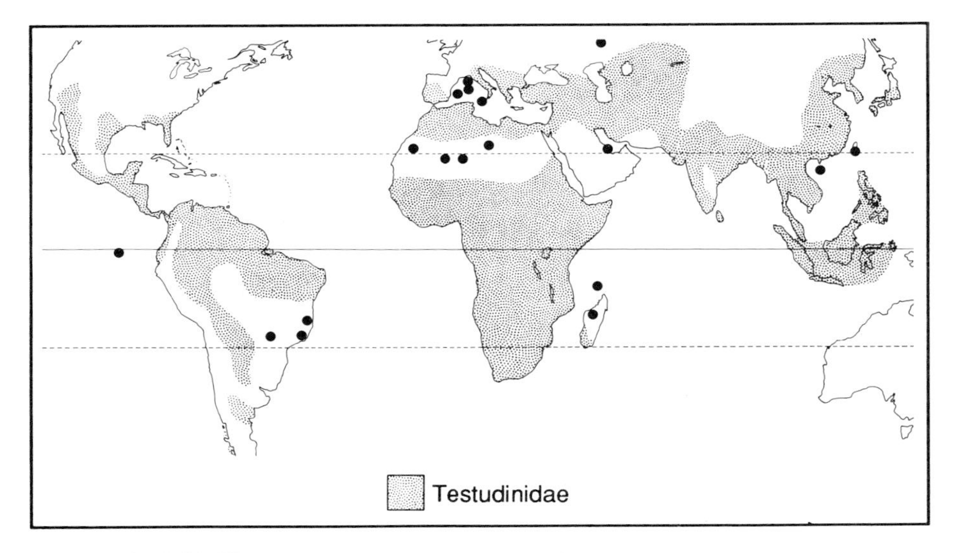

FIGURE 16.12 Distribution of the extant Testudinidae.

the broad overlap in body shapes and habits of these two families. The exception to this comment is the tortoises (Testudininae), one of the two testudinid subfamilies.

The tortoises are strictly terrestrial turtles (10–125 cm CL), and all possess columnar elephantine hindlimbs and usually high-domed carapaces and large plastrons. Tortoises are exposed to numerous types of predators. Hatchlings and juveniles must rely on crypsis and avoidance as the shell offers little protection to small, bite-sized turtles; however, for larger juveniles and adults, high-domed shells exceed the gape of many predators, and even if mouth-grasped, the concave surface induces tooth slippage rather than penetration. The completeness of the shell allows the full retraction of head, limbs, and tail beneath the shell margin (also possible because of the great volume of domed shells); the forelimbs are large and heavily scaled and so form a formidable barrier in front and the heavy, thick-skinned lower legs form a similar barrier to the rear.

Tortoises are mainly arid-adapted reptiles (North American *Gopherus*, South African *Homopus* and *Kinixys*, Euroasian *Testudo*, and others), although not exclusively (South American *Geochelone*, Asian *Indotestudo* and *Manouria*). Batagurines are principally semiaquatic and aquatic species; there are terrestrial species in a few genera (e.g., *Cuora*, *Rhinoclemmys*) but these genera also have aquatic or semiaquatic species as well. Batagurines have their greatest diversity in southern Asia (20+ genera) and occur elsewhere only in northern Africa and Eurasia (*Mauremys*) and tropical America (*Rhinoclemmys*). They range in size from the tiny *Geoemyda* (13 cm CL maximum) to the giant *Orlitia* (80 cm CL). There are several other large (60–70 cm) batagurines (e.g., *Batagur*, *Callagur*) and these large species are predominantly herbivores. Most other batagurines are moderate-sized (20–40 cm CL) omnivores or carnivores.

Phylogenetic Relationships of Turtles

The two major lineages, pleurodires and cryptodires, are sister-groups and represent a very old divergence, but a divergence that occurred after the establishment of the basic chelonian anatomy. Their shell osteology, even though not identical, shares such major features as the same number of trunk vertebrae and ribs and a similar architecture of carapace and plastral dermal plates. The divergence of these two lineages seems to reflect a vicariant event, because pleurodires are now exclusively Southern Hemisphere (their fossil record is also predominantly Southern Hemisphere) and cryptodires are mainly Northern Hemisphere with the exception of tortoises and softshells.

Within the pleurodires, the chelids and pelomedusids are each other's closest relatives, and our knowledge of the fossil record suggests that only these two pleurodiran families have existed since the mid-Cretaceous. Within each family, the relationships of the genera are strongly intercontinental. In the pelomedusids, the Madagascar *Erymnochelys* is the sister-group of the South American *Peltocepha-*

Painted roof turtle *Kachuga kachuga* (Testudinidae). (E. O. Moll)

lus and *Podocnemis*; the African *Pelomedusa* and *Pelusios* are likely sister-groups. In the chelids, the likely sequence of divergence has the Australian shortnecks as the sister-group to all the South American genera and suggests that *Chelodina* and *Hydromedusa* are a sister-pair and the most recent divergence event (see Fig. 12.9).

Relationships within the cryptodires are much more complex than those of the pleurodires. There are nine known extinct families, both outside and within the four main groups of living cryptodires; however, the emphasis here is on the relationships of living families. Evolutionarily, the sequence of divergence appears to be chelydroids, chelonioids, trionychoids, and finally testudinoids. Contrary to earlier phylogenetic hypotheses, *Platysternon* is a close relative to the two American chelydrid genera and likely represents a later origin from the same ancestral stock. Within the chelonioids, the cheloniids and dermochelyids are distantly related, and contrary to earlier phylogenetic interpretations, *Dermochelys* is neither a separate cryptodiran seaturtle lineage nor derived from cheloniines. The cheloniines and carettines are sister-groups. The extant trionychoids also include two distinct and old lineages, carettochelyids-trionychids and dermatemydids-kinosternids. The trionychoids and testudinoids appear to be a sister-pair, and testudinoids likely have the most recent origin of all extant turtles. Emydids and batagurines until very recently have been considered sister-pair

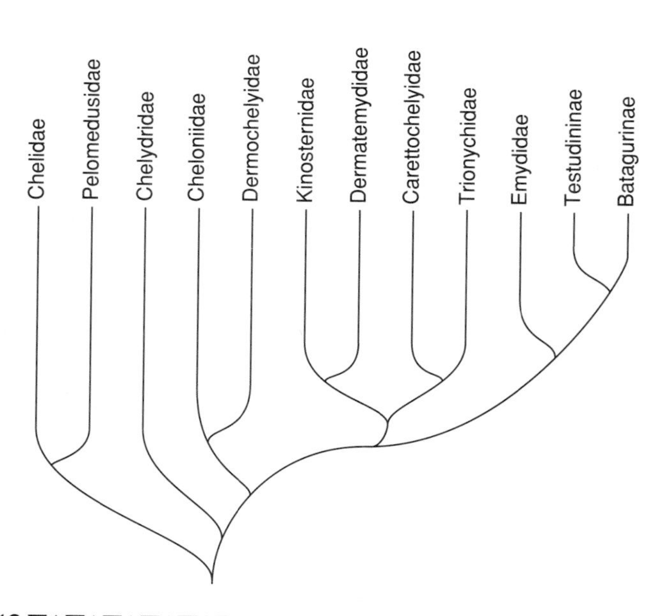

FIGURE 16.13 ▼▲
Dendrogram of presumed phyletic relationships among the living families of turtles.

families (or subfamilies) with testudinines as a basal divergence from their ancestoral stock. The dendrogram (Fig. 16.13) shows testudinines arising from within the batagurines, thus Testudinidae is a paraphyletic group. Such a classification is temporary and will change as more is discovered about the relationships of batagurine genera.

These relationships and the classification derived from it represent only one of several "competing" hypotheses. For the moment, it does have the strongest data base and the fewest inconsistencies.

CROCODYLIA

Crocodilians and birds are the sole reptilian survivors of the vast Mesozoic radiation of archosaurs. Crocodiles survived as the major freshwater predators of the tropics and subtropics. They hunt by stealth and ambush, capturing a wide variety of prey from small fish and invertebrates to birds and mammals. They are excellent swimmers, propelled slowly or rapidly by undulatory movements of powerful tails that are equal or longer than their attenuated bodies.

Crocodilians are the only ectothermic reptiles with a fully developed secondary palate and a four-chambered heart. Their bodies are armored dorsally and often ventrally by sheets of abutting osteoderms and a skin of thick, variable-sized, nonoverlapping scales; on the head, the thick skin is fused to bone. Although aquatic, their limbs are well developed; the hindlimbs are heavier and larger than the forelimbs and move nearly in a fore-and-aft swing. Crocodilians have high mobility on land, either via a high-walking gait or several running gaits including a high-speed gallop. Their forefeet are unwebbed with five digits, whereas the hindfeet are strongly webbed with four fully developed digits.

Like turtles, extant crocodilians are oviparous. Unlike turtles, crocodilians do not appear to have long-term sperm storage. Mating occurs shortly prior to egg-laying, usually within 1–3 weeks. Alligators, caimans, and a few crocodiles build mound-nests of vegetation and soil; females bite off pieces of the surrounding plants and with limbs and snout push these plant pieces, rotting detritus, and soil into a heap and lay the eggs within. Most crocodiles dig their nests in friable soils, also using their head and all limbs. Females of all crocodilians guard their nests and assist the young to escape from their eggshells and the nest.

Family Alligatoridae

Alligatorids are largely tropical American crocodilians (Fig. 16.14). All species are confined to freshwater habitats and all lack a lingual salt gland. Such glands might permit them to maintain a stable ionic and water balance in salt water. Alligatorids are easily distinguished from all other crocodilians by mandibular teeth that fit inside the upper jaw, so no teeth are visible when their mouths are closed.

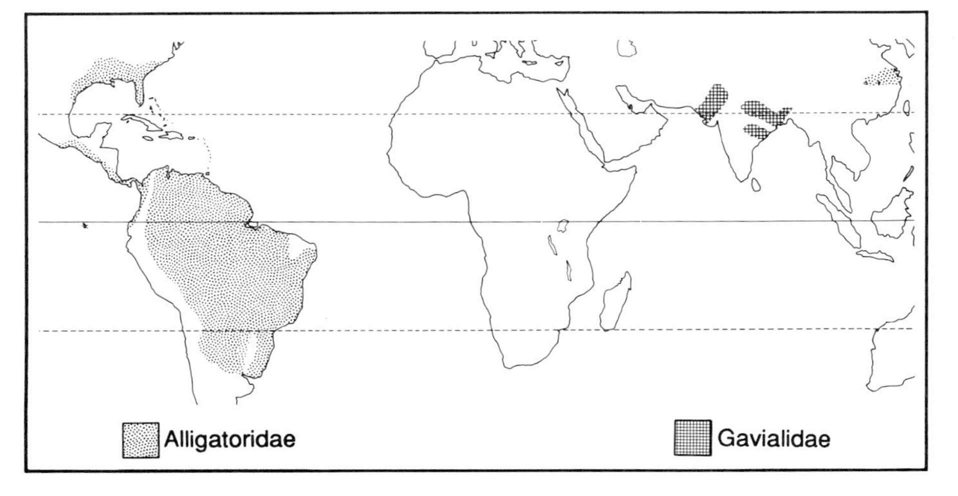

FIGURE 16.14 ▼▲
Distribution of the extant Alligatoridae and Gavialidae.

Alligatorids divide into two groups, the alligators (*Alligator*; two species) of warm temperate/subtropical China and southeastern North America and the caimans (*Caiman*, *Melanosuchus*, *Paleosuchus*; six species) of tropical America. Alligators are the only living crocodilians capable of living in areas of freezing temperatures. They do not, however, hibernate. Large juvenile and adult *A. mississippiensis* select steep-sided shorelines where they can lie with the tip of the snout above water and the body and tail in deeper, warmer water. If the water freezes, the alligator maintains an ice-free hole around its snout for it must have access to air for respiration.

Alligator mississippiensis (4 m TL maximum) occurs widely in lakes, swamps, marshes, and rivers of southeastern North America. In contrast, the Chinese alligator (*A. sinensis*; 3 m TL maximum) is an endangered species. Its original range was limited to lakes and marshes of the middle and lower Yangtse River and adjacent coastal freshwater marshes. Constant long-term persecution and

New Guinean freshwater crocodile *Crocodylus novaeguineae* (Crocodylidae). (C. A. Ross)

recent modifications of wetlands by the removal of marsh and riverine forest have largely eliminated suitable habitat.

Caimans are the major South American crocodilians, occupying all freshwater habitats from small rain forest streams (*Paleosuchus*), large swamps, rivers, and lakes (*Caiman latirostris*, *Melanosuchus*) to temporary savanna marshes and lakes (*Caiman crocodilus*). Their maximum sizes range from about 1.5 m TL for *Paleosuchus* to 5 m for *Melanosuchus*.

Family Crocodylidae

Crocodiles occur worldwide in tropical freshwater and nearshore-marine habitats (Fig. 16.15). They are moderate to large crocodilians with variable snout shapes, ranging from moderately short and blunt to long and pointed. Crocodiles are immediately recognizable by the fourth mandibular tooth on each side lying exposed on the outside of the upper jaw when the mouth is closed. All crocodiles, even freshwater species, have lingual salt glands, which permit them to maintain ionic and water balance in totally marine habitats.

Crocodiles include two subfamilies: Crocodylinae and Tomistominae. Tomistomines include the single species *Tomistoma schlegeli* (false gharial; 4 m TL maximum). It lives in the rivers and swamps of the Malay Peninsula and East Indies, and as its name implies, it is a gharial look-alike with a long, narrow snout and fish-eating habits. Unlike other crocodilians, both the first and fourth mandibular teeth on each side lie outside the closed mouth. The crocodylines have only the fourth mandibular tooth of each side lying externally. There are three crocodyline genera (*Crocodylus*, *Osteolaemus*, *Mecistops*). *Crocodylus* (10+ species) occurs worldwide, and in many areas, one species occupies the freshwater

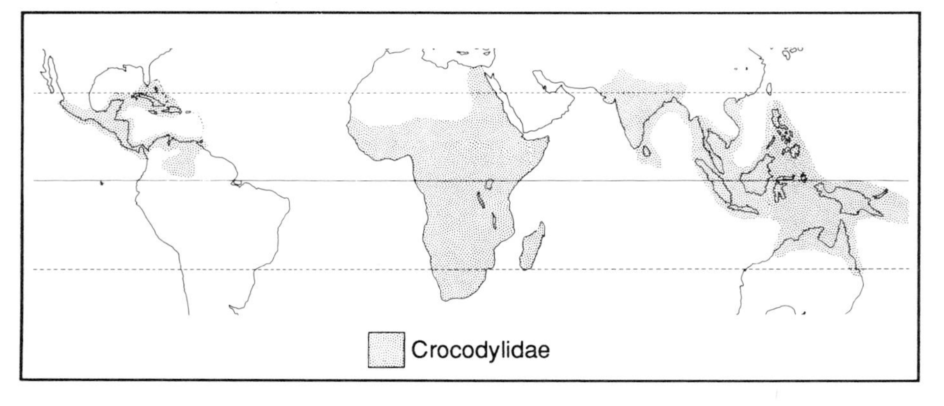

FIGURE 16.15 ▼▲▼

Distribution of the extant Crocodylidae.

habitats and another occurs in coastal estuaries and swamps and the tidal portions of rivers. In the Americas, *C. acutus* is the saltwater species and *C. intermedius*, *moreletii*, and *rhombifer* are the freshwater ones. In Australasia, *C. porosus* is the saltwater species with a half dozen freshwater species occupying contiguous areas. *Osteolaemus* is a dwarf crocodile (2 m TL maximum) from western Africa.

Family Gavialidae

The true gavial or gharial, *Gavialis gangeticus*, is the only living gavialid and now survives only in the upper reaches of the Indus, Ganges, Brahmaputra, and Mahanandi rivers (Figs. 16.14 and 16.16). Gharials prefer deep fast-flowing rivers, where adults congregate in the deep holes at river bends and the confluence of smaller streams. The juveniles select smaller side streams or river backwaters. Although the gharials are the most aquatic of extant crocodilians, they regularly bask each day in the cooler winter months.

The elongated crocodilian body of gharials is accentuated by an extremely long and slender snout. When the mouth is closed the fourth mandibular tooth and all teeth anterior to it lie on the outside of upper jaw, giving the tip of the snout a pincushion appearance. The lower-jaw symphysis extends backward almost to the end of the tooth row. The long and attenuate jaws have numerous, closely spaced teeth that form a proficient trap for catching fish, the gharial's major food. Gharial catch fish with quick sideward snaps of the jaw. With prey impaled on the teeth, the head is lifted out of the water and backward, and sideward head-jerks drop the fish headfirst into the esophagus. Frogs are another common prey, birds and mammals less so.

Male gharials usually reach maturity in about 13–14 yr and at about 3 m

FIGURE 16.16 Gharials *Gavialis gangeticus* (Gavialidae). (C. A. Ross)

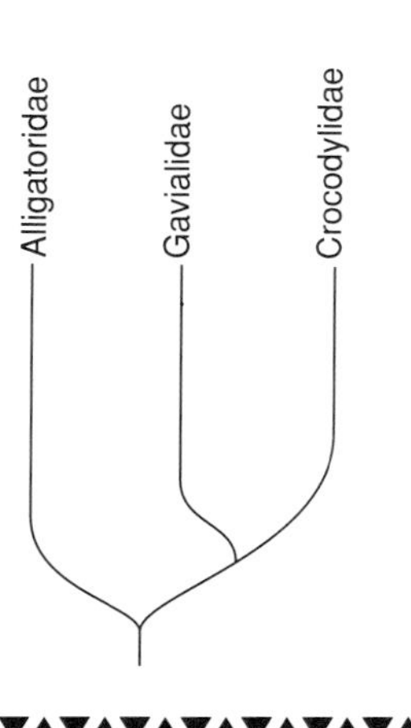

FIGURE 16.17 ▼▲▼▲▼▲▼▲▼▲▼▲▼▲▼▲▼▲▼▲▼▲▼▲▼▲▼▲▼

Dendrogram of presumed phyletic relationships among the living families of crocodilians.

(TL); females may mature at a slightly younger age and smaller size (2.6 m). Mature males develop a large, irregular growth on the tip of the snout. This bulbous boss grows progressively larger with age. Today most adult gharials are 4 m or less; previously they may have reached 8 m (TL), although the maximum verified length is 6.45 m.

Phylogenetic Relationships of Crocodilians

With only three families, only three relationship patterns are possible. Each of the possible patterns has been proposed at one time or another. Because several research groups are currently working on the problem and each has proposed a different pattern, there is no consensus on relationships. The pattern presented here (Fig. 16.17) is a conservative interpretation with the alligatorids as a sister-group of the crocodylids-gavialid pair, and with an early divergence between this pair. The alligatorids are monophyletic; this interpretation is repeatedly supported by new data sets. *Alligator* and caimans appear to have diverged early, then *Paleosuchus* from the *Caiman-Melanosuchus* line. The crocodylids and gavialids present a less certain situation.

The uncertainty arises from the placement of *Tomistoma*. Is it a crocodylid or a gavialid? The issue remains unresolved. Various sets of biochemical-molecular data suggest that the false gharial is a gavialid. Some morphological data suggest the preceding relationship, but as a whole, morphological data indicate that gavialids are a distinct line and link *Tomistoma* as a crocodylid but not closely to *Crocodylus* and *Osteolaemus*.

CHAPTER 17

Lizards, Amphisbaenians, and Tuataras

The Lepidosauria includes two groups of living reptiles, rhynchocephalians and squamates. Rhynchocephalians were once a diverse group, but today only two species (Sphenodontida) survive. Squamates, in contrast, are a dominant and speciose (>5000 species) group of vertebrates with a worldwide distribution.

Lepidosaurians share numerous characteristics including such features as small or absent lacrimal bone, pleurodont tooth attachment (sometimes acrodont), paired sternal plates fused, pelvic bones and astragalus-calcaneum fused in adults, and a transverse cloacal opening.

SPHENODONTIDA

Family Sphenodontidae

Sphenodon guentheri and *S. punctatus* (Fig. 17.1) are lizardlike, stout-bodied (19–28 cm adult SVL) reptiles with large heads and thick tails. They have a chisel-beaked upper jaw overhanging the lower jaw, a series of erect spines on the nape and back, and rudimentary hemipenes. Sphenodontidans share a mandible without splenial, no postorbital-parietal contact, no lacrimals, acrodont dentition, and other traits.

Tuataras now occur only on about 30 small islands off the New Zealand coast (Fig. 17.2; *Sphenodon guentheri* on only a single island). *Sphenodon punctatus* is most numerous on those islands shared with nesting seabirds. The seabird nesting activity yields abundant arthropod prey and burrows for daily shelter and winter hibernation. Tuataras forage principally at night, commonly at temperatures in the 12–16°C range; however, they are not exclusively nocturnal animals and, in warmer summer months, bask at their burrow entrances, retreating

FIGURE 17.1 ▼▲
Tuatara *Sphenodon punctatus* (Sphenodontidae). (R. W. Van Devender)

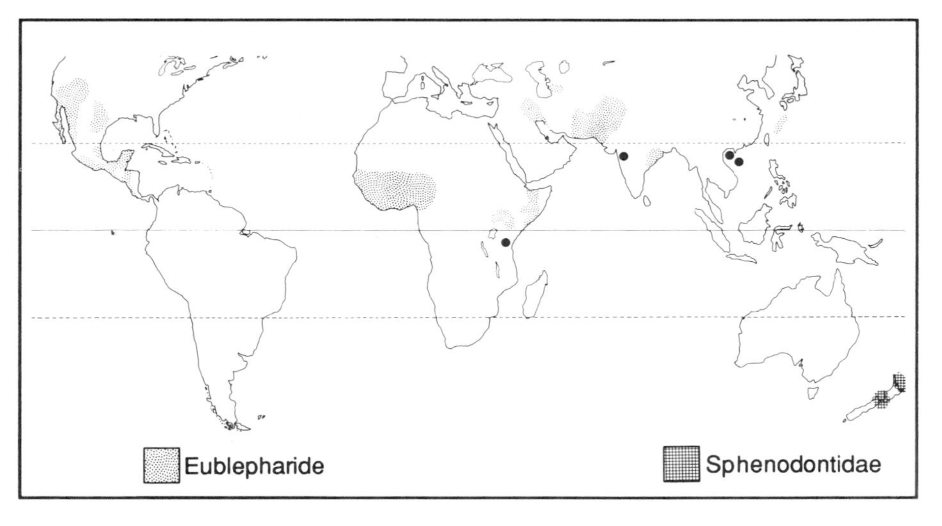

FIGURE 17.2 ▼▲▼▲▼▲▼▲▼▲▼▲▼▲▼▲▼▲▼▲▼▲▼▲▼▲▼▲▼▲▼▲▼▲▼▲▼▲
Distribution of the extant Sphenodontidae and Eublepharidae.

when they become too hot and reemerging after they cool. Their prey are predominantly insects, although they occasionally eat skinks, geckos, and hatchling seabirds.

Courtship and mating occur in January, but egg-laying is delayed until October–December, presumably indicating sperm storage rather than delayed development. The female digs a small nest cavity and deposits 8–15 eggs. Development is slow, stopping during the winter; hatching occurs 11–16 months after deposition. Optimal incubation temperatures in the laboratory are 18–22°C, the lowest known incubation temperatures of living reptiles. The eggs absorb moisture during incubation, so the mass of the hatchling is 1.2–1.3 times greater than the original egg mass.

SQUAMATA/LACERTILIA

The Squamata encompasses the lizards and all their derivatives—snakes, amphisbaenians, mosasaurs, and other extinct groups. A recent study identified more than 70 shared-derived traits for squamates in support of their monophyly. A few of these traits are: no vomerine teeth; no pterygoid-vomer contact; early fusion of exoccipital-opisthotic ossification centers; columella slender; tongue-in-groove ankle joint and reduction of tarsal bones; no gastralia; and well-developed hemipenes.

Squamate radiation has produced numerous distinct lineages and, because many of these subgroups are morphologically distinct, a plethora of names. An introduction to these names is necessary, for they are widely used and serve as convenient labels for the various monophyletic subgroups. The earliest divergence of squamates gave rise to the Iguania and Scleroglossa. Iguanians include the Iguanidae, Chamaeleonidae, and Agamidae, and scleroglossans encompass all other squamates. Gekkota and Autarchoglossa are the two primary groups of scleroglossans; the various gecko groups form the gekkotans, and all remaining lizards and their relatives comprise the autarchoglossans. Autarchoglossans include five subgroups: anguimorphs, scincomorphs, dibamids, amphisbaenians, and snakes. The origins of the latter three groups probably lie within either the anguimorphs or scincomorphs, but exactly where is unresolved. The familial content of the first two autarchoglossan subgroups is: scincomorphs—Cordylidae, Gymnophthalmidae, Lacertidae, Scincidae, Teiidae, and Xantusiidae; anguimorphs—Anguidae, Helodermatidae, Varanidae, and Xenosauridae.

Iguania

Family Agamidae

Agamids occur widely throughout Africa, Asia, and Australia (Fig. 17.3). Within each continent, agamids have had an independent radiation producing a variety of body shapes and sizes. Most agamids are moderate-sized (6–12 cm

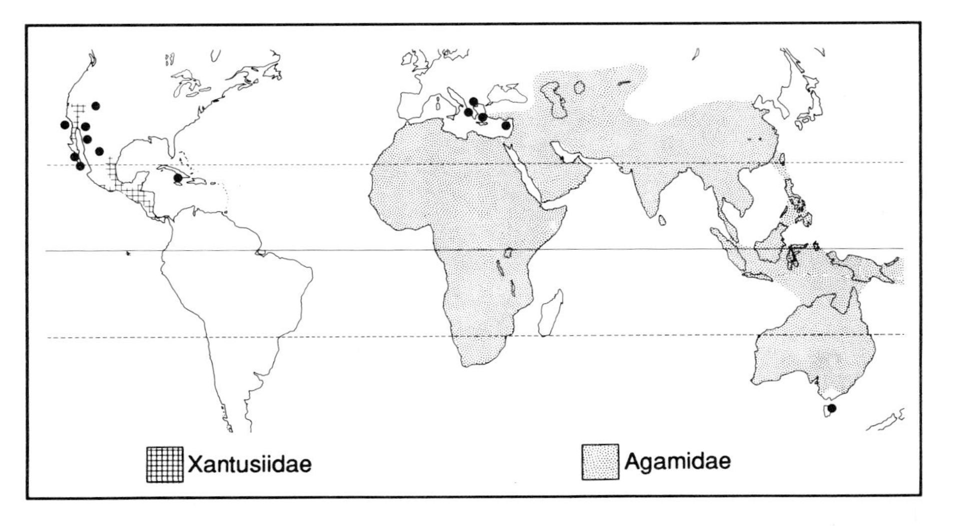

FIGURE 17.3 ▼▲▼▲▼▲▼▲▼▲▼▲▼▲▼▲▼▲▼▲▼▲▼▲▼▲▼▲▼▲▼▲▼▲▼▲
Distribution of the extant Agamidae and Xantusiidae.

SVL), but they range from the tiny *Amphibolorus microlepidotus* (14 mm SVL) to the large water dragon *Hydrosaurus amboinensis* (35 cm SVL, 110 cm TL). Body shape ranges from stout-bodied/tailed and short, heavy-limbed lizards (e.g., *Uromastyx*, *Moloch*) to more slender-bodied, long-tailed and -limbed forms (*Agama*, *Calotes*, *Amphibolurus*). All agamids have well-developed limbs. Many have keeled scales, middorsal crests, and throat flaps or fans. Agamids share an acrodont dentition (teeth fused to bone) with chameleons and differ thereby from all other lizard groups. Oviparity occurs in all except the live-bearing toad-headed lizards (*Phrynocephalus*).

Agamids are often partitioned into the agamines and leiolepidines (*Leiolepis*, *Uromastyx*). Although the agamines are monophyletic, the two leiolepidines appear to have no greater affinities to one another than either does to the agamines. *Leiolepis* is a terrestrial, tropical forest taxon, and *Uromastyx* is a terrestrial, arid-land one; both dig burrows for rest-retreats. The habits and habitats of agamines encompass those of the leiolepidines and include semiaquatic forms that use water for an escape refuge. A few Australasian species (e.g., *Hydrosaurus*) can run bipedally across the water surface; the modified scales on the hindfeet of *Hydrosaurus* form a weblike surface between the toes. *Draco* adopted aerial flight (gliding) as an escape device and they are unique among extant lizards in their modification of the rib cage to provide an aerofoil.

Family Chamaeleonidae

Chameleons are remarkable and peculiar lizards. They have strongly laterally compressed bodies, head casques covering their necks, zygodactylous feet (fusion of sets of two and three digits forming opposable, two-fingered, mittenlike feet; manus fusion 1-2-3 and 4-5, pes 1-2 and 3-4-5), projectile tongues, prehensile tails, and independently movable eyes with mufflerlike lids. They are highly arboreal animals, although a few species may forage on the ground. Chameleons avoid predator detection by metachrosis (color change) and slow movement.

Chameleons are mainly an African-Madagascan group (see Fig. 17.12) with independent radiations in each area. *Chamaeleo* is the most widespread (occurring everywhere within the familial distribution) and speciose of the chameleon genera (75+ species). *Chamaeleo* has the largest members (7–63 cm SVL) of the four chamaeleonid genera and has both oviparous and ovoviviparous species. *Bradypodion* (dwarf chameleons, 6–14 cm SVL) occur mainly in southern Africa; all its species are ovoviviparous. The leaf chameleons (*Brookesia*/Madagascar, *Rhampholeon*/Africa) are small, rain forest chameleons (25–55 mm SVL) and are oviparous.

Family Iguanidae

Iguanids are the dominant lizards of the Americas with outliers in Madagascar and the Southwest Pacific (Fig. 17.4). They display high diversity (50+ genera, 900± species), and this diversity reflects numerous adaptive radiations,

Chamaelinorops barbouri (Iguanidae). (G. R. Zug)

encompassing at least eight distinct lineages. The union of these lineages within the Iguanidae yields a paraphyletic family; however, this assemblage is retained until the origin of the agamids within the iguanids and the relationships among the iguanid subgroups gain stronger evidence. Recently, the iguanids were partitioned into nine, presumably monophyletic, families; these are used here as subfamilies to outline the broad diversity of the iguanids.

Iguanines are the true iguanas (seven genera) and are the largest iguanids; body sizes range from the small *Dipsosaurus* (11–14 cm SVL) to the large *Cyclura* (28–75 cm SVL). Iguanines occur from southwestern North America through the lowlands to central South America and widely in the Antilles and Galápagos. They are predominantly terrestrial and arid-adapted lizards (e.g., *Conolophus*, *Cyclura*, *Ctenosaura*). Water-conservation adaptations likely provided some of the physiological mechanisms for the marine iguana's (*Amblyrhynchus*) successful use of marine algae gardens. The Fijian iguana *Brachylophus* and the American *Iguana* are strongly arboreal species and descend to the ground only for egg-laying or moving to another patch of forest. All iguanas include a high proportion of vegetable matter in their diet, and all are oviparous.

Corytophanines and polychrines are largely arboreal lizards living in dry scrub forests to wet rain forests. Corytophanines (*Basiliscus*, *Corytophanes*, *Laemanctus*) are slender, long-limbed, and long-tailed lizards (9–20 cm SVL; nine species) of Central America. The latter two genera are strongly arboreal and live

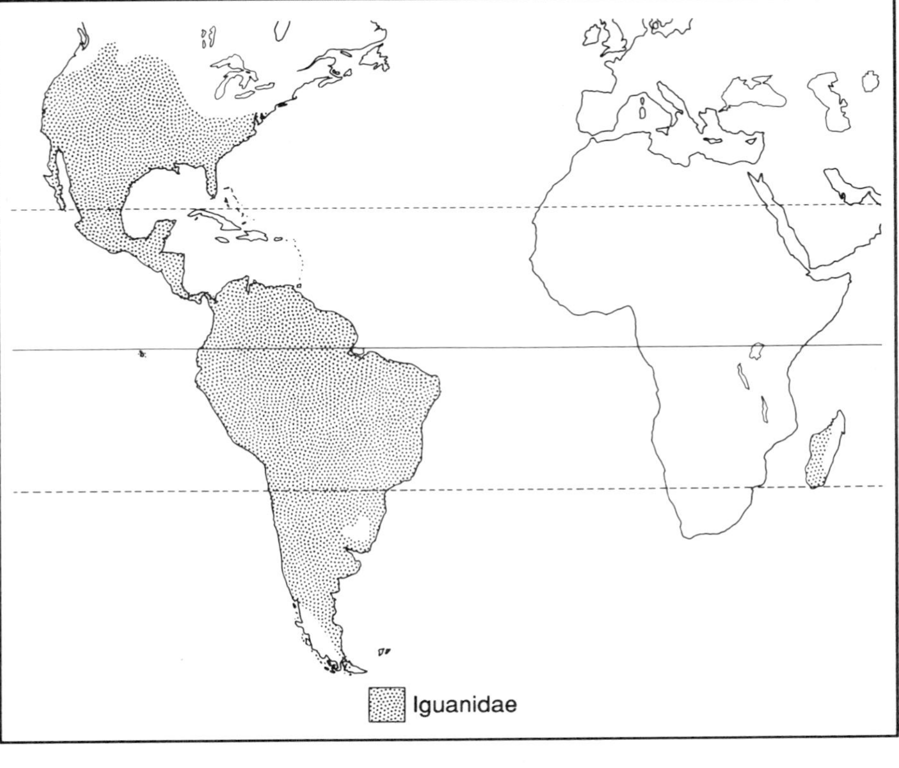

FIGURE 17.4 Distribution of the extant Iguanidae.

in the forest canopy. *Basiliscus* is a lower-level forest inhabitant and forages on the forest floor, frequently in the vicinity of water. One of its escape behaviors is to run bipedally across water surfaces. All corytophanines have casque heads and are oviparous, except the ovoviviparous *C. percarinata*. Polychrines (anoloids) are more widespread (southeastern North America to southern South America) and diverse (10+ genera; e.g., *Anolis*, *Enyalius*, *Urostrophus*; 650+ species). Most are small- to moderate-sized (25–120 mm SVL) and mainly arboreal. Habitus is typically slender-bodied, long-limbed, and long-tailed. All members are oviparous. *Anolis* lays one egg at a time with ovulation alternating between left and right ovaries.

Hoplocercines consist of three to five genera (e.g., *Hoplocercus*, *Enyaloides*, *Morunasaurus*) spottily distributed in tropical South America. They are moderate-sized, terrestrial and arboreal lizards; presumably, all are oviparous. Tropidurines are a far more diverse group of iguanids. They have had three major radiations: leiocephalans (10+ species) in the Antilles, tropidurans (60+ species) in the northern half of South America, and liolaemans (60+ species) in the southern half of South America. Members of all three radiations are small- to moderate-sized, usually spiny-scaled lizards. They are largely terrestrial, and many are arid-adapted species. Insectivory predominates, but some *Liolaemus* tend toward herbivory. All tropidurines are oviparous.

The remaining two American iguanid subfamilies (Crotaphytinae, Phrynosomatinae) are North American groups. Crotaphytines are the collared lizards (*Crotaphytes*, *Gambelia*; 10–14 cm SVL) of the central plains and southwestern deserts. Crotaphytines, especially *Gambelia*, are lizard predators, although they catch and eat a variety of other prey. In contrast to crotaphytines, phrynosomatines are diverse (9 genera; 100+ species). They are typically small- to moderate-sized lizards (most <10 cm SVL) and predominantly terrestrial-semiarboreal. *Phrynosoma*, horned lizards, matches no other phrynosomine lizards in appearance or for that matter any other iguanid. Their pancakelike bodies, often edged with spiny scales, and cranial crowns of horns or spines give them their name and camouflage them from predators. The other phrynosomines have moderately robust bodies, well-developed limbs and tails, and keeled-overlapping or granular-juxtaposed scales. *Sceloporus* epitomizes the spiny-lizard appearance; *Urosaurus* and *Holbrookia* have a smoother, more streamlined appearance. All the phrynosomines occur in arid habitats and, with the exception of *Sceloporus*, occur in the deserts and adjacent habitats of southwestern North America. *Sceloporus* (50+ species) occurs in a variety of arid to mesic habitats from the mid-Atlantic coastal forest of North America to western Panama. Some *Phrynosoma* and *Sceloporus* are ovoviviparous; all other phrynosomines are oviparous.

Oplurines are the Madasgascaran iguanids and include two genera, *Chalarodon* (one species; 6–9 cm SVL) and *Oplurus* (six species; 9–15 cm SVL; Fig. 17.5). In appearance and habits, they are very much like many phrynosomines and tropidurines. Indeed their most exceptional feature is their occurrence within the distribution of agamids and geographically far removed from other iguanids. They are a terrestrial, arid-adapted group, and all species are oviparous.

Gekkota

Family Eublepharidae

The eublepharid geckos are a small family of six genera (*Aeluroscalabotes*, *Coleonyx*, *Eublepharis*, *Goniurosaurus*, *Hemitheconyx*, *Holodactylus*). Eublepharids have eyelids, supratemporal bones in the skull, and angular bones in the mandible. With the exception of the arboreal *Aeluroscalabotes*, eublepharids are terrestrial geckos with narrow digits. All are nocturnal, moderate-sized (45–120 mm adult SVL), and predominantly arid-land geckos of the Northern Hemisphere (see Fig. 17.2) with three centers of diversity, sub-Saharan North Africa, Southwest Asia, and southwestern North America. The American radiation is within a

FIGURE 17.5 A Madagascaran spiny lizard *Oplurus sebae* (Iguanidae). (H. I. Uible)

single genus *Coleonyx* (six species); the two Central American species occur in mesic forest habitats, the other four species live in deserts (Fig. 17.6). Similarly, in Southwest Asia, *Eublepharis* (four species) occurs in arid habitats. In Africa, *Hemitheconyx* and *Holodactylus* (two species each) are also primarily scrub to desert inhabitants.

FIGURE 17.6 Texas banded gecko *Coleonyx brevis* (Eublepharidae). (R. W. Van Devender)

Eublepharids share a clutch size of two eggs with the gekkonids. They lay eggs with leathery shells that remain flexible throughout incubation and are usually deposited beneath objects or in rock crevices.

Family Gekkonidae

Gekkonids contain all the remaining geckos and a unique group of near-legless lizards. They differ from the eublepharid geckos by the absence of angulars, supratemporals (usually), and immovable eyelids. Their eyelids are fused, and the skin covering each eye forms a clear spectacle.

Gekkonids are a successful group, whether measured by generic (80+) or species (850+) diversity, or by local abundance (e.g., commonly four or more species are sympatric and often the density of at least one species exceeds 100 individuals ha^{-1}). They occur circumtropically on all continents and most oceanic islands (Fig. 17.7). They range in size from the tiny *Sphaerodactylus* (16 mm SVL) to the giant *Rhacodactylus leachianus* (25 cm SVL; and if extant, the 37-cm *Hoplodactylus delcourti*). They are terrestrial to arboreal and have a multitude of body shapes from broad-headed, robust-bodied and -limbed to slender, elongate, and nearly limbless. Most are nocturnal, although some commonly bask in late afternoon and others are strictly diurnal (e.g., *Phelsuma*/Indian Ocean islands, *Naultinus*/New Zealand). The arboreal taxa have dozens of foot and toe morphologies that permit them to climb vertical surfaces of various degrees

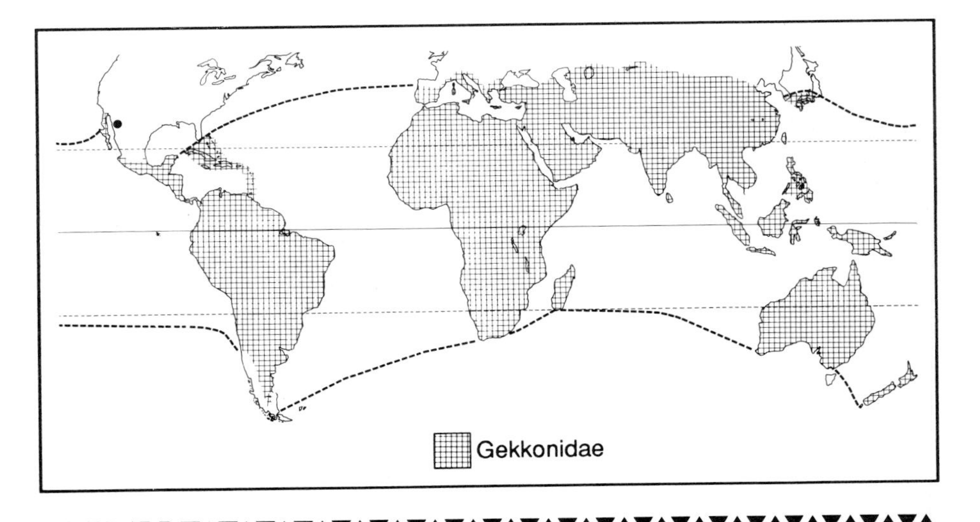

FIGURE 17.7 ▼▲▼▲▼▲▼▲▼▲▼▲▼▲▼▲▼▲▼▲▼▲▼▲▼▲▼▲▼▲▼▲▼▲▼▲

Distribution of the extant Gekkonidae. Most islands within the dashed line have one or more species of geckos.

of roughness. Their clinging/climbing abilities derive mainly from frictional adhesion via the claws and microvillous surfaces of their expanded digital lamellae. Within this broad diversity, most gekkonids have a fixed clutch of two eggs, although sphaerodactyls and a few other species have but one egg.

Two major groups, Gekkoninae and Pygopodinae, exist within the gekkonids. Pygopodines occur in Australia, New Zealand, and New Caledonia and include the limbed diplodactyls (e.g., *Bavayia*, *Carphodactylus*, *Hoplodactylus*, *Nephrurus*, *Oedura*, *Phyllurus*) and snakelike pygopods (*Delma*, *Lialis*); both groups are moderately diverse. Gekkonids include all other geckos. The two subfamilies differ by such traits as absence of a stapedial foramen or complete auditory meatal closure muscle in the pygopodines and the converse in most gekkonines. One prominent trait is eggshell structure. In gekkonines, eggs are leathery when laid but become hardened as they air-dry; in pygopodines, eggshells remain pliable and leathery. All but the New Zealand *Hoplodactylus* and *Naultinus* are oviparous.

Pygopodines are the dominant geckos of Australia. They occur over the entire continent from the arid interior (*Oedura*) into the moist temperate and tropical forests (*Phyllurus*). Diplodactyls are typically limbed geckos. The pygopods are functionally limbless; the forelimbs and pectoral girdle are absent, and the hindlimbs remain only as scaly flaps with a vestigial hindlimb and pelvic skeleton. All are slender, elongate species (7–25 cm SVL) and, with the exception of *Lialis*, they are semifossorial, foraging on the surface at night for arthropods and lizards. *Lialis* is a diurnal predator specializing on skinks. It hunts from ambush, bite-catching the skinks and sometimes holding them in a constrictorlike manner. Like some skink-eating snakes, *Lialis* has hinged teeth that are an adaptation for holding the smooth-scaled, hard-bodied skinks.

Gekkonine diversity in habitus greatly exceeds that of the pygopodines, except all gekkonines are limbed (Fig. 17.8). Although some gekkonines (e.g., *Sphaerodactylus*) have short limbs, their locomotion is propelled by the limbs. Gekkonines have major radiations in all tropical areas, even Australia, the center of a pygopodine radiation. In tropical America, the sphaerodactyl radiation (e.g., *Gonatodes*, *Lepidoblepharis*, *Sphaerodactylus*) yielded small (most <40 mm SVL), terrestrial, and possibly partly diurnal geckos. Other American geckos (*Aristelleger*, *Thecadactylus*, *Phyllodactylus*) are more typically nocturnal and arboreal. Multiple radiations occurred in Africa (*Lygodactylus* and *Pachydactylus*), Madagascar/Indian Ocean (*Phelsuma* and *Uroplatus*), Southwest Asia (terrestrial *Gymnodactylus*-*Cyrtodactylus* complex), Southeast Asia and Oceania (*Gekko*, *Gehyra*, *Hemidactylus*, and others), and Australia (*Heteronotia*). Several gekkonine genera (e.g., *Gehyra*, *Hemidactylus*, *Lepidodactylus*) contain bisexual and unisexual species; many of the latter species are polyploids, suggesting a hybrid origin from two bisexual species.

FIGURE 17.8 Moorish gecko *Tarentola mauritanica* (Gekkonidae). (R. G. Tuck)

Autarchoglossa

Family Anguidae

Anguids are small (*Elgaria parva*, 55–70 mm SVL) to large (*Ophisaurus apodus*, 50–52 cm SVL, 1.4 m TL maximum), limbed and limbless lizards. All are heavily armored with scales (largely nonoverlapping) underlain by rectangular osteoderms. Commonly, a longitudinal ventrolateral fold separates the dorsal and ventral armor on each side, thus maintaining the protective shield yet permitting diametric expansion for breathing and feeding. They are predominantly terrestrial-semifossorial lizards but not exclusively so (e.g., *Abronia* is arboreal with a prehensile tail). All anguids are carnivores with arthropod prey dominant, but small vertebrates, snails, and slugs are regular prey for some species. Prey availability and anguid body size are a large factor in determining the diet among the different anguid species. Anguids include oviparous and ovoviviparous species.

Anguids occur throughout the Americas and Eurasia (Fig. 17.9) and usually are divided into four subfamilies (Anguinae, Anniellinae, Diploglossinae, Gerrhonotinae). Anguines and anniellines are limbless, elongate forms with tails often twice the head-body length. *Anniella* with two species (30 cm TL maximum) is a slender-bodied taxon, living in arid habitats of California and Baja California.

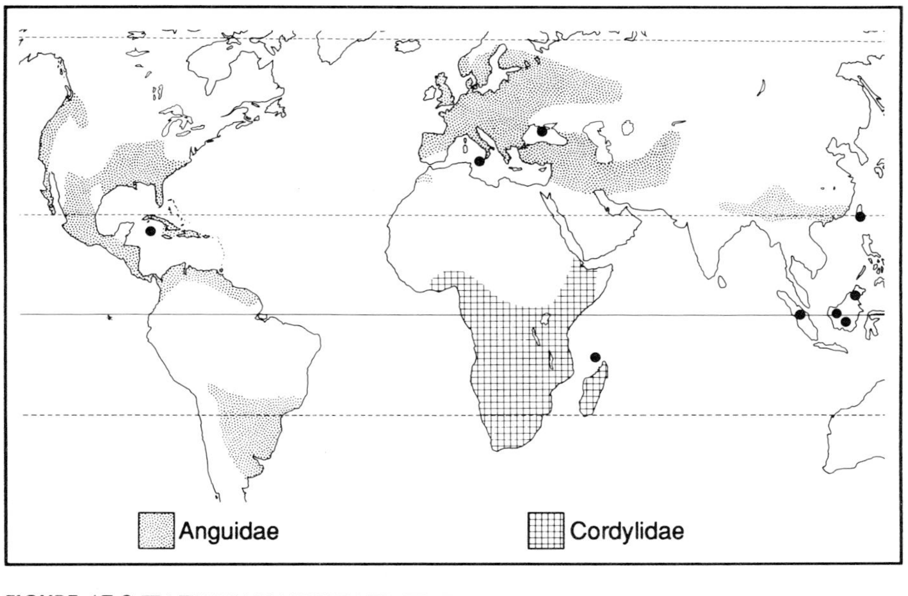

FIGURE 17.9 ▼▲▼▲▼▲▼▲▼▲▼▲▼▲▼▲▼▲▼▲▼▲▼▲▼▲▼▲▼▲▼▲
Distribution of the extant Anguidae and Cordylidae.

They burrow during the day and forage on the surface at night. Both species are live-bearers. Anguines (*Anguis*, 1 species; *Ophisaurus*, 10+ species) occur widely in more mesic habitats of North America and Eurasia and are mainly diurnal foragers. They are more thick-bodied and larger (1.5–3 times) than anniellines. *Anguis* is a live-bearer, *Ophisaurus* lays eggs (Fig. 17.10). Diploglossines (5 genera; e.g., *Diploglossus*, 30+ species, *Ophiodes*, 5 species) are also elongate, limbed lizards, although their limbs are often greatly reduced. They occur in the Antilles, Central America, and central South America. Gerrhonotines (5 genera; e.g., *Abronia*, 12+ species, *Elgaria*, 7 species) are broad-headed, heavy-bodied and -tailed lizards with short, strong limbs. These small- to moderated-sized lizards occur in xeric and mesic habitats from northwestern North America to western Panama and include both terrestrial and arboreal species as well as live-bearing and egg-laying species.

Family Cordylidae

Cordylids are the armored lizards of sub-Saharan Africa and Madagascar (see Fig. 17.9). These robust-bodied lizards have a well-developed osteoderm shield dorsally from head to tail; the scales may abut or overlap and frequently are strongly keeled. Their chests and bellies bear a lighter osteoderm shield and often abutting, smooth scales. Size ranges from a few small *Cordylus* (6–7 cm

FIGURE 17.10 ▼▲▼▲▼▲▼▲▼▲▼▲▼▲▼▲▼▲▼▲▼▲▼▲▼▲▼▲▼▲▼▲▼▲▼▲▼▲▼

Eastern glass lizard *Ophisaurus ventralis* (Anguidae). (R. W. Van Devender)

SVL) to the large *Gerrhosaurus validus* (30 cm SVL maximum, 70 cm TL). They are predominantly arid-land species of scrub forest to grassland and often occur in boulder fields and rocky outcrops. As a group, they are opportunistic omnivores, and both large and some small species regularly eat plant matter. Oviparity is the most frequent reproductive mode.

The two cordylid subfamilies (Cordylinae, Gerrhosaurinae) have largely overlapping distributions and preferred habitats in Africa. Each also includes stout, heavily armored taxa (cordyline *Cordylus* and *Pseudocordylus*, gerrhosaurine *Gerrhosaurus* and *Cordylosaurus*) and elongate, reduced-limbed taxa (*Chamaesaura* and *Tetradactylus*, respectively). The elongate taxa are mainly grassland inhabitants. *Anglosaurus skoogi* (gerrhosaurine) of the Namib Desert is an omnivore and regularly eats foliage. To escape predators and high daytime temperatures, it burrows in the dunes and moves beneath the surface via sand-swimming. *Tracheloptychus* and *Zonosaurus* (Fig. 17.11) are the Madagascan gerrhosaurines and somewhat less spiny versions of *Gerrhosaurus*. Cordylines are live-bearers, except the oviparous *Platysaurus*; gerrhosaurines lay eggs.

Family Dibamidae

The dibamids are small (5–20 cm SVL), elongate, nearly limbless lizards of Indomalaysia (*Dibamus*, nine species) and Mexico (*Anelytropsis*, one species)

FIGURE 17.11
Girdled lizard *Zonosaurus laticaudatus* (Cordylidae). (H. I. Uible)

(Fig. 17.12). They are smooth, shiny-scaled creatures and totally fossorial, but they do retain small eyes lying beneath head scales. Forelimbs and pectoral girdle are absent, and hindlimbs are reduced to small scaly flaps. *Dibamus* is a forest-floor dweller requiring moist soils; *Anelytropsis* is more arid-adapted and lives in drier upland forest and scrub. They may be ovoviviparous, but neither live-bearing nor egg-laying has been confirmed.

Family Gymnophthalmidae

These small (most <60 mm SVL) lizards are the microteiids, a name denoting an assumed relationship and their earlier assignment to teiids. They are not, however, miniature teiids in appearance. Semifossorial and fossorial habits match a more elongate habitus and shorter limbs (sometimes nearly vestigial, e.g., in *Bachia*, or absent as in *Calyptommatus*). Scalation patterns include an arrangement of abutting, polygonal scales subequal around the body (keeled, *Arthrosaura*; smooth, *Bachia*, *Tretioscincus*) and variously arranged rows of enlarged keeled scales on the back (*Leposoma*, *Neusticurus*).

Size and habits make them inconspicuous denizens of tropical American forests, even though they are a moderately diverse group (29 genera, 120+ species). *Gymnophthalmus* is the most studied gymnophthalmid; it includes about a half dozen bisexual and unisexual species. It and the other genera are oviparous.

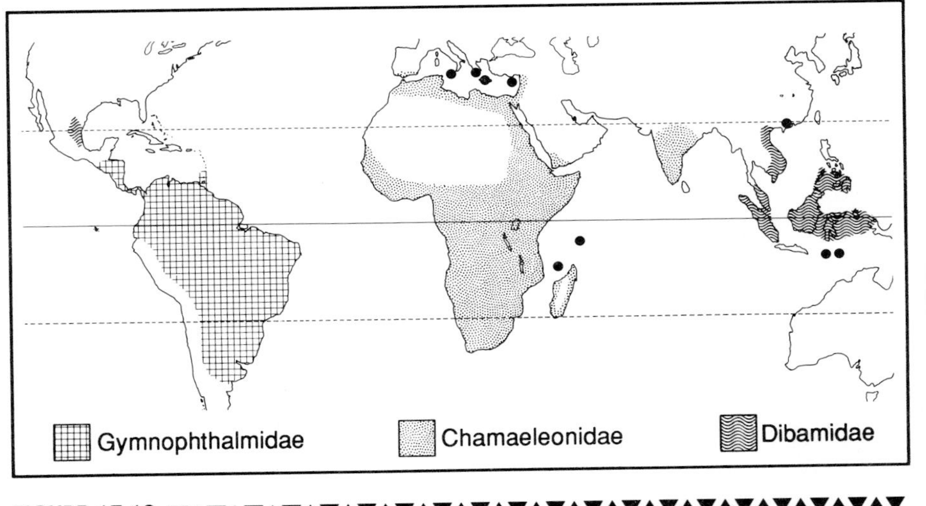

FIGURE 17.12 ▼▲▼▲▼▲▼▲▼▲▼▲▼▲▼▲▼▲▼▲▼▲▼▲▼▲▼▲▼▲▼▲▼▲▼▲▼
Distribution of the extant Chamaeleonidae, Dibamidae, and Gymnophthalmidae.

Family Helodermatidae

The Gila monster (*Heloderma suspectum*) and the Mexican beaded lizard (*H. horridum*) are the only lizards with grooved teeth and venom glands. The Gila monster lives in the deserts of the southwestern United States and northwestern Mexico. The beaded lizard occurs in arid central and southern Mexico and Guatemala (Fig. 17.13). Both species are large (30–35 cm SVL, 50 cm TL maximum for *suspectum*; 1 m TL for *horridum*), thick-bodied, stout-limbed, heavy-tailed, and strong-clawed lizards with a skin of small pebblelike scales.

They are slow-moving, diurnal lizards, which prey mainly on nestling rodents and lagomorphs, occasionally eating a lizard or bird and their eggs. They are solitary carnivores with good vision and excellent hearing. They methodically search above and below ground throughout their home ranges of several hectares, being both strong diggers and good climbers. Though *H. suspectum* are active throughout the warm months, they forage most intensely in late winter and early spring. When not foraging, they rest underground in burrows and similar retreats. Courtship and mating occur from late April to early June, and 2–12 eggs are laid in mid-July to mid-August. The eggs overwinter, hatching after about 10 months in May.

The broad, flat head of helodermatids contains a large mouth filled with sharp, slightly recurved teeth and strong muscular jaws. Some of the lower teeth are deeply grooved, particularly those nearest the openings of the venom ducts; none of the teeth are hollow fangs. The paired venom glands are multilobed, one on each side of the lower jaw; each lobe of the gland has its own duct. The

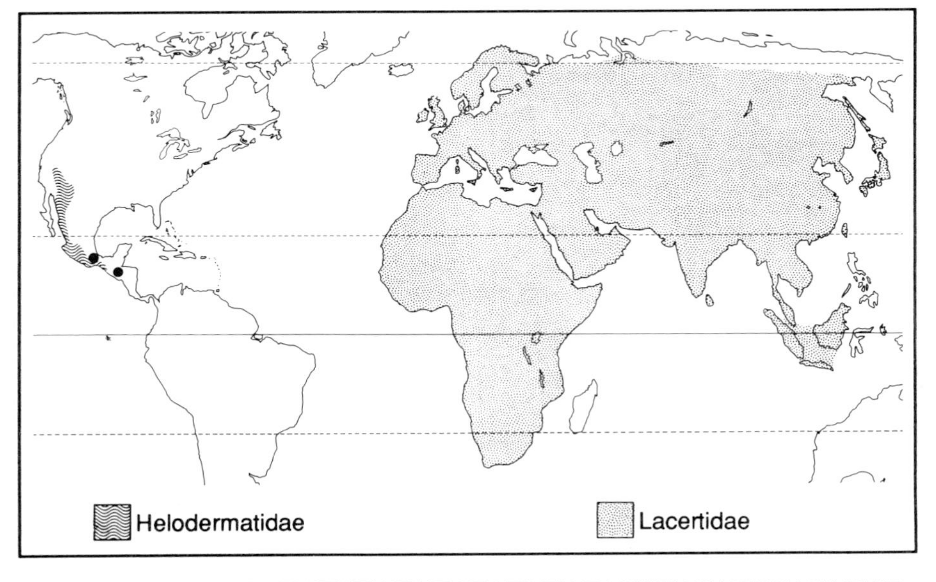

FIGURE 17.13
Distribution of the extant Helodermatidae and Lacertidae.

venom enters the wound by capillary action along the grooves of the teeth. Typically, *Heloderma* bites and retains a firm grasp while it chews, thus enhancing the entry of the venom into the wound. Death appears to result from shock associated with respiratory and cardiac paralysis.

Family Lacertidae

Lacertids are moderately diverse (20+ genera, 200+ species). This diversity exists within a remarkably uniform habitus; elongate lizards with conical heads on distinct necks, long and robust trunks, long moderately thick tails, and well-developed limbs (hind distinctly larger than fore). Scalation is also fairly uniform with large head scales, granular scales dorsally on the neck and trunk, and enlarged abutting scales ventrally; a few genera (e.g., *Algyroides*, *Ichnotropis*, *Takydromus*) have large, often keeled dorsal scales. Lacertid scalation and body forms are similar to those of the teiids, although lacertids are usually smaller. Most are less than 90 mm SVL, but a few species exceed 150 mm SVL (*Lacerta lepida*). All lacertids, except the ovoviviparous *Lacerta vivipara*, are oviparous.

Lacertids occur in Eurasia and Africa (Fig. 17.13). *Lacerta* and *Podarcis* contain most of the temperate European and Mediterranean species, living mainly in forest and scrub habitats. Eastward *Eremias* and *Ophisops* replace them in the grassland and desert habitats of Asia; however, in the Caucasia, the *Lacerta*

Corsican mountain lizard *Lacerta bedriagae* (Lacertidae). (G. R. Zug)

group has bisexual and unisexual populations. *Acanthodactylus*, *Algyroides*, and *Psammodromus* comprise small groups of Mediterranean scrub species. The long-tailed, cadaverous *Takydromus* is the only oriental species. A different group of lacertids occurs in sub-Saharan Africa. These lacertids include the desert-adapted *Meroles*, *Nucras*, and several other terrestrial genera, including the long-limbed gracile *Aporosaura anchietae*. *Aporosaura* is one of the few lizards that regularly eats seeds, not an unlikely food for a lizard of the harsh Namib Desert. *Holaspis* is one of the few arboreal lacertids, and its single species (*H. guentheri*) is a glider, although apparently a poor one using its broad tail and flattened body as an aerofoil.

Family Scincidae

Skinks occur worldwide (Fig. 17.14) and are the dominant (= most species) members of lizard faunas in many areas, for example, Australia, West Africa, and southeastern North America. Locally, their abundance regularly exceeds 1000 ha^{-1} (e.g., *Emoia cyanura* in Southwest Pacific islands). There are nearly 100 genera of skinks and over 1000 species. They range from the salt-sprayed littoral zone to near tree line in the Himalayas and from semiaquatic to arboreal habits. Most species are terrestrial or semifossorial and small- to moderate-sized (approximately 80% are 30–120 mm SVL maximum). A few skinks are larger, for example, *Tiliqua* is 9–31 cm SVL and *Corucia* is 35 cm SVL.

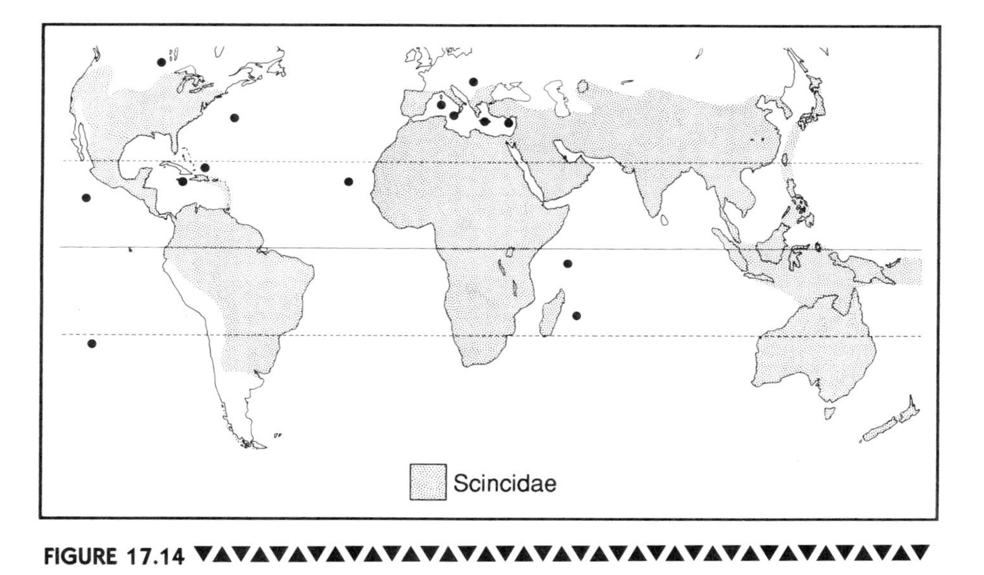

FIGURE 17.14 ▼▲▼▲▼▲▼▲▼▲▼▲▼▲▼▲▼▲▼▲▼▲▼▲▼▲▼▲▼▲▼▲▼▲▼▲▼
Distribution of the extant Scincidae.

Their great diversity encompasses a single clade; skinks are monophyletic. They uniquely share several cranial and vertebral characters, tongue surface morphology, cycloid scales, compound osteoderms dorsally and ventrally, and a secondary palate. The secondary palate ranges from partial to complete in skinks.

Regardless of habits or habitats, all but a few skinks (e.g., *Gnypetoscincus*, *Tribolonotus*) have cylindrical bodies, robust, medium-length tails, and conical heads, and all have smooth (to moderately keeled), shiny scales. Limb development varies from strong, moderately long limbs through vestigial rods to limbless (at least externally). Commonly, as limbs reduce and shorten, the body and tail elongate and movement becomes increasingly serpentine.

Reproduction in skinks includes all stages from oviparity to viviparity. Eggs are laid as single clutches (guarded or abandoned) or, infrequently, communally. Clutch size is commonly low (≤ six) and is limited to one or two eggs in some species groups (e.g., *Tribolonotus* and *Emoia*, respectively).

Skinks segregate into four subfamilies (Acontiniinae, Feyliinae, Lygosominae, Scincinae). Acontiniines (three genera, South Africa) and feyliines (two genera, Central Africa) are limbless, fossorial skinks. Members of both families are elongate, smooth-scaled lizards and live-bearers. Acontiniines (<55 cm SVL maximum; most <28 cm) are mainly grassland skinks living within bunch-grass or burrowing in sandy soils. Feyliniines (<30 cm SVL maximum) are also semifossorial skinks of grasslands and humid forest.

In contrast, lygosomines and scincines are geographically and morphologically diverse groups. Both include limbed to limbless, small to large, and oviparous to viviparous skinks. Lygosominae are predominantly Australasian skinks with major radiations in Australia, India-Indochina, and the Southwest Pacific. *Mabuya* is the only lizard genus with a circumtropical distribution, although less than a dozen species occur in the neotropics. *Cryptoblepharus*, the Indopacific snake-eyed skink, occurs on islands (Australia as well) from the east coast of Africa to the west coast of South America. Only two other lygosomines (*Scincella*, *Sphenomorphus*) occur in the Americas and with fewer than ten species. *Lygosoma*, *Riopa*, and *Scincella* are the principal Asian lygosomines, *Emoia* for the Pacific, and *Ctenotus*, *Egernia*, and *Lerista* in Australia. *Sphenomorphus* is the most speciose lygosomine (120+ species) with radiations in both Australia and Asia. A few speciose lygosomines (*Panaspis*, *Mabuya*) are African.

Scincines are predominantly African-Eurasian and have about half the genera (25+ versus 40+) and about a quarter of the species (150+ versus 500+) of the lygosomines. Only *Eumeces* and *Scelotes* are particularly diverse (>20 species). *Eumeces* is a subtropical-temperate group of the Northern Hemisphere, and the only skink to radiate significantly in the Americas (Fig. 17.15); *Neoseps* is the only other American scincine. *Scelotes* is an African group of small, semifossorial-fossorial skinks, and like some of the Australian lygosomine genera, its species show the entire spectrum from limblessness to well-developed limbs.

Family Teiidae

Teiids are a moderately diverse group (nine genera) of medium to large lizards (7–50 cm SVL, most <13 cm). They are predominantly terrestrial and

FIGURE 17.15 ▾▴▾▴▾▴▾▴▾▴▾▴▾▴▾▴▾▴▾▴▾▴▾▴▾▴▾▴▾▴▾▴▾▴▾
Prairie skink *Eumeces septentrionalis* (Scincidae). (R. W. Van Devender)

occur throughout the Americas (Fig. 17.16) in a wide variety of habitats from xeric scrub and sparse grasslands to rain forest. Body form is elongate; head pointed; neck, trunk, and tail long; and limbs well developed. This habitus correlates with a speedy motile-search life-style. Similarly most are eurythermic and appear only when diurnal temperatures are high and permit constant activity. Body scalation consists of dorsal granular scales and large, rectangular ventral plates; the head bears large plates.

Teiids divide into two size groups. One group contains *Tupinambis* (two species), a large (>30 cm SVL) carnivore that preys on vertebrates as well as arthropods. The other group contains the smaller taxa; they are mostly <16 cm SVL (<36 cm TL) and include *Ameiva*, *Cnemidophorus*, *Kentropyx*, and a few other genera. They are largely insectivores, although the heavier-bodied *Dracaena* specializes on snails.

Teiids are oviparous. The widespread *Cnemidophorus* (45 + species) includes unisexual and bisexual species (Fig. 17.17). The unisexual species are most prevalent in the desert areas of southwestern North America and usually sympatric with one or more bisexual species. Their high accessibility has fostered numerous studies, which have demonstrated (morphologically, karyotypically, and biochemically) that each unisexual species arose through the hybridization of parapatric or sympatric bisexual species or a unisexual and bisexual species. The unisexual species regularly thrive in disturbed habitats or ecotones, hence they have been called "weed species." Even though they may be locally abundant

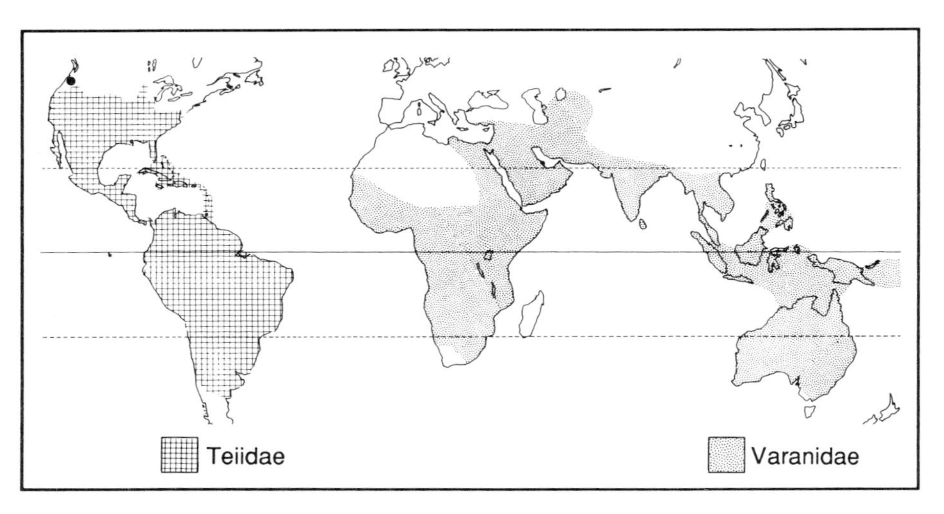

FIGURE 17.16 ▼▲▼▲▼▲▼▲▼▲▼▲▼▲▼▲▼▲▼▲▼▲▼▲▼▲▼▲▼▲▼▲▼
Distribution of the extant Teiidae and Varanidae.

FIGURE 17.17 ▼▲▼
Six-lined racerunner *Cnemidophorus sexlineatus* (Teiidae). (R. W. Van Devender)

and in some cases geographically widespread, they are likely evolutionary dead ends, lacking the genetic diversity and panmixia to adapt to major environmental changes. Teiid unisexuality has been demonstrated elsewhere only in *Kentropyx*.

Family Varanidae

Varanids include two dissimilar groups, Lanthanotinae and Varaninae. The former contain only the earless monitor (*Lanthonotus borneensis*) of Borneo. *Lanthonotus* is a slender helodermatidlike, semiaquatic lizard (30–40 cm TL) living in and near small forest streams and swamps. It appears to be nocturnal, spending its diurnal hours in burrows. At night, it forages on land and in the water, moving via undulatory locomotion in both habitats. It is oviparous.

All living varanines share a small head, long neck, sturdy body and limbs, and long, powerful tail, and all are included in the genus *Varanus* (35 + species)(Fig. 17.18). These goannas or monitor lizards are widespread throughout the Old World tropics (see Fig. 17.16). They have their greatest diversity in Australasia, where both the smallest and largest species occur. The pygmy goanna (*Varanus brevicauda*) is a small desert goanna of Australia (12 cm SVL maximum, 23 cm TL). The giant of all lizards, the Komodo dragon or ora (*V. komodensis*; 3.1 m TL maximum, 250 kg), occurs on the East Indies islands of Flores, Komodo, and Padar. All monitors are carnivores, although the Philippine butaans (*V. olivaceus*) seasonally eat fruit. The smaller species prey mainly on insects, small reptiles, and amphibians. With increasing body sizes, prey preference shifts

FIGURE 17.18

African savanna monitor *Varanus exanthematicus* (Varanidae). (R. G. Tuck)

increasingly to larger vertebrates, including mammals (e.g., *V. komodensis* kills and scavenges goats, deer, and wild cattle). Most species are terrestrial-arboreal predators, searching for prey in trees as well as on the ground. Some species (*V. indicus*, *V. niloticus*) regularly forage near water and occasionally in water. *Varanus mertensi* feeds and hides in water.

Varanus is strictly oviparous. Their courtship is often preceded by ritualized male combat, that is, an upright grappling/dancing posture.

Family Xantusiidae

Xantusiids are small- to moderate-sized lizards (<100 mm SVL) from southwestern North America, Middle America, and eastern Cuba (see Fig. 17.3). They are geckolike with soft granular-scaled back and limbs but large plates on the head and ventrally on the trunk. Their immovable eyelids form a spectacle over each eye, and like geckos, xantusiids clean their spectacles and face with flat, broad tongues. They are seldom seen and, with their elliptical pupils, were mistakenly assumed to be nocturnal and named "night lizards." Instead, they are strictly diurnal but seldom venture into the open. They move about slowly in and under ground litter, in rock crevices, or beneath the canopy of low, dense vegetation. All, whether desert or forest inhabitants, are probably sedentary with home ranges of only a few square meters.

Xantusiids contain three genera (*Cricosaura*, *Lepidophyma*, *Xantusia*). *Cricosaura* (one species) occurs only on an isolated peninsula in Cuba and, as yet, is little studied. *Lepidophyma* (14+ species) occurs widely in Central American lowland forest. *Xantusia* (4 species)(Fig. 17.19) inhabits the deserts and scrub of southwestern North America. Their diets are catholic and include plant matter.

All xantusiids are live-bearers, producing a few young each year or biennially. *Lepidophyma* has unisexual and bisexual species.

Family Xenosauridae

Xenosaurids include *Shinisaurus crocodilurus* (China) and three species of *Xenosaurus* (Mexico)(Fig 17.20). They are moderately large (<40 cm TL) lizards with a slender helodermatidlike habitus, although less broadly jowled and with slender, longer tails. Their backs are armored by rows of enlarged keeled scales, but the osteoderms are small and separate. Both are stenohydric taxa occurring in moist mountain forests. *Shinisaurus* (Fig. 17.21) is semiaquatic and forages in mountain streams for fish and other animal prey. *Xenosaurus* is a terrestrial denizen of cloud forests. All xenosaurids are live-bearers.

FIGURE 17.19 Granite recluse lizard *Xantusia henshawi* (Xantusiidae). (R. W. Van Devender)

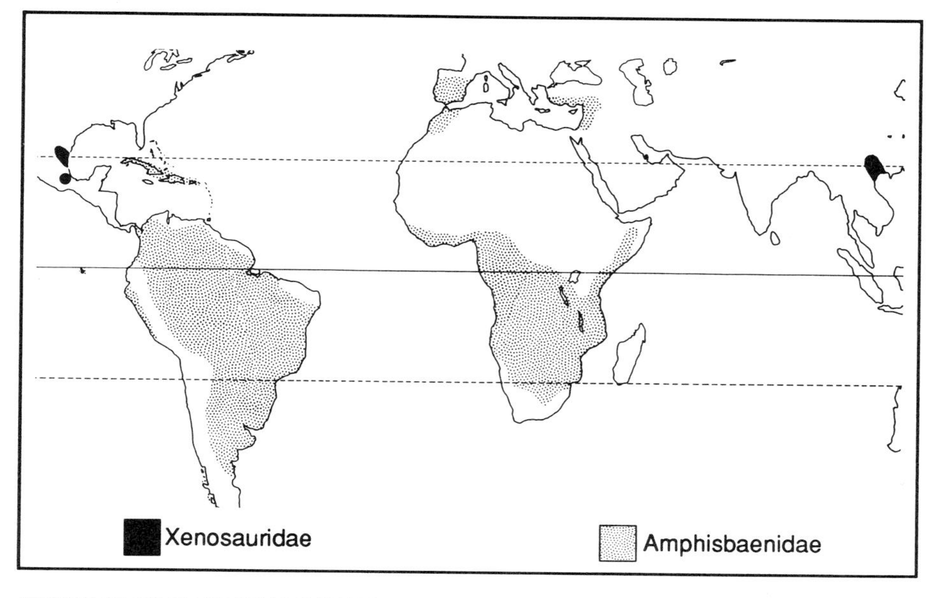

FIGURE 17.20 ▼▲▼
Distribution of the extant Amphisbaenidae and Xenosauridae.

FIGURE 17.21 ▼▲
Chinese crocodile lizard *Shinisaurus crocodilurus* (Xenosauridae). (R. W. Van Devender)

Amphisbaenia

The worm-lizards are recognized as a distinct group of lizards because of their earthwormlike appearance (Fig. 17.22) and associated anatomical modification for fossorial life-style. Their annulated appearance is unique for reptiles, but their origin is within the Lacertilia. As fossorial animals, amphisbaenians have compact skulls (the digging instrument), reduced but functional eyes beneath opaque head scales, no external ears, and no limbs or girdles (except *Bipes*). Their bodies are elongated through an increase in number of trunk vertebrae, perhaps at the expense of caudal vertebrae because their tails are short. Body elongation results in relocation, narrowing, and reduction of the viscera, for example, fewer loops in the intestine, slender gonads with the right one anterior to the left, and the liver long and narrow. As in snakes and other elongated and legless lizards, the development of one lung is suppressed; however, unlike any other vertebrate, it is the right lung.

FIGURE 17.22 ▼▲▼

An Argentinian wormlizard *Amphisbaena angustifrons (Amphisbaenidae)*. (C. Gans)

Family Amphisbaenidae

Amphisbaenids occur in Africa and South America (see Fig. 17.20). They contain some of the most generalized (*Blanus*, circum-Mediterranean) and specialized (*Ancylocranium*, Africa) forms, and they are the most diverse (15 genera, 100+ species) amphisbaenians. Most amphisbaenids are moderate-sized (25–40 cm SVL), but a few *Amphisbaena* species are larger (e.g., *A. alba* is approximately 70 cm TL).

Amphisbaenids display three general head shapes associated with different burrowing techniques. The blunt-cone or bullet-headed genera (e.g., *Amphisbaena*, South America; *Blanus* and *Cadea*, Cuba; *Zygaspis*, Africa) burrow by simple head-ramming. The spade-snouted taxa (*Leposternon*, South America; *Monopeltis*, Africa) tip the head downward, thrust forward, and then lift the head. The laterally compressed keeled-headed taxa (*Anops*, South America; *Ancylocranium*) ram their heads forward, then alternately swing it to the left and right. Reproduction appears to be oviparous in most amphisbaenids, but some *Loveridgea* and *Monopeltis* are live-bearers.

Family Bipedidae

Bipes (three species) is unique among the amphisbaenians. Remarkably, it has large molelike forelimbs and hands. These limbs assist *Bipes* to move about on the surface and to dig a small pit for reentry into the soil. *Bipes* are desert amphisbaenians of Mexico (Fig. 17.23) of small to moderate size (12–26 cm

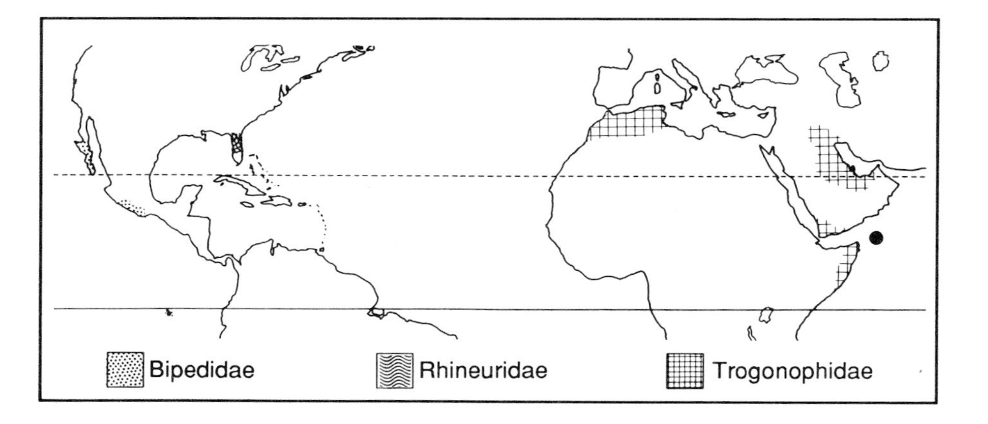

FIGURE 17.23 ▼▲▼▲▼▲▼▲▼▲▼▲▼▲▼▲▼▲▼▲▼▲▼▲▼▲▼▲▼▲▼▲
Distribution of the extant Bipedidae, Rhineuridae, and Trogonophidae.

SVL). They are blunt-headed and burrow by head ramming. All three species are oviparous and lay small clutches of one to four eggs.

Family Trogonophidae

Although rhineurids had many species in the North American Tertiary, only *Rhineura floridana* survives today. *Rhineura* is a moderate-sized (24–38 cm SVL), spaded-snouted amphisbaenian with a shieldlike tail, lacking caudal autonomy. It occurs in a variety of forest habitats from xeric scrub to mesic live-oak hammocks of central Florida (Fig. 17.23); all habitats have sandy soil. It is oviparous with clutches of two eggs.

Family Trogonophidae

Trogonophids are the most divergent amphisbaenians. They have an acrodont dentition (pleurodont in other families), lack caudal autonomy (present in most others), show a tendency to shorten the body by reduction of trunk vertebrae and fusion of cervical vertebrae, are triangular or I-shaped in cross section (circular in others), have a strongly flattened snout with slightly upturned edges, and use an oscillatory digging motion. With the exception of the dentition, the preceding characteristics correlate with their locomotion in the loose, dry sands of their homes in the Horn of Africa and the Arabian Peninsula. Unlike the other amphisbaenians, trogonophids use their tail as a brace to support the backward thrust of their digging. Their burrows match the shape of their noncircular bodies. They create their burrow by an alternating rotational movement of the head, which simultaneously shaves off the sides of the tunnel and compacts the wall.

Agamodon (three species) and *Trogonophis* (one species) are African; *Pachycalamus brevis* occurs on Socotra, an oceanic island in the Arabian Sea; and *Diplometopon* with a single species ranges from southwestern Arabia to Iran. All trogonophids are small to moderate size (8–24 cm SVL) and oviparous, except the live-bearing *Trogonophis*.

Phylogenetic Relationships of Lizards

Modern lizards divide into two major clades: Iguania and Scleroglossa. This divergence of the ancestral lizard stock is likely quite ancient, possibly Jurassic but certainly by mid-Cretaceous. The two lineages differ consistently (possession of synapomorphies) in numerous features, such as tongue structure and feeding behavior, brain morphology, abdominal muscles, and skull structure. While the taxonomic content of each group seems certain, the relationships among and within the families are less certain and still debated.

Iguania consists of ten or more presumably monophyletic groups. Eight of these groups are the iguanid subfamilial groups described earlier. The interrelationships of these subfamilies are uncertain, and where relationships have been proposed, the evidence is either equivocal or conflicting. One such relationship is the agamids-chamaeleonids to a hoplocercine-polychrine stock; however, it is also possible that agamids-chamaeleonids diverged from the iguanids prior to the radiation into the eight modern groups. Agamids and chamaeleonids are sister-groups, presumably with the chameleons diverging early from an agamid stock. Agamids and chameleons uniquely share an acrodont dentition on the dentary and posterior maxillary.

Scleroglossa (hard-tongued lizards) includes the sister-groups Gekkota and Autarchoglossa (Fig. 17.24). Both groups have flattened tongues with at least part of the tongue keratinized, and this condition appears to be associated with prey capture depending entirely on use of jaws, that is, there is no tongue manipulation of prey. Gekkotans possess numerous unique traits that distinguish them from all other lizards and indicate monophyly. Their ears are highly modified. The inner ear has a distinct pattern of hair cells on the auditory papillae, an elongated cochlear duct, and basilar membrane; the middle ear has an extracolumellar muscle and several other unique structural specializations. Only gekkotan hatchlings have a pair of egg teeth, a single egg tooth in all other squamates. Similarly several gekkotan subgroups (eublepharids, diplodactyls, pygopods, sphaerodactyls) are sharply defined; the first and the last are monophyletic.

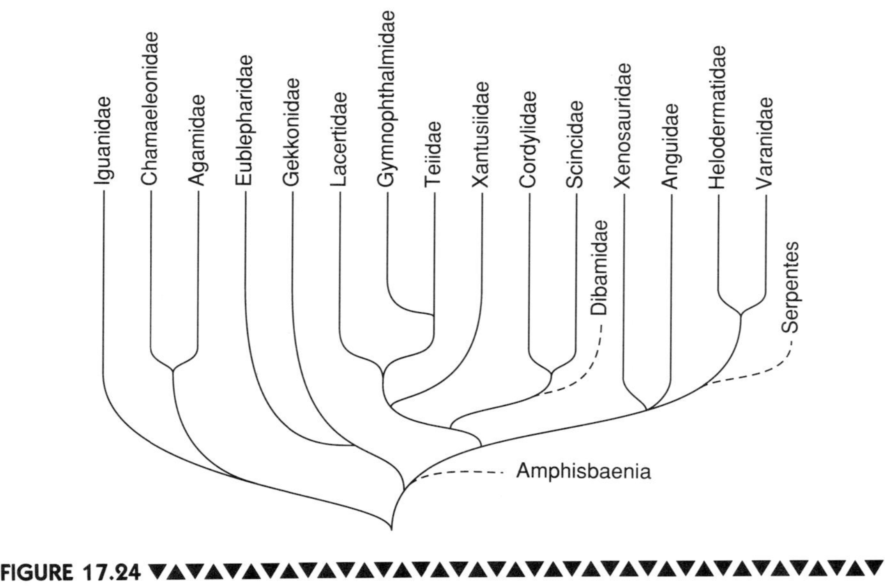

FIGURE 17.24 ▼▲▼

Dendrogram of presumed phyletic relationships among the living families of squamates, exclusive of snakes.

Diplodactyls are the Australian (including New Caledonia and New Zealand) legged geckos, and the pygopods are the Australian legless geckos. Previously pygopods were placed in their own family and all the legged geckos in the family Gekkonidae; however, a recent analysis demonstrates that diplodactyls and pygopods are sister-groups, and thus pygopods cannot be removed from the Gekkonidae without making the family paraphyletic.

The autarchoglossans have a greater diversity and are much more difficult to characterize. Currently only a single unique trait is known for all members, that is, the presence of a rectus abdominis lateralis muscle. Even lacking a greater number of unique traits, Autarchoglossa appears to be monophyletic. It contains two subclades (Scincomorpha, Anguimorpha) with certainty and three other subclades (Dibamidae, Amphisbaenia, Serpentes) with less certainty.

Scincomorphans are the lacertids, gymnophthalmids, teiids, xantusiids, cordylids, and scincids (Fig. 17.24). The last two families are a sister-pair (scincoids) and the sister-group to the other four families (lacertoids). Cordylids and scincids share a large number of features, such as cephalic, dorsal, and ventral trunk osteoderms, although some traits also occur in other lizard groups. A few others (e.g., compound osteoderms and tongue structure) are unique and support the close relationships of cordylids and scincoids. Lacertids and teiids are similar in general appearance and have a long history of presumed close relationship. The external similarity is also matched by shared internal traits, thus reenforcing the presumed close relationships. Gymnophthalmids were until recently considered teiids because of shared traits, and a close relationship seems highly probable. A recent, if unlikely, proposal relates teiids with iguanians. Xantusiids have been variously allied to gekkotans or anguimorphs. Aside from their eye-licking behavior and some primitive traits, they share little similarity to the geckos. Their similarities to anguimorphs are those possessed by many autarchoglossans; so among other traits, the teiid-lacertid body scalation supports their lacertoid relationships.

Anguimorphs are the anguids, xenosaurids, helodermatids, and varanids. Anguimorphs are another long-recognized group of families with numerous uniquely shared traits. One of the most prominent traits is the forked tongue in which the distal portion retracts into the proximal portion (base). This trait is also shared with snakes. Similarly, the sister-group relationship of helodermatids and varanids has a broad spectrum of characters in its support. The relationship among helodermatids-varanids, anguids, and xenosaurids is unresolved. Within the anguids, the interrelationships of the subfamilies are plagued by numerous conflicts of evidence. Anniellines as the sister-group to the anguines-diploglossines-gerrhonotines seems to be the best supported of the possible relationship patterns in anguids.

The ancestry of dibamids, amphisbaenians, and snakes remains unresolved in spite of numerous proposals. With each group, different sets of characters yield contrary interpretations of relationships. Suggestions for the origin of amphisbaenians have ranged from a basal divergence from the early squamates to derivation from snakes. Gekkotans and gymnophthalmids have also been proposed as sister-groups for amphisbaenians. Since amphisbaenians possess numerous scleroglossan characteristics it seems likely that their origins lie somewhere among the scleroglossan taxa. Dibamids are no less enigmatic and similarly share numerous scleroglossan traits; however, they share nearly equal numbers of derived traits with geckos, anniellines, and amphisbaenians, and only slightly fewer with felyiniines, acontines, and pygopods. Snake origins are somewhat less confused but no better resolved. Two propositions have nearly equal currency: snakes and lizards represent a basal divergence or snakes arose within the anguimorphs, likely from a varanoid stock.

CHAPTER 18

Snakes

SQUAMATA/LACERTILIA

Although snakes are morphologically distinct and easily distinguished from other lizards, snakes are lizards. They are no more or no less lizards than the amphisbaenians, because both snakes and amphisbaenians have their origins within specific groups of modern lizards. Only our uncertainty of which lizard group has a shared ancestry with snakes removes snakes from the lizard clade.

Serpentes

Limblessness is the most visible characteristic of snakes, but as noted in Chapter 17, it is a feature shared with several other groups of lizards. The forelimbs and pectoral girdle are totally absent in snakes; a vestige of the hindlimb and/or pelvic girdle occurs in most families of primitive snakes. A number of other characteristics are shared with scleroglossan lizards and most frequently with the varanoid-anguimorphs. The keratinized and strongly forked foretongue and its retraction into the hindtongue is characteristic of both snakes and varanoids. Reduction or loss of the left lung, loss of external and middle ears, lidless eyes protected by spectacles, simplification of the hyoid skeleton, and modifications of gonads and ducts are traits shared with one or more of the limbless scleroglossans. Several traits are unique to snakes and support the monophyly of snakes: dentaries of the lower jaw with a loose tendinous symphysis (except blindsnakes); exoccipitals and basioccipital form the foramen magnum; left systemic arch larger than right one (except some booids); and no ciliary-body muscles in the eye.

Limblessness has been no handicap for evolutionary success. Snakes as well as the other limbless scleroglossan lizards are speciose. The origin of limblessness is often, although not exclusively, associated with fossorial habits, and this association is vividly reflected in a comparison of the two basic groups (Scolecophidia, Alethinophidia) of living snakes. Scolecophidians, the blindsnakes, are look-alikes of several scleroglossan groups and exemplify a fossorial life-style, whereas the predominantly surface-dwelling alethinophidians have no scleroglossan twins, even among the presumably secondarily fossorial alethinophidian taxa.

Scolecophidians are uniformly slender, cylindrical-bodied snakes with blunt heads and short tails with apical spines. All have smooth, shiny, subequal, overlapping scales encircling the entire body. Their eyes are reduced and lie beneath large head scales. Uniquely, sebaceous glands are present; their vertebrae lack neural spines and hypapophyses; those taxa with vestigial pelvic girdles have no external spurs; their liver is multilobed; and the left oviduct is absent in *Leptotyphlops* and the Typhlopidae. Scolecophidians comprise three families: Anomalepididae; Leptotyphlopidae; and Typhlopidae.

Alethinophidians are a much more diverse group and include the remaining families of snakes. They are less easily characterized because of their great diversity. Alethinophidians divide into primitive (henophidians) and advanced (caenophidians) snakes. These two groups are somewhat easier to define by contrasting the retention of ancestral (= primitive) traits in the former and derived states in the latter. The so-called primitive groups are, however, often highly specialized in behavior and anatomy. Henophidians (Aniliidae, Boidae, Bolyeriidae, Loxocemidae, Pythonidae, Uropeltidae, Xenopeltidae) usually have teeth on the premaxillae, an open Meckel's groove on dentaries, left and right carotid arteries, an intercostal artery to each body segment, and vestigial hindlimbs and/or a pelvic girdle. Caenophidians (Atractaspididae, Colubridae, Elapidae, Viperidae) lack premaxillary teeth and vestiges of hindlimb or pelvic girdle, usually have a closed Meckel's groove, a single left carotid artery (except some viperids), and each intercostal artery serves several body segments. Acrochordids possess a mixture of henophidian and caenophidian characteristics.

Family Anomalepididae

Anomalepidids are small (<40 cm adult TL, most 15–30 cm) blindsnakes of South America (Fig. 18.1). Although anomalepidids are geographically the most restricted of the blindsnake groups, they show the greatest generic diversity (*Anomalepis*, *Helminthophis*, *Liotyphlops*, *Typhlophis*; 15+ species). In habits and habitats, they share a fossorial behavior and a preference for termite prey with typhlopids and leptotyphlopids. Furthermore, they are sympatric with both typhlopids and leptotyphlopids throughout their entire range, suggesting feeding differences or some other partitioning of resources among the three blindsnake families.

Anomalepidids have dentary and maxillary teeth, movable maxillae, one to three teeth on dentary, no left lung, a tracheal lung and left oviduct present (reduced or absent in *Anomalepis*), and usually no pelvic-girdle vestige. All anomalepidids appear to be oviparous, laying from 2 to 13 eggs.

Family Leptotyphlopidae

The wormsnakes or thread blindsnakes are small (40 cm SVL maximum, most 15–25 cm), distinctly slender blindsnakes. They have immovable edentulous maxillae and four to five teeth on each dentary. The left oviduct is absent; the tracheal and left lungs are absent; the pelvic girdle is rudimentary and occasionally rudimentary femurs occur, embedded in trunk musculature, rarely showing as external spurs.

Two genera (*Leptotyphlops*, *Rhinoleptus*; 80+ species) are recognized. *Leptotyphlops* occurs widely in the Americas, Africa, and Southwest Asia (Fig. 18.2); *Rhinoleptus* has a single species in West Africa. Leptotyphlopids feed nearly exclusively on termites and selectively on the abdomens of the larger termite species. Unlike many termite predators, they are capable of living in termite nests

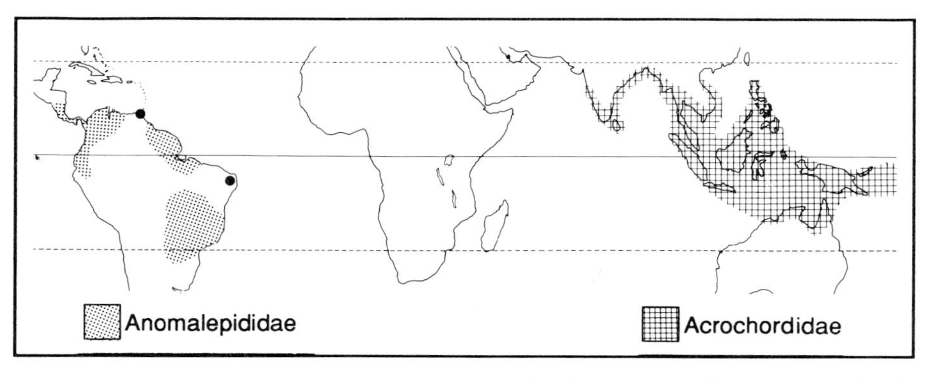

FIGURE 18.1 Distribution of the extant Acrochordidae and Anomalepididae.

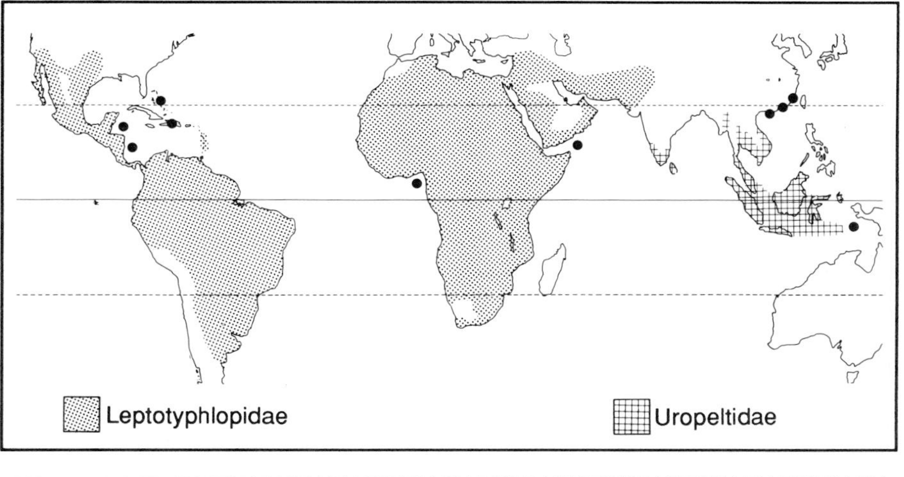

FIGURE 18.2 ▼▲
Distribution of the extant Leptotyphlopidae and Uropeltidae.

and do so regularly. They have evolved a secretion that repulses the attack of the soldier termites and ants and causes them to be recognized as nestmates and permit them to live undisturbed in the termite and ant galleries. Because they can move freely through the galleries, they have occasionally been found high above the ground in trees, a highly unexpected location for a subterranean animal.

Leptotyphlopids are oviparous, laying 1–12 small, elongate eggs. *Leptotyphlops dulcis*, the Texas blindsnake, coils around the incubating eggs; parental care may occur in other *Leptotyphlops* but has not been reported.

Family Typhlopidae

Typhlopids occur throughout the tropics and into temperate Asia and Australia (Fig. 18.3) and contain three genera. *Ramphotyphlops* (50± species) includes African, Asian, and Australian species, *Typhlops* (125+ species) occurs worldwide in the tropics, and *Rhinotyphlops* (20+ species) is a sub-Saharan African group. Typhlopids look like the other blindsnakes, although their size range is much broader, ranging from small species, such as *Ramphotyphlops braminus* (14–18 cm TL), to the largest, *Rhinotyphlops schlegelii* (95 cm TL maximum). All typhlopids have movable maxillae with teeth and toothless dentaries; all possess a tracheal lung (multichambered) and the left lung is commonly absent; only the right oviduct is present, as is a vestigial pelvic girdle.

Most typhlopids are oviparous, but *Typhlops diardi* is ovoviviparous. Clutch size varies with body size, ranging from 2–7 eggs (*Ram. braminus*) to 40–60

Jamaican blindsnake *Typhlops jamaicensis* (Typhlopidae). (R. G. Tuck)

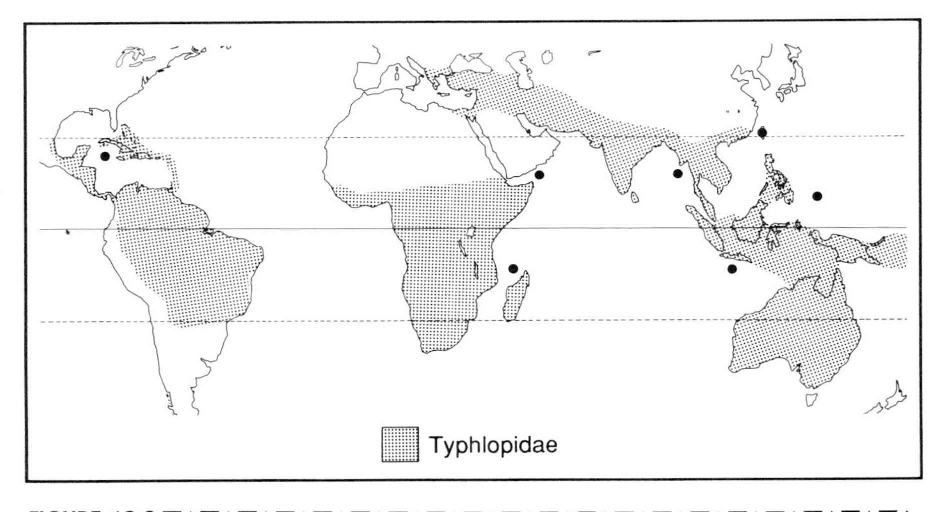

FIGURE 18.3
Distribution of the extant Typhlopidae, excluding the introduced populations of *Rhamphotyphlops braminus*.

eggs (*Rhin. schlegelii*). Eggs may be laid shortly after fertilization and incubated for 6–8 weeks or held within the oviducts and hatch within a week of laying (*Typhlops bibronii*). *Ram. braminus* is the only known unisexual species of snake. It has been accidently transferred in potted plants throughout the tropics from Africa through Indoaustralia to the Americas. As a parthenogen, only one individual needs to survive a trip to a new locality for the "flowerpot" blindsnake to establish a new population.

Family Acrochordidae

The wart, file, or elephant-trunk snakes acquired their common name from their unusual skin. The skin bears numerous small, granular scales giving it a rugose texture; further the scales do not overlap and the interstitial skin forms bristle-tipped tubercles. Acrochordids are heavy-bodied snakes, yet the skin appears several sizes too large, lying loose in folds. Such a skin would seem inappropriate for an aquatic snake, but all three *Acrochordus* species (*arafurae*, *granulatus*, *javanicus*) are totally aquatic, strong swimmers, and nearly incapable of terrestrial locomotion. Aquatic adaptations include eyes shifted dorsally on the head, valvular nostrils, a flap to close the lingual opening of the mouth, and small ventral scales. *Acrochordus granulatus* is an estuarine-marine species and has a laterally compressed tail and lingual salt glands. The other two species are mainly freshwater inhabitants.

Filesnakes are moderately large snakes (58–180 cm SVL) of the Indoaustralian region (see Fig. 18.1), and all three feed primarily on fish (living and dead). *Acrochordus arafurae* appears to be strictly piscivorous and consumes a variety of fish species. Prey capture usually requires the fish to touch the anterior part of the snake's body, and then *Acrochordus* entraps fish in body loops or coils with its spinose scales. The fish is quickly shifted forward and rapidly swallowed. All acrochordids are ovoviviparous with litters ranging from 2 to 32 neonates. Clutch size is positively correlated with body size in each species, and *Acrochordus arafurae* and *A. javanicus* are the most fecund.

Family Aniliidae

Anilius scytale is a moderate-sized (60–100 cm SVL) fossorial snake of tropical South America (Fig. 18.4) and the sole living aniliid. Because it is brightly banded in red and black, it is commonly called the false coral snake. It is not venomous and is phylogenetically far removed from the advanced colubroid snakes. *Anilius* is a henophidian and possesses such features as vestigial hindlimbs and girdle, premaxillary teeth, and a robust, tightly knit skull. The latter feature is associated with its burrowing behavior and includes the corresponding eye reduction (lying beneath large head scales), numerous small, smooth scales around the body, and a reduction in ventral-scale size. *Anilius* preys on other burrowing creatures, particularly caecilians and amphisbaenians. It is ovoviviparous and bears small litters (3–13 neonates).

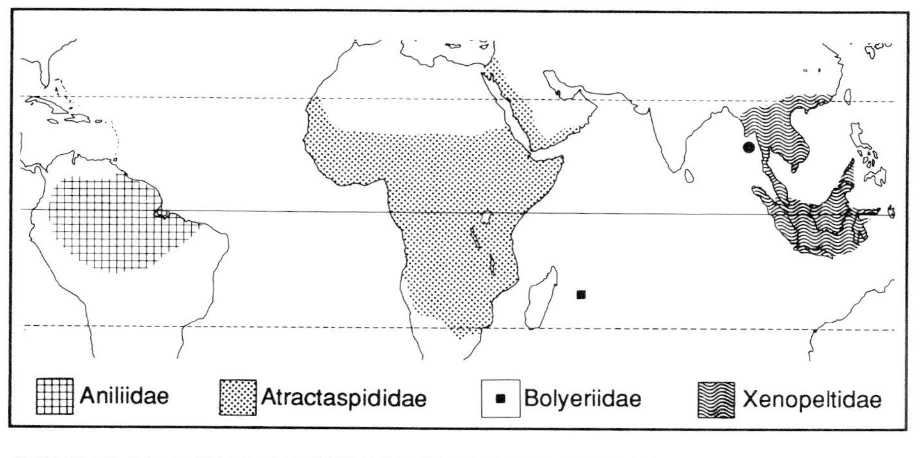

FIGURE 18.4 ▼▲
Distribution of the extant Aniliidae, Atractaspididae, Bolyeriidae, and Xenopeltidae.

Family Atractaspididae

The African burrowing asps consist of the single genus *Atractaspis* (16 species) in Africa and adjacent Eurasia (Fig. 18.4). They are blunt-headed, slender snakes (25–80 cm SVL) with short, blunt tails; all have very long maxillary fangs and venom glands. They live and feed subterraneanly, which prevents strike-bite envenomation. Instead, they crawl alongside their prey (mainly newborn rodents and burrowing reptiles), depress their lower jaw on one side exposing a fang, and with a backward stab envenomate the prey. Burrowing asps are oviparous and lay small clutches of 2–11 eggs.

Family Boidae

Boas are a widespread group (Fig. 18.5), with their greatest diversity in the tropics but with some species (erycines) living in cool temperate habitats. They range in size from tiny West Indies boas (22 cm SVL, *Tropidophis greenwayi*) to the giant anacondas (*Eunectes*, >10 m TL). Boas are aquatic to arboreal including burrowing forms, and occur in deserts to tropical rain forests. The diversity of boas and the likely paraphyly of the family make the group difficult to characterize; nonetheless, most boids possess vestigial hindlimbs and girdles displayed externally as cloacal spurs (usually larger in males), left and right lungs, premaxilla-maxilla articulation, premaxillary, maxillary, and palatine teeth, a coronoid element in the mandible, hypophyses on anterior trunk vertebrae, and ovoviviparity.

Boids comprise three subfamilies (Boinae, Erycinae, Tropidophiinae). Tropidophiines are small (<60 cm SVL, except *Tropidophis melanurus*) neotropical

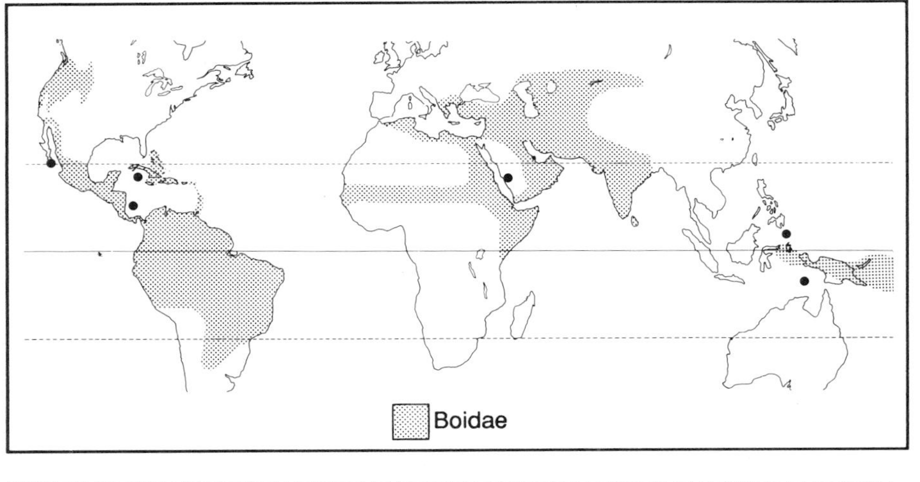

FIGURE 18.5 ▼▲
Distribution of the extant Boidae.

ground boas, typically nocturnal and terrestrial species. In contrast to most other boids, tropidophiines have no left lungs, a tracheal lung, a few large dorsal head scales, and $\leq$27 dorsal scales around midbody. *Trachyboa* (northwestern South America), *Ungaliophis* (Mexico to Ecuador), and *Boella* and *Exiliboa* (southern Mexico) are strictly continental taxa with one to two species each. *Tropidophis*, in contrast, has 15 species and is predominantly a Caribbean group (Bahamas and Greater Antilles), although two species occur in south-tropical South America. These ground boas prey mainly on small amphibians and reptiles. With the exception of oviparous *Trachyboa*, tropidodophiines are live-bearers, generally with small litters of two to six neonates.

Boines and erycines occur in both the Old and New World. Boines include both large (adults >1.5 m SVL; *Acrantophis*, *Boa*, *Epicrates*, *Eunectes*, *Sanzinia*) and small boas (<1.5 m SVL; *Candoia*, *Corallus*). They are principally neotropical, although *Acrantophis* and *Sanzinia* are Madagascaran endemics and *Candoia* are the Pacific island boas. Boines are mainly arboreal snakes, and many are bird and mammal predators, particularly the larger species. *Candoia carinata* shows niche displacement; it is arboreal when in sympatry with the terrestrial *C. aspera*, and terrestrial in sympatry with the arboreal *C. bibroni*. *Eunectes*, the anacondas, are aquatic, a habit not seen in any other boids. All boines are ovoviviparous.

Erycines are predominantly Old World and exclusively terrestrial and semifossorial taxa. They are moderate to small taxa (<1 m SVL) with small compact heads, stout cylindrical bodies, short tails, and small eyes. *Calabaria* (West Africa), *Charina,* and *Lichanura* (western North America) (Fig. 18.6) are monotypic in contrast to the widespread *Eryx* (10 species; Africa through Southwest

FIGURE 18.6
Rubber boa *Charina bottae* (Boidae). (R. W. Barbour)

Asia to India). Small reptiles and mammals are their major prey, and except for the oviparous *Calabaria*, they are live-bearers.

Family Bolyeriidae

The bolyeriids consist of *Casarea dussumieri* and *Bolyeria multocarinata* from the Mascarene islands in the western Indian Ocean (see Fig. 18.4). They are unique among vertebrates in having each maxilla jointed into a separate anterior and posterior element. Although these snakes are commonly linked with boas, they lack hindlimb-pelvic vestiges and their bodies appear more colubroid, slender with proportionately longer tails (80–138 cm TL) than the typical stout-bodied boids. The doubled upper jaw and the specialized "sensing posture" during prey hunting and capture indicate that at least *Casarea* shares its prey-capture behavior with lizard-eating colubroids. *Casarea* is oviparous, *Bolyeria* unknown.

Family Colubridae

The family Colubridae is often referred to as a "trash-can" taxon, one in which a genus or species is tossed when it fits nowhere else. Colubrids are advanced snakes and have all the characteristics of caenophidians. The multiple lineages, one of which includes the elapids, represent the sister-group of the viperids.

Owing to the composite nature of the Colubridae, it is a difficult group to characterize in any manner other than by the lack of the traits that delimit the elapids and viperids. Similarly as a composite, it possesses high diversity (roughly 1800 species or about two-thirds of the living snakes). Colubrids occur worldwide, although only marginally in Australia (Fig. 18.7). The multiple lineages are commonly partitioned into five subfamilies (Colubrinae, Homalopsinae, Lycodontinae, Natricinae, Xenodontinae). These subfamilies contain some monophyletic generic groups, but like the family, the subfamilies often represent an amalgam of genera, some of which are included for convenience rather than with certainty of relationship.

Colubrines (racers, ratsnakes, and their allies) include nearly 100 genera and 700 species of aglyphous (without fangs) and opisthoglyphous (fangs grooved, not canaliculate) snakes. Colubrines occur worldwide but have their greatest diversity in North America, Eurasia, and tropical Asia. Most are moderate-sized snakes, although they range from small (15–40 cm SVL, e.g., *Calamaria*, *Ficimia*, *Tantilla*) to large (>1.5 m SVL, *Boiga*, *Drymarchon*, *Masticophis*, *Ptyas*, *Spilotes*). They occupy a wide spectrum of habitats from tropical and temperate forests to deserts and have fossorial to arboreal habits. None are aquatic, although some prefer moist habitats and enter water freely. The burrowers occur predominantly in arid habitats (e.g., *Chionactis*, *Sonora*), but a few (*Stilosoma*) prefer mesic ones. Terrestrial species span the full spectrum of colubrine habitats; many

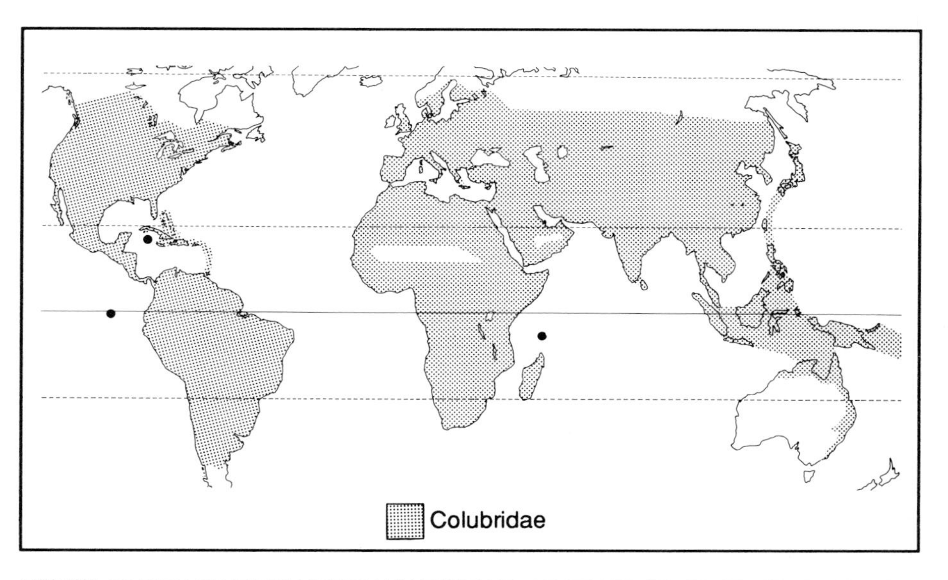

FIGURE 18.7 ▼▲▼▲▼▲▼▲▼▲▼▲▼▲▼▲▼▲▼▲▼▲▼▲▼▲▼▲▼▲▼▲▼▲▼▲▼▲

Distribution of the extant Colubridae.

have the slender racer habitus (*Chironius* , *Coluber*, *Dendrelaphis*, *Dendrophidion*, *Masticodryas*, *Telescopus*), whereas others have a stockier cylindrical body form (*Drymarchon*, *Lampropeltis*). Many of the slender terrestrial colubrines are good climbers (especially *Elaphe* and its allies); however, other colubrines are truly arboreal (*Ahaetulla*, *Chrysopelea*, *Thelotornis*), seldom descending to the ground other than for egg-laying. With few exceptions, colubrines are oviparous.

Colubrine prey encompasses most animal life, and many colubrines are dietary specialists, eating only one type or class of prey. *Dasypeltis* (the African egg-eating snakes; six species) specializes on bird eggs and possesses a number of anatomical specializations to swallow, crack, and empty eggs. Only a few other snake taxa (e.g., *Elaphe*) can process bird eggs with their rigid calcareous shells; these snakes are not egg specialists and, unlike *Dasypeltis*, consume a variety of other prey. *Cemophora* (scarlet snake) and other fossorial colubrines readily eat the eggs of reptiles. Mammals and birds are common prey for the larger terrestrial and arboreal species, although some of these taxa and many other colubrines prey on amphibians and reptiles. Invertebrates are not uncommon among the smaller taxa. *Opheodrys* (green snakes; two species) eats spiders and insects, particularly orthopterans.

One group of colubrines, the Boigini, is venomous. These opisthoglypous or rear-fanged snakes have enlarged and grooved teeth posteriorly on the maxillae. Typically, they must bite and chew to allow the venom to enter the wound. Since the venoms are often prey-specific, they are of variable toxicity to humans. Snakes (*Ahaetulla*, *Boiga*, *Telescopus*) preying mainly on reptiles may produce only a minor poisoning in humans; however, envenomation by the bird-eating *Thelotornis* (African vine snake) and *Dispholidus* (boomslang) is usually lethal.

Madagascaran hognose snake *Lioheterodon modestus* (Colubridae). (H. I. Uible)

Homalopsines are a small group (10 genera, 30+ species) of Asian freshwater and estuarine-marine snakes, ranging from Pakistan eastward to southeastern China and southward into the Philippines and northern Australia. They are predominantly moderate-sized snakes (30–100 cm adult TL). Although highly aquatic snakes, their bodies and tails tend to be more cylindrical than compressed, but they do have valvular nostrils and reduced ventral scales. All are opisthoglyphous with grooved posterior maxillary teeth. *Enhydris* (22± species) and *Cerberus* (3 species) occur throughout most of the subfamily's range. As aquatic snakes, homalopsines are predominantly piscivorous; however, a few taxa (e.g., *Fordonia*, *Myron*) are crustacean specialists. These specialists are marine and eat decapod crabs; they are able to crush the crabs' shells. All homalopsines are livebearers.

Lycodontines (50+ genera, 200+ species) are the major components of the African and tropical Asian colubrid snake fauna. They are mainly small- to moderate-sized snakes, although a few (e.g., *Malpolon*, *Psammophis*, *Pseudaspis*) are >1 m as adults. They occupy all African and Asian habitats from desert to forest; most are either fossorial or terrestrial, although a few are aquatic (*Lycodonomorphus*) or semiarboreal (*Philothamnus*). Lycodontines are predominantly oviparous.

Lycodontine prey includes invertebrates and vertebrates, with many dietary specialists particularly among the small-bodied taxa. *Pareas* is a slug-eating snake. *Apparalactus* specializes on centipedes and scorpions. Several groups of lycodontines are opisthoglypous (Aparallactini, Psammophiini), although their bites are not known to be lethal to humans.

Natricines are the major aquatic and semiaquatic colubrids (nearly 50 genera, 200+ species). Their distribution is largely confined to North America, Eurasia, and tropical Asia; none is known from South America and only marginally in Africa and Australia. Natricines consist mainly of small- to moderate-sized snakes, although a few (e.g., some *Nerodia*, *Xenochrophis*) reach or exceed 1 m as adults. Although the majority of natricines are aquatic, some taxa are semiaquatic (*Rhapdophis*, *Thamnophis*) or strictly terrestrial/semifossorial (*Aspidura*, *Storeria*, *Virginia*). Although most aquatic natricines are freshwater species, a few species and populations (e.g., *Nerodia fasciata compressicauda*) have adapted to estuarine conditions.

As expected for aquatic snakes, many natricines are piscivorous, but most supplement their diet with amphibians. Several species (e.g., *Regina alleni*, *R. septemvittata*) have become crustacean specialists, feeding on crayfish. They avoid the hard-shell problem by eating moulting crayfish. The semiaquatic and terrestrial taxa feed on amphibians and earthworms. All American natricines are livebearers (ovovivi- to viviparous) and often have large litters of 25 or more neonates. Most Eurasian natricines are oviparous.

Xenodontines are exclusively American snakes (80+ genera, 600± species) with their greatest diversity in the neotropics. Most are small- to moderate-sized snakes. Less than a dozen genera (*Alsophis*, *Clelia*, *Farancia*, *Hydrodynastes*,

Eastern ringneck snake *Diadophis punctatus* (Colubridae). (G. R. Zug)

Uromacer) contain species with adults exceeding 1 m SVL. Xenodontines occur in all but marine habitats; most species are forest (open and closed canopied) inhabitants. Further, most xenodontines are secretive, foraging and resting in the forest-floor litter or within the epiphytes on the trees. A few of the larger taxa (*Alsophis*, *Philodryas*) are diurnal racerlike snakes. The few aquatic taxa range from natricinelike *Helicops*, *Hydrops*, and *Tretanorhinus* to aniliidlike *Farancia* and *Pseuderyx*. Arboreal xenodontines also show two different body forms; the long-nosed, diurnal *Oxybelis* and *Uromacer* have the colubrinelike form, whereas the blunt-headed, nocturnal *Dipsas*, *Imantodes*, and *Leptodeira* have extremely slender and long bodies. These body forms contrast the stealth-stalking diurnal predators to the methodical nocturnal searchers. Amphibians, lizards, and other snakes are the major prey for xenodontines, including the larger species. The more aquatic xenodontines eat fish, and some terrestrial and arboreal species consume invertebrates. Live-bearing is uncommon in xenodontines.

Family Elapidae

Elapids are the erect-fanged venomous snakes. Unlike the venomous colubrids, each fang contains an enclosed passage extending the length of the tooth from an opening at the base for entry of the venom to an opening near the tip for injection into the prey. The fangs (usually one functional fang on each side) are permanently erect on the anterior ends of largely immovable maxillae and fit into grooved slots in the buccal floor when the mouth is closed. The general elapid habitus and scalation match the typical elongate, slender form and large, overlapping body scalation (usually without keels) of the colubrids. Their head scales are also colubridlike and they almost always lack a loreal scale between the nostril and eye. Elapids are predominantly diurnal snakes and range from small (*Ogmodon vitiensis*, 18–32 cm SVL) to large (*Ophiophagus hannah*, 6 m TL maximum).

Elapids (50+ genera) occur worldwide (Fig. 18.8), although they are largely absent from the northern temperate zone. Each of the southern continents and tropical Asia displays an adaptive radiation of elapids: coralsnakes in the Americas; cobras and mambas in Africa; cobras, kraits, and coralsnakes in Asia; seasnakes, tiger snakes, and others in Australia. The slender, brightly banded coralsnakes of each geographical area are not closely related (*Micrurus*; *Aspidelaps*; *Calliophis* and *Maticora*, respectively). As noted in Chapter 6, the coralsnake pattern serves for both crypsis and warning. The cobras (e.g., *Hemachatus*, *Naja*, *Walterinnesia*) appear to be a related group, most of which defensively display a hood. Four subfamilies (Elapinae, Hydrophiinae, Laticaudinae, Notechinae) are commonly recognized.

Elapines are a diverse group and include all non-Australian elapids. The wide distribution and the likely origin of Australian notechines from Asian

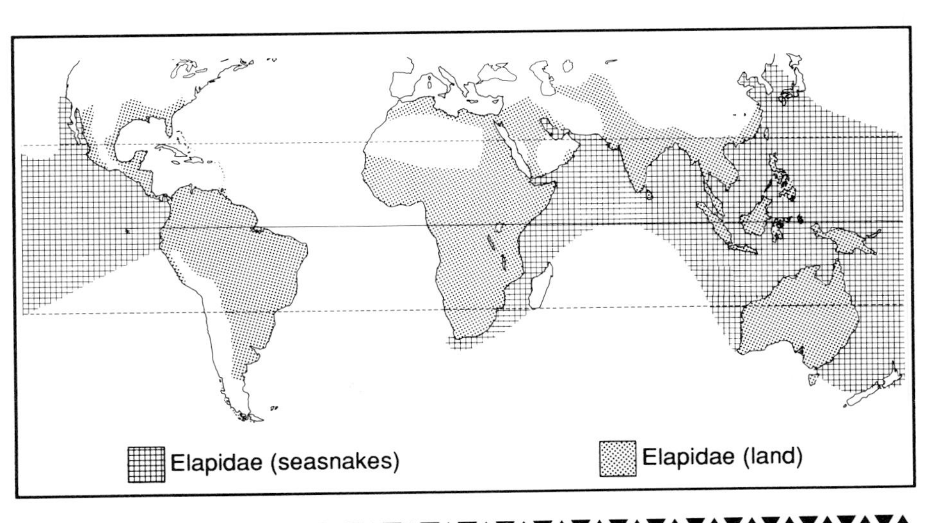

FIGURE 18.8 ▼▲▼▲▼▲▼▲▼▲▼▲▼▲▼▲▼▲▼▲▼▲▼▲▼▲▼▲▼▲▼▲▼▲▼▲

Distribution of the extant Elapidae. The cross hatching marks the distribution of seasnakes (Hydrophinae, Laticaudinae).

elapines indicate the paraphyly of Elapinae. In the Americas, elapines are all of one type, coralsnakes (*Micrurus* 50+ species, *Micruroides* 1 species; 1.6 m TL maximum, but most <1 m). American coralsnakes are semifossorial-terrestrial snakes, predominantly diurnal foragers of small reptiles. Most have a banded pattern of red, yellow, and black rings encircling the body and most commonly arranged in a red-yellow-black-yellow-red or a red-black-yellow-black-yellow-black-red sequence. All are oviparous, laying from two to nine eggs.

African elapines show a greater diversity of genera (9) but about half the number of species (26+). Their taxonomic diversity is reflected by a variety of life-styles. *Aspidelaps* (2 species) and *Elapsoidea* (7 species) are *Micrurus*-like in habits and size but more somberly banded. Mambas (*Dendroaspis*, 4 species) are slender, diurnally active snakes that prey on birds and mammals. Three of the four species are arboreal, and they glide quickly across branches and foliage. They are moderately large snakes, 1.5–2 m TL as adults, although the black mamba (*D. polylepis*) can attain lengths of more than 4 m. The other six genera of African elapines are cobras (e.g., *Naja*, 7 species; *Pseudohaje*, 2 species; usually <2 m TL). Most African cobras are nocturnal and opportunistic predators; however, the water cobras (*Boulengerina*, 2 species) are piscivorous. All African elapines are oviparous, except for the ovoviviparous rinkhals (*Hemachatus*).

Several species of *Naja* occur in India and Southeast Asia, but the kraits (*Bungarus*, 12+ species) and coral snakes (*Calliophis*, 10 species; *Maticora*, 2 species) are the major Asian elapines. These elapines are terrestrial species and usually nocturnal predators of amphibians and reptiles.

The diversity of the notechines matches that of the elapines, even though this diversity (25+ genera, 90+ species) arose on a single continent (Australia). Notechines are exclusively terrestrial to fossorial, and most are small (<80 cm SVL), diurnal, amphibian-reptile predators (Fig. 18.9). Taipans (*Oxyuranus*, 2 species, 1–3 m TL) are the largest Australian elapids; they are slender-bodied, quick-moving, savanna-scrub inhabitants of tropical Australia. *Pseudechis* (black snakes, 6 species) and *Demansia* (whip snakes, 6 species) are similarly slender-bodied, although smaller (1–2 m TL), snakes of open habitats. The death adders (*Acanthophis*, 2 species, 30–60 cm SVL) take their common name from their viperine appearance. They possess other viperine characteristics such as nocturnal behavior and vertical pupils, hunting by ambush, heavy-bodied, and hinged maxillary fangs. The smaller, semifossorial notechines (e.g., *Simoselaps*, *Unechis*) may also be nocturnal, although round pupils suggest a predominantly diurnal foraging. Ovoviviparity is common among the notechines (e.g., *Acanthophis*, *Cryptophis*, *Drysdalia*, *Glyphodon*, *Pseudechis*, *Tropidechis*) and some *Denisonia* may be viviparous. The tiger snake *Notechis* is especially prolific and may carry nearly a hundred embryos, although 20–40 embryos are more common. *Oxyuranus*, *Demansia*, and some of the smaller-bodied genera are oviparous.

Marine elapids segregate into two subfamilies, Hydrophiinae (true seasnakes) and the Laticaudinae (seakraits). Both subfamilies are adapted for a

FIGURE 18.9 ▼▲▼▲▼▲▼▲▼▲▼▲▼▲▼▲▼▲▼▲▼▲▼▲▼▲▼▲▼▲▼▲▼▲▼▲▼▲
Australian curl snake *Suta suta* (Elapidae). (T. Schwaner)

marine existence. They have laterally compressed bodies, paddlelike tails, valvular nostrils, low-permeable skin, lingual salt glands, and a variety of other anatomical, physiological, and behavioral adaptations. Laticaudines differ from the hydrophiines by the retention of enlarged ventral scales (gastrosteges) and oviparity, which brings them ashore to lay their eggs. Laticaudines consist of a single genus *Laticauda* (4 species, 70–100 cm SVL); their distribution encompasses the China Sea and the Southwest Pacific (Fig. 18.10). Hydrophiines comprise 14 genera and nearly 55 species ranging from the east coast of Africa to the west coast of tropical America (see Fig. 18.8). The yellow-bellied seasnake *Pelamis platurus* occurs throughout this huge distribution; the other seasnakes occur largely within the shallow, coastal waters of Indoaustralia. Hydrophines range from moderate-sized (0.5–1 m SVL), gracile taxa (*Aipysurus*, *Hydrophis*, *Pelamis*) to the large, robust *Astrotia stokesii* (1.6 m TL). Some are generalist piscivores, and others are specialists, eating only one or two types of fish or just fish eggs. All hydrophiines are live-bearers (ovoviviparous to viviparous).

Family Loxocemidae

Loxocemus bicolor of Central America (Fig. 18.11) is a moderate-sized (70 cm TL), semifossorial-terrestrial snake and the sole living representative of this family. It is an enigmatic snake having an assortment of primitive characteristics,

FIGURE 18.10
Yellow-lipped seakrait *Laticauda colubrina* (Elapidae). (G. R. Zug)

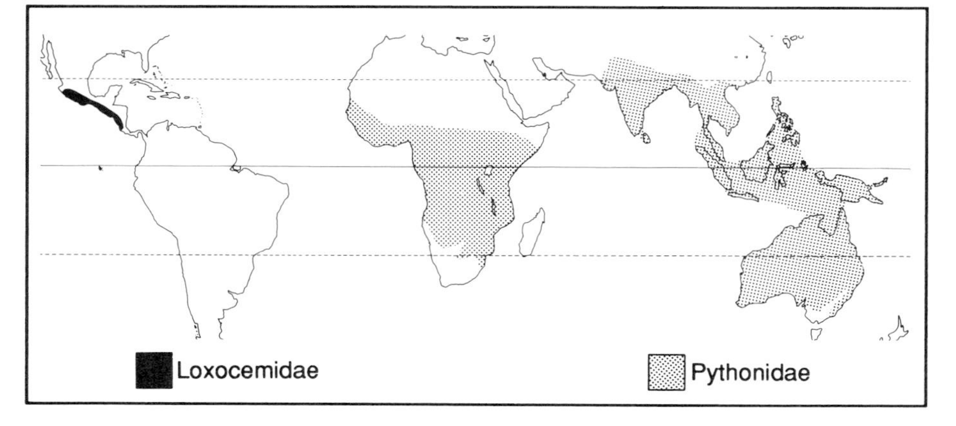

FIGURE 18.11
Distribution of the extant Loxocemidae and Pythonidae.

which have resulted in it being linked previously with aniliids, boids, and xenopeltids. *Loxocemus*'s biology is poorly known; it is oviparous and preys on small mammals and reptiles, including seaturtle and iguanine eggs.

Family Pythonidae

Pythons (3 genera, 30+ species) are the African and Indoaustralian (Fig. 18.11) equivalents of the neotropical boines and possess a number of boine traits, although they differ from boids in the structure of the bony orbit, several other cranial features, and paired subcaudal scales. Pythons are predominantly large snakes (adults >1.5 m SVL), although several of the Australian pythons (e.g., *Morelia boeleni*, *M. viridis*) attain sexual maturity at about half that length. Others (e.g., *M. amethistina*, *Python sebae*, *P. reticulatus*) are giants and commonly grow to >3 m SVL. Pythons occupy a variety of habitats from rain forest to arid scrub and savannas. Most are terrestrial, although *M. viridis* (green tree python) is strongly arboreal, and *M. fusca* (Australian water python) is semi-aquatic and always in or near water. Most pythons prey on endothermic prey, and most have labial infrared receptors and pits to assist in their location and capture of birds and mammals. All pythons are oviparous; several species guard and likely incubate their eggs by curling around them.

The African-Asian pythons and the Australian ones represent two separate lineages. All species of the former are classified as *Python*, and of the latter, *Aspidites* for the Australian black-headed pythons and *Morelia* for all other Australopapuan species.

Family Uropeltidae

Uropeltids consist of two Asian groups of fossorial snakes (see Fig. 18.2): the pipesnakes (Cylindrophiinae) and the shield-tailed snakes (Uropeltinae). These two henophidian groups share compact, firmly articulated (akinetic) skulls, smooth shiny scales, thick cylindrical bodies, short tails, and an array of henophidian traits. They differ in the presence (cylindrophiines) or absence (uropeltines) of pelvic-hindlimb vestiges and eye-covering spectacles.

Cylindrophiines include the oviparous *Anomochilus* (two species) of the Malayan peninsula and East Indies and the ovoviviparous *Cylindrophis* (eight species) of Sri Lanka, Burma through Indochina, and southward into the East Indies to Aru Island. They are mostly moderate-sized snakes (25–85 cm TL) that apparently forage nocturnally on and in forest-floor litter for caecilians, small snakes, and lizards and burrow for protection and rest.

Uropeltines include eight genera (e.g., *Plectrurus*, *Rhinophis*; 40+ species) of highly fossorial snakes from southern India and Sri Lanka. They seldom appear on the surface unless uncovered by searching predators or forced to the surface by waterlogged soils. They are almost exclusively forest inhabitants, occurring in open areas only where soils are highly friable permitting them to burrow deeply to avoid high soil-surface temperatures. They derive their colloquial name from an extremely short tail capped (in some genera) by a unique keratinous shield (Fig. 18.12), whose rough dorsal surface holds a thin layer of soil. The tail shield and its soil layer plug the burrow behind the snake. When exposed, uropeltids hide their head in body coils or beneath objects, exposing the armored tail to attack by predators. Their conical head and heavily muscled anterior quarter of the body are the digging mechanism. Digging begins with the head embedded in the tunnel wall and the muscular body folded into a series of loops within the skin envelope. The head is driven forward with the straightening of muscular loops; then the head anchors and pulls the body forward, creating a new series of internal loops to begin the next penetration cycle. This concertina-style burrowing is effective in moist and friable soils, and a uropeltid can disappear rapidly, even with a predator attacking its tail.

Uropeltines are small- to moderate-sized snakes (20–70 cm SVL). As totally subterranean creatures, their diet likely consists principally of earthworms but includes small burrowing vertebrates (e.g., lizards) as well. All uropeltines are ovoviviparous.

Family Viperidae

Viperids are the hinged-fanged, venomous snakes. The hinged-fang mechanism permits long fangs for deep penetration and envenomation and their storage against the roof of the mouth when not in use. The structure of the hinge

FIGURE 18.12 ▼▲▼

A Sri Lankan shield-tailed snake *Pseudotyphlops philippinus* (Uropeltidae). (C. Gans)

mechanism is unique; the lateral process of the palatine bone is absent, each maxilla is cuboidal and bears a single enlarged maxillary tooth (several replacement fangs lie behind the functional fang), the maxillary fang has a duct on its anterior face, and the elongate ectopterygoid serves as a lever to erect or depress the fang. Along with this hinged-fang mechanism, viperids typically have broad, triangular heads. Most are also heavy-bodied with relatively short, slender tails; the triangular head and stout body produce the viper habitus. Most vipers have numerous and heavily keeled body scales.

Vipers are a diverse group (20+ genera, 150+ species) occurring on all continents except Australia (Fig. 18.13). They include aquatic to arboreal species, range from sea level to the tree line, occur in near rainless deserts to evergreen rain forest, prey on invertebrates and vertebrates (the latter dominating the diets of most vipers), and are small (*Bitis schneideri*, 28 cm TL maximum) to large (*Lachesis muta*, 3.6 m TL maximum). This diversity divides among three subfamilies: Azemiopinae, Crotalinae, and Viperinae.

Azemiops feae is the sole member of its subfamily and occurs only in the mountains of southwestern China and adjacent Tibet and Burma. It is a

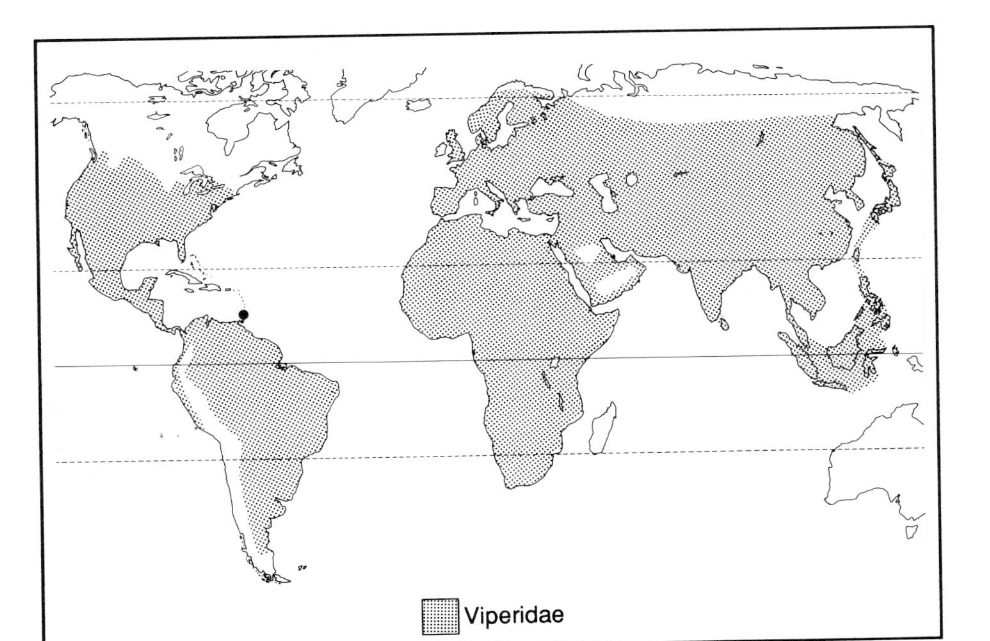

FIGURE 18.13 ▼▲▼

Distribution of the extant Viperidae.

moderate-sized, terrestrial viper, likely preying mainly on mammals. Unlike most other vipers, it has colubridlike head scales, smooth body scales, no tracheal lung, and no loreal pit-organs. It is unknown whether *Azemiops* is oviparous or ovoviviparous.

Crotalines are the pit vipers, uniquely characterized by a loreal pit-organ on each side of the head between the nostril and eye. The pit-organs are infrared receptors, usually with a "binocular" receptor field that permits the accurate tracking of prey whose body temperature differs from background temperatures. Crotalines are a highly successful group of 120+ species in temperate and tropical Asia (*Agkistrodon*, *Calloselasma*, *Deinagkistrodon*, *Hypnale*, *Trimeresurus*) and the temperate and tropical Americas (*Agkistrodon*, *Bothriechis*, *Bothriopsis*, *Bothrops*, *Crotalus*, *Lachesis*, *Ophryacus*, *Porthidium*, *Sistrurus*)(Fig. 18.14). Their adaptations span the full range of viperid adaptations outlined above. Most crotalines are live-bearers; however, *Calloselasma*, some *Trimeresurus*, and *Lachesis* are oviparous. Possibly all oviparous crotalines guard their eggs. Most crotalines are moderate-sized species (60–120 cm SVL) preying predominantly on vertebrates.

Viperines are less diverse (50± species) and less broadly distributed than crotalines. Viperines occur from Europe to India and throughout Africa. They are almost exclusively terrestrial (some burrow in desert sands, e.g., *Cerastes*, *Echis*), although *Atheris* is arboreal. Viperines range from forest to desert and

FIGURE 18.14 American copperhead *Agkistrodon contortrix* (Viperidae). (R. G. Tuck)

from subarctic to equatorial habitats. Most viperines are moderate-sized snakes; none is known to exceed 2 m SVL, although the Gaboon viper (*Bitis gabonica*) and its relatives are proportionately the heaviest living snakes. They prey mainly on vertebrates and, like crotalines, forage primarily at night. Viperines include both oviparous (e.g., *Causus*, *Echis coloratus*) and ovoviviparous groups (*Bitis*, *Echis carinatus*, most *Vipera*).

The geographical distributions of viperine genera occur in a layercakelike pattern. *Vipera* (16+ species) forms the northern tier from western Europe to Caucasia; the southern tier contains *Adenorhinus* (1 species), *Atheris* (9 species), *Bitis* (12+ species), *Causus* (6 species), and *Echis* (5 species) in sub-Saharan Africa. Sandwiched between these two tiers around the Mediterranean and eastward through arid Southwest Asia to India are *Cerastes* (2 species), *Eristicophis* (1 species), *Pseudocerastes* (4 species), and *Vipera* and *Echis* from the north and south, respectively.

Family Xenopeltidae

Xenopeltis unicolor is a moderate-sized (70–90 cm SVL), semifossorial snake of Indomalaysia and the East Indies (see Fig. 18.4). In some respects, the sunbeam snake is structurally similar to *Anilius*, such as possessing a tightly knit skull and premaxillary teeth, but the left lung is large, the pelvic girdle is absent, and the ventral scales are moderately enlarged. It also has more flexible jaws, a larger gape, and a broader diet including frogs, reptiles, and small mammals. Presumably *Xenopeltis* is oviparous, but its reproductive mode remains unconfirmed.

Phylogenetic Relationships of Snakes

Our understanding of snake evolution is a morass of unresolved lineage divergences. The relationship of one snake lineage to another is unclear in many instances, and similarly the monophyly of many snake subfamilies and families remains suspect. In the origin of snakes from a lizard ancestor, snakes passed through a morphological bottleneck, losing or reducing and reorganizing many structures. This "simplification" provided a limited base for subsequent evolution, and the same or similar morphological adaptations arose in unrelated lineages. These similarities are convergences, not shared-derived features, and cloud the recognition of phylogenetic relationships. More rigorous analysis of morphological data and the introduction of new data sets (e.g., karyotypic and molecular characters) are slowly improving our knowledge of snake phylogeny.

There is a fundamental division between the scolecophidian and the alethinophidian snakes. The differences are as visible externally as internally. Scolecophidians or blindsnakes with their smooth shiny scales with little dorsoventral differentiation look more like legless skinks than snakes. Alethinophidians with enlarged ventral scales (gastrosteges) have no lizard counterparts. In spite of the dissimilarities of these two groups, they uniquely share several cranial features that confirm the monophyly of snakes. Within the scolecophidians, anomalepidids and typhlopids share kinetic and toothed maxillae and a tracheal lung in contrast to the akinetic and toothless maxillae and no tracheal lung in leptotyphlopids. These differences in dentition and jaw mechanics, tracheal lungs, and other characteristics support the sister-pair relationship of the former two blindsnakes and a distant relationship to leptotyphlopids.

Origins and relationships within the alethinophidians remain equivocal, especially so for the primitive or henophidian groups. It is uncertain whether henophidians represent a single clade as shown in Fig. 18.15 or a grade with multiple independent origins. Multiple interpretations are possible, and advocates exist for many of these. A consensus does not yet exist. Most researchers accept the sister-pair relationship of the uropeltines and cylindrophines and much more tentatively their relationship to aniliids. The placement of xenopeltids is less certain. They may be linked to the former "anilioid" families, or possibly to loxocemids. Loxocemids, possibly xenopeltids, and the remaining henophidian families comprise the "booids." Previously all have been combined into a single family with each representing a subfamily. Current research is emphasizing the distinctiveness of each of the "subfamilies" and recognizing each as a distinct family; however, the sister-pair relationships of these booid families or subfamilies is ambiguous. Are the pythons the sister-group to the boines or to loxocemids? Do the bolyerids represent an early divergence from the ancestor boid stock

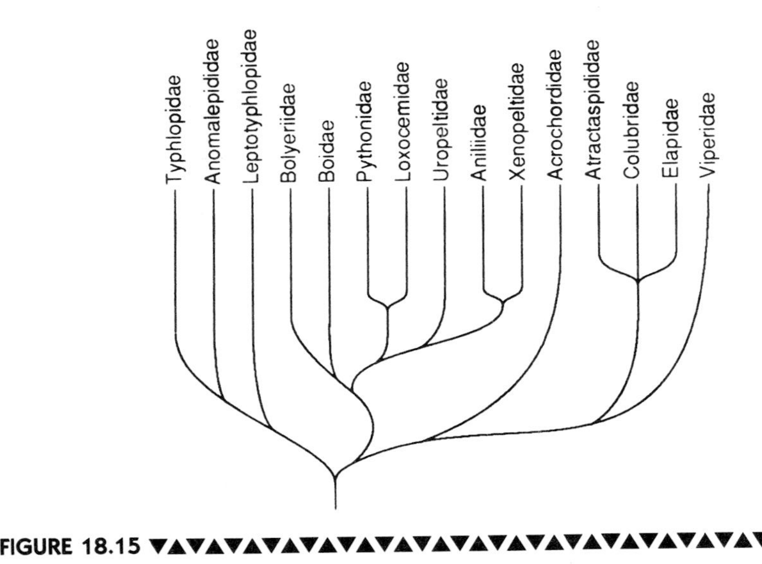

FIGURE 18.15 ▼▲▼
Dendrogram of presumed phyletic relationships among the living families of snakes.

or are they a sister-group of the caenophidians? Are the tropidophines boids or are they more closely related to caenophidians? Evidence exists for each of these alternate interpretations.

The Caenophidia including the acrochordids is supported by a number of shared-derived traits as is also the sister-pair relationship between acrochordids and the "colubroids." The distinctiveness of the vipers indicates an early divergence from the colubroid stock. The most primitive viper *Azemiops* shares many similarities with mainline colubroids. Presently, *Azemiops* is considered the most primitive extant viperid, mainly because of the similarities of its envenomation mechanisms with viperine and crotaline ones. Crotalines and viperines are sister-groups. The unique pit-organs identify the monophyly of crotalines.

Atractaspidids, colubrids, and elapids are each other's closest relatives, but their exact sister relationships remain unresolved. There is a hint of evidence that atractaspidids are an early divergence from the ancestral elapid stock. Elapids are unquestionably a monophyletic group and easily divisible into two evolutionary groups: Australian elapids + seasnakes + seakraits; and nonaustralian elapids. The Australian elapid-seasnake clade likely arose from an Asian elapid stock, but the identity of the ancestor is ambiguous. Further, it seems likely that the laticaudines diverged early from the ancestral Australian stock (or independently from an Asian stock), probably before the extensive elapid radiation within Australia. One of the radiations of Australian elapids gave origin to the hydrophiines. The pattern of divergence within the elapines is less clear. Micururini coralsnakes are definitely a neotropical radiation of elapids, but as yet their affinities to other elapids are not clear. Similarly, the interrelationships among the Afroasian elapines are uncertain, largely owing to absence of a comparative study of all taxa.

As noted earlier, our knowledge of colubrid relationships is uncertain, consisting of a morass of conjectures. Evidence is available to confirm the monophyly of numerous small groups of genera (e.g., Homalopsini, Thamnophiini, and others); however, data linking these tribal or suprageneric groups are not as robust or straightforward. These linkages are necessary before the current Colubridae can be partitioned into families reflecting phylogenetic relationships rather than similarity of appearance. Of the colubrids, the colubrines come the closest to being a monophyletic lineage. American colubrine genera all appear to be closely related to one another as well as to the Eurasian colubrines. *Elaphe* and *Lampropeltis* are paraphyletic with different species or species groups linking to other colubrine genera. The American natricines (Thamnophiini includes all American aquatic and terrestrial genera) are likely monophyletic and possibly a sister-group to the Asian *Sinonatrix*. The Eurasian natricines contain numerous groups (e.g., *Natrix*, *Afronatrix*, *Sinonatrix*, and the *Amphiesma* complex), whose relationships are uncertain. The homalopsines, although aquatic, are not closely related to any natricines; their affinity to other colubrids is obscure. Lycodontines include at least two distinct groups with the boodontines including many small African and Madagascaran colubrids and the "true" lycodontines including African and Asian genera. Xenodontines similarly include two distinct groups: mainly, but not exclusively, a Central American one and a South American one.

CHAPTER BIBLIOGRAPHIES

Books and articles are cited fully in the chapter in which they were first used. When used thereafter, the citation is "See Chap. o," referring to chapter bibliography with full citatation. Articles in edited, multiauthored books are cited in this bibliography by editors' names and op. cit. when in the same chapter bibliography or "See Chap. o," when in a subsequent chapter bibliography.

CHAPTER 1. Amphibians

Altig, R., and G. F. Johnston. 1986. Major characteristics of free-living anuran tadpoles. *Smithson. Herpetol. Inform. Serv.* No. 67.

Altig, R., and G. F. Johnston. 1989. Guilds of anuran larvae: relationships among developmental modes, morphologies and habitats. *Herpetol. Monogr.* **3:** 81–109.

Bagnara, J. T. 1986. Pigment cells. Pp. 136–149. *In* J. Bereiter-Hahn, A. G. Matoltsy, and K. S. Richards (eds.), "Biology of the Integument. 2. Vertebrates." Berlin: Springer-Verlag.

Bereiter-Hahn, J., A. G. Matoltsy, and K. S. Richards (eds.). 1986. "Biology of the Integument. 2. Vertebrates." Berlin: Springer-Verlag.

Capranica, R. R. 1976. Morphology and physiology of the auditory system. Pp. 551–575. *In* J. Bereiter-Hahn *et al.* (eds.). Op. cit.

Catton, W. T. 1976. Cutaneous mechanoreceptor. Pp. 629–642. *In* R. Llinás and W. Precht (eds.), "Frog Neurobiology. A Handbook." Berlin: Springer-Verlag.

Cope, E. D. 1889. The Batrachia of North America. *Bull. U.S. Natl. Mus.* **34:** 1–525.

Duellman, W. E., and L. Trueb. 1986. "Biology of Amphibians." New York: McGraw-Hill Book Co.

Elias, H., and J. Shapiro. 1957. Histology of the skin of some toads and frogs. *Am. Mus. Novit.* No. 1819: 1–27.

Fox, H. 1985. Changes in amphibian skin during larval development and metamorphosis. Pp. 59–87. *In* M. Balls and M. Bownes (eds.), "Metatmorphosis." Oxford: Clarendon Press.

Fox, H. 1986. Epidermis. Pp. 78–110. Dermis. Pp. 111–115. Dermal glands. Pp. 116–135. *In* J. Bereiter-Hahn *et al.* (eds.). Op. cit.

Fritzsch, B., and M. H. Wake. 1988. The inner ear of gymnophione amphibians and its nerve supply: a comparative study of regressive events in a complex sensory system. *Zoomorphology* **108:** 201–217.

Hetherington, T. E. 1987. Timing of development of the middle ear of Anura. *Zoomorphology* **106:** 289–300.

Hetherington, T. E., A. P. Jaslow, and R. E. Lombard. 1986. Comparative morphology of the amphibian opercularis system: I. General design features and functional interpretation. *J. Morphol.* **190:** 43–61.

Jaeger, C. B., and D. E. Hillman. 1976. Morphology of gustatory organs. Pp. 588–606. *In* R. Llinás and W. Precht (eds.), "Frog Neurobiology. A Handbook." Berlin: Springer-Verlag.

Krejsa, R. J. 1979. The comparative anatomy of the integumental skeleton. Pp. 112–191. *In* M. H. Wake (ed.), "Hyman's Comparative Vertebrate Anatomy," Third Edition. Chicago: Univ. of Chicago Press.

Lannoo, M. J. 1987. Neuromast topography in anuran amphibians. *J. Morphol.* **191:** 115–129.

Lannoo, M. J. 1987. Neuromast topography in urodele amphibians. *J. Morphol.* **191:** 247–263.

Llinás, R., and W. Precht (eds.). 1976. "Frog Neurobiology. A Handbook." Berlin: Springer-Verlag.

Northcutt, R. G. 1979. The comparative anatomy of the nervous system and the sense organs. Pp. 615–769. *In* M. H. Wake (ed.), "Hyman's Comparative Vertebrate Anatomy," Third Edition. Chicago: Univ. of Chicago Press.

Nussbaum, R. A., and B. G. Naylor. 1982. Variation in the trunk musculature of caecilians. *J. Zool.* **198:** 383–398.

Pang, P. K. T., and M. P. Schreibman (eds.). 1986. "Vertebrate Endocrinology: Fundamentals and Biomedical Implications. Vol. 1. Morphological Considerations." Orlando: Academic Press.

Ruibal, R., and V. Shoemaker. 1984. Osteoderms in anurans. *J. Herpetol.* **18:** 313–328.

Saint-Aubain, M. L. de. 1981. Amphibian ontogeny and its bearing on the phylogeny of the group. *Z. Zool. Syst. Evol.* **19:** 175–194.

Saint-Aubain, M. L. de. 1985. Blood flow patterns of the respiratory systems in larval and adult amphibians: functional morphology and phylogenetic significance. *Z. Zool. Syst. Evol.* **23:** 229–240.

Scalia, F. 1976. Structure of the olfactory and accessory olfactory systems. Pp. 213–233. *In* R. Llinás and W. Precht (eds.). Op. cit.

Spray, D. C. 1976. Pain and temperature receptors of anurans. Pp. 607–628. *In* R. Llinás and W. Precht (eds.). Op. cit.

Vial, J. L. (ed.). 1973. "Evolutionary Biology of the Anurans." Columbia, Missouri: Univ. of Missouri Press.

Wake, M. H. 1975. Another scaled caecilian (Typhlonectidae). *Herpetologica* **31:** 134–136.

Wake, M. H. (ed.). 1979. "Hyman's Comparative Vertebrate Anatomy," Third Edition. Chicago: Univ. of Chicago Press.

Wever, E. G. 1985. "The Amphibian Ear." Princeton: Princeton Univ. Press.

Whitear, M. 1977. A functional comparison between the epidermis of fish and of amphibians. *Symp. Zool. Soc. London* **39:** 291–313.

CHAPTER 2. Origin and Evolution of Amphibians

Benton, M. J. (ed.). 1988. "The Phylogeny and Classification of the Tetrapods. Vol. 1. Amphibians, Reptiles, Birds." Oxford: Clarendon Press.

Bishop, M. J., and A. E. Friday. 1987. Tetrapod relationships: the molecular evidence. Pp. 123–139. *In* C. Patterson (ed.), "Molecules and Morphology in Evolution: Conflict or Compromise?" Cambridge: Cambridge Univ. Press.

Bolt, J. R., and R. E. Lombard. 1985. Evolution of the amphibian tympanic ear and the origin of frogs. *Biol. J. Linn. Soc.* **24:** 83–99.

Carroll, R. L. 1977. Patterns of amphibian evolution: an extended example of the incompleteness of the fossil record. Pp. 405–437. *In* A. Hallam (ed.), "Patterns of Evolution As Illustrated by the Fossil Record." Amsterdam: Elsevier Sci. Publ. Co.

Carroll, R. L. 1988. "Vertebrate Paleontology and Evolution." New York: W. H. Freeman & Co.

Duellman, W. E., and L. Trueb. 1986. See Chap. 1.

Edwards, J. L. 1989. Two perspectives on the evolution of the tetrapod limb. *Am. Zool.* **29:** 235–254.

Estes, R. 1981. "Gymnophiona, Caudata. Handbuch der Paläoherpetologie." Stuttgart: Gustav Fischer Verlag.

Estes, R., and O. A. Reig. 1973. The early fossil record of frogs. A review of the evidence. Pp. 11–63. *In* J. L. Vial (ed.). See Chap. 1.

Gaffney, E. 1979. Tetrapod monophyly: a phylogenetic analysis. *Bull. Carnegie Mus. Nat. Hist.* **13:** 92–105.
Gauthier, J., A. Kluge, and T. Rowe. 1988a. Amniote phylogeny and the importance of fossils. *Cladistics* **4:** 105–209.
Gauthier, J., A. Kluge, and T. Rowe. 1988b. The early evolution of the Amniota. Pp. 103–155. *In* M. J. Benton (ed.). Op. cit.
Hanken, J. 1986. Developmental evidence for amphibian origins. *Evol. Biol.* **20:** 389–417.
Lambert, D., and Diagram Group. 1985. "The Field Guide to Prehistoric Life." New York: Facts on File Publ.
Milner, A. C. 1980. A review of the Nectridea (Amphibia). Pp. 377–405. *In* A. L. Panchen (ed.), "The Terrestrial Environment and the Origin of Land Invertebrates." New York: Academic Press.
Milner, A. C. 1988. The relationships and origin of living amphibians. Pp. 59–102. *In* M. J. Benton (ed.). Op. cit.
Panchen, A. L. (ed.). 1980. "The Terrestrial Environment and the Origin of Land Vertebrates." New York: Academic Press.
Panchen, A. L. 1980. The origin and relationships of the anthracosaur Amphibia from the Late Palaeozoic. Pp. 319–350. *In* A. L. Panchen (ed.). Op. cit.
Panchen, A. L., and T.R. Smithson. 1988. The relationships of the earliest tetrapods. Pp. 1–32. *In* M. J. Benton (ed.). Op. cit.
Patterson, C. (ed.). 1987. "Molecules and Morphology in Evolution: Conflict or Compromise?" Cambridge: Cambridge Univ. Press.
Piveteau, J. (ed.). 1955. Traité de Paléontologie. V. Amphibiens, Reptiles, Oiseaux. Paris: Masson et Cie Éditeurs.
Queiroz, K. de, and D. C. Cannatella. 1987. The monophyly and relationships of the Lissamphibia. *Am. Zool.* **27:** 60A. (Abstr.)
Rage, J.-C. 1984. Are the Ranidae known prior to the Oligocene? *Amphibia–Reptilia* **5:** 281–288.
Romer, A. S. 1960. "Vertebrate Paleontology." Chicago: Univ. of Chicago Press.
Schultze, H.-P., and L. Trueb (eds.). 1991. "Origins of the Higher Groups of Tetrapods. Controversy and Consensus." Ithaca, New York: Comstock Publ. Assoc.
Scott, A. C. 1980. The ecology of some Upper Palaeozoic floras. Pp. 87–115. *In* A. L. Panchen (ed.). Op. cit.
Spinar, Z. V. 1972. "Tertiary Frogs from Central Europe." The Hague: W. Junk N. V.
Thomson, K. S. 1980. The ecology of Devonian lobe-finned fishes. Pp. 187–122. *In* A. L. Panchen (ed.). Op. cit.
Trueb, L., and R. Cloutier. 1987. Historical constraints on Lissamphibia. *Am. Zool.* **27:** 33A. (Abstr.)
Whitear, M. 1977. See Chap. 1.

CHAPTER 3. Reptiles

Baird, I. L. 1970. The anatomy of the reptilian ear. Pp. 193–275. *In* G. Gans and T. S. Parsons (eds.), "Biology of the Reptilia," Vol. 2. New York: Academic Press.
Barghusen, H. R., and J. A. Hopson. 1979. The endoskeleton: the comparative anatomy of the skull and the visceral skeleton. Pp. 263–326. *In* M. H. Wake (ed.). See Chap. 1.
Bechtel, H. B., and E. Bechtel. 1991. Scaleless snakes and a breeding report of scaleless *Elaphe obsoleta linheimeri*. *Herpetol. Rev.* **22:** 12–14.
Bellairs, A. d'A. 1969. "The Life of Reptiles," Vols. 1 and 2. London: Weidenfeld & Nicolson.
Bellairs, A. d'A., and A. M. Kamal. 1981. The chondrocranium and the development of the skull in Recent reptiles. Pp. 1–263. *In* G. Gans and T. S. Parsons (eds.), "Biology of the Reptilia," Vol. 11. New York: Academic Press.
Bereiter-Hahn, J., A. G. Matoltsy, and K. S. Richards (eds.). 1986. See Chap. 2.

Burggren, W. W. 1987. Form and function in reptilian circulations. *Am. Zool.* **27:** 5–19.

Burke, A. C. 1989. Development of the turtle carapace: implications for the evolution of a novel bauplan. *J. Morphol.* **199:** 363–378.

Dowling, H. G., and W. E. Duellman. 1978. "Systematic Herpetology: A Synopsis of Families and Higher Categories." New York: HISS Publ.

During, M. von, and M. R. Miller. 1979. Sensory nerve endings of the skin and deeper structures. Pp. 407–441. *In* C. Gans, R. G. Northcutt, and P. Ulinski (eds.), "Biology of the Reptilia," Vol. 9. New York: Academic Press.

Edmund, A. G. 1969. Dentition. Pp. 117–200. *In* C. Gans, A. d'A. Bellairs, and T. S. Parsons (eds.), "Biology of the Reptilia," Vol. 1. New York: Academic Press.

Epple, A., and J. E. Brinn. 1986. Pancreatic islets. Pp. 279–317. *In* P. K. T. Pang and M. P. Schreibman (eds.). See Chap. 1.

Ernst, C. H., and R. W. Barbour. 1989. "Turtles of the World." Washington, D.C.: Smithsonian Inst. Press.

Estes, R., and G. Pregill (eds.). 1988. "Phylogenetic Relationships of the Lizard Families." Stanford: Stanford Univ. Press.

Fox, H. 1977. The urogenital system of reptiles. Pp. 1–157, 463–464. *In* C. Gans and T. S. Parsons (eds.), "Biology of the Reptilia," Vol. 6. New York: Academic Press.

Frazzetta, T. H. 1986. The origin of amphikinesis in lizards. A problem in functional morphology and the evolution of adaptive system. *Evol. Biol.* **20:** 419–461.

Gabe, M. 1970. The adrenal. Pp. 263–318. *In* C. Gans and T. S. Parsons (eds.), "Biology of the Reptilia," Vol. 2. New York: Academic Press.

Gaffney, E. S. 1979. Comparative cranial morphology of Recent and fossil turtles. *Bull. Am. Mus. Nat. Hist.* **164:** 65–376.

Gans, C., A. d'A. Bellairs, and T. S. Parsons (eds.). 1969. "Biology of the Reptilia," Vol. 1. New York: Academic Press.

Gans, C., R. G. Northcutt, and P. Ulinski (eds.). 1979. "Biology of the Reptilia," Vol. 9. New York: Academic Press.

Gans, C., and T. S. Parsons (eds.). 1970. "Biology of the Reptilia," Vol. 2. New York: Academic Press.

Gans, C., and T. S. Parsons (eds.). 1973. "Biology of the Reptilia," Vol. 4. New York: Academic Press.

Gans, C., and T. S. Parsons (eds.). 1977. "Biology of the Reptilia," Vol. 6. New York: Academic Press.

Gans, C., and T. S. Parsons (eds.). 1981. "Biology of the Reptilia," Vol. 11. New York: Academic Press.

Gasc, J.-P. 1981. Axial musculature. Pp. 355–435. *In* G. Gans and T. S. Parsons (eds.). Op. cit.

Gauthier, J. A., R. Estes, and K. de Queiroz. 1988. A phylogenetic analysis of Lepidosauromorpha. Pp.15–98. *In* R. Estes and G. Pregill (eds.). Op. cit.

Gauthier, J. A., A. G. Kluge, and T. Rowe. 1988a,b. See Chap. 2.

Grassé, P.-P. (ed.). 1970. "Traité de Zoologie. Tome XIV. Reptiles Caractéres Généraux et Anatomie." Paris: Masson.

Guibé, J. 1970. La peau et les productions cutanees. Pp. 6–32. *In* P.-P. Grassé (ed.). Op. cit.

Guibé, J. 1970. Le squelette céphalique. Pp. 78–143. *In* P.-P. Grassé (ed.). Op. cit.

Guibé, J. 1970. La musculature. Pp. 78–143. *In* P.-P. Grassé (ed.). Op. cit.

Haas, G. 1973. Muscles of the jaws and associated structures in the Rhynchocephalia and Squamata. Pp. 285–490. *In* C. Gans and T. S. Parsons (eds.). Op. cit.

Hoffstetter, R., and J.-P. Gasc. 1969. Vertebrae and ribs of modern reptiles. Pp. 201–310. *In* C. Gans and T. S. Parsons (eds.). Op. cit.

Iordansky, N. N. 1973. The skull of the Crocodilia. Pp. 201–262. *In* C. Gans and T. S. Parsons (eds.). Op. cit.

Irish, F. J., E. E. Williams, and E. Seling. 1988. Scanning electron microscopy of changes in epidermal structure occurring during the shedding cycle in squamate reptiles. *J. Morphol.* **197:** 105–126.

Komnick, H. 1986. Chloride cells and salt glands. Pp. 499–516. *In* J. Bereiter-Hahn *et al.* (eds.). See Chap. 2.

Kuhn-Schnyder, E. 1980. Observations on temporal openings of reptilian skulls and the classification of reptiles. Pp. 153–175. *In* L. L. Jacobs (ed.), "Aspects of Vertebrate History." Flagstaff, Arizona: Museum of Northern Arizona Press.

Landmann, L. 1975. The sense organs in the skin of the head of Squamata (Reptilia). *Isr. J. Zool.* **24:** 99–135.

Landmann, L. 1986. Epidermis and dermis. Pp. 150–187. *In* J. Bereiter-Hahn *et al.* (eds.). See Chap. 2.

Luppa, H. 1977. Histology of the digestive tract. Pp. 225–313. *In* G. Gans and T. S. Parsons (eds.). Op. cit.

Lynn, W. G. 1970. The thryoid. Pp. 201–234. *In* C. Gans and T. S. Parsons (eds.). Op. cit.

Maderson, P. F. A. 1965. The structure and development of the squamate epidermis. Pp. 96-102. *In* A. G. Lyne and B. F. Short (eds.), "Biology of Skin and Hair Growth." Sydney: Angus & Robertson.

Miller, M. R., and M. D. Lagios. 1970. The pancreas. Pp. 319–346. *In* C. Gans and T. S. Parsons (eds.). Op. cit.

Oelrich, T. M. 1956. The anatomy of the head of *Ctenosaura pectinata*. *Misc. Publ. Mus. Zool. Univ. Mich.* **94:** 1–122.

Oldham, J. C., H. M. Smith, and S. A. Miller. 1970. "A Laboratory Perspectus of Snake Anatomy." Champaign, Illinois: Stipes Publ. Co.

Ottaviani, G., and A. Tazzi. 1977. The lymphatic system. Pp. 315–462. *In* G. Gans and T. S. Parsons (eds.). Op. cit.

Pang, P. K., and M. P. Schreibman (eds.). 1986. See Chap. 1.

Parson, T. S. 1970. The nose and Jacobson's organ. Pp. 99–191. *In* G. Gans and T. S. Parsons (eds.). Op. cit.

Perry, S. F. 1983. Reptilian lungs. Functional anatomy and evolution. *Adv. Anat. Embryol. Cell Biol.* **29:** 1–81.

Quay, W. B. 1979. The parietal eye-pineal complex. Pp. 245–406. *In* C. Gans *et al.* (eds.). Op. cit.

Quay, W. B. 1986. Glands. Pp. 188–193. *In* J. Bereiter-Hahn *et al.* (eds.). See Chap. 2.

Ralph, C. L. 1983. Evolution of pineal control of endocrine function in lower vertebrates. *Am. Zool.* **23:** 597–605

Romer, A. S. 1956. "Osteology of the Reptiles." Chicago: Univ. of Chicago Press.

Ross, C. A. (ed.). 1989. "Crocodiles and Alligators." Silverwater, New South Wales: Golden Press Pty., Ltd.

Saint Girons, H. 1988. Les glandes céphaliques exocrines des reptiles. I. Données anatomiques et histologiques. *Ann. Sci. Nat. Zool., Paris* **9:** 221–255.

Saint Girons, M.-C. 1970. Morphology of the circulating blood cells. Pp. 73–91. *In* C. Gans and T. S. Parsons (eds.). Op. cit.

Schreibman, M. P. 1986. Pituitary gland. Pp. 11–55. *In* P. K. T. Pang and M. P. Schreibman (eds.). See Chap. 1.

Schumacher, G.-H. 1973. The head muscles and hyolaryngeal skeleton of turtles and crocodilians. Pp. 101–199. *In* C. Gans and T. S. Parsons (eds.). Op. cit.

Schwenk, K. 1985. Occurrence, distribution and functional significance of taste buds in lizards. *Copeia* **1985:** 91–101.

Underwood, G. 1970. The eye. Pp. 1–97. *In* G. Gans and T. S. Parsons (eds.). Op. cit.

Wake, D. B. 1979. The endoskeleton: the comparative anatomy of the vertebral column and ribs. Pp. 192–237. *In* M. H. Wake (ed.). See Chap. 1.

Walker, W. F., Jr. 1973. The locomotor apparatus of Testudines. Pp. 1–100. *In* C. Gans and T. S. Parsons (eds.). Op. cit.

Wever, E. G. 1978. "The Reptile Ear. Its Structure and Function." Princeton: Princeton Univ. Press.

Zimmerman, K., and H. Heatwole. 1990. Cutaneous photoreception: a new sensory mechanism for reptiles. *Copeia* **1990:** 860–862.

CHAPTER 4. Origin and Evolution of Reptiles

Bellairs, A. d'A., and C. B. Cox (eds.). 1976. "Morphology and Biology of Reptiles." London: Linnean Soc. London.

Benton, M. J., and J. M. Clark. 1988. Archosaur phylogeny and the relationships of the Crocodylia. Pp. 295–338. *In* M. J. Benton (ed.). See Chap. 2.

Carroll, R. L. 1982. Early evolution of reptiles. *Annu. Rev. Syst. Ecol.* **13:** 87–109.

Carroll, R. L. 1988. See Chap. 2.

Carroll, R. L., and D. Baird. 1972. Carboniferous stem-reptiles of the family Romeriidae. *Bull. Mus. Comp. Zool.* **143:** 321–363.

Charig, A. J., and C. Gans. 1990. Two new amphisbaenians from the Lower Miocene of Kenya. *Bull. Br. Mus. Nat. Hist. (Geol.)* **46:** 19–26.

Estes, R. 1983. "Sauria terrestria, Amphisbaenia. Handbuch der Paläoherpetologie, 10A." Stuttgart: Gustav Fischer Verlag.

Estes, R. 1983. The fossil record and early distribution of lizards. Pp. 365–398. *In* A. G. J. Rhodin and K. Miyata (eds.), "Advances in Herpetology and Evolutionary Biology. Essays in Honor of Ernest E. Williams." Cambridge, Massachusetts: Museum of Comparative Zoology.

Evans, S. E. 1988. The early history and relationships of the Diapsida. Pp. 221–260. *In* M.Benton (ed.). See Chap. 2.

Gaffney, E. S. 1986. Triassic and Early Jurassic turtles. Pp. 183–187. *In* K. Padian (ed.), "The Beginnings of the Age of Dinosaurs." Cambridge: Cambridge Univ. Press.

Gaffney, E. S. 1990. The comparative osteology of the Triassic turtle *Proganochelys*. *Bull. Am. Mus. Nat. Hist.* **194:** 1–263.

Gauthier, J. A. 1986. Saurischian monophyly and the origin of birds. *Mem. Calif. Acad. Sci.* **8:** 1–55.

Gauthier, J. A., A. G. Kluge, and T. Rowe. 1988a,b. See Chap. 2.

Hecht, M. A. 1982. The vertebral morphology of the Cretaceous snake, *Dinilysia patagonica* Woodward. *Neue. Jahrb. Geol. Palaeontol. Monatsh.* **9:** 523–532.

Hotton, N., P. D. MacLean *et al.* (eds.). 1986. "The Ecology and Biology of Mammal-like Reptiles." Washington, D.C.: Smithsonian Inst. Press.

Martin, L. D., and B. M. Rothschild. 1989. Paleopathology and diving mosasaurs. *Am. Zool.* **77:** 460–467.

Mlynarski, M. 1976. "Testudines. Handbuch der Paläoherpetologie, 7." Stuttgart: Gustav Fischer Verlag.

Olson, E. C. 1976. The exploitation of land by early tetrapods. Pp. 1–30. *In* A. d'A. Bellairs and C. B. Cox (eds.). Op. cit.

Packard, G. C., and M. J. Packard. 1980. Evolution of the cleidoic egg among reptilian antecedents of birds. *Am. Zool.* **20:** 351–362.

Piveteau, J. (ed.). 1955. See Chap. 2.

Rage, J.-C. 1984. "Serpentes. Handbuch der Paläoherpetologie, 11." Stuttgart: Gustav Fischer Verlag.

Rage, J.-C. 1988. The oldest known colubrid snakes. The state of the art. *Acta Zool. Cracov.* **31:** 457–474.

Rage, J.-C. 1988. Un serpent primitif (Reptilia, Squamata) dans le Cénomanien (base du Crétacé supérieur). *C. R. Acad. Sci., Paris* Sér. II **307:** 1027–1032.

Reisz, R. R., and M. Laurin. 1991. *Owenetta* and the origin of turtles. *Nature* (*London*) **349:** 324–326.

Rhodin, A. G. J., and K. Miyata (eds.). 1983. "Advances in Herpetology and Evolutionary Biology. Essays in Honor of Ernest E. Williams." Cambridge, Massachusetts: Museum of Comparative Zoology.

Romer, A. S. 1960. See Chap. 2.

Ross, C. A. (ed.). 1989. See Chap. 2.

Steel, R. 1973. "Crocodylia. Handbuch der Paläoherpetologie, 16." Stuttgart: Gustav Fischer Verlag.

Tarsitano, S., and J. Reiss. 1982. Plesiosaur locomotion—underwater flight versus rowing. *Neue. Jahrb. Geol. Palaeontol. Abh.* **164:** 188–192.

Wood, R. C. 1984. Evolution of the pelomedusid turtles. *Stvd. Geol. Salmanticensia Espec.* **1:** 269–282.

CHAPTER 5. Diet and Feeding

Auffenberg, W. 1978. Social and feeding behavior in *Varanus komodoensis*. Pp. 301–331. *In* N. Greenberg and P. D. MacLean (eds.), "Behavior and Neurology of Lizards." Poolesville, Maryland: Natl. Inst. Mental Health.

Auffenberg, W. 1982. Feeding strategy of the Caicos ground iguana, *Cyclura carinata*. Pp. 84–116. *In* G. M. Burghardt and A. S. Rand, (eds.), "Iguanas of the World. Their Behavior, Ecology, and Conservation." Park Ridge, New Jersey: Noyes Publ.

Auffenberg, W. 1988. "Gray's Monitor Lizard." Gainesville, Florida: Univ. of Florida Press.

Bennett, A. F. 1982. The energetics of reptilian activity. Pp. 155–199. *In* C. Gans and F. H. Pough (eds.), "Biology of the Reptilia," Vol. 13. New York: Academic Press.

Bjorndal, K. A. 1980. Nutritional and grazing behavior of the green turtle, *Chelonia mydas*. *Mar. Biol.* **56:** 147–154.

Bjorndal, K. A. 1985. Nutritional ecology of sea turtles. *Copeia* **1985:** 736–751.

Bjorndal, K. A. 1987. Digestive efficiency in a temperate herbivorous reptile, *Gopherus polyphemus*. *Copeia* **1987:** 714–720.

Bjorndal, K. A. 1989. Flexibility of digestive responses in two generalist herbivores, the tortoises *Geochelone carbonaria* and *Geochelone denticulata*. *Oecologia* **78:** 317–321.

Burghardt, G. M., and A. S. Rand (eds.). 1982. "Iguanas of the World. Their Behavior, Ecology, and Conservation." Park Ridge, New Jersey: Noyes Publ.

Cock Buning, T. de 1985. Thermal sensitivity as a specialization for prey capture and feeding in snakes. *Am. Zool.* **23:** 363–75.

Congdon, J. D., A. E. Dunham, and D. W. Tinkle. 1982. Energy budgets and life histories of reptiles. Pp. 233–271. *In* C. Gans and F. H. Pough (eds.), "Biology of the Reptilia," Vol. 13. New York: Academic Press.

Cooper, W. E., Jr. 1990. Prey odour discrimination by lizards and snakes. Pp. 533–538. *In* D. W. MacDonald *et al.* (eds.), "Chemical Signals in Vertebrates 5." Oxford: Oxford Univ. Press.

Cooper, W. E., Jr., and G. M. Burghardt. 1990. Vomerolfaction and vomodor. *J. Chem. Ecol.* **16:** 103–104.

Coulson, R. A., and T. Hernandez. 1983. "Alligator Metabolism. Studies on Chemical Reactions *In Vivo*." Oxford: Pergamon Press; see also *Comp. Biochem. Physiol. B* **74B:** i–iii, 1–182.

Cundall, D. 1987. Functional morphology. Pp. 106–140. *In* R. A. Seigel, J. T. Collins, and S. S. Novak (eds.), "Snakes: Ecology and Evolutionary Biology." New York: Macmillan Publ. Co.

Cundall, D., J. Lorenz-Elwood, and J. D. Groves. 1987. Asymmetric suction feeding in primitive salamanders. *Experientia* **43:** 1229–1231.

Elvers, I. 1977. Flower-visiting lizards on Madeira. *Bot. Not.* **130:** 231–234.

Ernst, C. H., and R. W. Barbour. 1989. See Chap. 3.

Feder, M. E., and G. V. Lauder (eds.). 1986. "Predator–Prey Relationships. Perspectives and Approaches from the Study of Lower Vertebrates." Chicago: Univ. of Chicago Press.

Fellers, G. M., and C. A. Drost. 1991. Ecology of the island night lizard, *Xantusia riversiana*, on Santa Barbara Island, California. *Herpetol. Monogr.* **5:** 28–78.
Fialho, R. F. 1990. Seed dispersal by a lizard and a treefrog—effect of dispersal site on seed survivorship. *Biotropica* **22:** 423–424.
Gans, C. 1986. Functional morphology of predator–prey relationships. Pp. 6–23. *In* M. E. Feder and G. V. Lauder (eds.). Op. cit.
Gans, C., and D. Crews (eds.). 1992. "Biology of the Reptilia," Vol. 18. Chicago: Univ. of Chicago Press.
Gans, C., and K. A. Gans (eds.). 1978. "Biology of the Reptilia," Vol. 8. New York: Academic Press.
Gans, C., and G. C. Gorniak. 1982. Functional morphology of lingual protrusion in marine toad (*Bufo marinus*). *Am. J. Anat.* **163:** 195–222.
Gans, C., and F. H. Pough (eds.). 1982. "Biology of the Reptilia," Vol. 13. New York: Academic Press.
Glodek, G. S., and H. K. Voris. 1982. Marine snake diets: prey composition, diversity and overlap. *Copeia* **1982:** 661–666.
Greene, H. W. 1982. Dietary and pheonotypic diversity in lizards: why are some organisms specialized? Pp. 107–128. *In* D. Mossakowski and G. Roth (eds.), "Environmental Adaptation and Evolution." Stuttgart: Gustav Fischer Verlag.
Greene, H. W., and G. M. Burghardt. 1978. Behavior and phylogeny: constriction in ancient and modern snakes. *Science* **200:** 74–77.
Greer, A. E. 1976. On the evolution of the giant Cape Verde scincid lizard *Macroscincus coctei. J. Nat. Hist.* **10:** 691–712.
Guard, C. L. 1980. The reptilian digestive system: general characteristics. Pp. 43–51. *In* K. Schmidt-Nielsen *et al.* (eds.), "Comparative Physiology: Primitive Mammals." Cambridge: Cambridge Univ. Press.
Halpern, M. 1992. Nasal chemical senses in reptiles: structure and function. Pp. 423–523. *In* C. Gans and D. Crews (eds.). Op. cit.
Hamilton, J., and M. Coe. 1982. Feeding, digestion and assimilation of a population of giant tortoises (*Geochelone gigantea*) on Aldabra atoll. *J. Arid Environ.* **5:** 127–144.
Heatwole, H. F., and J. Taylor. 1987. "Ecology of Reptiles." Chipping Norton, New South Wales: Surrey Beatty & Sons, Ltd.
Hetherington, T. E. 1985. Role of the opercularis muscle in seismic sensitivity in the bullfrog *Rana catesbeiana*. *J. Exp. Zool.* **235:** 27–34.
Hetherington, T. E. 1987. See Chap. 1.
Hetherington, T. E., A. P. Jaslow, and R. E. Lombard. 1986. See Chap. 1.
Holmberg, A. R. 1957. Lizard hunts on the north coast of Peru. *Fieldiana, Anthropol.* **36:** 203–220.
Huey, R. B., and E. R. Pianka. 1981. Ecological consequences of foraging mode. *Ecology* **62:** 991–999.
Huey, R. B., E. R. Pianka, and T. W. Schoener (eds.). 1983. "Lizard Ecology. Studies of a Model Organism." Cambridge: Harvard Univ. Press.
Iverson, J. B. 1982. Adaptations to herbivory in iguanine lizards. Pp. 60–76. *In* G. M. Burghardt and A. S. Rand (eds.). Op. cit.
Jaeger, R. G. 1978. Ecological niche dimensions and sensory functions in amphibians. Pp. 169–196. *In* M. A. Ali (ed.), "Sensory Ecology: Review and Perspectives." New York: Plenum Publ. Corp.
Kenny, J. S. 1969. Feeding mechanisms in anuran larvae. *J. Zool.* **157:** 225–246.
Kochva, E. 1978. Oral glands of the Reptilia. Pp.43–161. *In* C. Gans and K. A. Gans (eds.). Op. cit.
Lannoo, M. J. 1986. Vision is not necessary for size-selective zooplanktivory in aquatic salamanders. *Can. J. Zool.* **64:** 1071–1075.
Larsen, J. H., J. T. Beneski, and D. B. Wake. 1989. Hyolingual feeding systems of the Plethodontidae:

comparative kinematics of prey capture by salamanders with free and attached tongues. *J. Exp. Zool.* **252:** 25–33.

MacDonald, L. A., and H. R. Mushinsky. 1988. Foraging ecology of the gopher tortoise, *Gopherus polyphemus*, in a sandhill habitat. *Herpetologica* **44:** 345–353.

Mautz, W. J., and W. Lopez-Forment. 1978. Observations on the activity and diet of the cavernicolous lizard *Lepidophyma smithii*. *Herpetologica* **34:** 311–313.

Mautz, W. J., and K. A. Nagy. 1987. Ontogenetic changes in diet, field metabolic rate, and water flux in the herbivorous lizard *Dipsosaurus dorsalis*. *Physiol. Zool.* **60:** 640–658.

Moll, E. O. 1980. Natural history of the river terrapin, *Batagur baska* (Gray) in Malaysia. *Malays. J. Sci.* **6:** 23–62.

Moskovits, D. K., and K. A. Bjorndal. 1990. Diet and food preferences of the tortoises *Geochelone carbonaria* and *G. denticulata* in northwestern Brazil. *Herpetologica* **46:** 207–218.

O'Brien, W. J., H. Brownman, and B. I. Evans. 1990. Search strategies of foraging animals. *Am. Sci.* **78:** 152–160.

Pandian, T. J., and F. J. Vernberg (eds.). 1987. "Animal Energetics. Vol. 2. Bivalvia through Reptilia." San Diego: Academic Press.

Pough, F. H. 1973. Lizard energetics and diet. *Ecology* **54:** 837–844.

Pough, F. H. 1980. The advantages of ectothermy for tetrapods. *Am. Nat.* **115:** 92–112.

Rand, A. S. 1978. Reptilian arboreal folivores. Pp. 115–122. *In* G. G. Montgomery (ed.), "The Ecology of Arboreal Folivores." Washington, D.C.: Smithsonian Inst. Press.

Rand, A. S., B. A. Dugan *et al.* 1990. The diet of a generalized folivore: *Iguana iguana* in Panama. *Copeia* **1990:** 211–214.

Reilly, S. M., and G. V. Lauder. 1990. The evolution of tetrapod feeding behavior: kinematic homologies in prey transport. *Evolution* **44:** 1542–1557.

Robinson, M. D., and A. B. Cunningham. 1978. Comparative diets of two Namib Desert sand lizards. *Madoqua* **11:** 41–45.

Roth, G. 1986. Neural mechanisms of prey recognition: an example in amphibians. Pp. 42–68. *In* M. E. Feder and G. V. Lauder (eds.). Op. cit.

Ruppert, R. M. 1980. Comparative assimilation efficiencies of two lizards. *Comp. Biochem. Physiol. A* **67A:** 491–496.

Russell, F. E. 1983. "Snake Venom Poisoning." Great Neck, New York: Scholium Int., Inc.

Saint-Girons, H. 1988. Op. cit., Chap. 3.

Sakaluk, S. K., and J. J. Belwood. 1984. Gecko phonotaxis to cricket calling song: a case of satellite predation. *Anim. Behav.* **32:** 659–662.

Schoener, T. W. 1974. Theory of feeding strategies. *Annu. Rev. Syst. Ecol.* **2:** 369–404.

Schwenk, K. 1986. Morphology of the tongue in tuatara, *Sphenodon punctatus*, with comments on function and phylogeny. *J. Morphol.* **188:** 129–156.

Seale, D. B. 1987. Amphibia. Pp. 467–552. *In* T. J. Pandian and F. J. Vernberg (eds.). Op. cit.

Seigel, R. A., J. T. Collins, and S. S. Novak (eds.). 1987. "Snakes: Ecology and Evolutionary Biology." New York: Macmillan Publ. Co.

Shine, R., and T. Schwaner. 1985. Prey constriction by venomous snakes: a review and new data on Australian species. *Copeia* **1985:** 1067–1071.

Silva, R. S. M. da, and R. H. Migliorini. 1990. Effects of starvation and refeeding on energy-linked metabolic processes in the turtle (*Phrynops hilarii*). *Comp. Biochem. Physiol. A* **94A**: 415–419.

Skoczylas, R. 1978. Physiology of the digestive tract. Pp. 589–717. *In* C. Gans and K. A. Gans (eds.). Op. cit.

Spotila, J. R., and E. A. Standora. 1985. Energy budgets of ectothermic vertebrates. *Am. Zool.* **25:** 973–986.

Steyn, W. 1963. *Angolocaurus* [sic] *skoogi* (Andersson)—a new record from south west Africa. *Cimbebasia* **6:** 8–11.

Sylber, C. K. 1988. Feeding habits of the lizards *Sauromalus varius* and *S. hispidus* in the Gulf of California. *J. Herpetol.* **22:** 413–424.

Troyer, K. 1984. Behavioral acquisition of the hindgut fermentation system by hatchling *Iguana iguana*. *Behav. Ecol. Sociobiol.* **14:** 189–193.
Troyer, K. 1984. Microbes, herbivory and evolution of social behavior. *J. Theor. Biol.* **106:** 157–169.
Troyer, K. 1984. Structure and function of the digestive tract of a herbivorous lizard *Iguana iguana*. *Physiol. Zool.* **57:** 1–8.
Waldschmidt, S. R., S. M. Jones, and W. P. Porter. 1986. The effect of body temperature and feeding regime on activity, passage time, and digestive coefficient in the lizard *Uta stansburiana*. *Physiol. Zool.* **59:** 376–383.
Waldschmidt, S. R., S. M. Jones, and W. P. Porter. 1987. Reptilia. Pp. 553–619. *In* T. J. Pandian and F. J. Vernberg (eds.). Op. cit.
Wassersug, R. 1972. The mechanism of ultraplanktonic entrapment in anuran larvae. *J. Morphol.* **137:** 279–288.
Whitaker, A. H. 1968. The lizards of the Poor Knights Islands, New Zealand. *N. Z. J. Sci.* **11:** 623–651.
Wirot, N. 1979. "The Turtles of Thailand." Bangkok: SIAMFARM Zool. Gard.
Zug, G. R., and P. B. Zug. 1979. The marine toad, *Bufo marinus*: a natural history resumé of native populations. *Smithson. Contrib. Zool.* No. 284: 1–58.

CHAPTER 6. Defense and Escape

Arnold, E. N. 1988. Caudal autotomy as a defense. Pp. 235–273. *In* C. Gans and R. B. Huey (eds.), "Biology of the Reptilia," Vol. 16. New York: Alan R. Liss, Inc.
Bauer, A. M., A. P. Russell, and R. E. Shadwick. 1989. Mechanical properties and morphological correlates of fragile skin in gekkonid lizards. *J. Exp. Biol.* **145:** 79–102.
Brodie, E. D. 1983. Anitpredator adaptations of salamanders: evolution and convergence among terrestrial species. Pp. 109–133. *In* N. S. Margaris *et al.* (eds.), "Plant, Animal, and Microbial Adaptations to Terrestrial Environments." New York: Plenum Publ. Corp.
Brodie, E. D., and D. R. Formanowicz. 1987. Antipredator mechanisms of larval anurans: protection of palatable individuals. *Herpetologica* **43:** 369–373.
Cocroft, R. B., and K. Hambler. 1989. Observations on a commensal relationship of the microhylid frog *Chiasmocleis ventrimaculata* and the burrowing theraphosid spider *Xenesthis immanis* in southeastern Peru. *Biotropica* **21:** 2–8.
Daly, J. W., C. W. Myers, and N. Whittaker. 1987. Further classification of skin alkaloids from neotropical poison frogs (Dendrobatidae), with a general survey of toxic/noxious substances in the Amphibia. *Toxicon* **25:** 1023–1095.
Daniels, C. B. 1990. The relative importance of host behaviour, method of transmission and longevity on the establishment of an acanthocephalan population in two reptilian host. *Mem. Queensl. Mus.* **29:** 367–374.
Duellman, W. E., and L. Trueb. 1986. See Chap. 1.
Endler, J. A. 1986. Defense against predators. Pp. 109–134. *In* M. E. Feder and G. V. Lauder (eds.). See Chap. 5.
Fernandez, P. J., and J. P. Collins. 1988. Effect of environment and ontogeny on color pattern variation in Arizona tiger salamanders (*Ambystoma tigrinum nebulosum*). *Copeia* **1988:** 928–938.
Formanowicz, D. R., and E. D. Brodie. 1982. Relative palabilities of members of a larval amphibian community. *Copeia* **1982:** 91–97.
Gans, C., and R. B. Huey (eds.). 1988. "Biology of the Reptilia," Vol. 16. New York: Alan R. Liss, Inc.
Greene, H. W. 1988. Antipredator mechanisms in reptiles. Pp. 1–152. *In* C. Gans and R. B. Huey (eds.). Op. cit.
Greene, H. W., and G. M. Burghardt. 1978. See Chap. 5.

Hallman, G. M., C. E. Ortega *et al.* 1990. Effect of bacterial pyrogen on three lizard species. *Comp. Biochem. Physiol. A* **96A:** 383–386.

Harkey, G. A., and R. D. Semlitsch. 1988. Effects of temperature on growth, development, and color polmorphism in the ornate chorus frog *Pseudacris ornata. Copeia* **1988:** 1001–1007.

Heyer, W. R., R. M. McDiarmid, and D. L. Weigmann. 1975. Tadpoles, predation, and pond habitats in the tropics. *Biotropica* **72:** 100–111.

Middendorf, G. A., and W. C. Sherbrooke. 1992. Canid elicitation of blood-squirting in a horned lizard (*Phrynosoma cornutum*). *Copeia* **1992:** 519–527.

Moreno, G. 1989. Behavioral and physiological differentiation between the color morphs of the salamander, *Plethodon cinereus. J. Herpetol.* **23:** 335–341.

Morey, S. R. 1990. Microhabitat selection and predation in the Pacific treefrog, *Pseudacris regilla. J. Herpetol.* **24:** 292–296.

Morin, P. J. 1983. Predation, competition and composition of larval anuran guilds. *Ecol. Monogr.* **53:** 119–138.

Mushinsky, H. R. 1987. Foraging ecology. Pp. 302–334. *In* R. A. Seigel *et al.* (eds.). See Chap. 5.

Pasteur, G. 1982. A classificatory review of mimicry systems. *Annu. Rev. Ecol. Syst.* **13:** 169–199.

Petranka, J. W. 1989. Response of toad tadpoles to conflicting chemical stimuli: predator avoidance verus "optimal" foraging. *Herpetologica* **45:** 283–292.

Pounds, J. A., and M. L. Crump. 1987. Harlequin frogs along a tropical montane stream: aggregation and the risk of predation by frog-eating flies. *Biotropica* **19:** 306–309.

Pough, F. H. 1988. Mimicry and related phenomena. Pp. 153–234. *In* C. Gans and R. B. Huey (eds.). Op. cit.

Rand, A. S., S. Guerrero, and R. M. Andrews. 1983. The ecological effects of malaria on populations of the lizard *Anolis limifrons* on Barro Colorado Island. Panama. Pp. 455–471. *In* A. G. J. Rhodin and K. Miyata (eds.). See Chap. 4.

Ryan, M. J. 1985. "The Túngara Frog. A Study in Sexual Selection and Communication." Chicago: Univ. of Chicago Press.

Schall, J. J. 1983. Lizard malaria: parasite–host ecology. Pp. 84–100. *In* R. B. Huey *et al.* (eds.). See Chap. 5.

Schall, J. J., and E. R. Pianka. 1980. Evolution of escape behavior diversity. *Am. Nat.* **115:** 551–566.

Scudder, R. M., and G. M. Burghardt. 1983. A comparative study of defensive behavior in three sympatric species of water snakes (*Nerodia*). *Z. Tierpsychol.* **63:** 17–26.

Sherbrooke, W. C., and S. K. Frost. 1989. Integumental chromatophores of a color-change, thermoregulating lizard, *Phrynosoma modestum. Am. Mus. Novit.* No. 2943: 1–14.

Sherbrooke, W. C., M. E. Hadley, and A. M. de L. Castrucci. 1988. Melanotropic peptides and receptors: an evolutionary perspective in vertebrate physiological color change. Pp. 175–189. *In* M. E. Hadley (ed.), "The Melanotropic Peptides. Vol. II. Biological Role." Boca Raton, Florida: CRC Press.

Sherbrooke, W. C., and R. R. Montanucci. 1988. Stone mimicry in the round-tailed horned lizard, *Phrynosoma modesturm* (Sauria: Iguanidae). *J. Arid Environ.* **14:** 275–284.

Sullivan, B. K. 1991. Parasites and sexual selection: separating causes and effects. *Herpetologica* **47:** 250–264.

Tinsley, R. C. 1990. The influence of parasite infection on mating success in spadefoot toad, *Scaphiopus couchii. Am. Zool.* **30:** 313–324.

Wake, D. B., and I. G. Dresner. 1967. Functional morphology and evolution of tail autotomy in salamanders. *J. Morphol.* **122:** 265–306.

Wassersug, R. J. 1973. Aspects of social behavior in anuran larvae. Pp. 273–297. *In* J. L. Vial (ed.). See Chap. 1.

Watkins, J. F., F. R. Gehlbach, and J. C. Kroll. 1969. Attractant-repellant secretions in blind snakes (*Leptotyphlops dulcis*) and army ants (*Neivamyrex nigrescens*). *Ecology* **50:** 1098–1102.

Weldon, P. J. 1990. Responses by vertebrates to chemicals from predators. Pp. 500–521. *In* D. W. MacDonald *et al.* (eds.), "Chemical Signals in Vertebrates 5." Oxford: Oxford Univ. Press.

CHAPTER 7. Modes of Reproduction and Development

Altig, R., and G. F. Johnston. 1989. See Chap. 1.

Andrén, C., and G. Nilson. 1987. The copulatory plug of the adder, *Vipera berus*: does it keep sperm in or out? *Oikos* **49:** 230–232.

Andrews, R. M. 1976. Growth rate in island and mainland anoline lizards. *Copeia* **1976:** 477–482.

Andrews, R. M. 1982. Patterns of growth in reptiles. *In* C. Gans and F. H. Pough (eds.). See Chap. 5.

Auffenberg, W. 1981. "The Behavioral Ecology of the Komodo Monitor." Gainesville, Florida: Univ. of Presses Florida.

Baker, C. L. 1945. The natural history and morphology of amphiumae. *Rep. Reelfoot Lake Biol. Stn.* **9:** 55–91.

Barbault, R., and M. T. Rodrigues. 1979. Observations sur la reproduction et la dynamique des populations de quelques anoures tropicaux. III. *Arthroleptis poecilonotus. Trop. Ecol.* **20:** 64–77.

Bell, B. D. 1985. Development and parental-care in the endemic New Zealand Frogs. Pp. 269–278. *In* G. Grigg, R. Shine, and H. Ehlmann (eds.), "Biology of Australasian Frogs and Reptiles." Chipping Norton, New South Wales: Surrey Beatty & Sons, Ltd.

Bishop, S. C. 1941. The salamanders of New York. *N.Y. State Mus. Bull.* No. 324: 1–365.

Blackburn, D. G., L. J. Vitt, and C. A. Beuchat. 1984. Eutherian-like reproductive specializations in a viviparous reptile. *Proc. Natl. Acad. Sci. U.S.A.* **81:** 4860–4863.

Bogart, J. P., R. P. Elinson, and L. E. Licht. 1989. Temperature and sperm incorporation in polyploid salamanders. *Science* **246:** 1032–1034.

Bogart, J. P., and L. E. Licht. 1986. Reproduction and the origins of polyploids in hybrid salamanders of the genus *Ambystoma. Can. J. Genet. Cytol.* **28:** 605–617.

Bogart, J. P., L. E. Licht *et al.* 1985. Electrophoretic identification of *Ambystoma laterale* and *Ambystoma texanum* as well as their diploid and triploid interspecific hybrids on Pelee Island, Ontario. *Can. J. Zool.* **63:** 340–347.

Bokermann, W. C. A. 1974. Observacoes sobre desenvolvimento precoce em *Sphaenorhynchus bromelicola. Rev. Bras. Biol.* **34:** 35–41.

Bourne, D., and M. Coe. 1978. The size, structure and distribution of the giant tortoise population of Aldabra. *Philos. Trans. R. Soc. London, Ser. B* **282:** 139–175.

Bruce, R. C. 1988. An ecological life table for the salamander *Eurycea wilderae. Copeia* **1988:** 15–26.

Bruce, R. C. 1988. Life history variation in the salamander *Desmognathus quadramaculatus. Herpetologica* **44:** 218–227.

Bull, J. J. 1980. Sex determination in reptiles. *Q. Rev. Biol.* **55:** 3–21.

Caldwell, D. K. 1959. The loggerhead turtles of Cape Romain, South Carolina. *Bull. Fla. State Mus.* No. 4: 319–348.

Castanet, J., and M. Baez. 1991. Adaptation and evolution in *Gallotia* lizards from the Canary Islands: age, growth, maturity and longevity. *Amphibia–Reptilia* **12:** 81–102.

Castanet, J., D. G. Newman, and H. Saint Girons. 1988. Skeletochronological data on the growth, age, and population structure of the tuatara, *Sphenodon punctatus*, on Stephens and Lady Alice Islands, New Zealand. *Herpetologica* **44:** 25–37.

Cole, C. J. 1979. Chromosome inheritance in parthenogenetic lizards and evolution of allopolyploidy in reptiles. *J. Hered.* **70:** 95–102.

Congdon, J. D. 1987. Parental investment in reptiles: an important component of reproduction. *Abstr. SSAR HL 1987* p. 64.

Cott, H. B. 1961. Scientific results of an inquiry into the ecology and economic status of the Nile crocodile (*Crocodilus niloticus*) in Uganda and northern Rhodesia. *Trans. Zool. Soc. London* **29:** 211–356.

Crews, D. 1987. Courtship in unisexual lizards: a model for brain evolution. *Sci. Am.* **257:** 116–121.

Daniel, J. C. 1983. "The Book of Indian Reptiles." Bombay: Bombay Nat. Hist. Soc.

Dawley, R. M., and J. P. Bogart (eds.). 1989. "Evolution and Ecology of Unisexual Vertebrates." Albany: Bull. New York State Mus.

Duellman, W. E., and L. Trueb. 1986. See Chap. 1.

Elinson, R. P. 1987. Changes in developmental patterns: embryos of amphibians with large eggs. Pp. 1–21. *In* R. Raff and E. Raff (eds.), "Development as a Evolutionary Process." New York: Alan R. Liss, Inc.

Ernst, C. H. 1971. Population dynamics and activity cycles of *Chrysemys picta* in southeastern Pennsylvania. *J. Herpetol.* **5:** 151–160.

Fellers, G. M., and C. A. Drost. 1991. See Chap. 5.

Fitch, H. S. 1970. Reproductive cycles of lizards and snakes. *Univ. Kans. Mus. Nat. Hist. Misc. Publ.* No. 52: 1–247.

Fitch, H. S. 1975. A demographic study of the ringneck snake (*Diadophis punctatus*) in Kansas. *Univ. Kans. Mus. Nat. Hist. Misc. Publ.* No. 62: 1–53.

Fitch, H. S. 1989. A field study of the slender glass lizard, *Ophisaurus attenuatus*, in northeastern Kansas. *Univ. Kans. Mus. Nat. Hist. Occas. Pap.* No. 125: 1–50.

Forester, D. C. 1981. Parental care in the salamander *Desmognathus ochrophaeus*: female activity pattern and trophic behavior. *J. Herpetol.* **15:** 29–34.

Fox, H. 1983. "Amphibian Morphogenesis." Clifton, New Jersey: Humana Press.

Frazer, N. B., J. W. Gibbons, and J. L. Greene. 1990. Life tables of a slider turtle population. Pp. 183–200. *In* J. W. Gibbons (ed.). Op. cit.

Gans, C., and F. Billet (eds.). 1985. "Biology of the Reptilia," Vol. 15. New York: John Wiley & Sons.

Gans, C., F. Billet, and P. F. A. Maderson (eds.). 1985. "Biology of the Reptilia," Vols. 14. New York: John Wiley & Sons.

Gibbons, J. W. 1987. Why do turtles live so long?. *BioScience* **37:** 262–269.

Gibbons, J. W. (ed.). 1990. "Life History and Ecology of the Slider Turtle." Washington, D.C.: Smithsonian Inst. Press.

Gist, D. H., and J. M. Jones. 1989. Sperm storage within the oviduct of turtles. *J. Morphol.* **199:** 379–384.

Graf, J.-D., and M. P. Pelaz. 1989. Evolutionary genetics of the *Rana esculenta* complex. Pp. 289–302. *In* R. M. Dawley and J. P. Bogart (eds.). Op. cit.

Grassé, P.-P. 1986. La fécondation. Pp. 56–75. *In* P.-P. Grassé and M. Delsol (eds.), Op. cit.

Grassé, P.-P., and M. Delsol (eds.). 1986. "Traité de Zoologie. Anatomie, Systématique, Biologie. Tome XIV. Batraciens." Paris: Masson.

Grigg, G., R. Shine, and H. Ehlmann (eds.). 1985. "Biology of Australasian Frogs and Reptiles." Chipping Norton, New South Wales: Surrey Beatty & Sons, Ltd.

Grubb, P. 1971. The growth, ecology and population structure of giant tortoises on Aldabra. *Philos. Trans. R. Soc. London, Ser. B* **260:** 327–372.

Günther, R. 1990. "Die Wasserfrösche Europas." Wittenberg Lutherstadt, Germany: A. Ziemsen Verlag.

Guillette, L. J. 1989. The evolution of vertebrate viviparity: morphological modifications and endocrine control. Pp. 219–233. *In* D. B. Wake and G. Roth (eds.), "Complex Organismal Functions: Integration and Evolution in Vertebrates." London: John Wiley & Sons.

Guillette, L. J., S. L. Fox, and B. D. Palmer. 1989. Oviductal morphology and egg shelling in the oviparous lizards *Crotaphytus collaris* and *Eumeces obsoletus*. *J. Morphol.* **201:** 145–159.

Halliday, T. R., and P. A. Verrell. 1984. Sperm competition in amphibians. Pp. 487–508. *In* R. Smith (ed.), "Sperm Competition and the Evolution of Animal Mating Systems." Orlando: Academic Press.

Hanken, J. 1989. Development and evolution in amphibians. *Am. Sci.* **77:** 336–343.

Hardy, L. M., C. J. Cole, and C. R. Townsend. 1989. Parthenogenetic reproduction in the neotropical unisexual lizard, *Gymnophthalmus underwoodi*. *J. Morphol.* **201:** 215–234.

Howard, R. D. 1978. The evolution of mating strategies in bullforgs, *Rana catesbeiana*. *Evolution* **32:** 850–871.

Iverson, J. 1979. Behavior and ecology of the rock iguana *Cyclura carinata*. *Bull. Fla. State Mus., Biol. Sci.* No. 24: 175–358.

Jackson, D. R. 1988. Reproductive strategies of sympatric freshwater emydid turtles in northern peninsular Florida. *Bull. Fla. State Mus., Biol. Sci.* No. 33: 113–158.

Janzen, F. J., and G. L. Paukstis. 1991. Environmental sex determination in reptiles: ecology, evolution, and experimental design. *Q. Rev. Biol.* **66:** 149–179.

Joanen, T. 1969. Nesting ecology of alligators in Louisiana. *Proc. Annu. Conf. Southeast. Assoc. Game Fish Comm., 23rd* pp. 141–151.

Kalb, H. J., and G. R. Zug. 1990. Age estimates for a population of American toads, *Bufo americanus*, in northern Virginia. *Brimleyana* **16:** 79–86.

Klosterman, L. L. 1987. Ultrastructural and quantitative dynamics of the granulosa of ovarian follicles of the lizard *Gerrhonotus coeruleus*. *J. Morphol.* **192:** 125–144.

Kluge, A. G. 1981. The life history, social organization, and parental behavior of *Hyla rosenbergi* Boulenger, a nest-building gladiator frog. *Misc. Publ. Mus. Zool. Univ. Mich.* No. 160: 1–170.

Kok, D., L. H. du Preez, and A. Channing. 1989. Channel construction by the African bullfrog: another anuran parental care strategy. *J. Herpetol.* **23:** 435–437.

Lowcock, L. A. 1989. Biogeography of hybrid complexes of *Ambystoma*: interpreting unisexual–bisexual genetic data in space and time. Pp. 180–208. *In* R. M. Dawley and J. P. Bogart (eds.). Op. cit.

Macartney, J. M., and P. T. Gregory. 1988. Reproductive biology of female rattlesnakes (*Crotalus viridis*) in British Columbia. *Copeia* **1988:** 47–57.

Maeda, N., and M. Matsui. 1989. "Frogs and Toads of Japan." Tokoyo: Bun-ichi Sogo Shuppan Co.

McDiarmid, R. W. 1978. Evolution of parental care in frogs. Pp. 127–147. *In* G. M. Burghardt and M. Beckoff (eds.), "The Development of Behavior: Comparative and Evolutionary Aspects." New York: Garland STPM Press.

Medica, P. A., and F. B. Turner. 1984. Natural longevity of lizards in southern Nevada. *Herpetol. Rev.* **15:**34–35.

Metter, D. E. 1964. A morphological and ecological comparison of two populations of the tailed frog, *Ascaphus truei* Stejneger. *Copeia* **1964:** 181–195.

Morris, M. A., and R. A. Brandon. 1984. Gynogenesis and hybridization between *Ambystoma platineum* and *Ambystoma texanum* in Illinois. *Copeia* **1984:** 324–357.

Nussbaum, R. A. 1985. The evolution of parental care in salamanders. *Misc. Publ. Mus. Zool. Univ. Mich.* No. 169: 1–50.

Oliver, J. A. 1956. Reproduction in the king cobra, *Ophiophagus hannah* Cantor. *Zoologica* **41:** 145–152.

Organ, J. A. 1961. Studies of the local distribution, life history, and population dynamics of the salamander genus *Desmognathus* in Virginia. *Ecol. Monogr.* **31:** 189–220.

Ouboter, P. E., and L. M. R. Nanhoe. 1987. Notes on nesting and parental care in *Caiman crocodilus crocodilus* in northern Suriname and an analysis of crocodilian nesting habitats. *Amphibia–Reptilia* **8:** 331–348.

Packard, G. C., and M. J. Packard. 1988. The physiological ecology of reptilian eggs and embryos. Pp. 523–605. *In* C. Gans and R. B. Huey (eds.). See Chap. 6.

Packard, M. J., and K. F. Hirsch. 1986. Scanning electron microscopy of eggshells of contemporary reptiles. *Scanning Electron Microsc.* **4:** 1581–1590.

Parker, W. S., and W. S. Brown. 1980. Comparative ecology of two colubrid snakes, *Masticophis t. taeniatus* and *Pituophis melanoleucus deserticola*, in northern Utah. *Milwaukee Public Mus., Publ. Biol. Geol.* No. 7: 1–104.

Peterson, C. L., R. F. Wilkinson *et al.* 1983. Age and growth of the Ozark hellbender (*Cryptobranchus alleganiensis bishopi*). *Copeia* **1983:** 225–231.

Pieau, C. 1985. Déterminisme du sexe chez les reptiles; influence de facteurs épigénétique. *Bull. Soc. Zool. Fr.* **110:** 97–111.

Reinert, H. K., and W. R. Kodrich. 1982. Movements and habitat utilization by the massasauga, *Sistrurus catenatus catenatus*. *J. Herpetol.* **16:** 162–171.

Salthe, S. N. 1963. The egg capsules in the Amphibia. *J. Morphol.* **113:** 161–171.

Schuett, G. W., and J. C. Gillingham. 1986. Sperm storage and multiple paternity in the copperhead, *Agkistrodon contortrix*. *Copeia* **1986:** 807–811.

Sebens, K. P. 1987. The ecology of indeterminate growth in animals. *Annu. Rev. Ecol. Syst.* **18:** 371–407.

Semlitsch, R. D., and H. M. Wilbur. 1987. Artificial selection for paedomorphosis in the salmander *Ambystoma talpoideum*. *Evolution* **43:** 105–112.

Shine, R. 1987. The evolution of viviparity: ecological correlates of reproductive mode within a genus of Australian snakes (*Pseudechis*: Elapidae). *Copeia* **1987:** 551–563.

Shine, R. 1988. Parental care in reptiles. Pp. 275–329. *In* C. Gans and R. B. Huey (eds.). See Chap. 6.

Shoop, C. R. 1960. The breeding habits of the mole salamander, *Ambystoma talpoideum*, in southeastern Louisiana. *Tulane Stud. Zool.* **8:** 65–82.

Shoop, C. R. 1965. Aspects of reproduction in Louisiana *Necturus* populations. *Am. Midl. Nat.* **74:** 357–367.

Smith, B. G. 1907. The life history and habits of *Cryptobranchus allegheniensis*. *Biol. Bull.* **13:** 5–39.

Somma, L. A. 1990. A categorization and bibliogrpahic survey of parental behavior in lepidosaurian reptiles. *Smithson. Herpetol. Inform. Serv.* **81:** 1–53.

Stewart, J. R., and D. G. Blackburn. 1988. Reptilian placentation: structural diversity and terminology. *Copeia* **1988:** 839–852.

Tinkle, D. W. 1967. The life and demography of the side-blotched lizard, *Uta stansburiana*. *Misc. Publ. Mus. Zool. Univ. Mich.* No. **132:**1–182.

Tinkle, D. W., and J. W. Gibbons. 1977. The distribution and evolution of viviparity in reptiles. *Misc. Publ. Mus. Zool. Univ. Mich.* No. **154:** 1–55.

Townsend, D. S., and M. M. Stewart. 1985. Direct development in *Eleutherodactylus coqui*: a staging table. *Copeia* **1985:** 423–436.

Townsend, D. S., M. M. Stewart, and F. H. Pough. 1984. Male parental care and its adaptive significance in a neotropical frog. *Anim. Behav.* **32:** 421–431.

Trivers, R. L. 1972. Parental investment and sexual selection. Pp. *In* B. Campbell (ed.), "Sexual Selection and the Descent of Man." Chicago: Aldine.

Turner, F. B., P. A. Medica *et al.* 1969. A demographic analysis of fenced populations of the whiptailed lizard, *Cnemidophorus tigris,* in southern Nevada. *SW Nat.* **14:**189–202.

Van Gansen, P. 1986. Ovogenése des amphibiens. Pp. 21–55. *In* P.-P. Grassé and M. Delsol (eds.). Op. cit.

Vaz-Ferreira, R., and A. Gehrau. 1975. Comportamiento epimelético de la rana común, *Leptodactylus ocellatus*. I. Atención ed la cria y actividades alimentarias y agresivas relaciónadas. *Physis* **34B:** 1–14.

Vinegar, A., V. H. Hutchison, and H. G. Dowling. 1970. Metabolism energetics, and thermoregulation during brooding of snakes of the genus *Python*. *Zoologica* **55:** 19–48.

Vitt, L. J., and W. E. Cooper. 1989. Maternal care in skinks (*Eumeces*). *J. Herpetol.* **23:** 29–34.

Vrijenhoek, R. C., *et al.* 1989. A list of the known unisexual vertebrates. Pp. 19–23. *In* R. Dawley and J. Bogart (eds.). Op. cit.

Wake, M. H. 1982. Diversity within a framework of constraints. Amphibian reproductive modes. Pp. 87–106. *In* D. Mossakowski and G. Roth (eds.), "Environmental Adaptation and Evolution." Stuttgart: Gustav Fischer Verlag.

Wassarman, P. 1987. The biology and chemistry of fertilization. *Science* **235:** 553–560.
Wells, K. D. 1981. Parental behavior of male and female frogs. Pp. 184–197. *In* R. D. Alexander and D. W. Tinkle (eds.), "Natural Selection and Social Behavior: Recent Research and New Theory." New York: Chiron Press.
Werner, E. E. 1986. Amphibian metamorphosis: growth rate, predation risk, and the optimal size at transformation. *Am. Nat.* No. 128: 319–341.
Weygoldt, P. 1987. Evolution of parental care in dart poison frogs (Dendrobatidae). *Z. Zool. Syst. Evol.* **25:** 51–67.
Whali, W., I. B. Dawid *et al.* 1981. Vitellogenesis and the vitellogenin gene family. *Science* **212:** 298–304.
Wilbur, H. M. 1975. The evolutionary and mathematical demography of the turtle *Chrysemys picta. Ecology* **56:** 64–77.

CHAPTER 8. Dynamics of Reproduction

Arnold, S. J. 1977. The evolution of courtship behavior in New World salamanders with some comments on Old World salamanders. Pp. 141–183. *In* D. H. Taylor and S. I. Guttman (eds.), "The Reproductive Biology of Amphibians." New York: Plenum Press.
Böhme, W., R. Hutterer, and W. Bings. 1985. Die Stimme der Lacertidae, speziell der Canareneidechsen. *Bonn. Zool. Beitr.* **36:** 337–354.
Bradshaw, S. D. 1986. "Ecophysiology of Desert Reptiles." North Ryde, New South Wales: Academic Press Australia.
Carpenter, C. C. 1961. Patterns of social behavior in the desert iguana, *Dipsosaurus dorsalis. Copeia* **1961**:396–405.
Carpenter, C. C., and G. W. Ferguson. 1977. Variation and evolution of stereotyped behavior in reptiles. Pp. 335–554. *In* C. Gans and D. W. Tinkle (eds.), "Biology of the Reptilia," Vol. 7. New York: Academic Press.
Cooper, W. E. 1988. Aggressive behavior and courtship rejection in brightly and plainly colored female keeled earless lizards (*Holbrookia propinqua*). *Ethology* 77: 265–278.
Cooper, W. E., and N. Greenberg. 1992. Reptilian coloration and behavior. Pp. 298–422. *In* C. Gans and D. Crews (eds.). See Chap. 5.
Cooper, W. E., and L. J. Vitt. 1985. Lizard pheromones: behavioral responses and adaptive significance in skinks of the genus *Eumeces*. Pp. 323–340. *In* D. Duvall, D. Müller-Schwarze, and R. M. Silverstein (eds.), "Chemical Signals in Vertebrates. 4. Ecology, Evolution, and Comparative Biology." New York: Plenum Press.
Crews, D., and M. C. Moore. 1986. Evolution of mechanisms controlling mating behavior. *Science* **231:** 121–125.
Crump, M. L. 1988. Aggression in harlequin frogs: male–male competition and a possible conflict of interest between sexes. *Anim. Behav.* **36:** 1064–1077.
Drewery, G. E., W. R. Heyer, and A. S. Rand. 1982. A functional analysis of the complex call of the frog *Physalaemus pustulosus. Copeia* **1982:** 636–645.
Duellman, W. E. 1967. Courtship isolating mechanisms in Costa Rican hylid frogs. *Herpetologica* **23:** 169–183.
Duellman, W. E., and L. Trueb. 1986. See Chap. 1.
Duvall, D., L. J. Guillette, and R. E. Jones. 1982. Environmental control of reptilian reproductive cycles. Pp. 201–231. *In* C. Gans and F. H. Pough (eds.). See Chap. 5.
Duvall, D., D. Müller-Schwarze, and R. M. Silverstein (eds.). 1985. "Chemical Signals in Vertebrates. 4. Ecology, Evolution, and Comparative Biology." New York: Plenum Publ. Corp.
Forester, D. C., and D. V. Lykens. 1986. Significance of satellite males in a population of spring peepers (*Hyla crucifer*). *Copeia* **1986:** 719–724.
Frazier, J., and G. Peters. 1981. The call of the Aldabra tortoise (*Geochelone gigantea*). *Amphibia–Reptilia* **2:** 165–179.

Fritzsch, B., M. J. Ryan *et al.* (eds.). 1988. "The Evolution of the Amphibian Auditory System." New York: John Wiley & Sons.
Gans, C. 1973. Sound production in the Salientia: mechanisms and evolution of the emitter. *Am. Zool.* **13:** 1179–1194.
Gans, C., J. C. Gillingham, and D. L. Clark. 1984. Courtship, mating and male combat in tuatara, *Sphenodon punctatus. J. Herpetol.* **18:** 194–197.
Gans, C., and P. F. A. Maderson. 1973. Sound producing mechanisms in recent reptiles: review and comment. *Am. Zool.* **13:** 1195–1203.
Gans, C., and D. W. Tinkle (eds.). 1977. "Biology of the Reptilia," Vol. 7. New York: Academic Press.
Garrick, L. D., J. W. Lang, and H. A. Herzog. 1978. Social signals of adult American alligators. *Bull. Am. Mus. Nat. Hist.* **160:** 153–192.
Gehlbach, F. R., and B. Walker. 1970. Acoustic behavior of the aquatic salamander, *Siren intermedia. BioScience* **20:** 1107–1108.
Gerhardt, H. C. 1988. Acoustic properties used in call recognition by frogs and toads. Pp. 455–483. *In* B. Fritsch *et al.* (eds.). Op. cit.
Gillingham, J. C. 1987. Social behavior. Pp. 184–209. *In* R. A. Seigel *et al.* (eds.). See Chap. 5.
Halliday, T. R. 1978. Sexual selection and mate choice. Pp. 180–213. *In* J. R. Krebs and N. S. Davies (eds.), "Behavioural Ecology: An Evolutionary Approach." Oxford: Blackwell Sci. Publ.
Houck, L. D. 1985. The evolution of salamander courtship pheromones. Pp. 173–190. *In* D. Duvall *et al.* (eds.). Op. cit.
Howard, R. D. 1988. Sexual selection on male body size and mating behaviour in American toads, *Bufo americanus. Anim. Behav.* **36:** 1796–1808.
Lamming, G. E. (ed.). 1984. "Marshall's Physiology of Reproduction. Vol. 1. Reproductive Cycle of Vertebrates." Edinburgh: Churchill Livingstone.
Lang, J. W. Social behavior. Pp. 102–117. *In* C. A. Ross (ed.). See Chap. 3.
Licht, P. 1984. Reptiles. Pp. 206–282. *In* G. E. Lamming (ed.). Op. cit.
Lofts, B. 1984. Amphibians. Pp. 127–205. *In* G. E. Lamming (ed.). Op. cit.
Martin, W. F., and C. Gans. 1972. Muscular control of the vocal tract during release signaling in the toad *Bufo valliceps. J. Morphol.* **137:** 1–28.
Maslin, T. P. 1950. The production of sound in caudate Amphibia. *Univ. Colo. Stud., Ser. Biol.* No. 1: 29–45.
Mason, R. T. 1992. Reptilian pheromones. Pp. 114–228. *In* C. Gans and D. Crews (eds.). See Chap. 5.
Mason, R. T., J. Chinn, and D. Crews. 1987. Sex and seasonal differences in the skin lipids of garter snakes. *Comp. Biochem. Physiol. B* **87B:** 999–1003.
Milton, T. H., and T. A. Jenssen. 1979. Description and significance of vocalizations by *Anolis grahami. Copeia* **1979:** 481–489.
Mitchell, S. L. 1990. The mating system genetically affects offspring performance in Woodhouse's toad (*Bufo woodhousei*). *Evolution* **44:** 502–519.
Moore, M. C., and J. Lindzey. 1992. The physiological basis of sexual behavior in male reptiles. Pp. 70–113. *In* C. Gans and D. Crews (eds.). See Chap. 5.
Norris, D., and R. Jones (eds.). 1987. "Hormones and Reproduction in Fishes, Amphibians, and Reptiles." New York: Plenum Publ. Corp.
Olson, D. H., A. R. Blaustein, and R. K. O'Hara. 1986. Mating pattern variability among western toad (*Bufo boreas*) populations. *Oecologia* **70:** 351–356.
Rand, A. S. 1985. Tradeoffs in the evolution of frog calls. *Proc. Indian Acad. Sci., Anim. Sci.* **94:** 623–637.
Rand, A. S. 1988. An overview of anuran acoustic communication. Pp. 415–431. *In* B. Fritzsch *et al.* (eds.). Op. cit.
Ritke, M. E., and R. D. Semlitsch. 1991. Mating behavior and determinants of male mating success in the gray treefrog, *Hyla chrysoscelis. Can. J. Zool.* **69:** 246–250.

Ryan, M. J. 1985. See Chap. 6.
Ryan, M. J., and A. S. Rand. 1990. The sensory basis of sexual selection for complex calls in the túngara frog, *Physalaemus pustulosus*. *Evolution* **44:** 303–314.
Saint Girons, H. 1982. Reproductive cycles of male snakes and their relationships with climate and female reproductive cycles. *Herpetologica* **38:** 5–16.
Saint Girons, H. 1984. Les cycles sexuels des lézard males et leurs rapports avec le climat et les cycles reproducterus des femelles. *Ann. Sci. Nat., Zool.* **6:** 221–243.
Schwartz, J. J. 1989. Graded aggressive calls of the spring peeper, *Pseudacris crucifer*. *Herpetologica* **45:** 172–181.
Seigel, R. A., and N. B. Ford. 1987. Reproductive ecology. Pp. 210–252. *In* R. A. Seigel *et al.* (eds.). See Chap. 5.
Stamp, J. A. 1977. Social behavior and spacing patterns in lizards. *In* C. Gans and D. W. Tinkle (eds.). Op. cit.
Stamp, J. A. 1983. Sexual selection, sexual dimorphism, and territoriality. Pp. 169–204. *In* R. B. Huey *et al.* (eds.). See Chap. 5.
Vitt, L. J., and J. S. Jacob (covenors). 1982. Reproductive Biology of Reptiles. *Herpetologica* **38:** 1–255.
Wells, K. D. 1977. The social behavior of anuran amphibians. *Anim. Behav.* **25:** 666–693.
Wells, K. D. 1988. The effect of social interaction on anuran vocal behavior. Pp. 433–454. *In* B. Fritsch *et al.* (eds.). Op. cit.
Wells, K. D., and T. L. Taigen. 1989. Calling energetics of a neotropical treefrog *Hyla microcephala*. *Behav. Ecol. Sociobiol.* **25:** 13–22.
Whitaker, R., and D. Basu. 1983. The gharial (*Gavialis gangeticus*): a review. *J. Bombay Nat. Hist. Soc.* **79:** 531–548.
Whittier, J. M, and D. Crews. 1987. Seasonal reproduction: patterns and control. Pp. 385–409. *In* D. O. Norris and R. E. Jones (eds.). Op. cit.
Whittier, J. M., and R. R. Tokarz. 1992. Physiological regulation of sexual behavior in female reptiles. Pp. 24–69. *In* C. Gans and D. Crews (eds.). See Chap. 5.
Woolbright, L. L., and M. M. Stewart. 1987. Foraging success of the tropical frog, *Eleutherodactylus coqui*: the cost of calling. *Copeia* **1987:** 69–75.

CHAPTER 9. Spacing, Movements, and Orientation

Adler, K. 1970. The role of extraoptic photoreceptors in amphibian rhythms and orientation: a review. *J. Herpetol.* **4:** 99–112.
Adler, K. 1976. Extraocular photoreception in amphibians. *Photochem. Photobiol.* **23:** 275–298.
Andrews, R. M., and B. S. Kenney. 1990. Diel patterns of activity and of selected ambient temperature of the sand-swimming lizard *Sphenops sepsoides*. *Isr. J. Zool.* **37**: 65–73.
Ashton, R. E. 1975. A study of movement, home range, and winter behavior of *Desmognathus fuscus*. *J. Herpetol.* **9:** 85–91.
Auffenberg, W. 1978. See Chap. 5.
Auffenberg, W. 1988. See Chap. 5.
Barbour, R. W., M. J. Harvey, and J. W. Hardin. 1969. Home range, movements, and activity of the eastern worm snake, *Carphophis amoenus amoenus*. *Ecology* **50:** 470–476.
Bell, E. L. 1955. An aggregation of salamanders. *Proc. Pa. Acad. Sci.* **24:** 265–66.
Boersma, P. D. 1982. The benefits of sleeping aggregations in marine iguanas, *Amblyrhynchus cristatus*. Pp. 292–299. *In* G. M. Burghardt and A. S. Rand (eds.). See Chap. 5.
Breden, F. 1987. The effect of post-metamorphic dispersal on the population genetic structure of Fowler's toad, *Bufo woodhousei fowleri*. *Copeia* **1987:** 386–395.
Carpenter, C. C. 1953. A study of hibernacula and hibernating associations of snakes and amphibians in Michigan. *Ecology* **34:** 74–80.

Carpenter, C. C. 1957. Hibernation, hibernacula and associated behavior of three toed box turtle (*Terrapene carolina triunguis*). *Copeia* **1957:** 278–282.

Carpenter, F. L. (ed.). 1987. Territoriality: Conceptual Advances in Field and Theoretical Studies. *Am. Zool.* **27:** 223–409.

Congdon, J. D., and R. E. Gatten. 1989. Movements and energetics of nesting *Chrysemys picta*. *Herpetologica* **45:** 94–100.

Crump, M. L. 1986. Homing and site fidelity in a neotropical frog, *Atelopus varius* (Bufonidae). *Copeia* **1986:** 438–444.

Cunningham, J. D. 1960. Aspects of the ecology of the Pacific slender salamander, *Batrachoseps pacificus*, in southern California. *Ecology* **41:** 88–99.

Davies, N. B., and A. I. Houston. 1984. Territory economics. Pp. 148–169. *In* J. R. Krebs and N. B. Davies (eds.), "Behavioral Ecology." Oxford: Blackwell Sci. Publ.

Duellman, W. E. 1967. Courtship isolating mechanisms in Costa Rican hylid frogs. *Herpetologica* **23:** 169–183.

Dundee, H. A., and M. C. Miller. 1968. Aggregative behavior and habitat conditioning by the prairie ringneck snake, *Diadophis punctatus arnyi*. *Tulane Stud. Zool. Bot.* **15:** 41–58.

Duvall, D., M. J. Goode *et al.* 1990. Prairie rattlesnake vernal migration: field experimental analyses and survival value. *Natl. Geogr. Res.* **6:** 457–469.

Duvall, D., M. B. King, and K. J. Gutzwiller. 1985. Behavioral ecology of the prairie rattlesnake. *Natl. Geogr. Res.* **1:** 80–111.

Ernst, C. H. 1976. Ecology of the spotted turtle, *Clemmys guttata* in southeastern Pennsylvania. *J. Herpetol.* **10:** 25–33.

Fellers, G. M., and C. A. Drost. 1991. See Chap. 5.

Ferguson, G. W., J. L. Hughes, and K. L. Brown. 1983. Food availability and territorial establishment of juvenile *Sceloporus undulatus*. Pp. 134–148. *In* R. B. Huey *et al.* (eds.). See Chap. 5.

Forester, D. C. 1985. The recognition and use of chemical signals by a nesting salamander. Pp. 205–219. *In* D. Duvall *et al.* (eds.). See Chap. 8.

Gauthreaux, S. A. (ed.). 1980. "Animal Migration, Orientation, and Navigation." New York: Academic Press.

Gibbons, J. W. (ed.). 1990. See Chap. 7.

Gibbons, J. W., J. L. Greene, and J. D. Congdon. 1990. Temporal and spatial movement patterns of sliders and other turtles. Pp. 201–215. *In* J. W. Gibbons (ed.). See Chap. 7.

Gibbons, J. W., and R. D. Semlitsch. 1987. Activity patterns. Pp. 396–421. *In* R. A. Seigel *et al.* (eds.). See Chap. 5.

Glandt, D. 1986. Die saisonalen Wanderungen der mitteleuropäischen Amphibien. *Bonn. Zool. Beitr.* **37:** 211–228.

Graves, B. M., D. Duvall *et al.* 1985. Initial den location by neonatal prairie rattlesnakes: functions, causes, and natural history in chemical ecology. Pp. 285–304. D. Duvall *et al.* (eds.). See Chap. 8.

Gregory, P. T., J. M. Macartney, and K. W. Larsen. 1987. Spatial patterns and movements. Pp. 366–395. *In* R. A. Seigel *et al.* (eds.). See Chap. 5.

Healy, W. R. 1975. Terrestrial activity and home range in efts of *Notophthalmus viridescens*. *Am. Midl. Nat.* **93:** 131–138.

Heatwole, H. 1977. Habitat selection in reptiles. Pp. 137–155. *In* C. Gans and D. W. Tinkle (eds.). See Chap. 8.

Heatwole, H., and J. Taylor. 1987. See Chap. 5.

Humphries, R. L. 1956. An unusual aggregation of *Plethodon glutinosus* and remarks on its subspecific status. *Copeia* **1956:**122–123.

Huntley, A. C. 1987. Electrophysiological and behavioral correlates of sleep in the desert iguana, *Dipsosaurus dorsalis* Hallowell. *Comp. Biochem. Physiol. A* **86A:** 325–330.

Jaeger, R. G. 1985. Pheromonal markers as territorial advertisement by terrestrial salamanders. Pp. 191–203. *In* D. Duvall *et al.* (eds.). See Chap. 8.

Jameson, D. L. 1955. The population dynamics of the cliff frog, *Syrrhophus marnocki*. *Am. Midl. Nat.* **54:** 342–381.

Johnson, C. R. 1969. Aggregation as a means of water conservation in juvenile *Limnodynastes* from Australia. *Herpetologica* **25:** 275–276.

Joly, J. 1968. Données écologiques sur la salamandre tachetée *Salamandra salamandra*. *Ann. Sci. Nat., Zool.* **10:** 301–366.

Kropach, C. 1971. Sea snake (*Pelamis platurus*) aggregations on slicks in Panama. *Herpetologica* **27:** 131–135.

Lang, J. W. 1976. Amphibious behavior of *Alligator mississippiensis*: roles of a circadian rhythm and light. *Science* **191:** 575–577.

Larsen, K. W. 1987. Movements and behavior of migratory garter snakes, *Thamnophis sirtalis*. *Can. J. Zool.* **65:** 2241–2247.

Lescure, J. 1986. La vie sociale de l'adulte. Pp. 525–537. *In* P.-P. Grassé (ed.). See Chap. 7.

Macartney, J. M., P. T. Gregory, and K. W. Larsen. 1988. A tabular survey of data on movements and home ranges of snakes. *J. Herpetol.* **22:** 61–73.

Madsen, T. 1984. Movements, home range size and habitat use of radio-tracked grass snakes (*Natrix natrix*) in southern Sweden. *Copeia* **1984:** 707–713.

Martin, E., P. Jolly, and P. Bovet. 1989. Diel pattern of activity in the alpine newt (*Triturus alpestris*) during the aquatic phase. *Biol. Behav.* **14:** 116–131.

Martof, B. 1953. Home range and movements of the green frog, *Rana clamitans*. *Ecology* **34:** 529–543.

Meylan, A. B., B. W. Bowen, and J. C. Avise. 1990. A genetic test of the natal homing versus social facilitation models for green turtle migration. *Science* **248:** 724–727.

Noble, G. K., and H. J. Clausen. 1936. The aggregation behavior of *Storeria dekayi* and other snakes with especial reference to the sense organs involved. *Ecol. Monogr.* **6:** 269–316.

Nuland, G. J. v., and H. Strijbosch. 1981. Annual rhythmics of *Lacerta vivipara* Jacquin and *Lacerta agilis agilis* (Sauria, Lacertidae) in the Netherlands. *Amphibia–Reptilia* **2:** 83–95.

Owens, D., D. C. Comuzzie, and M. Grassman. 1985. Chemoreception in the homing and orientation behavior of amphibians and reptiles, with special reference to sea turtles. Pp. 341–355. *In* D. Duvall *et al.* (eds.). See Chap. 8.

Pechmann, J. H. 1986. Diel activity patterns in the breeding migrations of winter-breeding anurans. *Can. J. Zool.* **64:** 1116–1120.

Philips, C. A., and O. J. Sexton. 1989. Orientation and sexual differences during breeding migrations of the spotted salamander *Ambystoma maculatum*. *Copeia* **1989:** 17–22.

Phillips, J. B. 1986. Two magnetoreception pathways in a migratory salamander. *Science* **233:** 765–767.

Rand, A. S. 1968. A nesting aggregation of iguanas. *Copeia* **1968:** 552–561.

Ritke, M. E., and J. G. Babb. 1991. Behavior of the gray treefrog (*Hyla chrysoscelis*) during the nonbreeding season. *Herpetol. Rev.* **22:** 5–6, 8.

Rodda, G. H. 1984. The orientation and navigation of juvenile alligators: evidence of magnetic sensitivity. *J. Comp. Physiol.* **154A:** 649–658.

Rodda, G. H. 1985. Navigation in juvenile alligators. *Z. Tierpsychol.* **68:** 65–77.

Rodda, G. H., and J. B. Phillips. 1992. Navigational systems develop along similar lines in amphibians, reptiles, and birds. *Ethol. Ecol. Evol.* **4:** 43–51.

Rose, B. 1981. Factors affecting activity in *Sceloporus virgatus*. *Ecology* **62:** 706–716.

Rose, F. L., and F. W. Judd. 1975. Activity and home range size of the Texas tortoise, *Gopherus berlandieri*, in south Texas. *Herpetologica* **31:** 448–456.

Ruby, D. E., and A. E. Dunham. 1987. Variation in home range size along an elevational gradient in the iguanid lizard *Sceloporus merriami*. *Oecologia* **71:** 473–480.

Salvador, A. 1987. Actividad del lagarto verdinegro (*Lacerta schreiberi*). *Mediterranea, Ser. Biol.* **9:** 41–56.

Sampedro-M., A., V. Berovides-A., and O. Torres-F. 1982. Hábitos alimentarios y actividad de *Bufo peltocephalus* Tschudi en el Jardín Botánico de Cienfuegos. *Poeyana* **233:** 1–14.
Savage, R. M. 1961. "The Ecology and Life History of the Common Frog." London: Pitman & Sons.
Schubauer, J. P., J. W. Gibbons, and J. R. Spotila. Home range and movement patterns of slider turtles inhabiting Par Pond. Pp. 223–232. *In* J. W. Gibbons (ed.). Op. cit.
Schwartz, E. R., C. W. Schwartz, and A. R. Kiester. 1984. The three-toed box turtle in central Missouri, part II: a nineteen-year study of home range, movements and population. *Mo. Dep. Conserv., Terrest. Ser.* **12:** 1–28.
Sexton, O. J., C. Phillips, and J. F. Bramble. 1990. The effects of temperature and precipitation on the breeding migration of the spotted salamander (*Ambystoma maculatum*). *Copeia* **1990:** 781–787.
Shine, R. 1979. Activity patterns in Australian elapid snakes. *Herpetologica* **35:** 1–11.
Shine, R., and R. Lambeck. 1985. A radiotelemetric study of movements, thermoregulation and habitat utilization of Arafur filesnakes. *Herpetologica* **41:** 351–361.
Sinsch, U. 1990. Migration and orietation in anuran amphibians. *Ethol. Ecol. Evol.* **2:** 65–79.
Sinsch, U. 1991. The orientation behavior of amphibians. *Herpetol. J.* **1:** 541–544.
Stamps, J. A. 1977. Social behavior and spacing patterns in lizards. Pp. 265–334. *In* C. Gans and D. W. Tinkle (eds.). See Chap. 8.
Stamps, J. A. 1983. Sexual selection, sexual dimorphism, and territoriality. Pp. 169–204. *In* R. B. Huey *et al.* (eds.). See Chap. 5.
Thomas, R. 1965. A congregation of the blind snake, *Typhlops richardi. Herpetologica* **21:** 309.
Tiebout, H. M., and J. R. Cary. 1987. Dynamic spatial ecology of the water snake, *Nerodia sipedon. Copeia* **1987:** 1–18.
Twitty, V. C. 1966. "Of Scientists and Salamanders." San Francisco: W. H. Freeman & Co.
Underwood, H. 1992. Endogenous rhythms. Pp. 229–297. *In* C. Gans and D. Crews (eds.). See Chap. 5.
Van den Elzen, P. 1975. Contribution á connaissance de *Pelodytes punctatus* étudié en Camarque. *Bull. Soc. Zool. Fr.* **100:** 691–692.
Wassersug, R. J. 1973. Aspects of social behavior in anuran larvae. Pp. 273–297. *In* J. L. Vial (ed.). See Chap. 1.

CHAPTER 10. Homeostasis: Air, Heat, and Water

Avery, R. A. 1982. Field studies of body temperature and themoregulation. Pp. 93–166. *In* C. Gans and F. H. Pough (eds.), "Biology of the Reptilia," Vol. 12. New York: Academic Press.
Balinsky, J. B. 1981. Adaptations of nitrogen metabolism to hyperosmotic environment in Amphibia. *J. Exp. Zool.* **215:** 335–350.
Bartholomew, G. A. 1982. Physiological control of body temperatures. Pp. 167–211. *In* C. Gans and F. H. Pough (eds.), "Biology of the Reptilia," Vol. 12. New York: Academic Press.
Bennett, A. F. 1982. See Chap. 5.
Bennett, A. F. 1983. Ecological consequences of activity metabolism. Pp. 11–23. *In* R. B. Huey *et al.* (eds.). See Chap. 5.
Bennett, A. F., and R. B. Huey. 1990. Studying the evolution of physiological performance. Pp. 251–284. *In* D. J. Futuyma and J. Antonovics (eds.), "Oxford Surveys in Evolutionary Biology," Vol. 7. Oxford: Oxford Univ. Press.
Bennett, A. F., and H. John-Alder. 1986. Thermal relations of some Australian skinks. *Copeia* **1986:** 57–64.
Bradshaw, S. D. 1986. See Chap. 8.
Brattstrom, B. H. 1963. A preliminary review of the thermal requirements of amphibians. *Ecology* **44:** 238–255.

Brattstrom, B. H. 1979. Amphibian temperature regulation studies in the field and laboratory. *Am. Zool.* **19:** 345–356.

Caldwell, J. P., and P. T. Lopez. 1989. Foam-generating behavior in tadpoles of *Leptodactylus mystaceus*. *Copeia* **1989:** 498–502.

Campbell, J. W., J. E. Vorhaben, and D. D. Smith. 1987. Uricoteley: its nature and origin during the evolution of tetrapod vertebrates. *J. Exp. Zool.* **243:** 349–363.

Christian, K. A., I. E. Clavijo *et al.* 1986. Thermoregulation and energetics of a population of Cuban iguanas (*Cyclura nubila*) on Isla Magueyes, Puerto Rico. *Copeia* **1986:** 65–69.

Costanzo, J. P. 1989. A physiological basis for prolonged submergence in hibernating garter snakes *Thamnophis sirtalis*: evidence for an energy-sparing adaptation. *Physiol. Zool.* **62:** 580–592.

Coulson, R. A. 1987. Aerobic and anaerobic glycolysis in mammals and reptiles *in vivo*. *Comp. Biochem Physiol. B* **87B:** 207–216.

Dawson, W. R., B. Pinshow *et al.* 1989. What's special about the physiological ecology of desert organisms? *J. Arid Environ.* **17:** 131–143.

Duellman, W. E., and L. Trueb. 1986. See Chap. 1.

Ernst, C. H. 1982. Environmental temperatures and activities in wild spotted turtles, *Clemmys guttata*. *J. Herpetol.* **16:** 112–120.

Etheridge, K. 1990. Water balance in estivating sirenid salamanders (*Siren lacertina*). *Herpetologica* **46:** 400–406.

Feder, M. E. 1982. Thermal ecology of neotropical lungless salamanders: environmental temperatures and behavioral responses. *Ecology* **63:** 1665–1674.

Feder, M. E., and W. W. Burggren. 1985. Skin breathing in vertebrates. *Sci. Am.* **253:** 126–142.

Feder, M. E., *et al.* (covenors). 1988. Cutaneous Exchange of Gases and Ions. *Am. Zool.* **28:** 939–1045.

Feder, M. E., J. F. Lynch *et al.* 1982 Field body temperature of tropical and temperate zone salamanders. *Smithson. Herpetol. Inform. Serv.* **52:** 1–23.

Frair, W., R. G. Ackman, and N. Mrosovsky. 1972. Body temperature of *Dermochelys coriacea*: warm turtle from cold water. *Science* **177:** 791–793.

Freda, J. 1986. The influence of acidic pond water on amphibians: a review. *Water Air Soil Pollut.* **30:** 439–450.

Gans, C. 1970. Strategy and sequences in evolution of the external gas exchangers of ectothermal vertebrates. *forma funct.* **3:** 61–104.

Gans, C., and B. D. Clark. 1978. Air flow in reptilian ventilation. *Comp. Biochem. Physiol. A* **60A:** 453–457.

Gatten, R. E. 1985. The uses of anaerobiosis by amphibians and reptiles. *Am. Zool.* **25:** 945–954.

Gatten, R. E. 1987. Cardiovascular and other physiological correlates of hibernation in aquatic and terrestrial turtles. *Am. Zool.* **27:** 59–68.

Gatten, R. E., A. C. Echternacht, and M. A. Wilson. 1988. Acclimatization versus acclimation of activity metabolism in a lizard. *Physiol. Zool.* **61:** 322–329.

Graves, B. M., and D. Duvall. 1987. An experimental study of aggregation and thermoregulation in prairie rattlesnakes (*Crotalus viridis viridis*). *Herpetologica* **43:** 259–264.

Gregory, P. T. 1982. Reptilian hibernation. Pp. 53–154. *In* C. Gans and F. H. Pough (eds.). See Chap. 5.

Heatwole, H. F., and J. Taylor. 1987. See Chap. 5.

Huey, R. B. 1982. Temperature, physiology, and ecology of reptiles. Pp. 25–91. *In* C. Gans and F. H. Pough (eds.). See Chap. 5.

Huey, R. B., and A. F. Bennett. 1987. Phylogenetic studies of coadaptation: preferred temperatures versus optimal performance temperatures of lizards. *Evolution* **41:** 1098–1115.

Huey, R. B., and R. D. Stevenson. 1979. Integrating thermal physiology and ecology of ectotherms: a discussion of approaches. *Am. Zool.* **19:** 357–366.

Hughes, G. M. (ed.). 1976. "Respiration of Amphibious Vertebrates." New York: Academic Press.

Huntley, A. C. Electrophysiological and behavioral correlates of sleep in the desert iguana, *Dipsosaurus dorsalis*. *Comp. Biochem. Physiol. A* **86A:** 325–330.

Hutchison, V. H. 1982. Physiological ecology of the telmatobiid frogs of Lake Titicaca. *Natl. Geogr. Res. Rep.* **14:** 357–361.

Hutchison, V. H., and L. G. Hill. 1976. Thermal selection in the hellbender, *Cryptobranchus alleganiensis*, and the mudpuppy, *Necturus maculosus*. *Herpetologica* **32:** 327–331.

Hutchison, V. H., and J. R. Ritchart. 1989. Annual cycle of thermal tolerance in the salamander, *Necturus maculosus*. *J. Herpetol.* **23:** 73–76.

Johnson, C. R. 1971. Thermal relations in some southern and eastern Australian anurans. *Proc. R. Soc. Queensl.* **82:** 87–94.

Katz, U. 1986. The role of amphibian epidermis in osmoregulation and its adaptive response to changing environment. *In* J. Bereiter-Hahn *et al.* (eds.). See Chap. 1.

Komnick, H. 1986. Chloride cells and salt glands. Pp. 499–516. *In* J. Bereiter-Hahn *et al.* (eds.). See Chap. 1.

Legler, J. M. 1960. Natural history of the ornate box turtle, *Terrapene ornata ornata*. *Mus. Nat. Hist., Univ. Kans. Publ.* No. 11: 527–669.

Lillywhite, H. B., and W. W. Burggren (eds.). 1987. Cardiovascular adaptation in reptiles. *Am. Zool.* **27:** 1–131.

Lillywhite, H. B., and P. F. A. Maderson. 1988. The structure and permeability of integument. *Am. Zool.* **28:** 945–962.

Mautz, W. J. 1982. Patterns of evaporative water loss. Pp. 443–481. *In* C. Gans and F. H. Pough (eds.). See Chap. 5.

Minnich, J. E. 1982. The use of water. Pp. 325–395. *In* C. Gans and F. H. Pough (eds.). See Chap. 5.

Nagy, K. A., and A. A. Degen. 1988. Do desert geckos conserve energy and water by being nocturnal? *Physiol. Zool.* **61:** 495–499.

Packard, G. M., and M. J. Packard. 1990. Patterns of survival at subzero temperatures by hatchling painted turtles and snapping turtles. *J. Exp. Zool.* **254:** 233–236.

Packard, M. J., G. C. Packard, and T. J. Boardman. 1982. Structure of eggshells and water relations of reptilian eggs. *Herpetologica* **38:** 136–155.

Paladino, F. V. (ed.). 1985. Animal energetics: amphibians, reptiles, and birds. *Am. Zool.* **25:** 929–1018.

Paladino, F. V., M. P. O'Connor, and J. R. Spotila. 1990. Metabolism of leatherback turtles, gigantothermy, and thermoregulation of dinosaurs. *Nature* (*London*) **344:** 858–860.

Pierce, B. A. 1985. Acid tolerance in amphibians. *BioScience* **35:** 239–244.

Pough, F. H. 1980. The advantages of ectothermy for tetrapods. *Am. Nat.* No. 115: 92–112.

Putnam, R. W., and A. F. Bennett. 1981. Thermal dependence of behavioural performance of anuran amphibians. *Anim. Behav.* **29:** 502–509.

Seale, D. B. 1987. See Chap. 5.

Sexton, O. J., and K. R. Marion. 1981. Experimental analysis of movements by prairie rattlesnakes, *Crotalus viridis*, during hibernation. *Oecologia* **51:** 37–41.

Seymour, R. S. 1982. Physiological adaptations to aquatic life. Pp. 1–51. *In* C. Gans and F. H. Pough (eds.). See Chap. 5.

Shoemaker, V. H. 1987. Osmoregulation in amphibians. Pp. 109–120. *In* P. Dejours *et al.* (eds.), "Comparative Physiology: Life in Water and on Land." Padua: Liviana Press.

Shoemaker, V. H., M. A. Baker, and J. P. Loveridge. 1989. Effect of water balance on thermoregulation in waterproof frogs (*Chiromantis* and *Phyllomedusa*). *Physiol. Zool.* **62:** 133–146.

Shoemaker, V. H., L. L. McClanahan *et al.* 1987. Thermoregulatory response to heat in the waterproof frogs *Phyllomedusa* and *Chiromantis*. *Physiol. Zool.* **60:** 365–372.

Sievert, L. M., and V. H. Hutchison. 1988. Light versus heat: thermoregulatory behavior in a nocturnal lizard (*Gekko gecko*). *Herpetologica* **44:** 266–273.

Spotila, J. R., and E. A. Standora. 1985. Environmental constraints on the thermal energetics of sea turtles. *Copeia* **1985:** 694–702.

Standora, E. A., J. R. Spotila *et al.* 1984. Body temperatures, diving cycles, and movement of a subadult leatherback turtle, *Dermochelys coriacea. Herpetologica* **40:** 169–176.

Stevenson, R. D. 1985. Body size and limits to the daily range of body temperature in terrestrial ectotherms. *Am. Nat.* No. 125: 102–117.

Stevenson, R. D. 1985. The relative importance of behavioral and physiological adjustments controlling body temperature in terrestrial ectotherms. *Am. Nat.* No. 126: 362–386.

Storey, K. B. 1986. Freeze tolerance in vertebrates: biochemical adaptations of terrestrially hiberating frogs. Pp. 131–138. *In* H. C. Heller *et al.* (eds.), "Living in the Cold. Physiological and Biochemical Adaptations." New York: Elsevier Sci. Publ. Co.

Storey, K. B., and J. M. Storey. 1987. Persistence of freeze tolerance in terrestrially hibernating frogs after spring emergence. *Copeia* **1987:** 720–726.

Storey, K. B. and J. M. Storey. 1988. Freeze tolerance in animals. *Physiol. Rev.* **68:** 27–84.

Swingland, I. R., and J. G. Frazier. 1979. The conflict between feeding and overheating in the Aldabran giant tortoise. Pp. 611–615. *In* C. J. Amlaner and D. W. MacDonald (eds.), "A Handbook on Biotelemetry and Radio Tracking." Oxford: Pergamon Press.

Taigen, T. L., S. B. Emerson, and F. H. Pough. 1982. Ecological correlates of anuran exercise physiology. *Oecologia* **52:** 49–56.

Taigen, T. L., and F. H. Pough. 1985. Metabolic correlates of anuran behavior. *Am. Zool.* **25:** 987–997.

Taigen, T. L., F. H. Pough, and M. M. Stewart. 1984. Water balance of terrestrial anuran (*Eleutherodactylus coqui*) eggs: importance of parental care. *Ecology* **65:** 248–255.

Taplin, L. E. 1988. Osmoregulation in crocodiles. *Biol. Rev.* **63:** 333–337.

Thompson, M. B. 1987. Water exchange in reptilian eggs. *Physiol. Zool.* **60:** 1–8.

Tome, M. A., and F. H. Pough. 1982. Responses of amphibians to acid precipitation. Pp. 245–254. *In* R. E. Johnson (ed.), "Acid Rain and Fisheries," Proc. Int. Symp. Bethesda, Maryland: Am. Fisheries Soc.

Tracy, C. R. 1976. A model of the dynamics exchanges of water and energy between a terrestrial amphibian and its environment. *Ecol. Monogr.* **46:** 293–326.

Turner, J. S. 1987. The cardiovascular control of heat exchange: consequences of body size. *Am. Zool.* **27:** 69–79.

Ultsch, G. R. 1988. Blood gases, hematocrit, plasma ion concentrations, and acid-base status of musk turtles (*Sternotherus odoratus*) during simulated hibernation. *Physiol. Zool.* **61:** 78–94.

Waldschmidt, S. R., *et al.* 1987. See Chap. 5.

Whitford, W. G. 1973. The effects of temperature on respiration in the Amphibia. *Am. Zool.* **13:** 505–512.

Wilson, M. A., and A. C. Echternacht. 1987. Geographic variation in the critical thermal minimum of the green anole, *Anolis carolinensis*, along a latitudinal gradient. *Comp. Biochem. Physiol. A* **87A:** 757–760.

Withers, P. C., S. S. Hillman, and R. C. Drewes. 1984. Evaporative water loss and skin lipids of anuran amphibians. *J. Exp. Zool.* **232:** 11–17.

Wygoda, M. 1989. A comparative study of heating rates in arboreal and nonarboreal frogs. *J. Herpetol.* **23:** 141–145.

CHAPTER 11. Populations Structure and Dynamics

Andrews, R. M. 1979. Evolution of life histories: a comparison of *Anolis* lizards from matched island and mainland habitats. *Breviora* **454:** 1–51.

Andrews, R. M., and J. D. Nichols. 1990. Temporal and spatial variation in survival rates of the tropical lizard *Anolis limifrons. Oikos* **57:** 215–221.

Auffenberg, W. 1978. See Chap. 5.

Ballinger, R. E. 1983. Life-history variations. Pp. 241–260. *In* R. B. Huey *et al.* (eds.). See Chap. 5.

Barbault, R. 1972. Les peuplements d'amphibiens des savanes de Lamto (Cote-d'Ivoire). *Ann. Univ. Abidjan, Ser. E* **5:** 59–142.

Barbault, R. 1975. Les peuplements de lézards des savanes de Lamto (Cote-d'Ivoire). *Ann. Univ. Abidjan, Ser. E* **8:** 147–221.

Barbault, R. 1976. Population dynamics and reproductive patterns of three African skinks. *Copeia* **1976:** 483–490.

Barbault, R. 1987. Rapid-aging in males, a way to increase fitness in a short-lived tropical lizard. *Oikos* **46:** 258–260.

Barbault, R., and Y.-P. Mou. 1988. Population dynamics of the common wall lizard, *Podarcis muralis*, in southwestern France. *Herpetologica* **44:** 38–47.

Barbault, R., and M. T. Rodrigues. 1978. Observations sur la reproduction et la dynamique des populations de quelques anoures tropicaux. II. *Phrynobatrachus plicatus*. *Geo-Eco-Trop.* **2:** 455–466.

Barbault, R., and M. T. Rodrigues. 1979. Op. cit., Chap. 7.

Berven, K. A. 1987. The heritable basis of variation in larval developmental patterns within populations of the wood frog (*Rana sylvatica*). *Evolution* **41:** 1088–1097.

Blaustein, A. R. 1988. Ecological correlates and potential functions of kin recognition and kin association in anuran larvae. *Behav. Genet.* **18:** 449–464.

Bourn, D., and M. Coe. 1978. The size, structure and distribution of the giant tortoise population of Aldabra. *Philos. Trans. R. Soc. London, Ser. B* **282:** 139–175.

Breden, F. 1988. Natural history and ecology of Fowler's toad, *Bufo woodhousei fowleri* in the Indiana Dunes National Lakeshore. *Fieldiana, Zool.* **49:** 1–16.

Brown, W. S., and W. S. Parker. 1984. Growth, reproduction and demography of the racer, *Coluber constrictor mormon*, in northern Utah. *Mus. Nat. Hist., Univ. Kans. Spec. Publ.* No. 10: 13–40.

Bruce, R. C. 1988. An ecological life table for the salamander *Eurycea wilderae*. *Copeia* **1988:** 15–26.

Burton, T. M., and G. E. Likens. 1975. Salamander populations and biomass in the Hubbard Brook Experimental Forest, New Hampshire. *Copeia* **1975:** 541–546.

Caldwell, J. P. 1987. Demography and life history of two species of chorus frogs in South Carolina. *Copeia* **1987:** 114–127.

Case, T. J. 1982. Ecology and evolution of the insular gigantic chuckawallas, *Sauromalus hispidus* and *Sauromalus varius*. Pp. 184–212. *In* G. Burghardt and A. Rand (eds.). See Chap. 5.

Chapleau, F., P. H. Johnansen, and M. Williamson. 1988. The distinction between pattern and process in evolutionary biology: the use and abuse of the term 'strategy.' *Oikos* **53:** 136–138.

Congdon, J. D. 1989. Proximate and evolutionary constraints on energy relations of reptiles. *Physiol. Zool.* **62:** 356–373.

Congdon, J. D., and J. W. Gibbons. 1987. Morphological constraint on egg size: a challenge to optimal egg size theory. *Proc. Natl. Acad. Sci. U.S.A.* **84:** 4145–4147.

Congdon, J. D., and J. W. Gibbons. 1990. The evolution of turtle life histories. Pp. 45–54. *In* J. W. Gibbons (ed.). See Chap. 7.

Danstedt, R. T. 1975. Local geographical variation in demographic parameters and body size of *Desmognathus fuscus*. *Ecology* **56:** 1054–1067.

Dunham, A. E., D. B. Miles, and D. N. Reznick. 1988. Life history patterns in squamate reptiles. Pp. 441–522. *In* C. Gans and R. B. Huey (eds.). See Chap. 6.

Fellers, G. M., and C. A. Drost. 1991. See Chap. 5.

Ferguson, G. W., C. H. Bohlen, and H. P. Woolley. 1980. *Sceloporus undulatus*: comparative life history and regulation of a Kansas population. *Ecology* **61:** 313–322.

Fitch, H. S. 1960. Autecology of the copperhead. *Mus. Nat. Hist., Univ. Kans. Publ.* No. 13: 85–288.

Frazer, N. B., J. W. Gibbons, and J. L. Greene. 1990. See Chap. 7.

Gibbons, J. W., J. L. Greene, and J. D. Congdon. 1983. Drought-related responses of aquatic turtle populations. *J. Herpetol.* **17:** 242–246.
Gill, D. E. 1978. The metapopulation ecology of the red-spotted newt, *Notophthalmus viridescens. Ecol. Monogr.* **48:** 145–166.
Godley, J. S. 1980. Foraging ecology of the striped swamp snake, *Regina alleni*, in southern Florida. *Ecol. Monogr.* **50:** 411–436.
Grenot, C., and R. Vernet. 1973. Sur une population d'*Uromastix acanthinurus* isolée au mileu du Grand Erg Occidental (Sahara algérien). *C. R. Acad. Sci., Ser. D* **276:** 1349–1352.
Hairston, N. G. 1983. Growth, survival and reproduction of *Plethodon jordani*: trade-offs between selective pressures. *Copeia* **1983:** 1024–1035.
Harris, R. N. 1987. An experimental study of population regulation in the salamander, *Notophthalmus viridescens dorsalis. Oecologia* **71:** 280–285.
Heatwole, H. F., and J. Taylor. 1987. See Chap. 5.
Hutchinson, G. E. 1978. "An Introduction to Population Ecology." New Haven: Yale Univ. Press.
Jasieński, M. 1988. Kinship ecology of competition: size hierarchies in kin and nonkin laboratory cohorts of tadpoles. *Oecologia* **77:** 407–413.
Leclair, R., and J. Castanet. 1987. A skeletochronological assessment of age and growth in the frog *Rana pipiens* from southwestern Quebec. *Copeia* **1987:** 361–369.
Mitchell, J. C. 1988. Population ecology and life histories of the freshwater turtles *Chrysemys picta* and *Sternotherus odoratus* in an urban lake. *Herpetol. Monogr.* **2:** 40–61.
Nussbaum, R. A. 1987. Parental care and egg size in salamanders: an examination of the safe harbor hypothesis. *Res. Popul. Ecol.* **29:** 27–44.
Parker, W. S., and M. V. Plummer. 1987. Population ecology. Pp. 253–301. *In* R. A. Seigel *et al.* (eds.). See Chap. 5.
Pechmann, J. H. K., D. E. Scott *et al.* 1991. Declining amphibian populations: the problem of separating human impacts from natural fluctuations. *Science* **253:** 892–895.
Pianka, E. R. 1970. On 'r' and 'K' selection. *Am. Nat.* No. 104: 592–597.
Pianka, E. R. 1983. "Evolutionary Ecology," Third Edition. New York: Harper & Row.
Pilorge, T. 1987. Density, size structure, and reproductive characteristics of three populations of *Lacerta vivipara. Herpetologica* **43:** 345–356.
Plummer, M. V. 1977. Activity, habitat and population structure in the turtle, *Trionyx muticus. Copeia* **1977:** 431–440.
Plummer, M. V. 1985. Demography of green snakes, *Opheodrys aestivus. Herpetologica* **41:** 373–381.
Schoener, T. W., and A. Schoener. 1980. Densities, sex ratios, and population structure in four species of Bahamian *Anolis* lizards. *J. Anim. Ecol.* **49:** 19–35.
Semlitsch, R. D. 1980. Geographic and local variation in population parameters of the slimy salamander *Plethodon glutinosus. Herpetologica* **36:** 6–16.
Shine, R. 1988. Constraints on reproductive investment: a comparison between aquatic and terrestrial snakes. *Evolution* **42:** 17–27.
Sinervo, B., and P. Licht. 1991. Proximate constraints on the evolution of egg size, number, and total clutch mass in lizards. *Science* **252:** 1300–1302.
Stearns, S. C. 1976. Life-history tactics: a review of the ideas. Q. Rev. Biol. **51:** 3–47.
Stearns, S. C. 1977. The evolution of life history traits: a critique on the theory and a review of the data. *Annu. Rev. Ecol. Syst.* **8:** 145–172.
Stewart, M. M., and F. H. Pough. 1983. Population density of tropical forest frogs: relation to retreat sites. *Science* **221:** 570–571.
Swingland, I. R., and M. J. Coe. 1979. The natural regulation of giant tortoises populations on Aldabra Atoll: recruitment. *Philos. Trans. R. Soc. London, Ser. B* **286:** 177–188.
Tilley, S. G. 1980. Life histories and comparative demography of two salamander populations. *Copeia* **1980:** 806–812.

Tinkle, D. W. 1967. See Chap. 7.
Tinkle, D. W., J. D. Congdon, and P. C. Rosen. 1981. Nesting frequency and success: implications for the demography of painted turtles. *Ecology* **62:** 1426–1432.
Turner, F. B. 1977. The dynamics of populations of squamates, crocodilians and rhynchocephalians. Pp. 157–264. *In* C. Gans and D. W. Tinkle (eds.). See Chap. 8.
Van Devender, R. W. 1982. Comparative demography of the lizard *Basiliscus basiliscus*. *Herpetologica* **38:** 189–208.
Vial, J. L. 1968. The ecology of the tropical salamander, *Bolitoglossa subpalmata*, in Costa Rica. *Rev. Biol. Trop.* **15:** 13–115.
Vial, J. L., T. J. Berger, and W. T. McWilliams. 1977. Quantitative demography of copperheads, *Agkistrodon contortrix*. *Res. Popul. Ecol.* **18:** 223–234.
Vitt, L. J. 1986. Reproduction tactics of sympatric gekkonid lizards with a comment on the evolutionary and ecological consequences of invariant clutch size. *Copeia* **1986:** 773–786.
Vitt, L. J., and R. A. Seigel. 1985. Life history traits of lizards and snakes. *Am. Nat.* No. 125: 480–484.
Voris, H. K. 1985. Population size estimates for a marine snake (*Enhydrina schistosa*) in Malaysia. *Copeia* **1985:** 955–961.
Wake, D. B. 1991. Declining amphibian populations. *Science* **253:** 860.
Waldman, B. 1985. Chemical ecology of kin recognition in anuran amphibians. Pp. 225–242. *In* D. Duvall *et al.* (eds.). See Chap. 8.
Waldman, B. 1988. The ecology of kin recognition. *Annu. Rev. Ecol. Syst.* **19:** 543–571.
Wilbur, H. M. 1975. The evolutionary and mathematical demography of the turtle *Chrysemys picta*. *Ecology* **56:** 64–77.
Wilbur, H. M. 1980. Complex life cycles. *Annu. Rev. Ecol. Syst.* **11:** 67–93.
Wilbur, H. M., and P. J. Morin. 1988. Life history evolution in turtles. Pp. 387–439. *In* C. Gans and R. Huey (eds.). See Chap. 6.
Williamson, I., and C. M. Bull. 1989. Life history variation in a population of the Australian frog *Ranidella signifera*: egg size and early development. *Copeia* **1989:** 349–356.
Wilson, E. O., and W. H. Bossert. 1971. "A Primer of Population Biology." Stamford, Connecticut: Sinauer Assoc., Inc.
Zug, G. R., and P. B. Zug. 1979. See Chap. 5.
Zweifel, R. G., and C. H. Lowe. 1966. The ecology of a population of *Xantusia vigilis*, the desert night lizard. *Am. Mus. Novit.* No. 2247: 1–57.

CHAPTER 12. Population and Species Interactions

Arnold, S. J. 1977. Species densities of predators and their prey. *Am. Nat.* No. 106: 220–236.
Barbault, R. 1976. Notes sur la composition et la diversité spécifiques d'une herpétocénose tropicale (Bouaké, Cote d'Ivoire). *Bull. I.F.A.N.* **38**A: 445–456. *Inst. Fondam, Afr. Noire, Ser. A*
Bourquin, O., and S. G. Sowler. 1980. The vertebrates of Vernon Crookes Nature Reserve: 1. *Lammergeyer* **28:** 20–32.
Brown, J. H., and A. C. Gibson. 1983. "Biogeography." St. Louis: C. V. Mosby Co.
Brown, W. S., and W. S. Parker. 1982. Niche dimensions and resource partitioning in a Great Basin Desert snake community. *U.S. Fish Wildl. Serv. Wildl. Res. Rep.* No. 13: 59–81.
Bury, R. B., and P. S. Corn. 1988. Douglas-fir forests in the Oregon and Washington Cascades: abundance of terrestrial herpetofauna. *USDA For. Serv. Tech. Rep.* **RM166:** 11–22.
Bury, R. B., and J. A. Whelan. 1984. Ecology and management of the bullfrog. *U.S. Fish Wildl. Serv. Resour. Publ.* No. 155: 1–23.
Busack, S. D. 1977. Zoogeography of amphibians and reptiles in Cádiz Province, Spain. *Ann. Carnegie Mus.* No. 46: 285–316.

Busack, S. D., and S. B. Hedges. 1984. Is the peninusular effect a red herring? *Am. Nat.* No. 123: 266–275.

Campbell, H. W., and S. P. Christman. 1982. The herpetological components of Florida sandhill and sand pine scrub associations. *U.S. Fish Wildl. Serv. Wildl. Res. Rep.* No. 13: 163–171.

Carothers, J. H., and F. M. Jaksić. 1984. Time as a niche difference: the role of interference competition. *Oikos* **42:** 403–406.

Case, T. J. 1983. Sympatry and size similarity in *Cnemidophorus*. Pp. 297–326. *In* R. Huey *et al.* (eds.). See Chap. 5.

Case, T. J. 1990. Invasion resistance arises in strongly interacting species-rich model competition communities. *Proc. Natl. Acad. Sci. U.S.A.* **87:** 9610–9614.

Cody, M. L., and J. M. Diamond (eds.). 1975. "Ecology and Evolution of Communities." Cambridge: Harvard Univ. Press.

Cortwright, S. A., and C. E. Nelson. 1990. An examination of multiple factors affecting community structure in an an aquatic amphibian community. *Oecologia* **83:** 123–131.

Creusere, F. M., and W. G. Whitford. 1982. Temporal and spatial resource partitioning in a Chihuahuan Desert lizard community. *U.S. Fish Wildl. Serv. Wildl. Res. Rep.* No. 13: 121–127.

Diamond, J., and T. J. Case (eds.). 1986. "Community Ecology." New York: Harper & Row.

Duellman, W. E. 1978. The biology of an equatorial herpetofauna in Amazonian Ecuador. *Univ. Kans. Mus. Nat. Hist Misc. Publ.* No. 65: 1–352.

Duellman, W. E. 1988. Patterns of species diversity in anuran amphibians in the American tropics. *Ann. Mo. Bot. Gard.* **75:** 79–104.

Dunham, A. E. 1983. Realized niche overlap, resource abundance, and intensity of interspecific competition. Pp. 261–280. *In* R. B. Huey *et al.* (eds.). See Chap. 5.

Endler, J. A. 1977. "Geographic Variation, Speciation, and Clines." Princeton: Princeton Univ. Press.

Endler, J. A. (covenor). 1982. Alternate Hypotheses in Biogeography. *Am. Zool.* **22:** 349–471.

Fitch, H. S. 1965. The University of Kansas Natural History Reservation in 1965. *Univ. Kans. Mus. Nat. Hist. Misc. Publ.* No. 42: 1–60.

Fitch, H. S. 1985. Variation in clutch and litter size in New World reptiles. *Univ. Kansas Mus. Nat. Hist. Misc. Publ.* No. 76: 1–76.

Fittkau, E.-J. 1970. Role of caimans in the nutrient regime of mouth-lakes of Amazon affluents (an hypothesis). *Biotropica* **2:** 138–142.

Fittkau, E.-J. 1973. Crocodiles and the nutrient metabolism of Amazonian waters. *Amazonia* **4:** 103–133.

Formanowicz, D. R. 1986. Anuran tadpole/aquatic insect predator-prey interactions: tadpoles size and predator capture success. *Herpetologica* **42:** 367–373.

Gaffney, E. S. 1977. The side-necked turtle family Chelidae: a theory of relationships using shared derived characters. *Am. Mus. Novit.* No. 2620: 1–28.

Gibbons, J. W., and R. D. Semlitsch. 1991. "Guide to the Reptiles and Amphibians of the Savannah River Site." Athens, Georgia: Univ. of Georgia Press.

Hairston, N. G. 1980. The experimental test of an analysis of field distributions: competition in terrestrial salamanders. *Ecology* **61:** 817–826.

Hairston, N. G. 1980. Evolution under interspecific competition. Field experiments on terrestrial salamanders. *Evolution* **34:** 409–420.

Hairston, N. G. 1983. Alpha selection in competing salamanders: experimental verification of an a priori hypothesis. *Am. Nat.* No. 122: 105–113.

Hairston, N. G. 1987. "Community Ecology and Salamander Guilds." Cambridge: Cambridge Univ. Press.

Hairston, N. G., K. C. Nishikawa, and S. L. Stenhouse. 1987. The evolution of competing species of terrestrial salamanders: niche partitioning or interference? *Evol. Ecol.* **1:** 247–262.

Heatwole, H. 1982. A review of structuring in herpetofaunal assemblages. U.S. Fish Wildl. Serv. Wildl. Res. Rep. No. 13: 1–19.

Heatwole, H., and J. Taylor. 1987. See Chap. 5.

Heyer, W. R. 1976. Studies in larval amphibian partitioning. *Smithson. Contrib. Zool.* No. 242: 1–27.

Heyer, W. R., and L. R. Maxson. 1982. Neotropical frog biogeography: paradigms and problems. *Am. Zool.* **22:** 397–410.

Heyer, W. R., A. S. Rand *et al.* 1988. Decimations, extinctions, and colonizations of frog populations in Southeast Brazil and their evolutionary implications. *Biotropica* **20:** 230–235.

Inger, R. F., and R. K. Colwell. 1977. Organization of contiguous communities of amphibians and reptiles in Thailand. *Ecol. Monogr.* **47:** 229–253.

Inger, R. F., H. B. Shaffer *et al.* 1984. A report on a collection of amphibians and reptiles from Ponmudi, Kerala, south India. *J. Bombay Nat. Hist. Soc.* **81:** 406–427, 551–570.

Jaksić, F. M. 1981. Abuse and misuse of the term "guild" in ecological studies. *Oikos* **37:** 397–400.

James, F. C., and W. J. Boecklen. 1984. Interspecific morphological relationships and the density of birds. Pp. 458–471. *In* D. R. Strong, D. Simberloff *et al.* (eds.), "Ecological Communities. Conceptual Issues and the Evidence." Princeton: Princeton Univ. Press.

Laurent, R. F. 1954. Aperçu de la biogéographie des batraciens et des reptiles de la région des grands lacs. *Bull. Soc. Zool. Fr.* **79:** 290–310.

Laurent, R. F., and E. M. Teran. 1981. Lista de los anfibios y reptiles de la Provincia de Tucumán. *Fundac. Miguel Lillo Misc.* **71:** 1–15.

Lloyd, M., R. F. Inger, and F. W. King. 1968. On the diversity of reptile and amphibian species in a Bornean rain forest. *Am. Nat.* No. 102: 497–515.

Loschenkohl, A. 1986. Niche partitioning and competition in tadpoles. *Stud. Herpetol.* **1986:** 399–402.

MacArthur, R. H. 1972. "Geographical Ecology. Patterns in the Distribution of Species." New York: Harper & Row.

Morin, P. J. 1983. Predation, competition, and the composition of larval anuran guilds. *Ecol. Monogr.* **53:** 119–138.

Morin, P. J. 1987. Predation, breeding asynchrony, and the outcome of competition among treefrog tadpoles. *Ecology* **68:** 675–683.

Morton, S. R., and C. D. James. 1988. The diversity and abundance of lizards in arid Australia: a new hypothesis. *Am. Nat.* No. 132: 237–256.

Myers, C. W., and A. S. Rand. 1969. Checklist of amphibians and reptiles of Barro Colorado Island, Panama, with comments on faunal change and sampling. *Smithson. Contrib. Zool.* No. 10: 1–11.

Nussbaum, R. A. 1985. Amphibian fauna of the Seychelles Archipelago. *Natl. Geogr. Soc. Res. Rep.* **1977:** 53–62.

Otte, D., and J. A. Endler (eds.). 1989. "Speciation and Its Consequences." Sunderland, Massachusetts: Sinauer Assoc., Inc.

Pianka, E. R. 1983. See Chap. 11.

Pianka, E. R. 1986. "Ecology and Natural History of Desert Lizards." Princeton: Princeton Univ. Press.

Pough, F. H., E. M. Smith *et al.* 1987. The abundance of salamanders in forest stands with different histories of disturbance. *For. Ecol. Manage.* **20:** 1–9.

Rand, A. S., and S. S. Humphrey. 1968. Interspecific competition in the tropical rain forest: ecological distribution among lizards at Belém, Pará. *Proc. U.S. Natl. Mus.* **125:** 1–17.

Ricklefs, R. E. 1987. Community diversity: relative roles of local and regional processes. *Science* **235:** 167–171.

Roughgarden, J. 1986. Overview: the role of species interactions in community ecology. Pp. 333–343. *In* J. Diamond and T. J. Case (eds.). Op. cit.

Roughgarden, J. 1986. A comparison of food-limited and space-limited animal competition communities. Pp. 492–516. *In* J. Diamond and T. J. Case (eds.). Op. cit.
Roughgarden, J., and S. Pacala. 1989. Taxon cycle among *Anolis* lizard populations: review of evidence. Pp. 403–432. *In* D. Otte and J. A. Endler (eds.). See Chap. 11.
Savage, J. M. 1982. The enigma of the Central America herpetofauna: dispersals or vicarance? *Ann. Mo. Bot. Gard.* **69:** 464–547.
Schoener, T. W. 1986. Overview: kinds of ecological communities—ecology becomes pluralistic. Pp. 467–479. *In* J. Diamond and T. J. Case (eds.). Op. cit.
Schwartz, A., and R. W. Henderson. 1991. "Amphibians and Reptiles of the West Indies." Gainesville, Florida: Univ. of Florida Press.
Scott, N. J. (ed.). 1982. Herpetological Communities. *U.S. Fish Wildl. Serv. Wildl. Res. Rep.* No. 13.
Shaldybin, S. L. 1981. Wintering and number of amphibians and reptiles of Lazo State Nature Reserve. Pp. 123–124. *In* L. J. Borkin (ed.), "Herpetological Investigations in Siberia and the Far East." Leningrad: Acad. Sci. USSR, Zool. Inst.
Simbleroff, D. 1988. The contribution of population community biology to conservation science. *Annu. Rev. Ecol. Syst.* **19:** 473–511.
Simbotwe, M. P., and G. R. Friend. 1985. Comparison of the herpetofaunas of tropical wetland habitats from Lochinvar National Park, Zambia and Kadadu National Park, Australia. *Proc. Ecol. Soc. Aust.* **14:** 141–151.
Soulé, M. E. (ed.). 1986. "Conservation Biology: The Science of Scarcity and Diversity." Sunderland, Massachusetts: Sinauer Assoc., Inc.
Strong, D. R., D. Simberloff *et al.* (eds.). 1984. "Ecological Communities. Conceptual Issues and the Evidence." Princeton: Princeton Univ. Press.
Tiedemann, F. (ed.). 1990. "Lurche und Kriechtiere Wiens." Vienna: J & V Edition Wien.
Toft, C. A. 1985. Resource partitioning in amphibians and reptiles. *Copeia* **1985:** 1–21.
Vitt, L. J. 1987. Communities. Pp. 335–365. *In* R. A. Seigel *et al.* (eds.). See Chap. 5.
White, M. J. D. 1978. "Modes of Speciation." San Francisco: W. H. Freeman.
Wilbur, H. M. 1984. Complex life cycles. *Annu. Rev. Ecol. Syst.* **11:** 67–93.
Wilbur, H. M. 1984. Complex life cycles and community organization in amphibians. Pp. 195–224. *In* P. W. Price *et al.* (eds.), "A New Ecology. Novel Approaches to Interactive Systems." New York: John Wiley & Sons.
Wilbur, H. M. 1987. Regulation of structure in complex systems: experimental temporary pond communities. *Ecology* **68:** 1437–1452.
Wiley, E. O. 1988. Vicariance biogeography. *Annu. Rev. Ecol. Syst.* **19:** 513–542.
Williams, E. E. 1983. Ecomorphs, faunas, island size, and diverse end points in island radiation of *Anolis*. Pp. 326–370. *In* R. Huey *et al.* (eds.). See Chap. 5.
Woinarski, J. C. Z. 1989. The vertebrate fauna of broombush *Melaleuca unicinata* vegetation in northwestern Victoria. *Aust. Wildl. Res.* **16:** 217–238.

CHAPTER 13. Systematics: Theory and Practice

Abbott, L. A., R. A. Bisby, and D. J. Rogers. 1985. "Taxonomic Analysis in Biology. Computers, Models, and Databases." New York: Columbia Univ. Press.
Buth, D. G. 1984. The application of electrophoretic data in systematic studies. *Annu. Rev. Ecol. Syst.* **15:** 501–522.
"Catalogue of American Amphibians and Reptiles." 1963 et seq. Oxford, Ohio: Soc. Study Amphibians & Reptiles.
de Queiroz, K., and M. J. Donoghue. 1988. Phylogenetic systematics and the species problem. *Cladistics* **4:** 317–338.

Goodman, M., M. M. Miyamoto, and J. Czelusniak. 1987. Pattern and process in vertebrate phylogeny revealed by coevolution of molecules and morphologies. Pp. 141–176. *In* C. Patterson (ed.). See Chap. 2.

Green, D. M., and S. K. Sessions (eds.) 1990. "Amphibian Cytogenetics and Evolution." New York: Academic Press.

Guttman, S. I. 1985. Biochemical studies of anuran evolution. *Copeia* **1985:** 292–309.

Hennig, W. 1966. "Phylogenetic Systematics." Chicago: Univ. of Chicago Press.

Hillis, D. M., A. Larson, *et al.* 1990. Nucleic acid III: sequencing. Pp. 318–370. *In* D. M. Hillis and C. Moritz (eds.), "Molecular Systematics." Sunderland, Massachusetts: Sinauer Assoc., Inc.

Hillis, D. M., and C. Moritz (eds.). 1990. "Molecular Systematics." Sunderland, Massachusetts: Sinauer Assoc., Inc.

International Commission of Zoological Nomenclature. 1985. "International Code of Zoological Nomenclature." London: Int. Trust Zool. Nomencl.

James, F. C., and C. E. McCullough. 1985. Data analysis and the design of experiments in ornithology. Pp. 1–63. *In* R. F. Johnston (ed.), "Current Ornithology," Vol. 2. New York: Plenum Publ. Corp.

James, F. C., and C. E. McCullough. 1990. Multivariate analysis in ecology and systematics: panacea or Pandora's box? *Annu. Rev. Ecol. Syst.* **21:** 129–166.

Kachigan, S. K. 1982. "Multivariate Statistical Analysis. A Conceptual Introduction." New York: Radius Press.

Kluge, A. G., and J. S. Farris. 1969. Quantitatives phyletics and the evolution of anurans. *Syst. Zool.* **18:** 1–32.

Maxson, L. R., and R. D. Maxson. 1990. Proteins II. Immunological techniques. Pp. 127–155. *In* D. M. Hillis and C. Moritz (eds.). Op. cit.

Maxson, R. D., and L. R. Maxson. 1986. Micro-complement fixation: a quantitative estimator of protein evolution. *Mol. Biol. Evol.* **3:** 375–388.

Mayr, E., E. G. Linsley, and R. L. Usinger. 1953. "Methods and Principles of Systematic Zoology." New York: McGraw-Hill Book Co.

Mertens, R., and H. Wermuth. 1960. "Die Amphibien und Reptilien Europas." Frankfurt am Main: Verlag Waldemar Kramer.

Moritz, C., T. E. Dowling, and W. M. Brown. 1987. Evolution of animal mitochondrial DNA: relevance for population biology and systematics. *Annu. Rev. Ecol. Syst.* **18:** 269–292.

Murphy, R. W., J. W. Sites *et al.* 1990. Proteins I: isozyme electrophoresis. Pp. 45–126. *In* D. M. Hillis and C. Moritz (eds.). Op. cit.

Oosterbroek, P. 1987. More appropriate definitions of paraphyly and polyphyly, with a comment on the Farris 1974 model. *Syst. Zool.* **36:** 103–108.

Richardson, B. J., P. R. Baverstock, and M. Adams. 1986. "Allozyme Electrophoresis. A Handbook for Animal Systematics and Population Studies." Academic Press.

Schmitt, M. 1989. Claims and limits of phylogenetic systematics. *Z. Zool. Syst. Evol.-Forsch.* **27:** 181–190.

Simpson, G. G. 1961. "Principles of Animal Taxonomy." New York: Columbia Univ. Press.

Sneath, P. H. A., and R. R. Sokal. 1973. "Numerical Taxonomy." San Francisco: W. H. Freeman & Co.

Sokal, R. R. 1986. Phenetic taxonomy: theory and methods. *Annu. Rev. Ecol. Syst.* **17:** 423–442.

Templeton, A. R. 1989. The meaning of species and speciation: a genetic perspective. Pp. 3–27. *In* D. Otte and J. A. Endler (eds.). See Chap. 12.

Thorpe, R. S. 1976. Biometric analysis of geographic variation and racial affinities. *Biol. Rev.* **51:** 407–452.

Thorpe, R. S. 1987. Geographic variation: a synthesis of cause, data, pattern and congruence in analysis and phylogenesis. *Boll. Zool.* **54:** 3–11.

Watrous, L. E., and Q. D. Wheeler. 1981. The outgroup comparison method of character analysis. *Syst. Zool.* **30:** 1–11.

Werman, S. D., M. S. Springer, and R. J. Britten. 1990. Nucleic acids I: DNA-DNA hybridization. Pp. 204–249. *In* D. M. Hillis and C. Moritz (eds.). Op. cit.

Wiley, E. O. 1979. Ventral gill arch muscles and the interrelationships of gnathostomes, with a new classification of the Vertebrata. *Zool. J. Linn. Soc.* **67:** 149–179.

Wiley, E. O. 1981. "Phylogenetics. The Theory and Practice of Phylogenetic Systematics." New York: John Wiley & Sons.

Wiley, E. O., *et al.* 1991. The compleat Cladist. A primer of phylogenetic procedures. *Univ. Kans. Mus. Nat. Hist. Spec. Publ.* No. 19: 1–158.

CHAPTER 14. Caecilians and Salamanders

Adler, K., and E. Zhao. 1990. Studies on hynobiid salamanders, with description of a new genus. *Asiat. Herpetol. Res.* **3:** 37–45.

Arnold, E. N., and J. A. Burton. 1978. "A Field Guide to the Reptiles and Amphibians of Britain and Europe." London: Collins.

Arnold, S. J. 1977. See Chap. 8.

Conant, R., and J. T. Collins. 1991. "A Field Guide to Reptiles and Amphibians. Eastern and Central North America." Boston: Houghton Mifflin.

Dowling, H. G., and W. E. Duellman. 1978. See Chap. 3.

Dubois, A. 1984. Miscellanea nomenclatorica batrachologica (IV). *Alytes* **3**(3): 103–110.

Duellman, W. E., and L. Trueb. 1986. See Chap. 1.

Durand, J. P. 1983. Données et hypothéses sur l'évolution des Proteidae. *Bull. Soc. Zool. Fr.* **108:** 617–630.

Edwards, J. L. 1976. Spinal nerves and their bearing on salamander phylogeny. *J. Morphol.* **148:** 305–328.

Engelmann, W.-E., J. Fritsche, R. Günther, and F. J. Obst. 1986. "Lurche und Kriechtiere Europas." Stuttgart: F. Enke Verlag.

Estes, R. 1981. See Chap. 2.

Frost, D. R. (ed.). 1985. "Amphibians Species of the World." Lawrence, Kansas: Allen Press & Assoc. Syst. Collns.

Good, D. A. 1989. Hybridization and cryptic species in *Dicamptodon*. *Evolution* **43:** 728–744.

Good, D. A., G. Z. Wurst, and D. B. Wake. 1987. Patterns of geographic variation in allozymes of the Olympic salamander, *Rhyacotriton olympicus*. *Fieldiana, Zool.* **32:** 1–15.

Hecht, M. K., and J. L. Edwards. 1977. The methodology of phylogenetic inference above the species level. Pp. 3–51. *In* M. K. Hecht et al. (eds.), "Major Patterns in Vertebrate Evolution." New York: Plenum Publ. Corp.

Larson, A. 1991. A molecular perspective on the evolutionary relationships of the salamander families. *Evol. Biol.* **25:** 211–277.

Lombard, R. E., and D. B. Wake. 1986. Tongue evolution in the lungless salamanders, Family Plethodontidae. IV. Phylogeny of plethodontid salamanders and the evolution of feeding dynamics. *Syst. Zool.* **35:** 532–551.

Naylor, B. G. 1980. Radiation of the Amphibia Caudata: are we looking too far into the past? *Evol. Theory* **5:** 119–126.

Nickerson, M. A., and C. E. May. 1973. The hellbenders: North American "Giant Salamanders." *Milwaukee Public Mus., Publ. Biol. Geol.* **1:** 1–106.

Nussbaum, R. A. 1979. The taxonomic status of the caecilian genus *Uraeotyphlus* Peters. *Occas. Pap. Mus. Zool. Univ. Mich.* No. 687: 1–20.

Nussbaum, R. A., and M. Wilkinson. 1989. On the classification and phylogeny of caecilians (Amphibia: Gymnophiona), a critical review. *Herpetol. Monogr.* **3:** 1–42.

Pfingsten, R. A., and F. L. Downs. 1989. Salamanders of Ohio. *Bull. Ohio Biol. Surv.* **7:** 1–315.
Reagan, N. L., and P. A. Verrell. 1991. The evolution of plethodontid salamanders: did terrestrial mating facilitate lunglessness? *Am. Nat.* No. 138: 1307–1313.
Ruben, J. A., and A. J. Boucot. 1989. The origin of the lungless salamanders (Plethodontidae). *Am. Nat.* **134:** 161–169.
Sessions, S. K., and J. E. Wiley. 1985. Chromosome evolution in salamanders of the genus *Necturus. Brimleyana* **10:** 37–52.
Tanaka, K. 1989. Mating strategy of male *Hynobius nebulosus.* Pp. 437–448. *In* M. Matsui *et al.* (eds.), "Current Herpetology in East Asia." Kyoto: Herpetol. Soc. Japan.
Taylor, E. H. 1968. "The Caecilians of the World. A Taxonomic Review." Lawrence, Kansas: Univ. of Kansas Press.
Wake, D. B. 1987. Adaptive radiation of salamanders in Middle American cloud forests. *Ann. Mo. Bot. Gard.* **74:** 242–264.
Wake, D. B., and N. Ozeti. 1969. Evolutionary relationships in the family Salamandridae. *Copeia* **1969:** 124–137.
Wake, M. H. 1977. The reproductive biology of caecilians: an evolutionary perspective. Pp. 73–101. *In* D. H. Taylor and S. I. Guttman (eds.). See Chap. 8.
Wake, M. H. 1986. A perspective on the systematics and morphology of the Gymnophiona (Amphibia). *Mém. Soc. Zool. Fr.* **43:** 21–38.
Zhao, E., Q. Hu *et al.* 1988. "Studies on Chinese Salamanders." Oxford, Ohio: Soc. Study Amphibians Reptiles.

CHAPTER 15. Frogs

Amiet, J. L. 1990. Images d'amphibiens camerounais. II. L'enfouissement et la phonation bouche ouverte chez *Conraua crassipes. Alytes* **8:** 99–104.
Arnold, E. N., and J. A. Burton. 1978. See Chap. 14.
Beebee, T. J. C. 1983. "The Natterjack Toad." Oxford: Oxford Univ. Press.
Bell, B. D. 1985. See Chap. 7.
Cannatella, D. C. 1986. A new genus of bufonid from South America, and phylogenetic relationships of the Neotropical genera. *Herpetologica* **42:** 197–205
Cannatella, D. C., and L. Trueb. 1988. Evolution of pipoid frogs: intergeneric relationships of the aquatic frog family Pipidae (Anura). *Zool. J. Linn. Soc.* **94:** 1–38.
Cannatella, D. C., and L. Trueb. 1988. Evolution of pipoid frogs: morphology and phylogenetic relationships of *Pseudhymenochirus. J. Herpetol.* **22:** 439–456.
Cogger, H. G. 1983. "Reptiles and Amphibians of Australia." Sanibel, Florida: Ralph Curtis Books.
Conant, R., and J. T. Collins. 1991. See Chap. 14.
Cruz, C. A. G. 1990. Sobre as relaçoes intergenéricas de Phyllomedusinae da floresta atlantica. *Rev. Bras. Biol.* **50:** 709–726.
Deuchar, E. M. 1975. "*Xenopus*: The South African Clawed Frog." New York: John Wiley & Sons.
Dowling, H. G., and W. E. Duellman. 1974–1978. See Chap. 3.
Drewes, R. C. 1984. A phylogenetic analysis of Hyperoliidae (Anura): treefrogs of Africa, Madagascar, and the Seychelles Islands. *Occas. Pap. Calif. Acad. Sci.* No. 139: 1–70.
Dubois, A. 1981. Liste des genres et sous-genres nominaux de Ranoidea du monde, avec identification de leurs espéces-types: conséquences nomenclaturales. *Monit. Zool. Ital. NS* **15** (N.S. Suppl. 15): 225–284.
Dubois, A. 1984. La nomenclature supragénérique des amphibiens anoures. *Mem. Mus. Natl. Hist. Natur., Zool.* **131:** 1–64.
Dubois, A. 1986. Miscellanea taxinomica batrachologica (I). *Alytes* **5:** 7–95.
Dubois, A. 1986. Living amphibians of the world: a first step towards a comprehensive checklist. *Alytes* **5:** 99–149.

Duellman, W. E., and L. Trueb. 1986. See Chap. 1.

Emerson, S. B. 1988. The giant tadpole of *Pseudis paradoxa*. *Biol. J. Linn. Soc.* **34:** 93–104.

Engelmann, W.-E., *et al.* 1986. See Chap. 14.

Frost, D. R. (ed.). 1985. See Chap. 14.

Gill, B. 1986. "Collins Handguide to the Frogs and Reptiles of New Zealand." Auckland: Collins.

Green, D. M., T. E. Sharbel *et al.* 1989. Genetic variation in the genus *Leiopelma* and relationships to other primitive frogs. *Z. Zool. Syst. Evol.-Forsch.* **27:** 65–79.

Guibé, J. 1978. Les batraciens de Madagascar. *Bonn. Zool. Monogr.* **11:** 1–140.

Heyer, W. R. 1975. A preliminary analysis of the intergeneric relationships of the frog family Leptodactylidae. *Smithson. Contrib. Zool.* **199:** 1–55.

Heyer, W. R., and D. S. Liem. 1976. Analysis of the intergeneric relationships of the Australian frog family Myobatrachidae. *Smithson. Contrib. Zool.* No. 233: 1–29.

Heyer, W. R., A. S. Rand *et al.* 1990. Frogs of Boracéia. *Arq. Zool.* **31:** 231–410.

Hutchinson, M. N., and L. R. Maxson. 1987. Biochemical studies on the relationships of the gastricbrooding frogs, genus *Rheobatrachus*. *Amphibia–Reptilia* **8:** 1–11.

Hutchinson, M. N., and L. R. Maxson. 1987. Phylogenetic relationships among Australian tree frogs: an immunological approach. *Aust. J. Zool.* **35:** 61–74.

Inger, R. F. 1966. The systematics and zoogeography of the Amphibia of Borneo. *Fieldiana, Zool.* **52:** 1–402.

Lamotte, M., and J. Lescure. 1989. Les tetards rhéophiles et hygropetriques de l'Ancien et du Nouveau Monde. *Ann. Sci. Nat., Zool.* **10:** 125–144.

Laurent, R. F. 1979. Exquisse d'une phylogenése des anoures. *Bull. Soc. Zool. Fr.* **104:** 397–422.

Laurent, R. F. 1984. La phylogenese des Ranoidea et le cladisme. *Alytes* **3:** 97–101.

Liem, S. S. 1970. The morphology, systematics, and evolution of the Old World treefrogs. (Rhacophoridae and Hyperoliidae). *Fieldiana, Zool.* **57:** 1–145.

Lynch, J. D. 1973. The transition from archaic to advanced frogs. Pp. 133–182. *In* J. A. Vial (ed.). See Chap. 1.

Maxson, L. R., and C. H. Daugherty. 1980. Evolutionary relationships of the monotypic toad family Rhinophrynidae: a biochemical perspective. *Herpetologica* **36:** 275–280.

Maxson, L. R., and C. W. Myers. 1985. Albumin evolution in tropical poison frogs (Dendrobatidae): a preliminary report. *Biotropica* **17:** 50–56.

McCranie, J. R., L. D. Wilson, and K. L. Williams. 1989. A new genus and species of toad (Anura: Bufonidae) with an extraordinary stream-adapted tadpole from norther Honduras. *Occas. Pap. Mus. Nat. Hist. Univ. Kans.* No. 129: 1–18.

Myers, C. W. 1987. New generic names for some neotropical poison frogs. *Pap. Avul. Zool.* **36:** 301–306.

Nussbaum, R. A. 1980. Phylogenetic implications of amplectic behavior in sooglossid frogs. *Herpetologica* **36:** 1–5.

Nussbaum, R. A., E. D. Brodie, and R. M. Storm. 1983. "Amphibians and Reptiles of the Pacific Northwest." Moscow, Idaho: Univ. of Idaho Press.

Passmore, N. I., and V. C. Carruthers. 1979. "South African Frogs." Johannesburg: Witwatersrand Univ. Press.

Perret, J.-L. 1988. Sur quelques genres d'Hyperoliidae (Anura) restés en question. *Bull. Soc. Neuchatelise Sci. Nat.* **111:** 35–48.

Ruiz-C., P. M., and J. D. Lynch. 1991. Ranas Centrolenidae de Colombia I. Propuesta de una nueva clasificación genérica. *Lozania* **57:** 1–30.

Sokol, O. M. 1975. The phylogeny of the anuran larvae: a new look. *Copeia* **1975:** 1–24.

Sokol, O. M. 1977. A subordinal classification of frogs. *J. Zool.* **182:** 505–508.

Tyler, M. J. 1976. "Frogs." Sydney: William Collins, Ltd.

Tyler, M. J. 1989. "Australian Frogs." Ringwood, Victoria: Viking O'Neil.

Wake, M. H. 1978. The reproductive biology of *Eleutherodactylus jasperi*, with comments on the evolution of live-bearing systems. *J. Herpetol.* **12:** 121–133.
Zweifel, R. G. 1986. A new genus and species of microhylid frog from the Cerro de la Neblina region of Venezuela and a discussion of relationships among New World microhylid genera. *Am. Mus. Novit.* No. 2863: 1–24.

CHAPTER 16. Turtles and Crocodilians

Benton, M. J., and J. M. Clark. 1988. See Chap. 2.
Bickham, J. W., and J. L. Carr. 1983. Taxonomy and phylogeny of the higher categories of cryptodiran turtles based on a cladistic analysis of chromosomal data. *Copeia* **1983:** 918–932.
Buffetau, E. 1985. The place of *Gavialis* and *Tomistoma* in eusuchian evolution: a reconciliation of palaeontological and biochemical data. *Neue Jahrb. Geol. Palaeontol. Monatsh.* **12:** 707–716.
Chen, B. 1990. The past and present situation of the Chinese alligator. *Asiat. Herpetol. Res.* **3:** 129–136.
Densmore, L. D., and P. S. White. 1991. The systematics and evolution of the Crocodilia as suggested by restriction endonuclease analysis of mitochondrial and nuclear ribosomal DNA. *Copeia* **1991:** 602–615.
Ernst, C. H., and R. W. Barbour. 1972. "Turtles of the United States." Lexington, Kentucky: Univ. of Kentucky Press.
Ernst, C. H., and R. W. Barbour. 1989. See Chap. 3.
Gaffney, E. S. 1984. Historical analysis of theories of chelonian relationships. *Syst. Zool.* **33:** 283–301.
Gaffney, E. S., J. H. Hutchison *et al.* 1987. Modern turtle origins: the oldest known cryptodire. *Science* **237:** 289–291.
Gaffney, E. S., and P. A. Meylan. 1988. A phylogeny of turtles. Pp. 157–219. *In* M. J. Benton (ed.). See Chap. 2.
Iverson, J. B. 1990. Phylogenetic hypotheses for the evolution of modern Kinosterninae turtles. *Herpetol. Monogr.* No. 4: 1–27.
Iverson, J. B. 1992. Species richness maps of the freshwater and terrestrial turtles of the world. *Smithson. Herpetol. Inform. Serv.* **88:** 1–18.
Iverson, J. B. 1992. "A Revised Checklist with Distribution Maps of the Turtles of the World." Richmond, Indiana: Privately printed.
King, F. W., and R. L. Burke. 1989. "Crocodilians, Tuatara, and Turtle Species of the World." Washington, D.C.: Assoc. Syst. Colln.
Legler, J. M. 1985. Australian chelid turtles: reproductive patterns in wide-ranging taxa. Pp. 117–123. *In* G. Grigg *et al.* (eds.). See Chap. 7.
Meylan, P. A. 1987. The phylogenetic relationships of soft-shelled turtles (family Trionychidae). *Bull. Am. Mus. Nat. Hist.* **186:** 1–101.
Meylan, P. A., and E. S. Gaffney. The skeletal morphology of the Cretaceous cryptodiran turtle, *Adocus*, and the relationships of the Trionychoidea. *Am. Mus. Novit.* No. 2941: 1–60.
Mrosovsky, N. 1983. "Conserving Sea Turtles." London: British Herpetol. Soc.
Pritchard, P. C. H., and P. Trebbau. 1984. "The Turtles of Venezuela." Athens, Ohio: Soc. Study Amphibians Reptiles.
Ross, C. A. (ed.). 1989. See Chap. 8.
Symposium. 1989. Biology of the Crocodilia. *Am. Zool.* **29:** 823–1054.
Tarsitano, S. F., E. Frey, and J. Riess. 1989. The evolution of the Crocodilia: a conflict between morphological and biochemical data. *Am. Zool.* **29:** 843–856.
Whitaker, R., and D. Basu. 1983. See Chap. 8.
Zangerl, R. 1969. The turtle shell. Pp. 311–339. *In* C. Gans *et al.* (eds.). See Chap. 3.

CHAPTER 17. Lizards, Amphisbaenians, and Tuataras

Arnold, E. N., and J. A. Burton. 1978. See Chap. 14.

Auffenberg, W. 1981. See Chap. 7.

Auffenberg, W. 1988. See Chap. 5.

Bauer, A. M. 1990. Phylogenetic systematics and biogeography of the Carphodactylini. *Bonn. Zool. Monogr.* **30:** 1–28.

Beck, D. D., and C. H. Lowe. 1991. Ecology of the beaded lizard, *Heloderma horridum*, in a tropical dry forest in Jalisco, México. *J. Herpetol.* 25: 395–406.

Bezy, R. L. 1988. The natural history of the night lizards, family Xantusiidae. Pp. 1–12. *In* H. F. De Lisle *et al.* (eds.), *Proc. Conf. Calif. Herpetol.* Van Nuys, California Southwestern: Herpetol. Soc.

Böhme, W. 1989. Zur systematishchen Stellung der Amphisbäen, mit besonderer Berüchsichtigung der Morphologie des Hemipenis. *Z. Zool. Syst. Evol.-Forsch.* **27:** 330–337.

Branch, B. 1988. "Field Guide to the Snakes and Other Reptiles of Southern Africa." Sanibel, Florida: Ralph Curtis Books.

Brygoo, E.-R. 1971. Reptiles sauriens Chamaeleonidae genre *Chamaeleo. Faune Madagascar* **33:** 1–318.

Brygoo, E.-R., and R. Roux-Estéve. 1983. *Feylinia*, genre de lézards africains de la famille des Scincidae, sous-famille des Feyliniinae. *Bull. Mus. Natl. Hist. Nat.* Ser, 4ᵉ 5: 307–341.

Burghart, G. M., and A. S. Rand. 1982. See Chap. 5.

Crother, B. I., M. M. Miyamoto, and W. F. Presch. 1986. Phylogeny and biogeography of the lizard family Xantusiidae. *Syst. Zool.* **35:** 37–45.

Daugherty, C. H., A. Cree *et al.* 1990. Neglected taxonomy and continuing extinctions of tuatara (*Sphenodon*). *Nature* (*London*) **347:** 177–179.

Dowling, H. G., and W. E. Duellman. 1978. See Chap. 3.

Engelmann, W.-E., *et al.* 1986. See Chap. 14.

Estes, R., K. de Queiroz, and J. Gauthier. 1988. Phylogenetic relationships within Squamata. Pp. 119–281. *In* R. Estes and G. Pregill (eds.). See Chap. 3.

Etheridge, R., and K. de Queiroz. 1988. A phylogeney of Iguanidae. Pp. 283–367. *In* R. Estes and G. Pregill (eds.). See Chap. 3.

Fitch, H. S. 1970. See Chap. 7.

Frost, D. R., and R. Etheridge. 1989. A phylogenetic analysis and taxonomy of iguanian lizards. *Misc. Publ. Mus. Nat. Hist., Univ. Kans.* No. 81: 1–65.

Gans, C. 1978. The characteristics and affinities of the Amphisbaenia. *Trans. Zool. Soc. London* **34:** 347–416.

Gill, B. 1986. See Chap. 15.

Good, D. A. 1987. An allozyme analysis of anguid subfamilial relationships. *Copeia* **1987:** 696–703.

Good, D. A. 1987. A phylogenetic analysis of cranial osteology in the gerrhonotine lizards. *J. Herpetol.* **21:** 285–297.

Greer, A. E. 1970. A subfamilial classification of scincid lizards. *Bull. Mus. Comp. Zool.* **139:** 151–184.

Greer, A. E. 1985. The relationships of the lizard genera *Anelytropsis* and *Dibamus. J. Herpetol.* **19:** 116–156.

Greer, A. E. 1989. "The Biology and Evolution of Australian Lizards." Chipping Norton, New South Wales: Surrey Beatty & Son, Ltd.

Grismer, L. L. 1988. Phylogeny, taxonomy, classification, and biogeography of eublepharid geckos. Pp. 369–469. *In* R. Estes and G. Pregill (eds.). See Chap. 3.

Hoogmoed, M. S. 1973. Notes on the herpetofauna of Surinam IV. The lizards and amphisbaenians of Surinam. *Biogeographica* **4:** 1–419.

Joger, U. 1991. A molecular phylogeny of agamid lizards. *Copeia* **1991:** 616–622.

King, M. 1987. Monophyleticism and polyphyleticism in the Gekkonidae: a chromosomal perspective. *Aust. J. Zool.* **35:** 641–654.

Kluge, A. G. 1976. Phylogenetic relationships in the lizard family Pygopodidae: an evaluation of theory, methods and data. *Misc. Publ. Mus. Zool. Univ. Mich.* No. 152: 1–72.

Kluge, A. G. 1987. Cladistic relationships in the Gekkonoidea. *Misc. Publ. Mus. Zool. Univ. Mich.* No. 373: 3–54.

Lang, M. 1991. Generic relationships within Cordyliformes. *Bull. Inst. R. Sci. Nat. Belg., Biol.* **61:** 121–188.

Losos, J. B., and H. W. Greene. 1988. Ecological and evolutionary implications of diet in monitor lizards. *Biol. J. Linn. Soc.* **35:** 379–407.

Lowe, C. H., C. R. Schwalbe, and T. B. Johnson. 1986. "The Venomous Reptiles of Arizona." Phoenix: Arizona Game & Fish Dep.

McCoy, C. J. 1968. Reproductive cycles and viviparity in Guatemalan *Corythophanes percarinatus. Herpetologica* **24:** 175–178.

Newman, D. G. 1987. Burrow use and population densities of tuatara (*Sphenodon punctatus*) and how they are influenced by fairy prions (*Pachyptila turtur*) on Stephens Island, New Zealand. *Herpetologica* **43:** 336–344.

Papenfuss, T. J. 1982. The ecology and systematics of the amphisbaenian genus *Bipes. Occas. Pap. Calif. Acad. Sci.* No. 136: 1–42.

Pregill, G. K., J. A. Gauthier, and H. W. Greene. 1986. The evolution of helodermatid squamates, with description of a new taxon and an overview of Varanoidea. *Trans. San Diego Soc. Nat. Hist.* **23:** 367–202.

Presch, W. 1974. Evolutionary relationships and biogeography of the marcoteiid lizards. *Bull. South Calif. Acad. Sci.* **73:** 23–32.

Presch, W. 1980. Evolutionary history of the South American microteiid lizards. *Copeia* **1980:** 36–56.

Presch, W. 1983. The lizard family Teiidae: is it a monophyletic group? *Zool. J. Linn. Soc.* **77:** 189–197.

Presch, W. 1988. Cladistic relationships within the Scincomorpha. Pp. 471–492. *In* R. Estes and G. Pregill (eds.). See Chap. 3.

Rage, J.-C. 1982. La phylogénie des lépidosauriens: une approche cladistique. *C. R. Acad. Sci.* Sér. 2, **294:** 563–566.

Rieppel, O. 1980. "The Phylogeny of Anguimorph Lizards." Basel: Birkhaeuser Verlag.

Rieppel, O. 1987. The phylogenetic relationships within the Chamaeleonidae, with comments on some aspects of cladistic analysis. *Zool. J. Linn. Soc.* **89:** 41–62.

Russell, A. P. 1988. Limb muscles in relation to lizard systematics: a reappraisal. Pp. 493–568. *In* R. Estes and G. Pregill (eds.). See Chap. 3.

Schwartz, A., and M. Carey. 1977. Systematics and evolution in the West Indian iguanid genus *Cyclura. Stud. Fauna Curacao Caribb. Isl.* **53:** 16–97.

Schwartz, A., and R. W. Henderson. 1991. See Chap. 12.

Schwenk, K. 1988. Comparative morphology of the lepidosaur tongue and its relevance to squamate phylogeny. Pp. 569–598. *In* R. Estes and G. Pregill (eds.). See Chap. 3.

Shcherbak, N. N., and M. L. Golubev. 1986. "Gekkony Fauny SSSR I Sopredelbnykh Stran." Kiev: Naukova Dumka.

Vanzolini, P. E., A. M. Ramos-Costa, and L. J. Vitt. 1980. "Repteis das Caatingas." Rio de Janeiro: Acad. Bras. Cienc.

CHAPTER 18. Snakes

Ashe, J. S., and H. Marx. 1988. Phylogeny of the viperine snakes: part II. Cladistic analysis and major lineages. *Fieldiana, Zool.* **52:** 1–23.

Branch, B. 1988. See Chap. 17.

Brodmann, P. 1987. "Die Giftschlangen Europas und die Gattung *Vipera* in Afrika und Asien." Bern: Kümmerly & Frey.

Bullock, D. J. 1986. The ecology and conservation of reptiles on Round Island and Gunner's Quoin, Mauritius. *Biol. Conserv.* **37:** 135–156.

Cadle, J. E. 1984. Molecular systematics of neotropical xenodontine snakes. III. Overview of xenodontine phylogeny and the history of New World snakes. *Copeia* **1984:** 641–652.

Cadle, J. E. 1987. Geographic distribution: problems in phylogeny and zoogeography. Pp. 77–105. *In* R. A. Seigel *et al.* (eds.). See Chap. 5.

Campbell, J. A., and W. W. Lamar. 1989. "The Venomous Reptiles of Latin America." Ithaca, New York: Comstock Publ. Assoc.

Cogger, H. G. 1975. "Reptiles and Amphibians of Australia." Sydney: A. H. & A. W. Reed.

Cundall, D., and F. J. Irish. 1986. Aspects of locomotor and feeding behaviour in the Round Island boa *Casarea dussumieri.* Dodo, *J. Jersey Wildl. Preserv. Trust* **23:** 108–111.

Deraniyagala, P. E. P. 1955. "A Colored Atlas of Some Vertebrates from Ceylon. III. Serpentoid Reptilia." Colombo: Govern. Press of Ceylon.

Dessauer, H. C., J. E. Cadle, and R. Lawson. 1987. Patterns of snake evolution suggested by their proteins. *Fieldiana, Zool.* **34:** 1–34.

Dowling, H. G., and W. E. Duellman. 1978. See Chap. 3.

Dowling, H. G., R. Highton *et al.* 1983. Biochemical evaluation of colubrid phylogeny. *J. Zool.* **201:** 309–329.

Dunson, W. A. (ed.). 1975. "The Biology of Sea Snakes." Baltimore: Univ. Park Press.

Fitch, H. S. 1970. See Chap. 7.

Gans, C. 1976. Aspects of the biology of uropeltid snakes. *Linn. Soc. Symp. Ser.* **3:** 193–204.

Gasperetti, J. 1988. Snakes of Arabia. *Fauna Saudi Arabia* **9:** 169–450.

Gloyd, H. K., and R. Conant. 1990. "Snakes of the *Agkistrodon* Complex. A Monographic Review." Oxford, Ohio: Soc. Study Amphibians Reptiles.

Golay, P. 1985. "Checklist and Keys to the Terrestrial Proteroglyphs of the World." Geneva: elapsoïdea.

Groombridge, B. C. 1984. The facial carotid artery in snakes: variations and possible cladistic significance. *Amphibia–Reptilia* **5:** 145–155.

Gyi, K. K. 1970. A revision of colubrid snakes of the subfamily Homalopsinae. *Univ. Kans. Publ. Mus. Nat. Hist.* No. 20: 47–223.

Heatwole, H. 1987. "Sea Snakes." Kensington, New South Wales: New South Wales Univ. Press.

Irish, F.J. 1989. The role of heterochrony in the origin of a novel bauplan: evolution of the ophidian skull. *Geobios, Mem. Spec.* **12:** 227–233.

Kardong, K. V. 1983. The evolution of the venom apparatus in snakes from colubrids to viperids and elapids. *Mem. Inst. Butantan* **46:** 105–118.

Kardong, K. V. 1986. Observations on live *Azemiops feae*, Fea's viper. *Herpetol. Rev.* **17:** 81–82.

Kluge, A. G. 1991. Boine snake phylogeny and research cycles. *Misc. Publ. Mus. Zool. Univ. Mich.* No. 178: 1–58.

Lemen, C. A., and H. K. Voris. 1981. A comparison of reproductive strategies among marine snakes. *J. Anim. Ecol.* **50:** 89–101.

Mattison, C. 1986. "Snakes of the World." New York: Facts on File.

McCarthy, C. J. 1985. Monophyly of elapid snakes. An assessment of the evidence. *Zool. J. Linn. Soc.* **83:** 79–93.

McCarthy, C. J. 1986. Relationships of the laticaudine sea snakes. *Bull. Br. Mus. Nat. Hist.* (*Zool.*) **50:** 127–161.

McDowell, S. B. 1979. A catalogue of the snakes of New Guinea and the Solomons, with special reference to those in the Bernice P. Bishop Museum. Part III. Boinae and Acrochordioidea. *J. Herpetol.* **13:** 1–92.

SCIENTIFIC NAME INDEX

Illustrations of taxa are indicated by boldface page numbers.

SUBJECT INDEX